Atmospheric and Oceanic Circulation

Robert V. Rohli · Anthony J. Vega ·
Keith G. Henderson

Atmospheric and Oceanic Circulation

An Explanation of Earth's Climate Patterns

Springer

Robert V. Rohli
Department of Oceanography and Coastal Sciences
Louisiana State University
Baton Rouge, LA, USA

Anthony J. Vega
Department of Geosciences
East Tennessee State University
Johnson City, TN, USA

Keith G. Henderson
Department of Geography and the Environment
Villanova University
Villanova, PA, USA

ISBN 978-3-031-66141-9 ISBN 978-3-031-66142-6 (eBook)
https://doi.org/10.1007/978-3-031-66142-6

Cover illustration: Image 1. A cloud-filled sky, rugged islands, and turbulent air joined to create fanciful designs in the atmosphere in late July 2019. The Moderate Resolution Imaging Spectroradiometer (MODIS) on board NASA's Terra satellite acquired a true-color image of cloud vortices off of the Canary Islands on July 22. Image 1. Global image of the M0D09CMG.061 data product from January 14, 2021. Image 2: NASA/Goddard Space Flight Center Scientific Visualization Studio. Image 3: A drought-parched landscape.

This Springer imprint is published by the registered company Springer Nature Switzerland AG
The registered company address is: Gewerbestrasse 11, 6330 Cham, Switzerland

Acknowledgments

We are grateful to our wives and the rest of our families for their unending patience and support, especially during the time that we worked on this project, and to our many dedicated teachers and mentors who have provided so much guidance and knowledge assembled here. We appreciate the patience of our research collaborators and friends as the deadlines for our various projects coincided. We also thank our colleagues, manuscript reviewers, universities, and especially our students, for always challenging us to do our best work and for providing countless useful insights and perspectives that have molded the way that we think about the material covered in this book and about science and knowledge in general. We are particularly grateful to the students in Oceanography and Coastal Sciences 3013 ("Atmospheric and Oceanic Circulation") at Louisiana State University, who highlighted the need for this book. We very much appreciate the willingness of Vanessa van Heerden (with Geaux Geospatial), Siarah Hall, and Suzanne Rohli for sharing their talents in so many of the book's illustrations. We are grateful to so many talented artists employed in federal agencies, both in the USA and abroad, who provided public domain graphics used in this book, especially NOAA and its various constituent agencies. We thank Former Dean Christopher D'Elia, who led the effort to have Louisiana State University admitted into the University Corporation for Atmospheric Research (UCAR). This membership gave us access to UCAR's "MetEd" library of graphics, so many of which are used in this book. We thankfully acknowledge other copyright holders who granted permission to use their figures. And we also thank our colleagues at Springer Nature for offering this opportunity to us. Of course, we take responsibility for all errors and oversights.

Contents

Part VI Regional Responses in the Tropics and Southern Hemisphere

Part I
Let's Go with the Flow

Chapter 1
Introduction to the Atmospheric and Oceanic Sciences

Abstract In this Information Age, understanding the world around us through a scientific lens has become more important than ever. People have studied the atmospheric and oceanic circulation using an established process called the Scientific Method. This chapter reviews several examples of the scientific genius of some of the most important figures in the development of modern meteorology and oceanography along with their discoveries. Then the major themes in the development of the sciences of meteorology and oceanography are presented from the Age of Discovery to the modern Information Age. The weather and climate that results from the atmospheric and oceanic circulation have effects on various sectors of economic and social systems that include primary and secondary impacts. The atmospheric and oceanic circulation is a critical component of the climate system that affects, and is affected by conditions within and exchanges of energy, matter, and momentum between the atmosphere hydrosphere, lithosphere, cryosphere, and biosphere. There is an important distinction and interplay between the local environmental setting—the site—and the broader environmental features, including the atmospheric and oceanic circulation—the situation. Finally, the goals of the book are presented.

Keywords Atmosphere · Bergen School · Biosphere · Climate system · Climatography · Cryosphere · Environmental determinism · Environmental possibilism · Hydrosphere · Lithosphere · Scientific Method · Spatial and temporal variability, extremes, and change · Structuration theory · Techno-fix era

The Nature of Modern Science

You are undoubtedly aware that we live in an Information Age. We are often, or maybe even usually, overloaded with information. Every day, we get information from many different sources. Scientists are no different. Scientists must be able to process that information to be able to put it to its best use (or its best non-use, in some cases). This includes scientists who study the atmospheric and

R. V. Rohli et al., *Atmospheric and Oceanic Circulation*,
https://doi.org/10.1007/978-3-031-66142-6_1

oceanic circulation—**meteorologists** and **oceanographers**—and the Earth-ocean–atmosphere system in general—climatologists, geographers, ecologists, geologists, and other environmental scientists.

The best discoveries in the atmospheric and oceanic sciences, as with those in all sciences, result from correct and careful implementation of a series of steps in a process known as the **Scientific Method**, which has its roots in ancient times and has evolved through the Enlightenment and into the Modern Age. The Scientific Method starts with an interesting, testable observation that comes from a scientist's experience, readings, or intuition. The scientist then formulates a testable statement, or **hypothesis**, that explains how or why the observation is the way that it is. The hypothesis is then tested fairly and objectively through data collection, which are typically tested using quantitative testing procedures. The test results are analyzed to determine whether the hypothesis is supported or rejected. The conclusions are then reported to other expert scientists on that topic in the form of a scientific report or journal article to undergo review before it is determined whether the report is fit to be published. If accepted for publication, scientific research can then be accessed by other scientists or, increasingly, by the general public. Similar experiments on the same or very similar topic may be repeated in many different studies, with the preponderance of the evidence eventually leading to a more generalized "conclusion" concerning the overlying processes being observed. At some point, after extensive consideration and debate, with few unknowns about the phenomenon and consensus about the explanation is reached, the collection of verified hypotheses eventually becomes a **theory**. A scientific **law** is entirely different; it is an absolute and predictable truth with no known variations.

We often are asked to "trust the science." The more appropriate adages may be, "Trust the Scientific Method," "Think critically about the extent to which the Scientific Method is being implemented correctly," and "Participate in the science." Exercise those principles when you read research in the atmospheric and oceanic sciences.

A Modern History of the Science of Atmospheric and Oceanic Circulation

Like most sciences, the early development of scientific understanding of the atmospheric and oceanic circulation during ancient times by brilliant people who were centuries ahead of their time is astounding. Nevertheless, the developments leading to modern understanding of atmospheric and oceanic circulation can be considered to have begun around the time of the Age of Discovery in the 1400s through 1600s, when explorers were commissioned by governments to sail to lands that were not yet known in their societies.

Superimposed on the Age of Discovery was the Enlightenment, an intellectual revolution in the 1600s and 1700s that pervaded Western thought. Francis Bacon was already introduced as a key Enlightenment figure. He advocated for stringent unbiased testing of collected data as a means of discovery. Before Bacon, logic was the basis of understanding. Logic is limited by present understanding and prevailing thought seeped within a culture, and therefore is systematically biased. Bacon was influential in expanding the use of the less-biased Scientific Method which led to a more proper understanding of the natural world. Much discovery occurred a few decades after the Enlightenment when technology advanced to allow the precise measurement of natural phenomena. This created a need for more and better observations and the quantitative analysis of physical data. Rapid written and verbal communication of the analytical findings followed. The one-two punch of the Age of Discovery, the Enlightenment, and modern inventions brought the atmospheric and oceanic sciences, along with many other fields of study, into modernity. In the Western world, Benjamin Franklin takes his place among those who ushered in this new era, as he hypothesized that weather systems that passed overhead in Pennsylvania were the same systems that crossed over Boston a day or so later. He formulated this hypothesis without the benefit of telecommunication devices to test it, but of course, technology eventually proved him correct. Franklin also was among the first to observe and map the currents of the North Atlantic Ocean, including the **Gulf Stream** of the North Atlantic Ocean.

A second genius who brought atmospheric and oceanic research to modernity was Alexander von Humboldt (1769–1859), a Prussian polymath—a person who has numerous skills and wide knowledge and talents in various, often very different fields. von Humboldt is considered to be the one of the two founders (along with fellow Prussian Carl Ritter) of modern scientific physical geography. Much more than that, he influenced and was invited to visit many of the great thinkers of his time, including Thomas Jefferson. He has even been credited for giving Charles Darwin some of his ideas. He influenced many other major scientists too, especially Louis Agassiz—a Swiss glaciologist, geologist, and geographer, along with conservationists like George Perkins Marsh and John Muir, and even poets and writers like Ralph Waldo Emerson, Henry David Thoreau, and Walt Whitman.

von Humboldt also contributed substantially to meteorology, philosophy of science, biogeography, ecology, botany, geology, and many other fields. He argued for the "unity of nature" which forms the basis of today's ideas of interdisciplinarity to understand complicated natural phenomena. von Humboldt also observed the changes in Earth's magnetic field with distance to the poles and advocated for using geophysical measurements that formed the basis of modern geomagnetism, weather data networks, vulcanology, and more. His fieldwork in and near the Andes was also noteworthy for forming the basis of understanding the relationship between atmospheric pressure and elevation, and the theories of what allows air to rise or sink in ways that make weather disturbances more or less likely and destructive. For his work in the oceans, including his sailing on the *HMS Beagle* with Charles Darwin, the **Humboldt Current**—also known as the **Peru Current**—in the Pacific was named for him. von Humboldt also traveled extensively throughout Europe, revolutionizing scientific writing and traveling for fieldwork rather than just exploration and leisure.

Another genius of the same era but slightly later was Matthew Fontaine Maury (1806–1873). After being sent away to school, Maury eventually applied to the U.S. Naval Academy, and later fought for the Confederacy in the U.S. Civil War. Maury is credited with many advances in navigation and oceanography. For instance, he is believed to be the first to point out the "great circle principle"—that trips require less time when following a spherical rather than x–y coordinate "straight line" direction. He also became interested in wind and ocean currents and developed the first wind and current chart in 1842. His book, *Sailing Directions*, accompanied the chart, and he wrote articles on weather, currents, and related items. Maury was the first to describe several of the wind systems that will be covered in later chapters. His charts aided travel by cutting route times by days to months.

A generation later, the Swedish physicist-chemist Svante Arrhenius (1859–1927) became another influential scientist who made his mark on atmospheric circulation. Arrhenius was the first to propose (in 1896) that humans can cause warming globally, and he estimated the extent to which atmospheric CO_2 is linked to warming. However, unlike most scientists today, Arrhenius framed this idea in a positive light, as a means of feeding the world. Arrhenius' idea was based on the fact that CO_2 would contribute to more photosynthesis, with the extra warming contributing to longer growing seasons. He was criticized sharply by the famous Swedish physicist Knut Ångstrom (1857–1910), who thought that the atmosphere was already absorbing as much radiation as it could absorb, so adding more CO_2 would have little impact. For his efforts, Arrhenius received the Nobel Prize for Chemistry in 1903.

The debates between Arrhenius and Ångstrom are even more impressive because they happened in an era in which most mainstream atmospheric scientists did not believe that climates change at all. One visionary scientist who began to change the paradigm to one in which climates are known to change over time is Robert Ward (1867–1931)—an American climatologist. However, Ward's views on other matters, like eugenics, contributed to his obscurity.

Another major visionary in the atmospheric and oceanic sciences was Alfred Wegener (1880–1930), who was trained as a geologist, meteorologist, and physical geographer. His meteorology interests were related to polar weather and the upper atmosphere. His studies led to his world-record-holding ballooning expeditions. This interest allowed him to be ahead of his time in describing what we now know as the **polar front jet stream**. The existence of the polar front jet stream was controversial at the time and only became established during World War II when airplanes used less fuel and arrived sooner when traveling from west to east than when traveling from east to west. Wegener died when he fell into a glacial crevasse in Greenland while exploring glaciers and their impacts on the atmosphere. He is most famous for his Continental Drift hypothesis, which eventually formed the basis of the modern Theory of Plate Tectonics, but he was also a multidisciplinary scientist. Interestingly, science has seemed to come full-circle, with a recognition today that multi-disciplinary perspectives on team projects are so valuable.

Meanwhile, another impactful atmospheric scientist was blazing new territory in the form of atmospheric dispersion and weather modeling. Lewis Fry Richardson (1881–1953) was a British pacifist (opposed war and violence), so during World War

I, he developed the field of dispersion meteorology to save lives during poison gas attacks while he was an ambulance driver in the war. He pioneered many of the more advanced topics in atmospheric dispersion studies, including the vital, fundamental concept of atmospheric stability, which will be covered in more detail later in this book. Richardson also contributed mightily to micro-scale circulations in the air, such as atmospheric **turbulence**. Although atmospheric turbulence is at too small a scale to be covered in this book, think about how important Richardson's research was the next time you are on an airplane about to take off or land. Richardson also made the first legitimate attempt at numerical weather forecasting while occupying his mind between medical emergencies during wartime. His pencil and paper computations required far, far more time than the short forecast that he calculated, and moreover the forecast itself failed miserably, but in the process, Richardson laid the groundwork for numerical modeling that awaited the invention of the digital computer. Once enough computer power was available, Richardson's approach transitioned from theoretical to practical, but he did not have the chance to work in the true Computer Age. In Richardson's later years, he aligned his pacifist principles and his mathematical genius to attempt, unsuccessfully and unfortunately, to develop a numerical model to discern the causes of war.

Amazing breakthroughs in atmospheric and oceanic circulation in Scandinavia were not limited to Arrhenius and Ångstrom. The **Bergen School** in Norway, later transferred to the University of Chicago, made quite an impact. Vilhelm Bjerknes (1862–1951) is generally credited as being the founder of weather forecasting and some very important principles in advanced fluid dynamics, such as the "circulation theorem" that is beyond the scope of coverage in this text. The circulation theorem is used to explain fluid flow in both atmospheric science and engineering. Vilhelm's son, Jacob Bjerknes (1897–1975), continued the Bergen school and led his father and others in developing the Norwegian cyclone model—today's modern notion of weather fronts and the **mid-latitude wave cyclone** that will be covered in Chap. 4. Jacob emigrated to the USA in 1940, served as a Colonel in the U.S. Air Force, and worked on the meteorology behind setting dates for bombing campaigns in World War II, including those dropped on Hiroshima and Nagasaki. He founded UCLA's Department of Atmospheric and Oceanic Sciences, and in 1969 helped to link the long-known **El Niño** and **La Niña** oceanic phenomena with atmospheric circulation features including the **Southern Oscillation**.

Other Scandinavian scientists more loosely associated with the Bergen School played a pivotal role in atmospheric and oceanic circulation. For example, the Swedish oceanographer Vagn Walfrid Ekman (1874–1954), the son of an oceanographer, was recruited by Vilhelm Bjerknes into oceanography from physics, to study fluid dynamics. The Norwegian explorer, athlete, scientist, diplomat, humanitarian, politician, and statesman Fridtjof Nansen (1861–1930) was perhaps Norway's version of Benjamin Franklin. He developed the theory and mathematical proof of concept of the spiraling motion with increasing distance from the source of **friction** that slows the flow in both the atmosphere and ocean—the **Ekman spiral**—in 1905, while observing the direction of movement of icebergs. However, Ekman received credit for the discovery. Carl-Gustaf Rossby (1898–1957) was another influential

Swedish and later American meteorologist and oceanographer. Rossby characterized the polar front jet stream, the long waves that were later named for him, and turbulence, and later in his short life he developed the field of atmospheric chemistry. He was affiliated with prestigious research institutions, such as MIT, U.S. Weather Bureau, Woods Hole Oceanographic Institution, and (most famously) founded the Chicago School of meteorology. Rossby became the first meteorologist on the cover of a major magazine (12/17/1956 in *Time*) and considered atmospheric CO_2 as a cause of climatic warming.

In the U.S., the field of climatology was beginning to evolve from mere record-keeping of weather measurements into a legitimate, holistic science. The human-environmental geographer Carl Sauer (1889–1975) posited that cultural landscapes are made up of "the forms" superimposed on the physical landscape, and therefore, it is critical to understand the physical environment, including climate, as a prerequisite to understanding society. During Sauer's time, climatology's development was aided by the ready market of military-trained meteorologists after World War II. Innovations such as radar and collection of increasing amounts of data created a market of applied meteorologists who opened consulting firms. At the beginning of this era, climatology was still considered to be composed of average and extreme weather data, and very simple statistics on the data, devoid of explanations about "why" or "how" the data occurred—this is known as **climatography**.

However, since the 1950s, meteorology and climatology were both helped enormously by computers and modeling, and availability of "Big Data" processing techniques revolutionized the field. Today, climatology is generally respected, especially outside the United States. It takes its place as a major discipline in Britain, Russia, Canada, and elsewhere. It is now recognized that many contemporary world problems are weather-, ocean-, and climate-related. For example, half of all Americans live in counties bordering an ocean, and coastal storms are more impactful than ever before. Other examples of complicated climate-related problems include El Niño-Southern Oscillation (ENSO), sea-level rise, wildfires, drought-induced famine, water supply, and floods. The bottom line is that meteorology and oceanography are well-respected, and climatology is becoming an increasingly important discipline. One piece of evidence is that American meteorologist Warren Washington and American climatologist Michael Mann shared the "Nobel Prize of Environmental Science" in 2019.

Meteorological/Oceanographic/Climatological Thought

The way scientists and the public think about the nature of science, including what it can and cannot tell us and how we should react to it, changes over time. This is what is meant by "meteorological/oceanographic/climatological thought." The way people react to sciences like meteorology, oceanography, and climatology also changes. For example, as the data from high-powered computer output and more precise and abundant observations have improved, do we respond more seriously to weather forecasts now than we used to? And of course, the source of the information also matters. Do we trust politicians who talk about climate change more than we trust scientists who talk about climate change?

And as implied earlier, what is it about "trusting the science" anyway? Should we really "trust the science"? Should we follow the Russian axiom that says, "Trust but verify"? Does it matter if the "science" is atmospheric and oceanic circulation or if it is other forms of environmental science or medicine or the engineering behind a bridge or airplane or roller coaster? Why or why not?

One characteristic of anything that changes over time is that the change tends to proceed through "fads" or phases. The environmental sciences, including meteorology, oceanography, and climatology are no different. For instance, **environmental determinism** was an important concept in the early 1900s in environmental sciences. Environmental determinism holds that weather, climate, and other aspects of the environment determine society—in other words, that human achievement is tied so intricately to weather/climate and the environment as to make success dependent upon having favorable conditions. For example, an environmental determinist might argue that New England has more intelligent people because the weather patterns change frequently, stimulating human thought. Elsworth Huntington and William Morris Davis, the latter of whom was a geomorphologist and the first president of the Association of American Geographers (now American Association of Geographers), were two of the major proponents. Of course, today environmental determinism is (very understandably) frowned upon, and is viewed as a form of environmental racism or elitism.

Environmental possibilism rose after the fall of environmental determinism, around the 1930s, as a scientific paradigm in environmental sciences, and flourished for a few decades in mid-twentieth century. This philosophy holds that the environment sets broad limits but does not determine human achievement. For instance, it would be impossible to grow oranges in Alaska naturally, but they could be grown in a greenhouse in Alaska (if it were economically advantageous, of course). Environmental possibilism began in the 1930s and flourished after World War II. Ellen Churchill Semple was important in bringing the transition from determinism to possibilism, though earlier in her life she had many environmental deterministic ideas.

Around the 1950s, the **techno-fix era** tended to dominate scientific thought, including in the environmental sciences. This notion holds that we can and will conquer any difficulties that climate or other constraints put on us because we have science and technology. A classic example of the kind of thinking characteristic of the techno-fix era was putting a person on the Moon. Other techno-fix ideas that have been considered have been to bring water from the Great Lakes to nourish the deserts of Arizona and convert them to farmland. You can probably already see that these kinds of notions are generally naive and oversimplified; we now realize that there are unintended consequences when we engineer nature to suit our needs. But yet sometimes, even today, we still try to outsmart nature, out of necessity or perceived necessity. Leveeing the Mississippi River to (supposedly) prevent floods was a techno-fix response that was ahead of its time in the wake of the Great Flood of 1927 that changed the U.S. in so many ways. Of course, the techno-fix of leveeing comes at a cost. Without periodic flooding, the lack of new sediment to build and nourish wetlands resulted in coastal erosion, while deposition further downstream caused regional sinking, or **subsidence**, of land, making the delta more vulnerable to flooding anyway.

One way of thinking about modern human/environment relations is borrowed from the social sciences. **Structuration theory** says that there are overlying structures that enable us as a society to accomplish certain things, but that these structures also constrain us. Capitalism, Socialism, and Marxism are examples of the kinds of overlying structure in social science, while the climate system might be considered a structure in the Earth/environmental sciences. The climate of an area enables certain activities to take place there but also enacts constraints or limits. We, as free actors, can choose to pursue activities like irrigating crops or developing beaches which reinforce the way the climate structure affects us. We can also choose to act in ways that alter the climate structure like emitting greenhouse gases and altering land use. We have a choice as to which course of action to take based on our trust in the Scientific Method and our goals for society.

Atmospheric Impacts

If structuration theory is correct, then the climate that results from atmospheric and oceanic circulation patterns is important to understand because it provides the "structure" that ripples throughout the Earth-ocean-atmosphere-society system. Those ripples are called impacts. **Primary impacts** are components of the natural or human systems that are affected directly by variability or changes to the atmospheric and oceanic circulation and the resulting climate. Sometimes the cause-effect directions can even reverse, so that the primary impacts can become impactors on the atmospheric and oceanic circulation too. Perhaps the most important primary impact, and one which works in both cause-effect directions, is oceanic processes. Another important primary impact is agriculture; drought and flood impacts agriculture directly. Water resources, such as California's persistent and worsening water supply problem, are another primary impact. Another would be energy supply and demand; there is a large and growing demand for energy during heat waves and brutally cold winters. The construction and transportation industries are two other examples of primary impacts. Inclement weather will wreak havoc on both. Can you think of other examples?

Secondary impacts are those components of the natural or human systems that are indirectly affected by variability or changes in the atmospheric and oceanic circulation. Food supply is a secondary impact because other factors affect food supply more directly than the atmospheric and oceanic circulation, although the circulation and the climate that it affects are certainly important. Tourism and, especially in more localized areas or smaller countries, politics and the economy, would also be considered secondary impacts of variability in the atmospheric and oceanic circulation.

Regardless of whether the resulting impact of an atmospheric or oceanic circulation feature is primary or secondary, the departure from the average conditions is the stressor, rather than the average conditions. When variability or long-term changes in the circulation occur, natural and social systems must be able to adapt. Some individuals, species, communities, populations, ecosystems, families, cities, regions, and nations will have more **resilience**—ability to adapt—than others. All too often, only the average conditions of the atmospheric and oceanic circulation are considered, when in fact the extreme conditions and the variability about the average are far more important ecologically, environmentally, economically, politically, and socially.

There are two types of such extremes and variability—temporal (through time) and spatial (across space). **Temporal variability, extremes, and change**—under the broad umbrella of climate change—is the subject of much attention and ongoing research. But **spatial variability, extremes, and change** are also important to consider. In some years, bitter cold air is circulated from the poles to certain subtropical locations, resulting in agricultural damage, pipe bursts in homes, and other primary and secondary impacts. Yet in other years, no such extreme is experienced at the same location. In some years, the coastal ocean circulation allows for much near-surface upward motion called **upwelling**, which impacts near-surface nutrient availability and therefore has numerous ecological and economic implications, including fisheries, while in other years little such upwelling may occur at that same location. These are examples of spatial variability and extremes of the atmospheric and oceanic circulation. If, over time, the extreme phases of such spatial variability become more or less common and/or the direction of the circulation changes, then spatiotemporal changes are said to be occurring.

The topics of this book are largely restricted to spatial variability of the atmospheric and oceanic circulation. Throughout the book, ask yourself, "Where does the circulation happen?", "Why does it happen?", and "What are the implications of that phenomenon happening?" But all the while, keep in mind that the phenomenon is also likely changing in frequency, location, intensity, seasonality, and distribution within the day-night cycle over time as well. Much research is being invested in understanding such variability and changes better, because of the effects on the world's people in a wide array of primary and secondary ways.

The Climate System

Just as your circulatory system or the Solar System or the fuel injection system of your vehicle can be characterized by certain components, and certain processes and functions of those components that impact each other and can be measured, the Earth-ocean-atmosphere system, or **climate system**, is composed of components and processes/functions that can be measured. The climate system is a busy and complicated place!

As suggested previously, the climate system has several components that all interact, as shown by the schematic in Fig. 1.1. For all schematics like this one, spend as much time studying the image as you would have spent reading text that occupies the same amount of space. These figures are intended to convey ideas in much more intuitive and efficient ways than the brain can process the words used to describe them. Besides, most people are visual learners, meaning that they can learn from images like these much more efficiently if they take the time to try to understand them.

As suggested in Fig. 1.1, the **atmosphere** is the "system component" that consists of Earth's environment above the surface. Technically, "atmosphere" refers to just the gases, but "air" includes not only the atmosphere, but also solid and liquid particles called **aerosols**, also known as **particulates**. Solid aerosols include small particles of salt crystals from the ocean, dust from stirred-up land surfaces, soot from volcanic eruptions, sulfates and other pollutants, and other items. Liquid aerosols are primarily

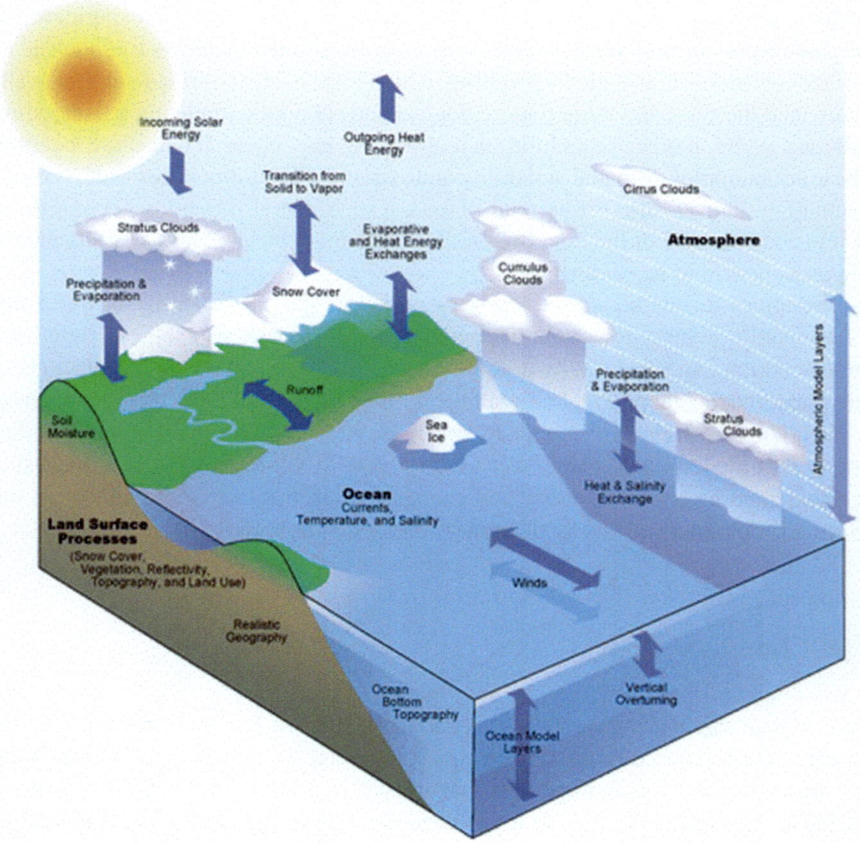

Fig. 1.1 Sample of interactions in the climate system. *Source* UCAR Image Library. Permission from Tara Torres

water in the form of cloud droplets. Clouds themselves are composed mostly of water (or ice)—not water vapor, although water vapor is in the presence of the cloud too.

The world ocean, or **hydrosphere**, is another climate system component. As suggested in the figure by the double-sided arrows, the hydrosphere and atmosphere interact via winds, which transfer **energy**, **matter**, and **momentum**—the three essential entities that the atmospheric and oceanic circulation transports—vertically back and forth within and between them, even as they move energy, matter, and momentum laterally, both in the air and in the ocean. Ocean currents serve this function within the hydrosphere, and also come into contact with the atmosphere and other spheres. The hydrosphere and atmosphere also transfer matter through **precipitation** and **evaporation**. All the while, processes within the ocean are occurring. In the climate system, the cards in the deck are constantly being shuffled!

The land surface—part of the **lithosphere**, comprising the Earth crust and the upper rigid underlying mantle—is much less extensive above sea level than the ocean surface itself, but includes many facets that impact the climate system, such as the shape of coastlines and topography. Figure 1.1 depicts the lithosphere-atmosphere and lithosphere-hydrosphere transfer of energy, matter, and momentum in several ways, with the double-sided arrows. If you did not notice these, pay attention to them as you review the figure. Notice also that the land surface also includes water bodies in the form of lakes and streams. Soil moisture and groundwater represent important land water sources as well. Even when combined, all land-based water is miniscule compared to the ocean, but such water plays a disproportionately important role in the system, especially for life. And the heterogeneous land surfaces complicate the understanding of the system because the different types of land cover support different amounts of energy, matter, and momentum transfer via the atmospheric and oceanic circulation.

Ice on or near Earth's surface, including **ice caps**, **ice sheets** which exceed 50,000 km^2 (30,000 mi^2) in area, floating **sea ice**, frozen inland water bodies, ice in soils, and glaciers are collectively known as the **cryosphere**. While the cryosphere can be considered to overlap the hydrosphere and lithosphere, it is useful to think of the cryosphere as its own system component because many important transfers of energy, matter, and momentum occur in and adjacent to the cryosphere that impact the climate system distinctly.

The final player in the climate system game is the **biosphere**—or life layer. The biosphere cross-cuts the atmosphere, hydrosphere, lithosphere, and even the cryosphere. The atmospheric composition and the energy transformations near Earth's surface result from photosynthesis and respiration in the biosphere.

The interactions between the atmosphere and the other climate system components—oceans and the rest of the hydrosphere, along with the lithosphere and cryosphere, all cross-cutting the biosphere—contribute to the dynamic nature of the climate system across space and time. Meteorologists tend to pay less attention to these interactions than they should—instead they primarily consider the processes occurring within the atmosphere or air. Climatologists are more likely to be concerned about the interactions of the atmosphere with the other "spheres" shown in Fig. 1.1, because those interactions often have their strongest impacts at longer time scales.

For example, tomorrow's weather forecast probably does not need to consider the fact that Lake Erie might have a little less of its surface frozen than it has today, because this difference probably will not cause weather conditions to change very much by tomorrow. However, at longer (climatological) time scales, if Lake Erie freezes less (say) in winters 50 years from now than it does today, then this change is likely to have an impact not only on processes in the climate system, but also on individual weather elements within the atmosphere or air that cannot be ignored.

Keep in mind that interactions occur across the full range of spatial and temporal scales shown in Fig. 1.2, with phenomena that occur on shorter spatial scales also tending to occur on shorter temporal scales, and vice versa. And these different phenomena at various scales interact with each other, influencing and being influenced by their temporal dimensions and spatial intensities. For example, short-duration microscale turbulence and updrafts in an individual thunderstorm cell can become a small part of the longer-lasting mesoscale (or regional-scale) circulation associated with a broader thunderstorm complex, which itself is steered by the broader, longer-persisting, synoptic-scale atmospheric flow. The synoptic-scale flow can make the mesoscale thunderstorm complex more or less likely to intensify, and in turn, the mesoscale system can promote or hinder the microscale turbulence and updrafts. However, for the atmospheric and oceanic circulation, the scales at the mesoscale and larger (Fig. 1.3) are usually emphasized, as in this book.

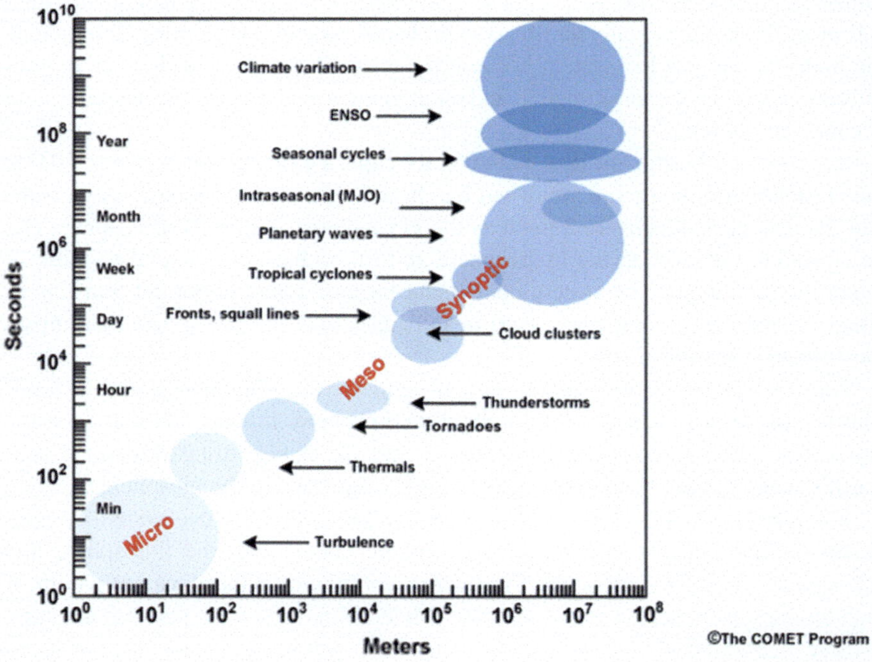

Fig. 1.2 Space–time correlation of atmospheric features. *Source* UCAR MetEd. Permission from Tara Torres

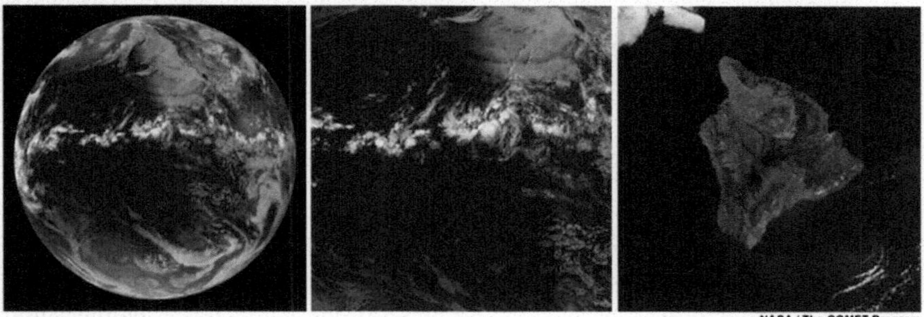

Fig. 1.3 Scales of interactions usually emphasized in examining the climate system. *Source* UCAR MetEd. Permission from Tara Torres

Weather Processes and Elements

Figure 1.1 suggests that processes occur within each component and across components of the climate system. These processes involve the atmospheric and oceanic circulation, which transport energy, matter, and momentum in ways that make the world the way that it is. Weather elements are the "evidence" that can be measured, as a means of representing the effectiveness of a weather process at a given place and time. Over long periods of time, patterns of the measured values of these elements, including not only the average values, but also the variability, extremes, seasonality, and distribution within the day-night cycle that characterize the climate and spatiotemporal changes therein. What are these processes and elements?

The transfer of energy takes three forms. One of these forms is the transfer of heat that can be sensed as "hot" to the touch—**sensible heat**, via the **sensible heat flux** ("flux" means "flow"). **Temperature** is the element that indexes the presence and distribution of sensible heat, measured by various types of **thermometers**. This would include various expressions of temperature, including average (or mean) annual temperature, monthly or seasonal mean temperature, temperature variability from one year to the next, ranges between the temperatures in summer vs. winter, and others. Virtually all of the energy that drives the atmospheric and oceanic circulation ultimately arrives from the Sun, through processes described in Chap. 2. The weather element that indicates the intensity and distribution of this solar **radiant energy** is solar radiation, which can be measured by **radiometers**.

Many types of matter are transferred in the climate system, through many processes. However, because water is so essential for life, it is the form of matter that receives the most attention. Evaporation and precipitation are the primary processes that involve the transport of water by atmospheric circulation. But because the water molecule contains energy that was used to break molecular bonds with other water molecules in the vaporization process, its transport via the atmospheric circulation also involves the second type of energy transfer even as it transfers matter too. That second type of energy is known as the **latent heat**, because "latent" means "hidden"—an apt description because evaporated water and the energy contained within its molecule cannot be seen or measured directly. The **latent heat flux** is the process that transports this latent energy contained within its elements—water molecules in solid, liquid, and gaseous states, both vertically and laterally. The atmospheric moisture in the gaseous state, which has absorbed latent energy when it evaporated, is called **humidity** and can be measured by various types of instruments and used to calculate the latent heat flux. The liquid and solid water in the air are the water molecules that have already given up their latent energy through transformation from vapor (or gaseous) to liquid state (i.e., **condensation**) or transformation from vapor to solid (i.e., **deposition**). These solid and liquid water elements can be measured by the proportion of sky covered by clouds and by other mechanisms. Chapter 2 also includes a deeper dive into the transfer of matter in the form of water. Other instruments are used to measure elements from processes related to other non-water types of matter transfer, such as pollutants, other aerosols, carbon dioxide (CO_2), and **ozone** (O_3).

Unlike the first two types of energy flux which both take place, at least partially, in a gaseous medium (the air), the third—the **substrate heat flux**—occurs in a liquid (the ocean) or solid (the lithosphere). Also unlike the first two types of fluxes, which occur by means of turbulent energy transfer in a gas or liquid—**convection**, the substrate heat flux transfers energy in a far less efficient manner. In solids, the energy transfers in a molecule-by-molecule fashion known as **conduction**, as there is no opportunity for turbulence to occur in a solid. Thus, the substrate heat flux is known as the **conductive heat flux** in the terrestrial Earth, while in the ocean, the substrate heat flux occurs via turbulence, except for at the interfaces between the ocean and the lithosphere.

Momentum transfer, or the **momentum flux**, in the climate system is by processes involving wind and ocean currents. Air moves faster aloft than near the surface, where it is slowed by **friction**, and ocean currents move faster away from the ocean bottom and the shorelines, where friction slows them. The transfer of momentum is always downward in the atmosphere, to move molecules that have more momentum toward the surface where momentum is lacking. When friction slows this air, the momentum is dissipated. In the ocean, the momentum transfer is more complicated because of the various sources of friction. The winds and ocean surface currents move at the speed and in the direction dictated by the combination of forces that are acting on air in the atmosphere and water in the oceans or hydrosphere. These forces will be explained more fully in Chap. 3.

Winds are measured in various ways, but the instruments all fall under the category of **anemometers**, though modern anemometers can also include precise remotely sensed technology like Doppler wind profilers. Traditionally, ocean currents have been measured by **current meters**. Modern global positioning system technology or satellite communications are now used along with Doppler profilers to measure ocean currents.

One way to think about weather/climate processes and elements is that the processes are functions that happen, while elements are the structure. Just as processes occur in an automobile that cause it to function, the functional processes in the atmosphere that cause the Earth-ocean-atmosphere system to function "properly" include the sensible heat flux, latent heat flux, substrate heat flux, and momentum flux. These are involved in circulating energy, matter, and momentum throughout the Earth-ocean-atmosphere system. Without them, the tropics, which receive abundant *radiant energy* from the Sun, would get hotter and hotter and hotter over time, while the polar parts of the system, which receive less energy from the Sun than they emit as radiation, would become colder and colder and colder over time. This would leave the planet uninhabitable except at the line where the energy budget balanced perfectly. It is good that the processes function properly.

And just as there is a structure to an automobile that consists of measurable elements, such as engine temperature, tire air pressure, and viscosity of the oil in the engine, the structural elements of the Earth-ocean-atmosphere system include temperature, humidity, precipitation, winds, and so forth. The basic principle is that it is easier and less dangerous to infer structure when the process is understood than to infer process when the structure is understood. For example, a physician's measurement of blood pressure, body temperature, and even cholesterol levels cannot diagnose heart disease, but having a look at processes such as via an EKG can. In the same way, the emphasis in the Earth-ocean-atmosphere system, including in this book, is on processes, with the resulting structural elements being reported as a response to those processes.

The rest of the book will include discussion of the most prominent functional processes and structural elements in different regions, but first it is important to understand the climate system. This will be done in Part II using a "concepts approach" in which the factors that influence the circulation of the atmosphere and ocean will be covered, including how the rest of the climate system affects and is affected by this circulation. Then, Part III uses a "regional approach" to examine the characteristic climatic features that result from the processes. This involves a trip around the world, with emphasis on the oceans and coastal locations, to determine the impacts and consequences of the weather and climates on life.

Throughout the book, but especially in the second half, emphasis will be placed on "where" in the climate system. "Where" can be described at various scales. Here they will be distinguished as "**site**" —what is the local environmental setting of that place?—versus "**situation**"—what larger-scale forces are acting upon, and interacting with, the climate at a place, including the atmospheric and oceanic circulation? Your present "site" is the **latitude/longitude** coordinates of your location, or your address. Your "situation" is, "within whichever section of whatever country you reside, where water from whatever major river near you is flowing in whatever direction toward whatever ocean, which has made it advantageous to have a few large cities that rely on trade from navigation on that important river." The advantages or disadvantages of a site and/or situation regarding one type of feature (such as weather/climate) may not be advantages or disadvantages regarding another type of feature (like trade, for example) for that place.

Goals of the Book

With this background, the goals of the book are threefold:

1. To explain the processes contributing to the broad-scale atmospheric and oceanic circulation;
2. To provide a better understanding of the forcing processes that produce climate (the "climate system");
3. To elucidate the local and regional effects of the atmospheric and oceanic circulation.

The focus is on the steering circulation, rather than local circulation such as in dust devils, tornado dynamics, upper-atmosphere and shallow water gravity waves, turbulence within the friction layer of the atmosphere and ocean, and tides. The circulation in the present-day world is described, but the same principles that drive the steering circulation in that system operate for any period of Earth's past, present, or future. Thus, models designed to simulate climates in the Earth's geologyic past or future must include representation of the processes described in the book.

Further Reading

Hulme, M. (2011). Reducing the future to climate: A story of climate determinism and reductionism. *Osiris, 26*, 245–266. https://doi.org/10.1086/661274

McLaughlin, P., & Dietz, T. (2008). Structure, agency, and environment: Toward an integrated perspective on vulnerability. *Global Environmental Change—Human and Policy Dimensions, 18*(1), 99–111. https://doi.org/10.1016/j.gloenvcha.2007.05.003

Nebeker, F. (1995). *Calculating the weather: Meteorology in the 20th century* (255 pp). Academic Press. 0-12-515175-6.

Popper, K. R. (1959). *The logic of scientific discovery*. Basic Books.

Thompson, R. D., & Perry, A. (eds). (1997). *Applied climatology: Principles and practice* (380 pp). Routledge. 978-0415141017.

Part II
Processes in the Air and Ocean

Chapter 2
The Basics of the Atmosphere

Abstract Solar energy powers the climate system, but energy received at the Earth's surface varies across space and time. Radiation laws quantify the amount of radiation emitted by wavelength on the electromagnetic spectrum. The position of the Earth, and any location on it, with respect to the Sun's most direct emitted (shortwave) radiation, along with the extent to which the solar radiation is absorbed, scattered, or reflected before it reaches the surface, all affect the radiant energy intensity incident at a place. In turn, each location on Earth emits longwave radiation. The accounting for shortwave and longwave radiation inputs and outputs at a location—net radiation (Q*)—is the radiation balance. Excess radiant energy input is expended through the convective (turbulent) fluxes of latent and sensible heat and the conductive substrate heat flux, which together comprise the energy balance. Positive Q* increases temperature, which, along with humidity, is a key component of local weather and climate. A stable atmosphere, including cases of a temperature inversion, is one in which a hypothetical air parcel will not rise spontaneously. In unstable atmospheres, spontaneous rising motion can support vertical cloud development, and, in the presence of abundant humidity, can cause clouds and precipitation.

Keywords Adiabatic and environmental lapse rates · Attenuation · Blackbody radiation laws · Counter-radiation (L↓) · Dew point temperature · Energy balance · Greenhouse effect · Longwave radiation · Net radiation (Q*) · Relative and specific humidity · Convective (turbulent) fluxes of sensible (Q_H) and latent (Q_E) heat · Shortwave radiation · Stability (atmospheric) · Substrate conductive heat flux (Q_G)

Introduction

Energy is needed to power the atmospheric and oceanic circulation. The Sun is the source of nearly all of that energy. Computer simulations show that without the Sun's *radiant energy*, the circulation would grind to a halt in a matter of weeks, in addition to all the other problems we would have. This chapter describes the nature of that

R. V. Rohli et al., *Atmospheric and Oceanic Circulation*,
https://doi.org/10.1007/978-3-031-66142-6_2

energy, how it is stored and distributed in various forms across the Earth, and how it is manifested as *temperature*.

Moisture is also a vital weather element that is transported by the atmospheric circulation. That moisture actually stores and releases energy within the water molecule that is ultimately derived from the Sun. This chapter also describes how moisture in the form of *humidity* works. The energy in the water molecule is stored and released, depending largely on the ability of the air in a given place to rise or sink. The atmospheric condition known as stability is what encourages or discourages air to rise or sink spontaneously. So atmospheric stability is also covered in this chapter.

Radiation, the heat it produces, the moisture, and the rising and sinking motions of the atmosphere all contribute to the **thermodynamic** state of the atmosphere—"thermo" means "heat" and "dynamic" implies movement. Thermodynamics of the atmosphere is contrasted with its **dynamics**, which largely involves broad-scale lateral motion in the air and ocean, which is the subject of Chap. 3. The processes involved in both atmospheric thermodynamics and dynamics can be explained relatively completely using higher-level mathematics. However, this book will include minimal mathematics and a largely qualitative treatment of these processes. Let's get started.

Earth-Sun Relationships

Energy from the Sun drives the entire *climate system*. You have no doubt noticed that this energy arrives to Earth in different intensities at different locations. More arrives in tropical locations than polar locations, more arrives near noon than near sunrise and sunset (and zero arrives at night) at a given location, and more arrives in summer than in winter (except in the tropical *latitudes* where energy receipt from the Sun remains rather consistent throughout the year).

But regardless of any of these factors, the solar radiation receipt (or incoming solar radiation, or **insolation**, or $K\downarrow$), is less at the Earth's surface than at the theoretical "top" of the atmosphere, at a given place. This difference is because of the weakening of the beam of radiation due to **absorption**, **scattering**, and **reflection** by matter, such as the air—collectively known as **attenuation**—which differs with the angle of the Sun above the horizon. When the Sun is lower in the sky (closer to the horizon, like at sunrise, sunset, and winter outside the tropics), the beam must travel through more distance in the atmosphere than when the Sun is higher in the sky (more overhead), so it has more opportunities to be absorbed, reflected, or scattered away from the surface by the air. Furthermore, the area of Earth's land or water surface heated by a "ray" of Sun also varies by the angle of the Sun above the horizon, as shown in Fig. 2.1. As areas near the tropics receive radiation at a higher angle (i.e., nearly 90° to the surface, or nearly overhead) for more of the year than near the poles, where the Sun's radiation arrives at an oblique angle, as depicted by the white lines. And of course, the more the sky is filled with clouds, water vapor, or pollutants, the more attenuation occurs and the less solar radiation reaches the surface.

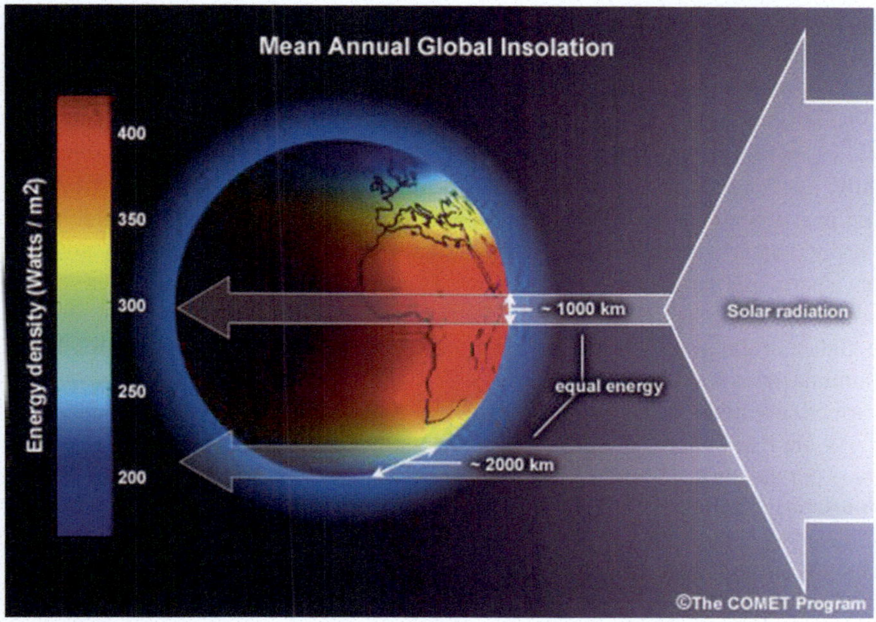

Fig. 2.1 Variation of solar radiation intensity with latitude and the path length of energy from the top of the atmosphere to the surface. *Source* UCAR MetEd. Permission from Tara Torres

As implied above, the attenuation from the radiation's trek through the atmosphere is also affected by the time of year. This is for two reasons. First, the Earth is actually at a slightly different distance from the Sun at different times of year. What is even more surprising is that the Earth is closer in January than it is in July! As you are probably aware, July is winter in the Southern Hemisphere, so to Southern Hemisphere residents, it makes more sense for the Earth to be farthest from the Sun in July and closest in January. This variation in Earth-Sun distance throughout the year changes very slowly through geologic time, so that at some times in Earth history, Earth is almost the same distance from the Sun throughout the year and at other times its distance varies more greatly with time of year. At some times it is closest in January and at other times it is closest in July. So, Earth-Sun distance is an important determinant of how much solar *radiant energy* is received at the Earth's surface.

The second way that the time of year affects attenuation is in terms of the number of daylight hours on a given day, which is a function of latitude. Because the Earth tilts on its axis as it makes its **revolution** around the Sun, the Northern Hemisphere part of the axis points toward the Sun in the half of the year from about March 21—the **March** (**vernal** to Northern Hemisphere inhabitants) **equinox**—through September 22—the **September** (**autumnal** to Northern Hemisphere inhabitants) **equinox**, with maximal tilt toward the Sun on or about June 21—the **June solstice**. Likewise, the Northern Hemisphere part of the axis tilts away from the Sun in the half of the year from about September 22 through March 21, with maximal tilt away from the Sun around December 21—the **December solstice** (Fig. 2.2, top). To Northern Hemisphere residents, the June solstice is also known as the **summer solstice**, and the December solstice is also known as the **winter solstice**.

Note from Fig. 2.2a–c that the tilt itself remains constant, at least on human time scales, but its orientation with respect to its position with respect to the Sun changes daily, reaching its peak toward the Sun near the June solstice, and away from the Sun near December 22–December solstice. In summer, the Sun appears higher in the sky than in winter, so there are more hours of daylight, as is shown by noticing that a particular parallel of latitude in Fig. 2.2 (top) will be illuminated for more than half of one complete rotation of the Earth on its axis. Furthermore, when the Sun is higher in the sky, the solar rays are more concentrated on a smaller area being illuminated, as was shown earlier in Fig. 2.1. This provides more intense energy and warmer conditions than in winter.

As implied above, latitudinal differences are related to the number of daylight hours per day throughout the year. Polar areas have almost no daylight in winter and almost no nighttime in summer. The poles experience six months of constant daylight followed by six months of constant night. Tropical regions have much more uniform daylight through the year with the **Equator** receiving 12 hours of daylight every day throughout the year. So the higher latitudes have more extreme differences in daylight between winter and summer and this effect lessens with decreasing latitude to the Equator. Despite the uneven distribution of radiant energy through the year, every line of latitude will ultimately receive a total of six months of daylight and six months of night each year.

The latitude that experiences the Sun's direct (i.e., overhead, perpendicular) rays changes by day of the year, as shown in Fig. 2.3. Notice that the June 21 label is on the 23.5° N line of latitude. This special line of latitude corresponds to the degree of tilt on Earth's axis. This northernmost line of latitude that can ever experience the direct rays of the Sun is called the **Tropic of Cancer**. The March 21 and September 22 labels (which can be on September 21 in some years) are on the 0° line of latitude (i.e., the Equator). The December 21 label is on the 23.5° S line of latitude. This southernmost line of latitude that can ever experience the direct rays of the Sun is called the **Tropic of Capricorn**. These are the latitudes that are experiencing the highest **Sun angle** on Earth on that day. The area between the Tropics of Cancer and the Tropic of Capricorn (i.e., 23.5° N to 23.5° S) are called simply the Tropics. The characteristic feature of the Tropics, or the tropical latitudes, is that the Sun passes overhead on two days per year, and the rays of sunshine are fairly well concentrated throughout the year.

Outside the tropics, the Sun's direct rays never pass overhead, the number of hours of daylight per 24-hour period is increasingly greater than 12 in summer and increasingly less than 12 in winter, and the rays of sunshine are increasingly more intense in summer compared to winter. Figure 2.4 depicts the number of daylight hours (i.e., day length) and amount of insolation by latitude and time of year.

Despite the fact that any location on Earth receives exactly six months of sunlight and six months of darkness through the year, the way that these daylight and darkness hours are distributed ranges tremendously with latitude. At the Equator every day

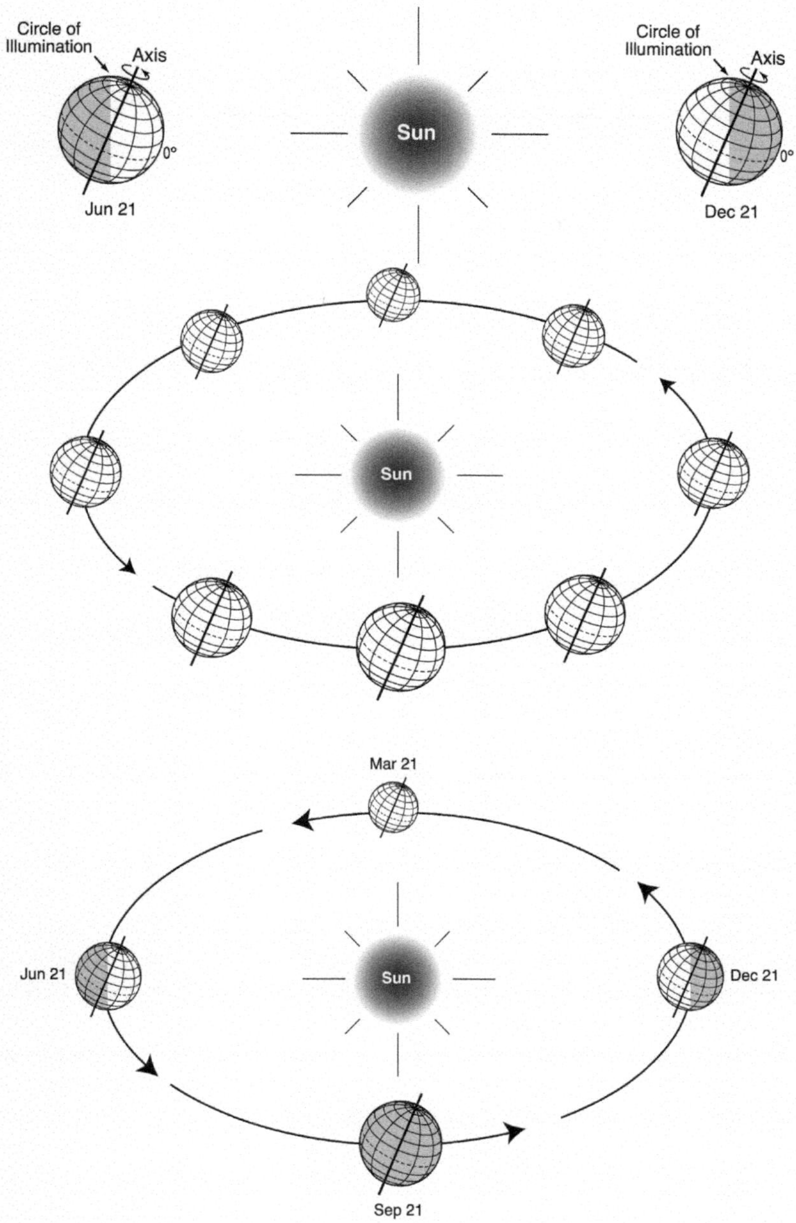

Fig. 2.2 Schematics depicting the contrast in daylight on the June versus December solstice (top). *Source* Figure 04.02 in Vega et al. (2013). Louisiana Weather and Climate. Jones and Bartlett; Earth's revolution around the Sun showing parallelism of axis (middle), and the position of the axis on the solstices and equinoxes (bottom). *Source* Figure 04.03 in Vega et al. (2013). Louisiana Weather and Climate. Jones and Bartlett, with all three showing the axial tilt, and top and bottom showing the amount of daylight

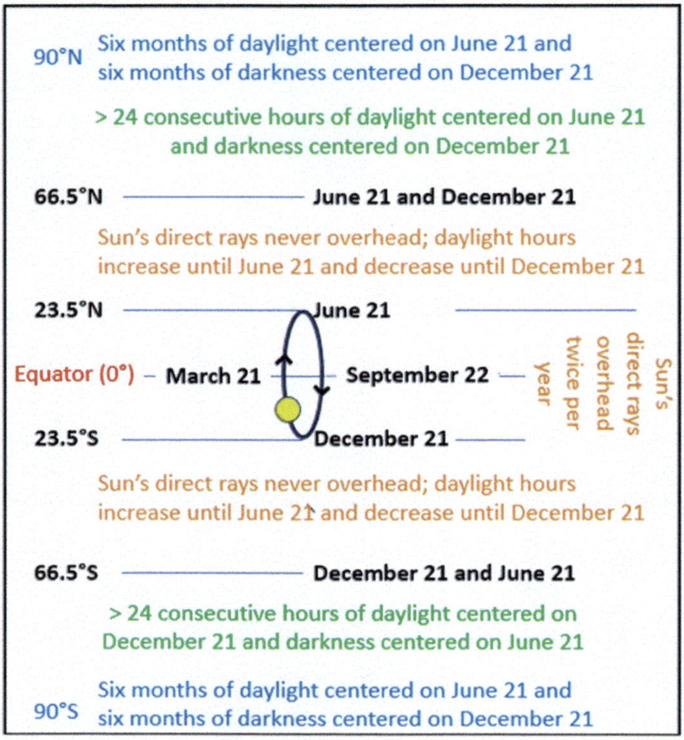

Fig. 2.3 Depiction of the latitude receiving the Sun's direct rays throughout the year. *Source* RVR

throughout the year experiences 12 hours of daylight and 12 hours of darkness. At the poles, all six months of daylight happen at one time, and then there are six consecutive months of darkness. At other locations, outside the tropics, as latitude increases (i.e., moving poleward), the number of daylight hours in a single 24-hour period increases in summer and decreases in winter. At the **Arctic Circle** and **Antarctic Circle**, at 66.5° N and 66.5° S, respectively, the number of daylight hours reaches 24 on the June solstice or December solstice, respectively. From the Arctic Circle to the North Pole, or Antarctic Circle to the South Pole, the number of consecutive daylight (or darkness) hours increases from 24 hours to six months.

The Nature of Electromagnetic Radiation

The presentation above verifies the reasons that different latitudes of the globe receive different intensities of radiation. But the differences even occur within a given latitude, because of the differences in surface type and weather elements from one place

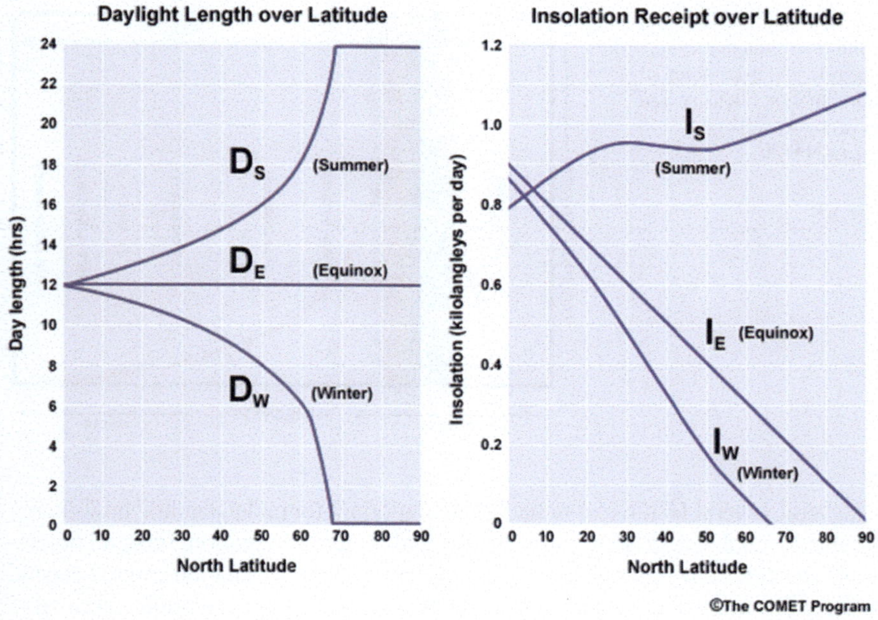

Fig. 2.4 Number of daylight hours and solar radiation receipt, by latitude. *Source* UCAR MetEd. Permission from Tara Torres

to another. This sets up variations in atmospheric and oceanic circulation, which in turn bring different temperature, *precipitation*, humidity, and other weather elements.

To understand how it all begins, it is first necessary to describe the nature of radiation, which can be thought of as traveling energy in wave form, oscillating as a combined electric and magnetic field. Such **electromagnetic waves** differ by the **wavelength** of the oscillation. Wavelength is the distance between adjacent crests, like on an ocean wave. However, the wavelengths of radiant energy emitted by the Sun are much shorter, on the order of millionths of meters, called **micrometers**, or microns, and abbreviated as μm. Shorter-wavelength radiation carries more energy than longer-wavelength radiation. **Electromagnetic radiation** can take several forms, including visible light, radio waves, microwaves, X-rays, gamma rays, and infrared (heat) rays, and others. All of these forms of radiation are characterized by their range of wavelengths on the **electromagnetic spectrum**—the full assemblage of all possible electromagnetic waves by wavelength.

Thus, radiation is related to light. Scientists usually think of light as consisting either of innumerable "wave packets" (characterized by specific wavelengths or colors) or "photons," particles of energy that carry no mass but travel at the speed of light (300,000 km per second). Our eyes perceive just a tiny fraction of the possible wavelengths of light (0.4–0.7 millionths of a meter, or μm)—centered on the peak wavelengths that the Sun emits (**visible light**; Fig. 2.5).

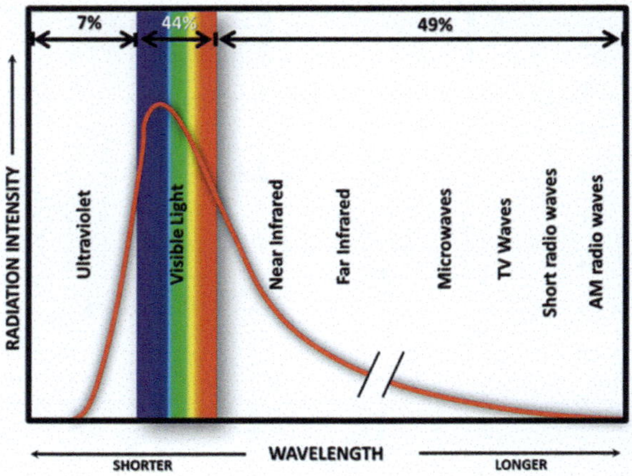

Fig. 2.5 Major subdivisions of the electromagnetic spectrum and the amount of total radiation intensity (red curve) by wavelength. *Source* NASA

Several general features of radiation are important for understanding how energy becomes available to drive the atmospheric and oceanic circulation. First, all objects above absolute zero emit radiation. The type and amount of radiation emitted depend on the temperature of the object. The hotter the object, the more radiation is emitted—the hot Sun emits lots of light. The hotter the object, the more energetic (i.e., the bluer) the radiation and the shorter the wavelength of emittance (see again Fig. 2.5).

But how does this work? The answer lies in the examination of some of the fundamental radiation laws. These laws work perfectly for hypothetical objects that absorb and emit radiation uniformly and with 100% efficiency. Such objects are called **blackbodies**. In the real world, nothing has perfectly blackbody radiation properties, though many common objects in nature behave nearly as blackbodies for at least some of the electromagnetic spectrum.

The first of these is **Wien's Law**, which tells the electromagnetic wavelength (λ) of peak emittance of a blackbody:

$$\lambda_{\max} = \frac{2897\ \mu\text{m K}}{T}$$

The μm in the equation above shows that the 2897 has units of micrometers. K is the **Kelvin temperature** unit that will cancel with the units of Kelvins in the denominator. Recall that temperature can be expressed in Kelvins, the metric unit of temperature, simply by adding 273.15 to its Celsius temperature. This is because the Kelvin temperature scale is at 0 when there is no molecular motion—the theoretical "absolute zero."

Therefore, Wien's Law says that shorter electromagnetic waves are associated with higher temperatures and vice versa. The wavelength of the peak radiance decreases linearly as the temperature increases. So hot things (like the Sun) emit electromagnetic radiation at shorter wavelengths than cooler things (like the air and the Earth itself). The Earth's temperature is controlled by the balance between the incoming **shortwave radiation** from the Sun (at wavelengths less than 4 μm, mostly as visible light according to Fig. 2.5) and the departing **outgoing longwave radiation**, mostly infrared light, as will be shown in greater detail later.

The second important radiation law is **Stefan-Boltzmann's Law**, which describes the total amount of radiation at all wavelengths emitted by a blackbody, not just the wavelength of peak radiation. Stefan-Boltzmann's Law involves using calculus to integrate the area under the curve depicted by the red lines in Fig. 2.6, across all wavelengths. Mathematically, Stefan-Boltzmann's Law is:

$$E = \sigma T^4$$

where E is the total energy emitted and σ (Greek lower-case letter "sigma") is the Stefan-Boltzmann constant (5.67×10^{-8} W m^{-2} K^{-4}). Here, W stands for **Watts**—the metric unit of power, T represents temperature, again in units of Kelvins (K), and m is for meters—the metric unit of length. Therefore, Stefan-Boltzmann's Law says that hot objects emit much more energy than cooler objects, since T is raised to the fourth power.

Planck's Law—also known as the "blackbody" formula—basically fills in the knowledge gap that Wien's Law and Stefan-Boltzmann's Law leave. It tells us the amount of blackbody emittance of radiant energy at any wavelength, not just the wavelength of peak emittance, and not just the total full-spectrum emittance like Wien's and Stefan-Boltzmann's Laws tell us, respectively. As with the other laws that were just introduced, Planck's Law works perfectly for objects that are perfect blackbodies. It also works very well for celestial bodies, because celestial bodies behave nearly as blackbodies. The equation is:

$$E_\lambda = \frac{a}{\lambda^5 \exp^{\left(\frac{b}{\lambda T} - 1\right)}}$$

where E_λ is the amount of radiant energy emitted at a given wavelength (called **monochromatic emittance**, since it is emitted at a single, given wavelength), in Watts per micrometer (μm), T is the temperature of the object in Kelvins, and a and b are known constants. As the equation shows, Planck's Law expresses that the spectrum of wavelengths emitted by a body at temperature T has a characteristic shape that is strongly dependent on the wavelength (to the inverse fifth power). An object's "signature" of energy emission as a function of wavelength describes the *spectral* distribution of radiation emitted by a blackbody. The red curves in Figs. 2.5 and 2.6 are derived by connecting the dots formed by using Planck's Law to calculate the emittance at each wavelength. As shown in Fig. 2.6 and verifiable using Planck's Law, the Sun emits abundant electromagnetic radiation at visible wavelengths. This explains why our eyes have evolved to be sensitive to the visible wavelength range, as these are the most abundant wavelengths from our Sun.

In summary, Wien's Law indicates that the hot Sun emits its peak radiation at a shorter wavelength than the cooler Earth, which actually has its peak emission in the infrared part of the electromagnetic spectrum. Stefan-Boltzmann's Law shows that the Sun emits much more energy than Earth. And Planck's Law tells the shape—or

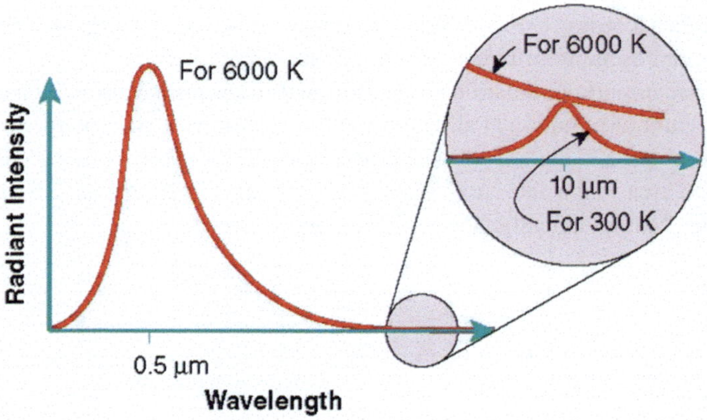

Fig. 2.6 Electromagnetic spectrum illustrating both Wien's Law—the Sun (shortwave radiation) emits at shorter peak wavelength than Earth (longwave radiation), and Stefan-Boltzmann's Law—the Sun emits much more radiation than Earth. *Source* Figure 02.07 in Rohli and Vega's Climatology (4/e), Jones and Bartlett

spectral signature—of the curves, thereby providing more information than Wien's and Stefan-Boltzmann's equation and including the information that Wien's and Stefan-Boltzmann's Laws provide.

Shortwave Radiation in the Atmosphere

The shortwave radiation flux reaching the Earth's surface (abbreviated as "K," with "K↓" being the shortwave radiation from the Sun that is emitted down to the surface of the Earth) strikes the top of the atmosphere at a constant rate of about 1366 W m^{-2}. But because of attenuation, the amount of radiation that reaches the surface will always be less. The additional path length of the Sun's rays through the atmosphere at more oblique angles near sunrise and sunset, for example, will attenuate so much radiation that it is harder to get sunburns and feel hot at those times of day. Likewise, at latitudes increasingly distant from the tropics, the Sun will be increasingly lower in the sky, especially in winter, thereby allowing for more attenuation. And even at the same location outside the tropics, the attenuation will increase in winter compared to summer because of the lower (i.e., more oblique) Sun angle. Inside the tropics, seasonal changes in attenuation get trickier, as there is generally less attenuation at the two times per year when the Sun is directly overhead, but those exact times of year depend on latitude. They are at the March and September equinoxes at the Equator, and both dates approach June with decreasing distance to the Tropic of Cancer and both approach December moving toward the Tropic of Capricorn. However, regardless of location, time of day, or time of year, a cloudy and/or polluted sky will increase attenuation of what would have occurred under a clear sky. And of course, regardless of location or time of year, attenuation is zero at night, because K↓ is zero.

Once shortwave radiation from the Sun reaches the "top" of the atmosphere (known as K_{EX}, for **extraterrestrial shortwave radiation**) how much of it goes into absorption, scattering, and reflection, at the surface and in the atmosphere? What percentage of K_{EX} becomes K↓? As might be expected, the answer depends on lots of things: latitude, time of day, time of year, cloud cover, and pollution have already been mentioned. Nevertheless, because of satellite observations, it is possible to compute a global annual average value of K↓, as well as a global annual average of each of the components of attenuation—absorption, scattering, and reflection.

Figure 2.7 answers these questions, and more. For now, consider only the orange arrows (which represent fluxes of shortwave radiation). Study carefully how many W m^{-2} of radiant energy flow into each of the various components of absorption, reflection, and emission. This diagram is simplified for your convenience, but of the three arrows labeled "reflected," only the ones bouncing up from the clouds and surface are really reflection, in the true sense of the word. The molecules in the air, like all gases, do not actually "reflect" shortwave radiation, because "reflection" implies that the entire flux "bounces" back in the same direction. Instead, the molecules in the air scatter (instead of just reflect) shortwave radiation in all directions, not just one, so think of that label as "reflection and scattering."

Notice that Fig. 2.7 implies that on a global annual average, K_{EX} is about 340 W m^{-2} (14 + 54 + 272 W m^{-2}) of shortwave radiation arriving at the "top" of the

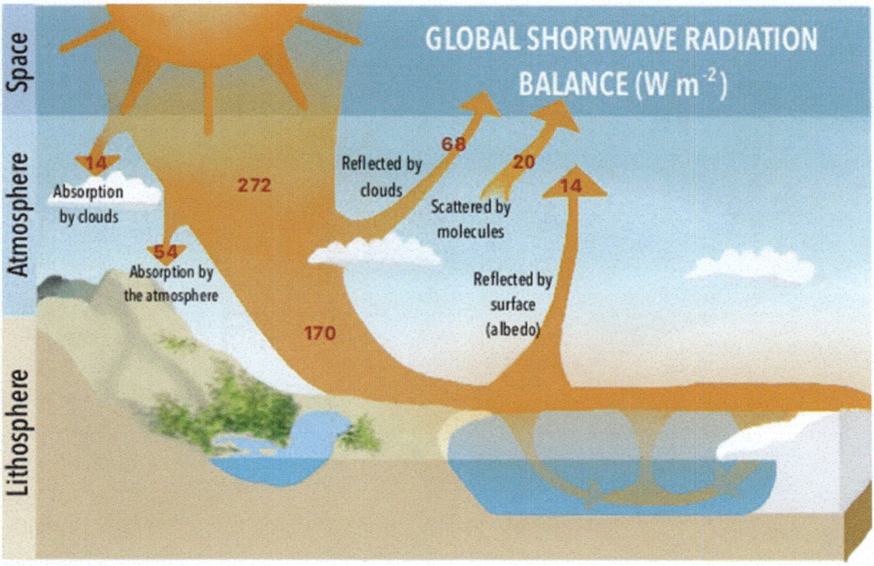

Fig. 2.7 Number of Watts per square meter on a global, annual average, involved in shortwave radiation gains and losses in the climate system. *Source* Modified from Australian Bureau of Meteorology (Siarah Hall)

atmosphere. But how can this be, since the solar constant is about 1366 W m^{-2}? The answer is that the solar radiant energy is intercepted by the spherical Earth in a "circle" of radiation of area πr^2, and that radiation arrives on the surface of the Earth, which is approximately a sphere of surface area $4\pi r^2$. Thus, the 1366 W m^{-2} must be divided by 4. Half of the Earth's surface receives no insolation at any given time because it is nighttime, and the insolation incident on increasingly high-latitude parts of the Earth is at increasingly oblique angles, which reduces radiation intensity at the surface.

Perhaps more importantly, Fig. 2.7 implies that the amount of radiation flux arriving at the surface as K↓ (170 W m^{-2}) far exceeds the amount that is attenuated in the atmosphere (14 + 54 W m^{-2}). Figure 2.7 also suggests that, of the 340 W m^{-2} of shortwave radiation that enters the atmosphere as K_{EX}, about 20% (68/340) W m^{-2} is absorbed by the clouds and atmosphere, and about 50% is absorbed at the surface as K↓, with about 102 (20 + 68 + 14) W m^{-2} or about 30%, reflected (or scattered by clouds) back upward as K↑.

In general, the percentage of K_{EX} that is reflected (by atmosphere as scattered radiation, or by clouds or by the surface) is known as **albedo** (K↑, to represent "shortwave reflected"). Albedo is color- and roughness-dependent, with lighter-colored and smoother objects having a greater albedo than darker rougher ones. Typical albedoes of some common surfaces are shown in Fig. 2.8.

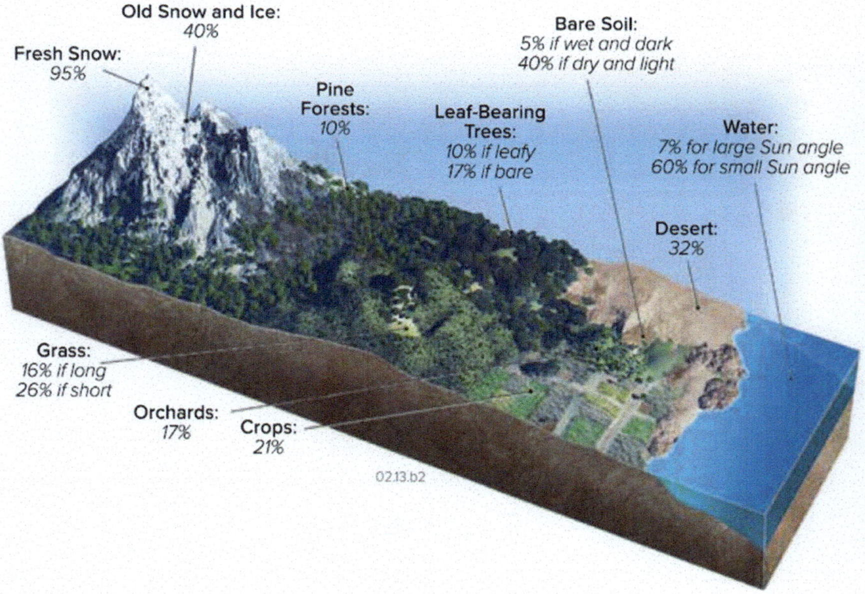

Fig. 2.8 Typical albedo of some common surfaces. *Source* Figure 02.13 from Exploring Physical Geography 3/e. Reynolds, Rohli, Johnson, Waylen, and Francek, Exploring Physical Geography, McGraw-Hill

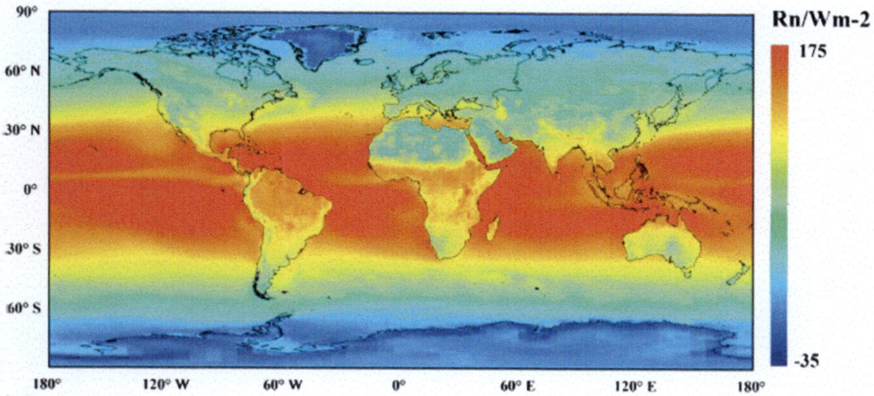

Fig. 2.9 Annual mean net shortwave radiation (K↓ − K↑, or K*) at the surface. *Source* Elsevier. Liang (2018)

At the surface, the difference in quantity of shortwave radiation that is absorbed vs. reflected (i.e., K↓ − K↑) is known as **net shortwave radiation** (**K***) at a place. K* is always zero at night, anywhere on Earth. Because no shortwave radiation is emitted downward from the Sun at night, none can be reflected from the surface as albedo. On a long-term average basis, the net shortwave radiation, collected by the European Centre for Medium-range Weather Forecasting, is shown in Fig. 2.9.

Longwave Radiation in the Atmosphere

Because Earth absorbs radiation, Stefan-Boltzmann's Law says that it must emit radiation. But this radiation is not as energetic as the radiation from the much-hotter Sun, so it is emitted at longer wavelengths than the energy that the Sun emits. This radiant energy emitted by Earth and its atmosphere is therefore known as longwave radiation (L), also termed **terrestrial radiation**, though the latter term is not preferred because marine and cryospheric surfaces also emit radiation at these wavelengths. Longwave radiation can be considered to have wavelengths exceeding 4 μm. To examine the sources and sinks (i.e., where the energy flux starts and where it goes) of longwave radiation emitted by Earth (L↑) and the atmosphere back down to Earth (L↓, also known as **counter-radiation**), look at the blue arrows added to Fig. 2.7, in Fig. 2.10, and study their magnitudes in W m^{-2} of radiant energy.

At any given place and time, the terrestrial radiation escaping back to space (which, on average, is 20 + 88 + 130 W m^{-2} of radiant energy, or 70% of K_{EX}, as shown in Fig. 2.10), is said to go through the **atmospheric window** out to space. The range of wavelengths of radiant energy from Earth that can escape to space at a given place and time (and therefore the value of the labels on the arrows) depends on the gases (especially water vapor) and liquids (especially liquid water, in the form of clouds)

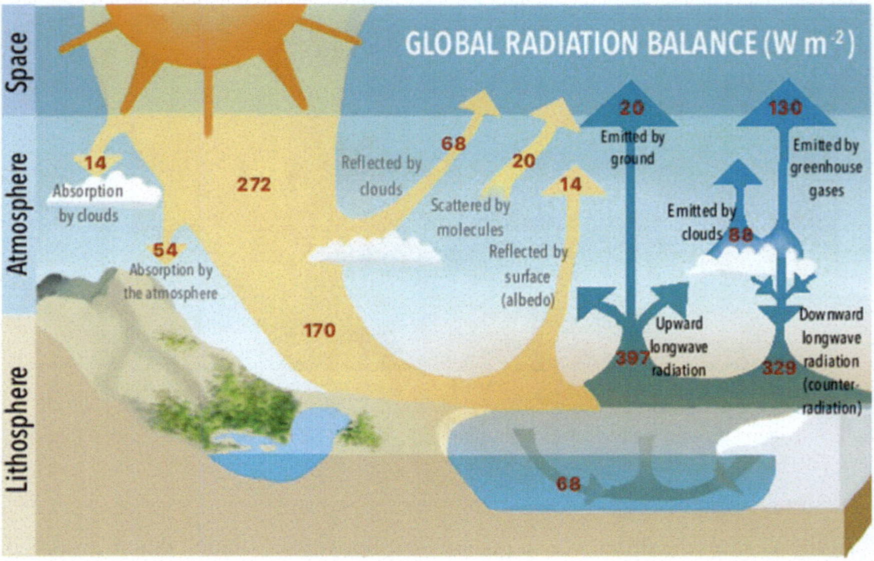

Fig. 2.10 Number of Watts per square meter of shortwave radiation on a global, annual average, involved in longwave radiation gains and losses in the climate system. Modified from Australian Bureau of Meteorology (Siarah Hall)

in the air at the time. The wavelength range that comprises the atmospheric window varies in space and time, primarily because of variation in water vapor and liquid water droplets (i.e., clouds) in the air.

Water vapor and liquid water droplets in air absorb L↑ well, so on a cloudy, humid night, the atmospheric window closes, keeping the energy associated with the longwave radiation in the air (where it is absorbed and then heats up the air), rather than letting it continue out to space. This would keep the nighttime temperatures warmer, since the radiation cannot escape the air so easily. By contrast, on a clear night there is less absorption of this L↑ in the air, so the atmospheric window remains open, allowing more L↑ to escape to space. This would allow the nighttime temperatures to drop. The coldest nights of the year are usually clear nights with weak or no winds. Winds would hinder this L↑ from escaping to space by moving it with more horizontal *turbulence* and with less vertically-oriented turbulence near the surface.

As is suggested by the blue arrows in Fig. 2.10, some L↑ is absorbed by the air and then re-radiated (remaining as longwave radiation) back to space (also as L↑) or back to the surface as counter-radiation (L↓), keeping the surface much warmer than it would be if no atmosphere surrounded Earth's surface. This is the well-known and mostly natural **greenhouse effect**. Over time, the buildup of **greenhouse gases** (most notably carbon dioxide and methane) would increase the amount of L↑ absorbed in the atmosphere and therefore reduce the amount of outgoing longwave radiation that leaves the Earth-ocean-atmosphere system through the atmospheric window, thereby warming the system more. This warming would occur by providing "extra" molecules in the air to absorb the L↑ and re-radiate longwave radiation back down to the surface as L↓, in the greenhouse effect—warming the Earth's surface further. The 2022 eruption of the undersea South Pacific Hunga Tonga-Hunga Ha'apai volcano near the island nation of Tonga has led some to wonder about the role of water vapor, which is by far the most common greenhouse gas, in the stratosphere as a potential contributor to warming, at least on short time scales.

Figure 2.10 suggests that more energy is involved in the longwave radiation than even enters the system as K_{EX}. This is because some longwave radiation is "recycled" within the Earth-ocean-atmosphere system, as suggested by Fig. 2.10. But also notice from Fig. 2.10 that for a global, annual average, the surface loses more energy from L↑ (397 W m^{-2}) than it gains from L↓ (329 W m^{-2}). This negative value for **net longwave radiation** (L*, or L↓ − L↑) makes sense because the surface is, on average, warmer than the air above it, thereby giving it more energy to emit than the atmosphere. This begs the question, "Why doesn't the surface continue getting warmer and warmer over time, at a much faster rate than Earth's surface is currently warming?" The answer to this question requires an understanding of how the radiation balance works.

The Radiation Balance

For the Earth-ocean-atmosphere system as a whole, K* and L* must balance in the atmosphere, or the Earth would warm or cool at fast rates. For the globe at any given instant, the actual amount of radiation a place on Earth's surface receives—the **net radiation** (Q*)—can be expressed using the radiation balance equation for the Earth as a whole:

$$Q^* = (K\downarrow - K\uparrow) + (L\downarrow - L\uparrow) = 0$$

The equation above implies that the calculation of Q* from measurement of all four variables above at a large, random sample of places across Earth's surface, and calculation of Q* should give an average Q* of zero, since some places are getting warmer while others are cooling off. The system is in balance.

However, at a given locality (i.e., at any one of those places) at a given moment in time, the terms are not likely to balance. This is because different places are experiencing different times of day and seasons (for a Northern vs. Southern Hemisphere location), at different latitudes. Africa is hotter than Greenland, for example, so an African weather station might usually have a larger Q* than a site in Greenland. At a given moment, Africa and/or places experiencing daylight might likely have a positive value of $(K\downarrow - K\uparrow) + (L\downarrow - L\uparrow)$, while Greenland and/or places experiencing nighttime usually have a negative value. This is because $(K\downarrow - K\uparrow)$ is always zero at a point on Earth experiencing nighttime and $(L\downarrow - L\uparrow)$ is usually (but not always) negative (as suggested by Fig. 2.10). The value of the equation above is probably negative in Alaska most of the time and positive over the tropical Pacific Ocean most of the time, because of the differences in Sun angle and number of daylight hours. In general, Q* increases (becomes more positive) from morning to afternoon and is at its largest value on a summer afternoon, and at its smallest value on a cold winter night. Figure 2.11 compares monthly Q* between January and July 2022.

The shortwave variables in the radiation balance are affected by several factors. K↓ is affected by latitude, time of year, time of day, and cloud and pollution cover, as described earlier. Albedo, or K↑, is locally site-specific, with smoother or lighter colored objects having greater albedo than darker, rougher surfaces, as described earlier. Average albedo values on a relative basis (low vs. high) over the terrestrial Earth are shown in Fig. 2.12.

Likewise, several factors affect the longwave components of the radiation balance equation. L↑ is the most straightforward of all four components in the equation. Local temperature governs L↑ through slight modification of Stefan-Boltzmann's Law to account for the fact that the surface is not a perfect blackbody.

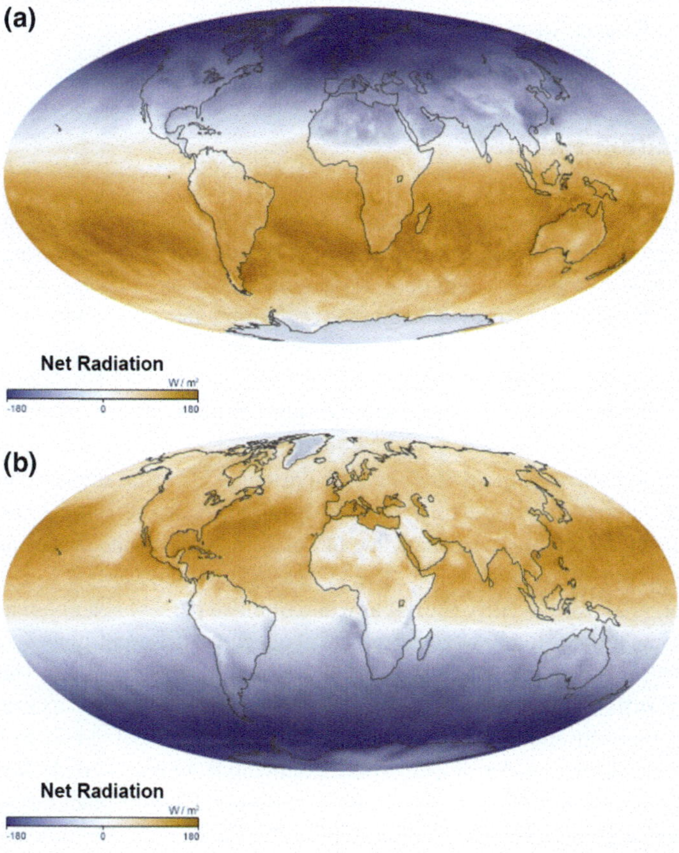

Fig. 2.11 Net radiation at the surface: **a** January 2022; and **b** July 2022. *Source* NASA. https://earthobservatory.nasa.gov/global-maps/CERES_NETFLUX_M

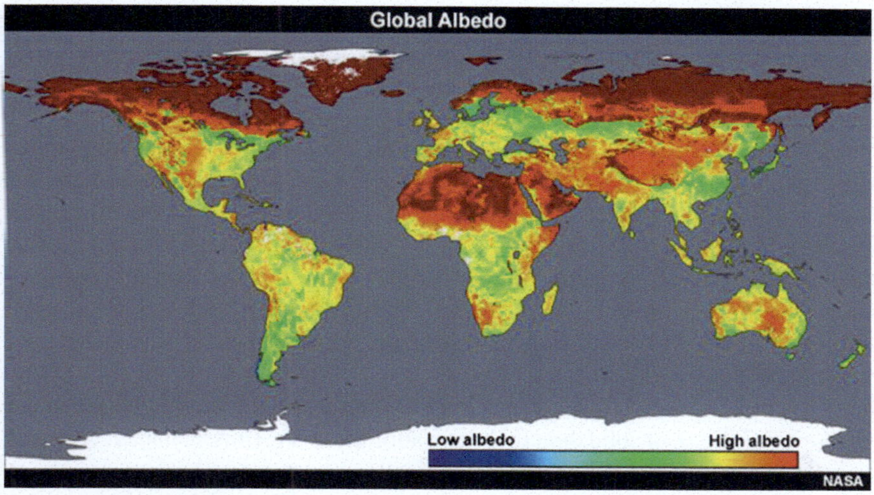

Fig. 2.12 Albedo on a global scale. Note that in the *cryosphere*, albedo is large though not shown here. *Source* NASA. https://earthobservatory.nasa.gov/images/2599/global-albedo

Counter-radiation (L↓) is affected by cloud cover and atmospheric circulation that brings air of different amounts of humidity and pollutant concentrations. In general, humid air absorbs L↑, and therefore re-radiates it as L↓, more effectively than drier air. Importantly, all four of these radiant energy fluxes (K↓, K↑, L↓, L↑) change at different rates.

The Energy Balance

Understanding the radiation fluxes is important because they contribute to the energy balance at a place. The local energy balance determines how much energy is available to drive winds, storms, ocean currents, **evapotranspiration**—the combined effect of evaporation and transpiration of water from the surface to the atmosphere via plant roots, and other characteristics of the atmospheric and oceanic circulation. In general, places near the Equator tend to have an energy surplus (positive Q*) more often than not, and places near the poles tend to have an energy deficit (negative Q*) more often than not, with 35° N and 35° S latitude being near the balance where Q* averages to zero over the course of time. Nevertheless, places poleward of 35° N and 35° S can certainly have a positive Q* at times, especially in summer, and places between 35° N and 35° S can certainly have a negative Q* at times, especially in winter. When a point on Earth's surface has a positive Q*, it has extra radiant energy, so it "spends" that extra radiant energy in one of three ways.

The first way that a radiant energy surplus can be "spent" is by evaporating water or by **transpiring** water from the surface into the atmosphere via the root systems of plants. These two processes are usually combined in meteorology as evapotranspiration. This type of energy transformation is known as **latent heating** (Q_E), although latent heat, *latent heat flux,* and latent energy are also used and refer to the same concept.

Latent energy is energy that is released from the molecular bonds of water to the surrounding environment when water changes phase from gas to liquid (*condensation*), liquid to solid (freezing), or gas directly to solid (*deposition*). This release of energy at the molecular scale is converted to **kinetic energy**—energy of motion at the molecular level—an unimaginable number of times per second that becomes manifested as the winds that drive storms! Thus, abundant water is needed to be condensed, frozen, or deposited in order to create a storm.

Latent heating also involves energy from the Sun that is absorbed into the molecular bonds of water, when water changes phase from solid to liquid (**melting**), liquid to gas (*evaporation*), or solid directly to gas (**sublimation**). Thus, abundant energy is needed, such as from the Sun in summer, to melt ice, evaporate a bucket of water, or sublimate ice. One reason that a swimmer feels cold after getting out of the pool, even on a very hot day, is that the Sun's energy is being "spent" evaporating the water from the skin rather than into warming the swimmer up as heat that they can sense. The concept of latent heating and latent heat release is a major concept in understanding meteorology, especially in coastal or marine environments, where phase changes of water are most abundant.

The amount of energy released in a given phase change is the same as the amount absorbed in that same phase change going in the opposite direction. For instance, condensing a given volume of water releases the same amount of energy as evaporating that same volume. However, the amount of energy absorbed or released is not the same for all types of phase changes. Instead, much more latent heat is released in condensation and absorbed in evaporation than is released in freezing or absorbed in melting. Said another way, much more energy is released when water vapor transforms into liquid water than when liquid water transforms into ice. And it requires a lot more energy to cause a liquid water molecule to "jump" into the air in evaporation than to make that same water molecule go from its solid to liquid phase on the surface. The gas to liquid or liquid to gas phase change is far more energetic than the liquid to solid or solid to liquid phase change. This principle is evident in at least some tangible ways, as ice begins to melt at only 0 °C (32 °F), but evaporating an equivalent amount of that water would require a lot more energy from the Sun.

The second way that surplus Q^* at the surface can be "spent" is by heating the surface. Heat that can be felt as an increase in temperature, as opposed to "hidden" in the phase changes of water as in latent heating, is known as *sensible heating*. The flow of more concentrated heat (i.e., with greater temperature) to a place with lesser temperature is known as the *sensible heat flux* (Q_H). When surplus energy is available from the Sun, evaporating water is the "first priority"—so latent heating usually occurs before sensible heating. After a rain event, for instance, the Sun evaporates the water from the surface, and as the water evaporates, more and more of the Sun's energy can go into heating the environment up again, thereby creating increasing sensations of warming. Likewise, after the swimmer finally dries off after getting out of the pool, the Sun's energy can then go into warming them up via sensible heating.

Many other everyday examples illustrate the concepts of latent vs. sensible heating. For instance, perspiration is a built-in protection mechanism that allows latent heating to protect against overheating. If a person is exposed to excessive heat, perspiration causes the Sun's energy to delay overheating them by moving water to the skin to be evaporated, at least until the person dehydrates, at which point they are in trouble because sensible heating takes place unimpeded.

As another example, spring is usually cooler than the day in autumn with the same noon Sun angle and daylight period, because more of the Sun's energy goes into latent heating from snow and water that could not evaporate during the low-energy input during winter. By contrast, in autumn, additional sensible heating occurs because the long, hot summer is likely to have evaporated much of the available water near the land surface. Similarly, oceans are usually cooler in summer than land masses at the same latitude because there is abundant water available to evaporate (more latent heating and less sensible heating than over land).

The ratio of sensible heating to latent heating is known as the **Bowen ratio**. In the tropics, typical Bowen ratios may be around 0.1. In the mid-latitude deserts, Bowen ratios often exceed 10.

Q* arrives to the surface from the Sun via the mechanism of energy transfer known as radiation. This is the only mechanism of heat transfer that can occur through the vacuum of space that separates the Sun from Earth. By contrast, the first two ways that surplus Q* at the surface can be "spent" (i.e., Q_E and Q_H) are through the mechanism of energy transfer known as *convection* or turbulence. The third mechanism, known as **substrate heating** or the *substrate heat flux* (Q_S), takes place through convection in water surfaces or *conduction* within land surfaces and involves the flow of heat from the surface downward into the *hydrosphere* or *lithosphere*, respectively. Convective and conductive energy transfers require matter to be present throughout the entire flux. In convection, the matter is a **fluid** (i.e., liquid or gas), and in conduction, the matter is a solid. Substrate heating over the oceans or other water bodies involves a downward convective flux from the water surface whenever the water surface is warmer than the water below it, and an upward convective flux whenever the deeper water happens to be warmer than the water above it. Likewise, substrate heating over land (sometimes called the **ground heat flux**, as a special case of the substrate heat flux) involves a downward conductive flux from the terrestrial surface whenever the surface is warmer than the soil below it and an upward conductive flux whenever the deeper lithosphere is warmer than the surface above it.

On a net basis over a long period of time at a place, the net Q_S will often be zero or near zero in the mid-latitudes. The **Second Law of Thermodynamics**—the principle that pervades all of natural science that energy flows from areas of higher concentration to areas of lesser concentration—requires that the flux be directed downward beneath the surface whenever the surface air is warmer than the substrate beneath it. This case is common on a summer afternoon, but it happens at other times too. The flux is directed upward back to the surface whenever the substrate is warmer than the air above it. This case occurs most typically on winter nights, but it happens at other times too. Regardless of the direction of the flux, its strength is proportional to the energy gradient as measured by temperature.

Because liquids and gases can move much more easily than solids, convection is much more efficient in transferring energy than conduction. Oceans can redistribute their energy relatively effectively, with water bodies having less temperature difference throughout them than land surfaces. By contrast, depending on the mineral content and moisture characteristics, a land surface may only transfer surplus radiant energy downward through Q_S by a meter or less during a hot summer afternoon. At some point in the evening hours, the direction of the flux reverses to upward as the land surface cools nocturnally and stored heat energy from below begins to send heat energy to the surface. On a seasonal basis, Q_S over land may heat down to a depth of about 15 m in summer before the direction reverses and the stored energy at depth begins to heat the land surface above it via conduction.

If the long-term net Q_S is greater than zero at a place, then the surplus heat energy transfers laterally to adjacent areas in the form of **advection**, the lateral transport of energy and matter. If the long-term net Q_S is less than zero at a place, then flow of heat must move laterally into the area from adjacent areas in the form of advection to satisfy the Second Law of Thermodynamics. Winds and ocean currents are the familiar examples of advection.

For example, the warm *Gulf Stream* advects warm surface waters from the tropical Atlantic Ocean poleward along the Atlantic coast of North America then across the North Atlantic to northwestern Europe. The situation transfers huge quantities of stored energy to relatively high latitudes. For this reason, coastal northwestern Europe is much warmer than most other places at similar latitudes.

When the sources of radiant energy are accounted for, the radiation balance (left side of the equation below) becomes the energy balance:

$$(K\downarrow - K\uparrow) + (L\downarrow - L\uparrow) = Q_H + Q_E + Q_S$$

or

$$(K\downarrow - K\uparrow) + (L\downarrow - L\uparrow) - Q_H - Q_E - Q_S = 0$$

The equation above is called the (atmospheric) **energy balance.**

As shown in Fig. 2.10, it appears that on a global annual average, the Earth's surface absorbs 170 W m^{-2} of shortwave radiation through K↓, and 329 W m^{-2} of longwave radiation through L↓, for a total of 499 W m^{-2}. The 14 W m^{-2} of K that are reflected back out from the surface are added and then promptly subtracted from this total. Figure 2.10 also suggests that Earth's surface emits only 397 W m^{-2} through L↑. How can it be that the Earth's surface gains more energy than it loses? The Earth's surface would be energized, and therefore warmed, by 102 W m^{-2} every year, if this would be the case.

Likewise, a close look at the atmosphere in Fig. 2.10 would suggest something just as scary. Specifically, the atmosphere appears to absorb 14 + 54 W m^{-2} of K↓ from above, through absorption by clouds and the atmosphere, respectively, and

397 W m^{-2} of $L\uparrow$ from below, for a total gain of 465 W m^{-2}. Yet Fig. 2.10 shows clearly that the atmosphere emits 20 + 88 + 130 W m^{-2} of $L\uparrow$ to space, from the surface, clouds, and greenhouse gases, respectively, and 329 W m^{-2} of $L\downarrow$ as counter-radiation, for a total loss of 567 W m^{-2}. Is the atmosphere really cooling by 102 W m^{-2} each year?

And what about this "coincidence" that the surface appears to be getting energized by the same amount (102 W m^{-2}) as the atmosphere is getting robbed of its energy? There must be something else happening to transfer the excess 102 W m^{-2} from the surface to the atmosphere! But that "something" must not be radiative transfer, because Fig. 2.10 includes all of the shortwave and longwave components of radiation. The radiative effects are out of balance.

But radiative transfer is only one means of transferring energy. The 102 W m^{-2} are transferred by convection, through the convective (or turbulent) fluxes of latent heat and sensible heat, or Q_E and Q_H. Figure 2.13 shows this "missing" energy as part of the "energy balance," rather than the "radiation balance" as was depicted by Fig. 2.10. Because the Earth is mostly water, Q_E is much larger than Q_H on a global, annual average.

But what about Q_S? On a global, annual average, Q_S balances out to zero, so it is not shown. The same amount of heat (energy) is conducted downward from the surface during the afternoon and summer as is conducted back upward to the surface during the night and winter.

Fig. 2.13 Number of Watts per square meter of shortwave radiation on a global, annual average, involved in shortwave and longwave radiation gains and losses in the climate system, along with the convective fluxes. *Source* Modified from Australian Bureau of Meteorology (Siarah Hall)

Temperature

As shown previously, when radiant energy interacts with a substance, it may reflect from the substance (albedo), scatter in all directions after hitting the substance, transmit through the substance, or be absorbed by the substance. The total of reflection, scattering, and absorption equals 100% of the incident energy. Different substances reflect, scatter, absorb, and transmit radiant energy differently, including dependencies on wavelength of the radiant energy, with the amount of radiant energy absorbed by a substance proportional to its heat content, and therefore, to its temperature. How that energy is manifested in the environment is also important in determining the temperature of a body.

Because they are composed of very different substances, continents and oceans have different tendencies to warm and cool. More specifically, continents heat and cool faster and to a greater extent than oceans. Solids, of which the continental lithosphere is largely composed, are primarily opaque rather than transparent to radiant energy. Liquids, such as in the oceans, are generally less opaque and more transparent. In clear water, visible light penetrates to 100 m (330 ft) or more. This spreads the same amount of energy striking and heating a land surface over a column of water 100 m (330 ft) thick. Thus, if the same amount of energy is incident on the same surface area of land and water, a higher percentage of that energy would be absorbed by the land surface than by the water surface. The land surface will heat much more, and more quickly, as a result.

Furthermore, during times of surplus local Q*, the energy incident at the continental surface would be transferred downward in accordance with the Second Law of Thermodynamics via conduction, while downward heat transport from the water surface could occur through the much more efficient mechanisms of convection and advection. During times when a deficit of local Q* sets in, such as in evening hours and in extratropical winter, the concentration of most of the energy near the land surface will facilitate more rapid and abundant cooling as heat is transferred to the atmosphere through latent and sensible convective (turbulent) heating. By contrast, the onset of a Q* deficit at the water surface will not result in cooling as quickly or completely until all of the energy stored in the water column can be convected or advected back upward to the surface, by which time the Q* may have returned to surplus conditions, especially if the heating during the previous Q* surplus was long and intense, and water body is deep. These features of energy transfer over land vs. water reinforce the tendency for inland surfaces to heat and cool rapidly and substantially compared to marine surfaces.

Another factor that aids in the rapid heating and cooling of land versus water surfaces is a physical property of matter. Different types of matter require different amounts of heat energy (in units of **Joules**) to be absorbed in order to change the temperature. The amount of energy (which is the same as saying the amount of heat or work) needed to produce a temperature change of one Kelvin (or 1 °C or 1.8 °F) is known as **specific heat**. Table 2.1 shows that the specific heat of water is much higher than that of solid substances. Thus, it normally takes at least five times more

Table 2.1 Specific heat of common substances

Substance	Specific heat (J kg^{-1} K^{-1})
Water (at 20 °C)	4182
Ice (at 0 °C)	2093
Water vapor (at 27 °C)	1864
Dry air at sea level	1005
Sandstone	710
Granite (intrusive igneous rock)	0.79
Iron	449
Lead	129

https://www.engineeringtoolbox.com/specific-heat-capacity-d_391.html

Joules of energy to raise the temperature of 1 kg of water by one Kelvin or 1 °C or 1.8 °F than for solid land surfaces.

The last factor that influences the speed and intensity of heating and cooling of land vs. water bodies is evaporation. During evaporation, the latent heat flux from the surface to the air (rather than sensible heat flux) buffers the rate and intensity of surface temperature increase. Naturally, terrestrial surfaces will be less likely to have abundant evaporation than marine surfaces, even when the movement of water from the roots of plants upward through the plant to the atmosphere—transpiration—is included with evaporation in the form of evapotranspiration. Over land, surface water in the forms of lakes, streams, and puddles, support evaporation and vegetated surfaces support transpiration. Evaporation of oceans occurs constantly during daytime hours, as long as the surface remains unfrozen. The Bowen ratio indicates that there is more sensible heating over continents and more latent heating over oceans.

The atmospheric and oceanic circulation, and therefore the climate types around the world respond to this difference. All other factors being equal, like latitude, elevation, and land cover, continental climates have more oppressive summers and harsher winters than coastal and (especially) marine areas, where temperatures are more moderate. Notice the differences between January temperatures and between July temperatures at the four cities at approximately the same latitude (Tijuana, Mexico; Midland, Texas; Hamilton, Bermuda; and Dera Gazi Khan, Pakistan) in Fig. 2.14.

Figure 2.15a, b shows the mean temperature for January and July, respectively, around the world. The difference between the temperature in the warmest month and that of the coldest month for the same place is the **annual temperature range**. The greater the difference between summer and winter temperature, the larger the annual temperature range. Figure 2.16 shows the annual temperature range. Inland locations are said to have greater **continentality** than coastal or marine places.

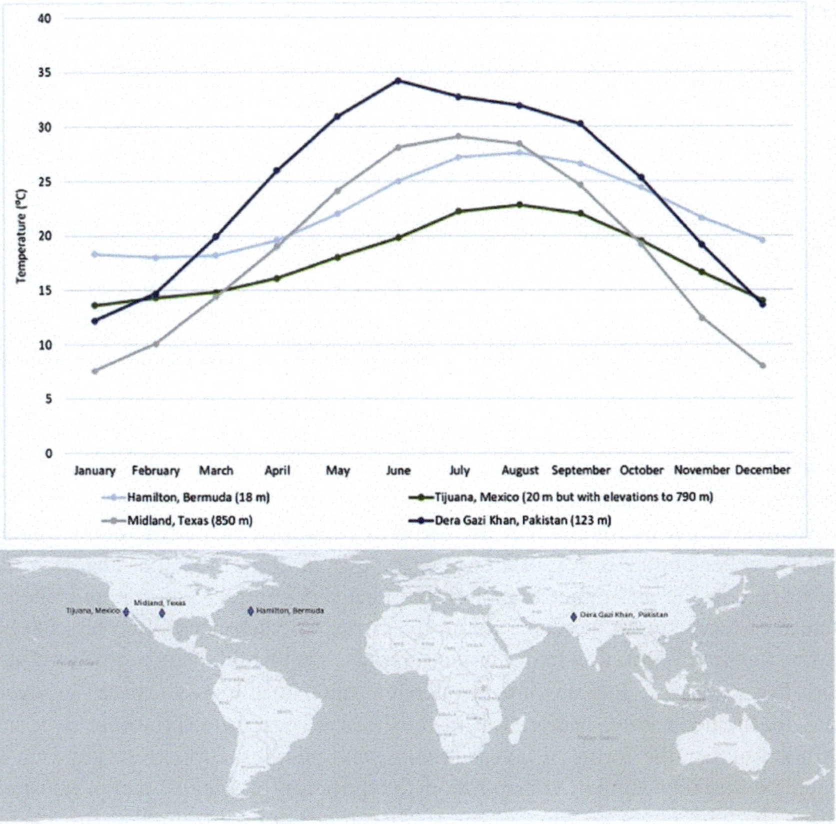

Fig. 2.14 Mean temperatures by month for four Northern Hemisphere cities at similar latitudes. *Source* Geaux Geospatial

Humidity

As latent heating increases, available energy in the form of Q* is used to drive evapotranspiration, transporting and transforming available liquid water from near the surface into water vapor in the air via the latent heat flux, increasing the humidity in the air. In particularly dry locations, all or nearly all of the Q* at the surface goes to the atmosphere via the sensible heat flux, and therefore little evapotranspiration and humidity occurs.

One way to tell whether a place is generally more humid or less humid is by comparing the water equivalent of precipitation to the water equivalent of evapotranspiration at that place. Humid places generally have more precipitation than evapotranspiration, even though evaporation may also be substantial. Arid places generally have more potential for evaporation than precipitation, even though actual

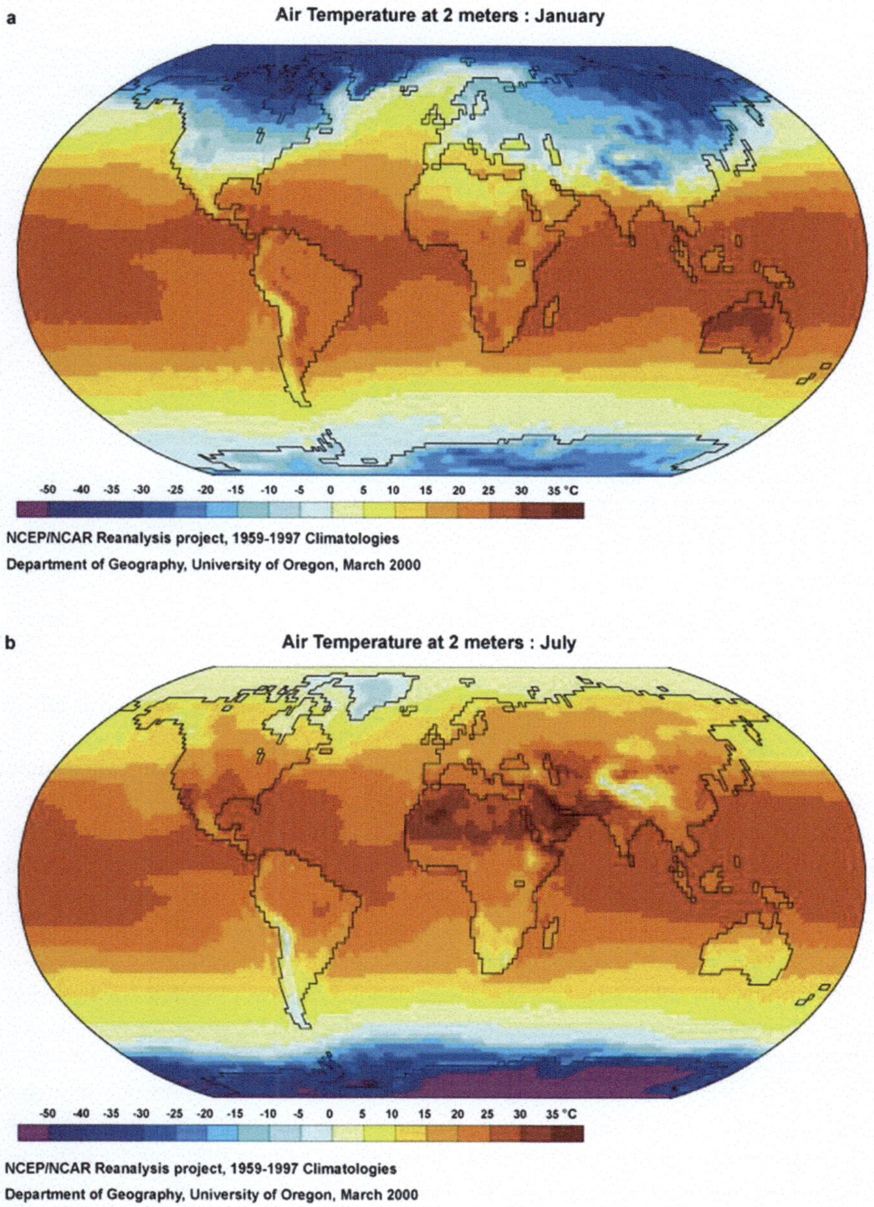

Fig. 2.15 Mean temperature for **a** January; and **b** July. *Source* UCAR MetEd. Permission from Tara Torres

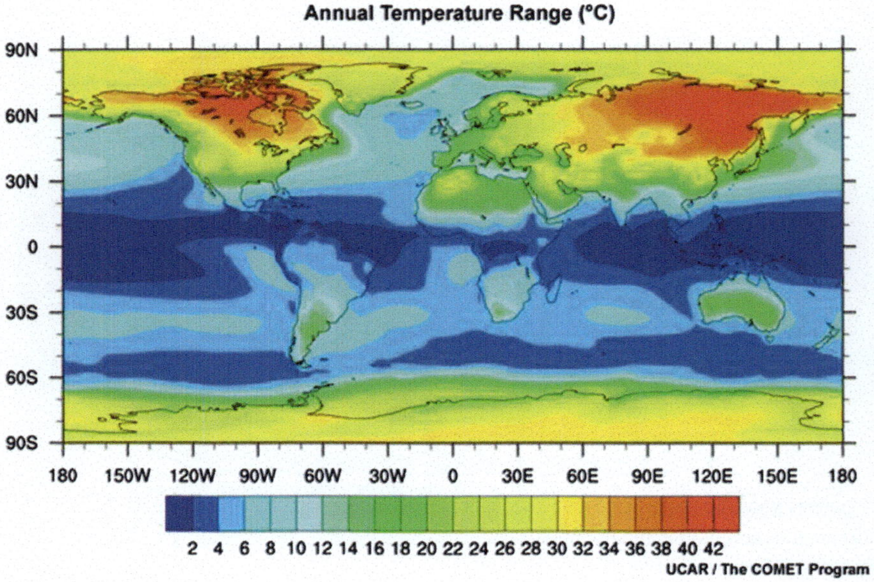

Fig. 2.16 Mean annual temperature range. *Source* UCAR MetEd. Permission from Tara Torres

evapotranspiration may be limited by a lack of water availability. This index tells a similar story as the Bowen ratio described previously.

Figure 2.17 shows the distribution of the difference between evaporation and precipitation. Despite often supporting abundant evapotranspiration and therefore humidity, oceanic areas near the Equator tend to have even more precipitation. Over the subtropical oceans, however, clear skies are dominant, with abundant local Q* and warm ocean currents in the area heating the surface and driving abundant evaporation. The lack of cloud cover means that little precipitation occurs in these areas. The reasons for the abundant precipitation in the equatorial areas, the lack of precipitation in the subtropical oceanic areas, and the patterns elsewhere in the world are related to preferred locations and intensities of atmospheric pressure systems that will be explained in the next chapter.

One of the keys to understanding atmospheric humidity is the concept of **water vapor capacity**, which is the maximum amount of humidity that can occur before the air becomes saturated. Water vapor capacity depends on temperature, with warmer air having a greater water vapor capacity than colder air. This relationship occurs because warmer air has more available energy (indexed by Q*) to evaporate water, allowing the water to enter gaseous (i.e., vapor) form. Cooler air has less energy available to drive the evapotranspiration process, and instead some of the vapor must condense back into a liquid or deposit directly into a solid. We see evidence of this condensation and deposition in clouds, which form after air has risen high enough to cool enough to reach its water vapor capacity and therefore begin to condense or deposit the water

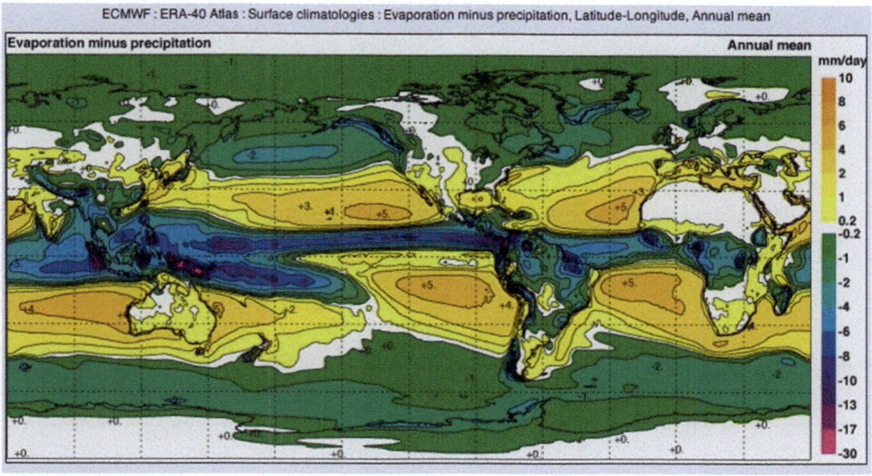

Fig. 2.17 Mean difference between evaporation and precipitation. Positive values indicate that evaporation exceeds precipitation and negative values denote precipitation that exceeds evaporation. *Source* ECMWF. https://sites.ecmwf.int/era/40-atlas/docs/section_B/charts/B12_LL_YEA.html

vapor present. Thick clouds usually indicate condensation, while wispy clouds with a smeared appearance, often higher in altitude, suggest that deposition has occurred.

Because warmer air has a greater water vapor capacity than cooler air, snow totals are likely to be greatest when temperature is near the freezing point (0 °C; 32 °F) rather than at lower temperatures, all other factors being equal. The snowiest places on Earth are not at the poles but instead in locations such as Buffalo, New York. Snowfall is usually more abundant during spring and fall than winter in many places for this reason too. The driest places on Earth, from an absolute humidity standpoint, are the coldest, because the temperatures are too low to allow much water vapor capacity.

Specific humidity (measured in grams of water vapor per kilogram of air, or g kg^{-1}) is a way to describe how much water vapor is in the air (i.e., atmospheric humidity). The maximum specific humidity on Earth is generally around 33 g kg^{-1} (or 0.033 kg kg^{-1}, or 3.3%), even over the tropical oceans. The smallest specific humidities are near zero, but not zero, over central Antarctica. Figure 2.18a, b shows average annual specific humidity for January and July, respectively.

Relative humidity is the ratio of the specific humidity to the water vapor capacity at that temperature. It is normally expressed as a percentage, with 100% relative humidity meaning that the air is saturated and actively condensing or depositing its water vapor. This does not indicate how humid the air is, but only shows how close the air is to saturation. In warmer environments, the greater water vapor capacity will drive up the denominator in the relative humidity calculation and therefore reduce the relative humidity, all other factors being equal. Thus, if the specific humidity—the actual water vapor content of that air—remains constant, as that air cools such as at

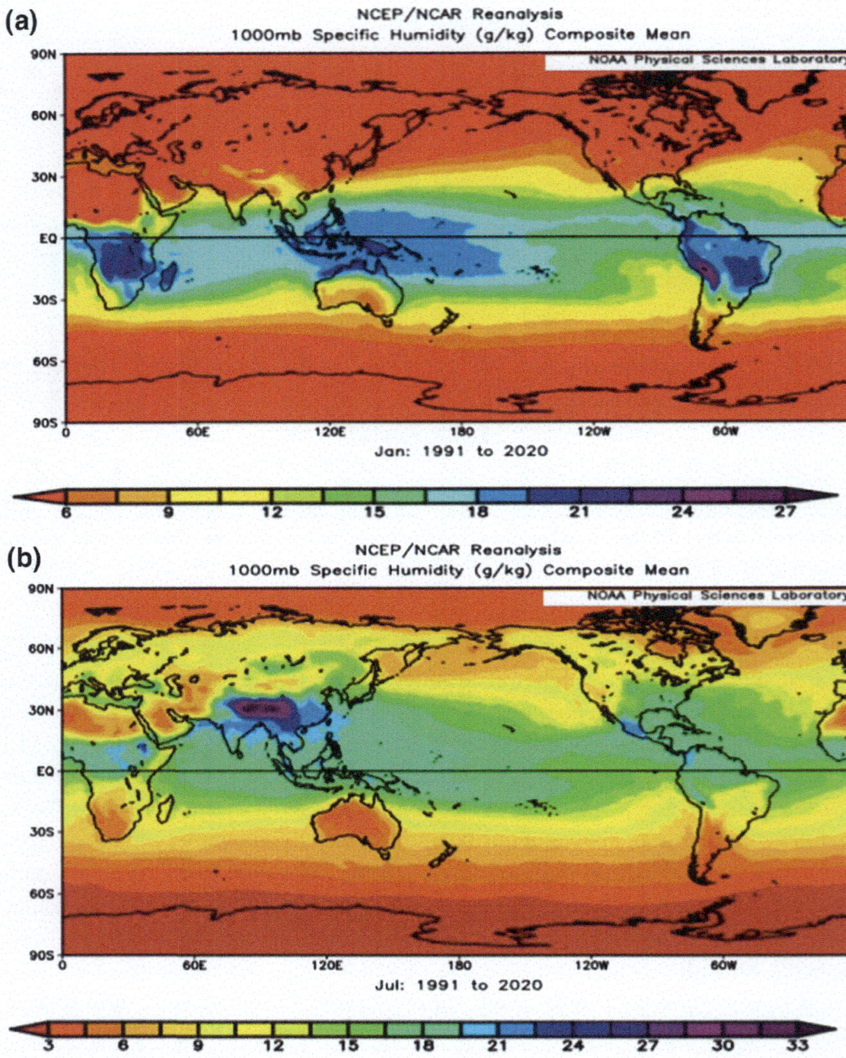

Fig. 2.18 Mean surface specific humidity (g kg^{-1}) in **a** January, and **b** July. *Source* NCEP/NCAR. https://psl.noaa.gov/cgi-bin/data/composites/printpage.pl

night, its relative humidity will increase. If it increases to 100%, then a layer of dew, fog, and/or frost will form. Relative humidity is inversely related to air temperature.

Assuming that all other factors—including the actual amount of water vapor in the air—are equal, relative humidity will be at its smallest during the warmest times. On a daily basis, this will usually occur in the afternoon, while the greatest relative humidity will happen during the coldest time of day—early morning. Seasonally, relative humidity is generally greater during winter, as colder air is more likely to be nearly saturated even if the specific humidity is small. The smallest average relative humidities will occur during summer afternoons, as warmer air has a large water vapor capacity that leaves it far from being saturated. While this concept may be confusing because summer is normally the time of oppressive humidity, the humidity experienced in summer is from the actual vapor content in the air—the specific humidity—rather than relative humidity. In winter, exhaling into cold air will put more vapor in the air, sometimes momentarily exceeding the water vapor capacity (i.e., saturation) and requiring condensation or deposition into the air, as the added water vapor instantly saturates the air near your face. This personal "face fog" then evaporates into the broader surrounding atmosphere.

Figure 2.19 shows satellite-derived mean relative humidity in the low latitudes. In the tropical parts of the Earth, relative humidity is smallest in the places where the temperatures are highest. Contrary to popular belief, the highest temperatures are usually not near the Equator but instead in the subtropics (fringe areas of the map), where persistent high-pressure patterns favor largely cloudless skies that can drive up temperatures. The equatorial areas (middle area in the map below) are generally too cloudy to have the highest temperatures and thus support greater relative humidities.

Dew point temperature is another important humidity variable. It is the temperature at which saturation occurs, thereby inducing condensation, and must be less than or equal to the air temperature. Although the dew point is a temperature, it is indicative of humidity values. The closer the dew point temperature is to the air temperature, the more humid the air. The **dew point depression**, which is the difference between the dew point and the surface air temperature, is used to indicate the humidity content of air. If the dew point temperature is below freezing, then it is

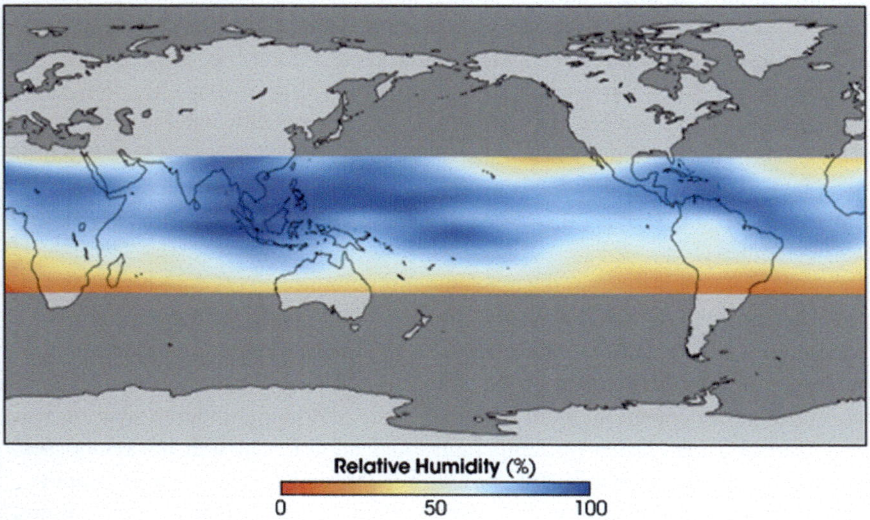

Fig. 2.19 Satellite-derived mean relative humidity. *Source* NASA https://eoimages.gsfc.nasa.gov/images/imagerecords/4000/4312/h2odata_540.jpg

called the **frost point temperature**, and deposition rather than condensation begins at that temperature.

Stability

Vertical motion is essential for weather and climate processes because it allows the humidity in the air to cool to the dew point temperature and condense and/or deposit, thereby forming clouds that grow vertically and perhaps precipitate. In some situations, the local air is more inclined to rise than in others. The tendency for a local parcel of air to rise is known as its **stability**. Specifically, an **unstable atmosphere** is one in which air locally has a tendency to rise spontaneously, and a **stable atmosphere** is one in which air locally does not have a tendency to rise, at least not spontaneously on its own. The stability condition of the atmosphere changes across space and time, and even in different layers of the atmosphere at the same place and time. In extremely unstable atmospheres, vigorous upward vertical motion called **updrafts** are present, which are associated with severe weather.

To understand the stability or **instability** conditions of the air on a given day and time, it is first necessary to understand the way that temperature changes in the atmosphere. The rate at which an imaginary parcel of air will cool if it ascends through the atmosphere without gaining or losing any energy from the surrounding air—i.e., **adiabatically** (known as an adiabatically-moving air parcel)—is the **unsaturated** (sometimes known as dry, even though air may not be completely dry when it is used) **adiabatic lapse rate** and is abbreviated by the upper-case Greek letter gamma with a subscripted "u" (Γ_u), or UALR. Every place on Earth, every day of the year, Γ_u has the value of about 9.8 K km^{-1}, which is the same as a cooling of 9.8 °C km^{-1}, or approximately 1 K per 100 m (which is the same as 1 °C per 100 m). By convention, all lapse rates are positive when temperature decreases with height, as is always the case with Γ_u. Air will always warm at the same rate if it descends through the atmosphere without gaining or losing any energy from the surrounding air—i.e., adiabatically. The UALR itself is derived mathematically from the fundamental gas laws in chemistry.

If at any point on its rising, the adiabatically-moving parcel of air becomes saturated with moisture because it has cooled to its dew point temperature, then it will always cool at a lesser rate, which varies based on temperature and moisture content. This lesser rate occurs because of the latent heat that is released during condensation, freezing, or deposition. Even though the rate of cooling under these saturation circumstances varies based on temperature and humidity, it will always be positive, because temperature will still cool with increasing height, even as latent heat is released during condensation. This rate is called the **saturated** (or wet or humid) **adiabatic lapse rate** (Γ_s or SALR). For instance, if the air temperature is 20 °C (68 °F) when it becomes saturated, it will begin cooling at a rate of about 4 °C km^{-1}, instead of the 9.8 °C km^{-1} rate of cooling with height when it was unsaturated. The SALR changes continuously with increasing height as its temperature and pressure changes. At successively lower temperatures as the parcel rises higher and higher, it will have less and less water in it, so it will have less and less latent heat to release during condensation, so its rate will get closer and closer to the UALR (9.8 CV km^{-1}) but it will never exceed that rate. If the saturated parcel of air would start to descend, it would warm at the UALR because it could not remain saturated, as its warming would instantaneously increase the water vapor capacity and make relative humidity less than 100%. Like the UALR, the SALR can be derived mathematically, for any starting temperature and humidity, from the fundamental gas laws in chemistry.

The third player in the game for assessing atmospheric stability is the actual rate at which temperature changes with height in the static atmosphere adjacent to that adiabatically-rising (or sinking) parcel of air. In most cases, the static air temperature decreases as height increases. There are two reasons for this tendency. First, air is less compressed by the weight of overlying air, with increasing height. This allows it to expand, and according to basic gas laws, its expansion will cause cooling, much like air released from a confined tire feels cool when it can expand into the free air under less pressure. Second, the energy balance has already shown that more units of radiant energy are absorbed by the surface than by the atmosphere. This implies that the surface is warmer than the air above it, on average.

This rate of temperature change with height in a static atmosphere varies in space and time. It is known as the **environmental lapse rate** and is abbreviated by the lower-case Greek letter gamma (γ) or ELR. ELR may be greater than, equal to, or less than Γ_u or Γ_S. It can be positive or negative, but on a global, annual average basis, it is about 6.4 K km^{-1} (which is the same as 6.4 °C km^{-1}). In other words, ascending a mountain that is 1 km above the starting point, a legitimate guess is that the temperature on its peak will be, on average, 6.4 °C lower than the temperature at the starting point. It is useful to note here that when comparing Fahrenheit versus Celsius versus Kelvin temperatures, without an absolute value of temperature, we multiply by 1.8 to go from C to F, so 6.4 °C is the same as 11.5 °F. So, a mountain with 1 km of local relief might have a temperature that is about 11.5 °F cooler than its base, but this can vary from day to day and place to place. It is useful to think of γ as the temperature gradient of the area around the adiabatically-moving parcel of air—NOT moving vertically, expressed in °C per kilometer. ELR is measured twice-daily around the world by weather balloons that have attached instrument panels and report telemetrically temperature and other readings at various heights above the surface.

To recap, three lapse rates must be considered as prerequisites to assessing stability of the air—its tendency to rise spontaneously, which might lead to vertical cloud growth and perhaps precipitation. The unsaturated lapse rate (Γ_u) or UALR is always the same, every place on Earth, every day of the year. The adiabatically-moving, unsaturated parcel of air will always cool at this rate of 9.8 °C km^{-1} of ascent, unless it becomes saturated somewhere along its ascent. If it started at a temperature of 10 °C (50 °F) by the time it reaches 1 km in altitude its temperature will always be 0 °C (32 °F) if it remains unsaturated throughout. However, if it becomes saturated at any point on the ascent, it begins to cool at Γ_S, and its rate of cooling with height is not constant. Instead, it changes with temperature and humidity, but for a given

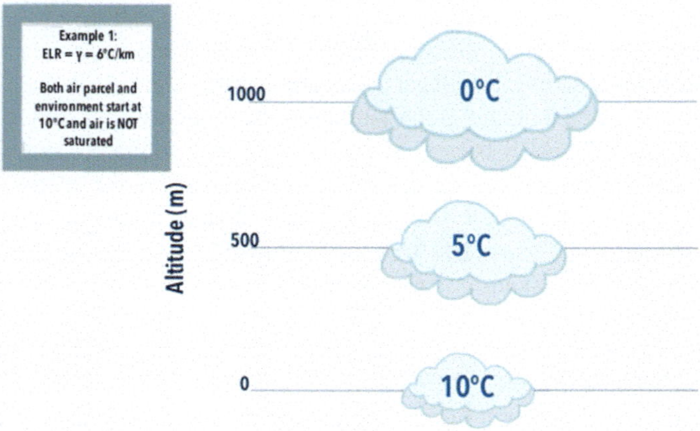

Fig. 2.20 Schematic example of a stable atmosphere; the puffy air represents an air parcel, not a cloud. *Source* Siarah Hall

temperature and humidity, SALR is the same, every place on Earth, every day of the year. And the ELR (γ) changes from place to place and from day to day, but its global annual average is 6.4 °C km^{-1}.

Determining the stability conditions of the atmosphere is as simple as comparing the adiabatically-moving air parcel's temperature to the temperature of its surrounding environment. In Fig. 2.20, the adiabatically-moving air parcel is represented by the parcel that expands as it rises, and it cools at Γ_u. But in this example, the surrounding environment cools by 6 °C km^{-1} (γ). So with the surface being 10 °C in this example, the air parcel's temperature will be 5 °C by the time it rises to 500 m, compared to the surrounding environment's temperature of 7 °C. By the time the air parcel rises to 1 km (1000 m), it will have cooled to 0 °C, compared to its surrounding environment's temperature of 4 °C. It is apparent that the air parcel will be cooler than its surrounding environment at either level considered. The atmosphere is stable in this case, because the air parcel will have no spontaneous rising.

A plot of Γ_u, if the parcel is unsaturated, or by Γ_s if the parcel is saturated, versus the temperature of its surrounding environment (shown by a plot of γ) can reveal the stability condition more quickly. If the air parcel is cooler than its surrounding environment at a given height, it will sink, and the atmosphere is stable. If the air parcel is warmer than its surrounding environment at that height, it will rise, and the atmosphere is unstable. If the air parcel is the same temperature as its surrounding environment at that height, it will want to stay where it is, and the atmosphere is neutral. In practice, meteorologists and climatologists have software that calculates stability conditions and simple indices that describe the stability conditions at a given place and time. The scenario described in Fig. 2.20 is graphed in Fig. 2.21. It is apparent that the air parcel, with a temperature shown by the red solid line, is cooler than its surrounding environment, with a temperature shown by the green dashed line, at any height on the graph. Thus, in this case the atmosphere is stable.

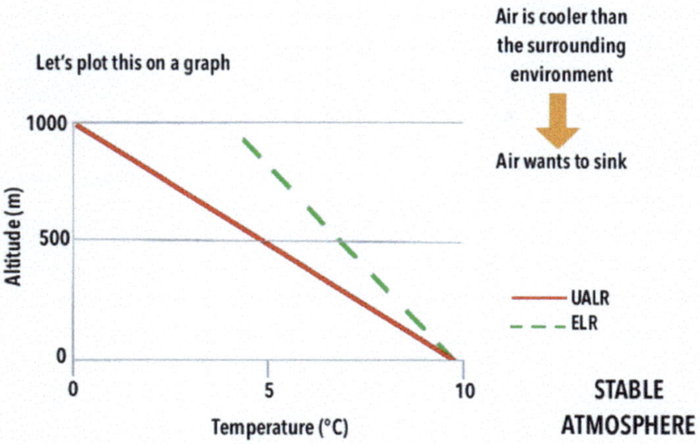

Fig. 2.21 Graph of the stable atmosphere shown in Fig. 2.20. *Source* Siarah Hall

In Example 2 (Fig. 2.22), γ actually is negative—temperature in the surrounding environment actually increases with height. Any temperature increase with increasing height, as in Example 2, is called a **temperature inversion**, or just an **inversion**. Cases of inversions represent extremely stable atmospheres.

Example 3 (Fig. 2.23) indicates another case—this time one in which saturation is reached by the adiabatically-rising air parcel on its ascent. As air ascends, it expands, which initiates cooling. As it cools, molecular energy decreases, which limits the air's ability to evaporate more water and forces some water vapor to condense back into the liquid phase. The air parcel then cools at a lesser rate (SALR), shown by the blue dotted line in Fig. 2.23, from that level upward. In some cases, the transformation

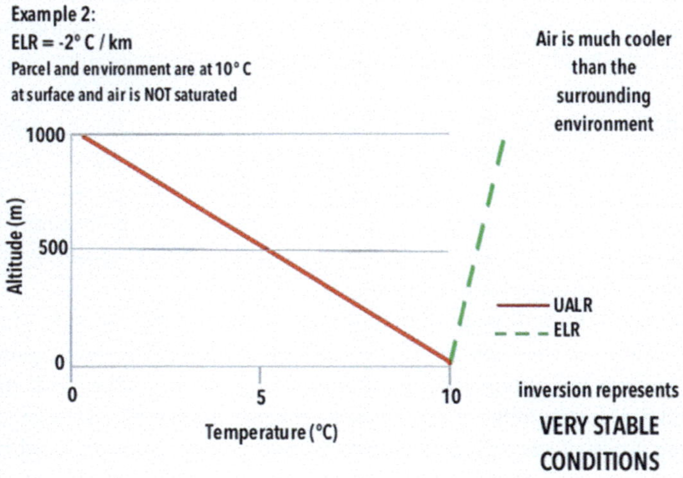

Fig. 2.22 Graph of an extremely stable atmosphere. *Source* Siarah Hall

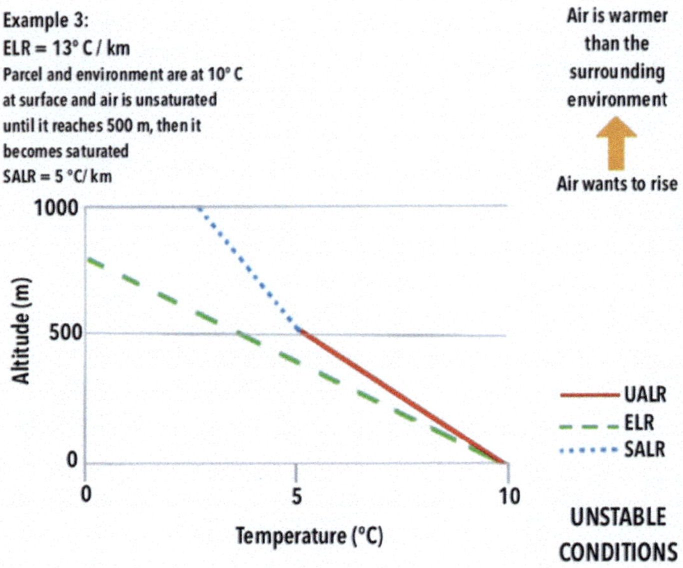

Fig. 2.23 Graph of an unstable atmosphere. *Source* Siarah Hall

from vapor (gas) directly to solid, bypassing the liquid phase—deposition—occurs. In physics, the process of skipping a phase, either from vapor to solid, or solid to vapor, is collectively termed sublimation. In the atmospheric sciences, the term deposition denotes water moving from a vapor to a solid as that form of sublimation occurs frequently in normal atmospheric processes.

Frost formation is an example of the deposition process. It typically occurs when water vapor molecules move along a very cold surface. This allows the water vapor to adhere to and freeze directly on the surface. Additional molecules repeat the process and the surface may become covered with the frost. Snow formation frequently involves deposition of water vapor within clouds with temperatures well below freezing.

Notice in Fig. 2.23 how saturation, which occurs at 500 m in this example, makes the air parcel even warmer than its surrounding environment at a given height. So if enough vapor is put into the air (like, perhaps having humid air from over the Gulf of Mexico being pushed into Louisiana by a southerly breeze), the atmosphere becomes more unstable. This is familiar, as we have all experienced increasingly humid air rising suddenly, creating a sudden downpour.

This property of humidity destabilizing the atmosphere makes the world the way that it is. If adding humid air would somehow make the atmosphere more stable instead of making the atmosphere more unstable as in the real world, then there would be much less precipitation, particularly inland, and our human activities, such as agriculture, hydroelectric power generation, and snow ski industry would be much more difficult and expensive.

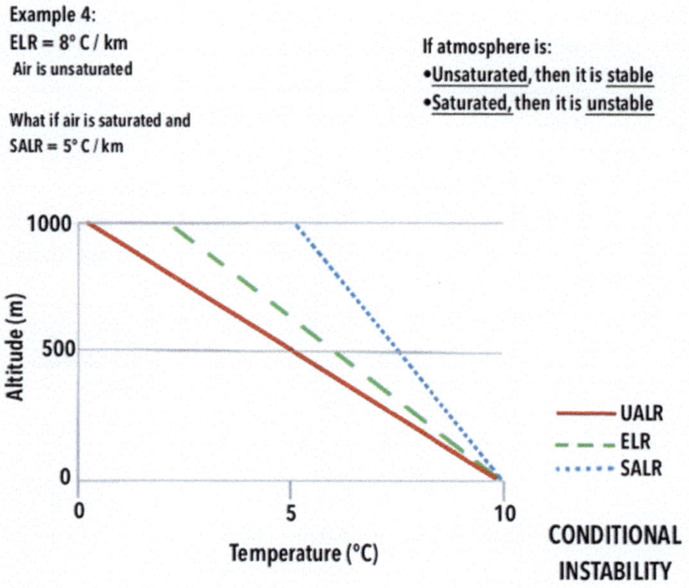

Fig. 2.24 Graph of a conditionally unstable atmosphere. *Source* Siarah Hall

Example 4 (Fig. 2.24) illustrates the case in which γ falls between the two adiabatic lapse rates. This is fairly common, because as shown before, the global average of γ is 6.4 °C km^{-1}, which is less than 9.8 and greater than what often happens in saturated atmospheres. This is the case of **conditional instability**. The "condition" is that stability depends on whether the atmosphere is saturated or not. Figure 2.25 serves as a reminder that γ can change with height, sometimes substantially, making some atmospheric layers stable while others are unstable at the same time.

Figure 2.26 shows a generally unstable situation where "fair weather cumulus" occurred. Fair weather cumulus is represented by stand-alone small clouds. Cumulus congestus is depicted to the left of the image while a towering **cumulonimbus** (aka thunderstorm) dominate the center of the image. The situation exists when surface heating causes localized convection and condensation associated with rising parcels. The thunderstorm does not require an overriding atmospheric low-pressure system associated with a broader stormy area, as will be described in the next chapter.

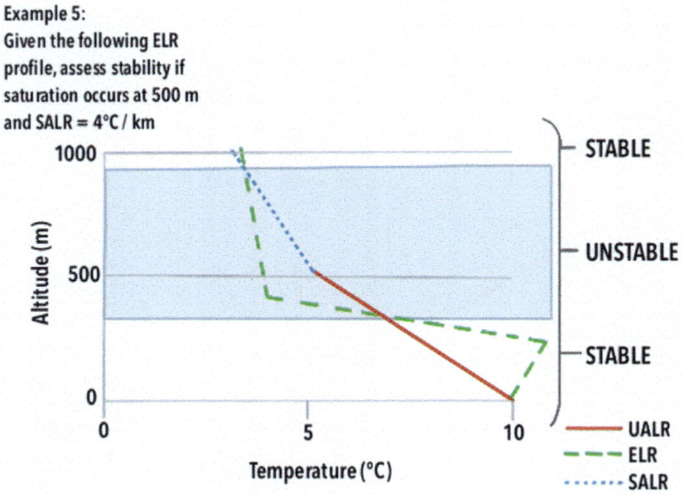

Fig. 2.25 Graph of an atmosphere with varying stability conditions by height. *Source* Siarah Hall

Fig. 2.26 Coastal thunderstorm over Biloxi, Mississippi, on an otherwise pleasant 17 July 2022, with Gulfport, Mississippi, in the foreground. *Source* Authors

Further Reading

Dai, A. (2006). Recent climatology, variability, and trends in global surface humidity. *Journal of Climate, 19*(15), 3589–3606. https://doi.org/10.1175/JCLI3816.1

Kapsch, M.-L. (2015). *The atmospheric contribution to arctic sea-ice variability*. Ph.D. Thesis, Stockholm University.

Liang, S. (2018). Volume 5 overview: Recent progress in remote sensing of Earth's energy budget. In S. Liang (Editor-in-Chief), *Comprehensive remote sensing*. Elsevier. 978-0-12-803221-3.

Reynolds, S. J., Rohli, R. V., Johnson, J. K., Waylen, P. R., & Francek, M. A. (2021). *Exploring physical geography* (Third edition) (736 pp). The McGraw-Hill Companies, Inc., New York, New York. 978-1-260-36499-6.

Rohli, R. V., & Vega, A. J. (2018). *Climatology* (Fourth edition, 418 pp). Jones and Bartlett, Sudbury, Massachusetts. 978-1-284-11998-5.

Trenberth, K., Fasullo, J. T., & Kiehl, J. (2009). Earth's global energy budget. *Bulletin of the American Meteorological Society, 90*(3), 311–323. https://doi.org/10.1175/2008BAMS2634.1

Chapter 3
Primary Atmospheric Circulation

Abstract The Sun's unequal heating of Earth creates spatial pressure differences that drive the primary atmospheric circulation. The pressure gradient force moves air from higher to lower pressure. The Coriolis effect (apparently, to us) deflects winds to the right (left) in the Northern (Southern) Hemisphere, and friction slows the wind and therefore weakens the Coriolis effect. Quasi-permanent pressure-derived features include the two Hadley cells, which have rising motion and low pressure known as the inter-tropical convergence zone (ITCZ) near the Equator and sinking motion with high pressure at the subtropical anticyclones near 30° N and 30° S latitude, with trade winds between them and the subtropical jet stream flowing aloft near the Hadley cells' poleward edges. Rising motion at the subpolar lows near 60° N and 60° S and sinking air and high pressure near both poles support the near-surface polar easterlies. The mid-latitudes of each hemisphere have near-surface westerly winds. West-to-east flow also prevails aloft over the extratropics of both hemispheres, with embedded poleward-bending ridges and equatorward-bending troughs within a continuous belt of Rossby waves and the polar front jet stream circumnavigating each pole. The ridge-to-trough side of the Rossby wave generally supports clear skies while the trough-to-ridge side supports surface cyclones and cloudy conditions.

Keywords Circumpolar vortex · Coriolis effect · Geostrophic and gradient wind · Hadley cells · Jet streams · Meridional versus zonal flow · Mid-latitude westerlies · Northeast and southeast trade winds · Polar cells · Polar easterlies · Pressure gradient force · Rossby wave ridges and troughs · Subpolar lows · Subtropical anticyclones

Winds and the Primary Circulation: An Overview

Air that has energy in the form of positive *net radiation* (Q*) and moisture in the form of *humidity* moves in order to redistribute that energy and matter, in accordance with the *Second Law of Thermodynamics*. The global atmospheric and oceanic circulation exists to try to redistribute energy that is not evenly received from the Sun, and in so

R. V. Rohli et al., *Atmospheric and Oceanic Circulation*,
https://doi.org/10.1007/978-3-031-66142-6_3

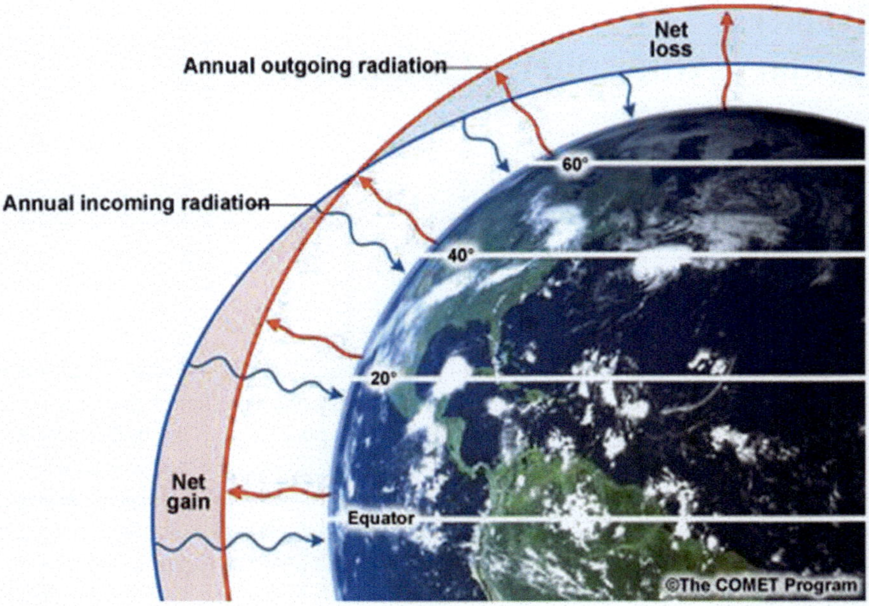

Fig. 3.1 Idealized latitudinal profile of positive and negative net radiation. *Source* UCAR MetEd

doing, the matter (i.e., humidity) moves with it. The atmosphere redistributes about 60% of the energy imbalance in the form of winds. The oceans redistribute about 25% in the form of ocean currents. And **latent heating** does the remaining 15% of the job. This part happens as surplus energy equatorward of about 40 degrees of latitude (Fig. 3.1) is used to sustain *evapotranspiration* and then the vapor moves to higher latitudes, transforms phase through *condensation* or *deposition*, and releases latent heat poleward of where it evaporated.

The so-called **primary circulation** of the atmosphere refers to the persistent, global-scale patterns. By contrast, the **secondary circulation** refers to the more short-lived and local-to-regional-scale flow patterns. For example, mid-latitude enclosed areas of high and low pressure that will be covered more comprehensively in the next chapter are secondary circulations.

Controls of Winds

Far above the surface, where *friction* can be neglected, the forces that determine wind speed and direction are relatively simple because only two factors are involved. The **pressure gradient force** affects direction and speed. It moves air laterally from higher to lower pressure, with wind speed proportional to the difference between the pressure per unit distance (i.e., the pressure gradient). The **Coriolis effect** then affects direction by apparently (to us) deflecting wind to the right in the Northern Hemisphere and to the left in the Southern Hemisphere. It is said to be "apparent" because the wind is being mapped with respect to a rotating planet. Figure 3.2 shows how winds, represented by the drawn line, on a rotating planet when viewed from above can have an apparent turn, even though they are actually moving only in a straight path in response to the pressure gradient force.

The Coriolis effect can be calculated as $f \times V$, where f is the **Coriolis parameter** such that

$$f = 2\Omega \sin \varphi$$

where Ω represents the angular velocity of the Earth ($7.27 \times 10^{-5}\ \mathrm{s}^{-1}$, which is essentially constant), φ is the latitude, and V represents the wind speed. Thus, according to the equation above, the Coriolis effect is a function of the wind speed and the latitude, with greater effects when winds are fast and latitude is large, and a zero value at the Equator, where φ is zero.

Figure 3.3 shows the effect of the pressure gradient force and Coriolis effect on airflow in the Northern Hemisphere. The lines drawn on the map each represent a line of constant pressure—called an **isobar**, but in official weather maps such lines on upper-level maps show **geopotential height**—called an **isohypse**. The labeled numerals on each isobar tell how much pressure, in hundreds of Pascals (or hPa), which is equivalent to the number of **millibars** (mb), exists. At sea level, an average of 1013 hPa or mb of surface pressure exists, so the numbers in Fig. 3.3 represent a height where pressure is about 60% of the sea level pressure. Upper-level maps that show geopotential heights tell how many meters high in the atmosphere above sea level that a fixed pressure, such as 500 millibars, is experienced. The greater the isohypse label, the higher in the sky that 500 mb of pressure is reached. Because pressure always increases moving downward from above, due to the additional weight and compression of the atmosphere from the overlying column, it can be inferred that higher geopotential heights correspond to higher pressures, and lower geopotential heights correspond to lower pressures.

Then why are these kinds of maps not simply labeled with the pressure value, as in Fig. 3.3, instead of the geopotential height value? The main reason is that much of

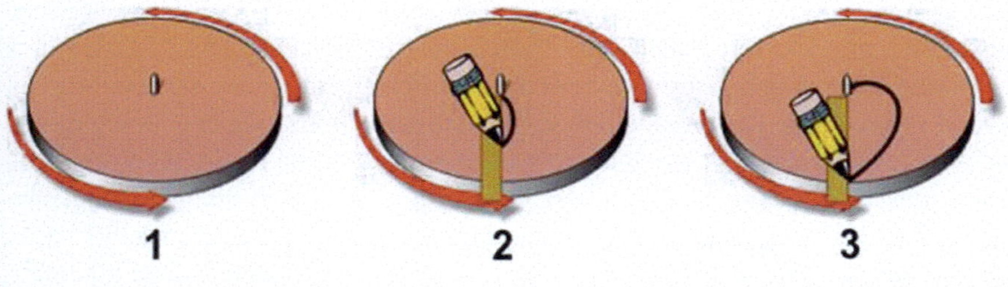

Fig. 3.2 Schematic depicting the "apparent" nature of the Coriolis effect, even though the pencil is moving in a straight line. *Source* NWS Teaching Resources

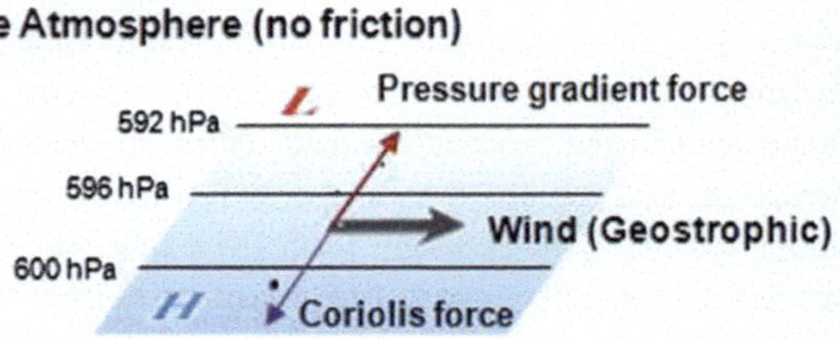

Fig. 3.3 Effect of the pressure gradient force and Coriolis effect on wind direction. *Source* UCAR MetEd

the developmental roots of meteorology were in the aviation industry. Aircraft tend to need to maintain constant pressure, not constant height. So a map of the height at which a constant pressure, like 500 mb—the level where approximately half of the atmospheric mass lies above—will tell aviators how many meters they must move vertically to remain safe. Maps are made for a range of constant pressure surfaces.

The top of Fig. 3.3, which represents the poleward latitudes, has lower pressure, and therefore lower geopotential height. The bottom, which represents the equatorward latitudes, has higher pressure and therefore higher geopotential height. Notice what happens to the airflow after the pressure gradient force gets air moving across the isobars or isohypses from higher to lower pressure, and the Coriolis effect deflects it to the right. The airflow parallels the isobars or isohypses, with higher pressure to the right of the flow (in the Northern Hemisphere) and lower pressure to the left of the flow (in the Northern Hemisphere).

This airflow that results from a balance between the pressure gradient force and the Coriolis effect is called the **geostrophic wind**. In the Northern Hemisphere, the geostrophic wind has higher geopotential heights (and also higher pressure) to the right of the flow, when facing the direction that the air is moving. Try it by studying the diagram above. In the Southern Hemisphere, the geostrophic wind has higher pressure to the left of the flow, when facing the direction that the air is moving. This is because the Coriolis effect apparently turns the air to the left in the Southern Hemisphere, but the pressure gradient force still works to move the air from higher to lower pressure, regardless of hemisphere. Furthermore, regardless of hemisphere, faster winds exist over steeper pressure gradients, which correspond to densely-packed isohypses. Slower winds are associated with gentler pressure gradients, which correspond to having widely-spaced isohypses.

These principles remain true near the surface, but surface winds are somewhat more complicated near the surface, where they are the result of a balance between the pressure gradient force, the Coriolis effect, and friction. The effect of friction is to slow the wind, which in turn weakens the Coriolis effect because it is proportional to wind speed. Thus, geostrophic flow cannot exist where friction is non-negligible. The result is that near-surface wind crosses the isobars at an angle, with the angle proportional to the amount of friction. Study Fig. 3.4, which compares the flow without friction being considered—the geostrophic case—to the case when friction is important too, like at the surface. Notice that surface weather maps, including Fig. 3.4, nearly always show pressures rather than geopotential heights. There is no need to show geopotential heights because aircraft cannot fly at the surface!

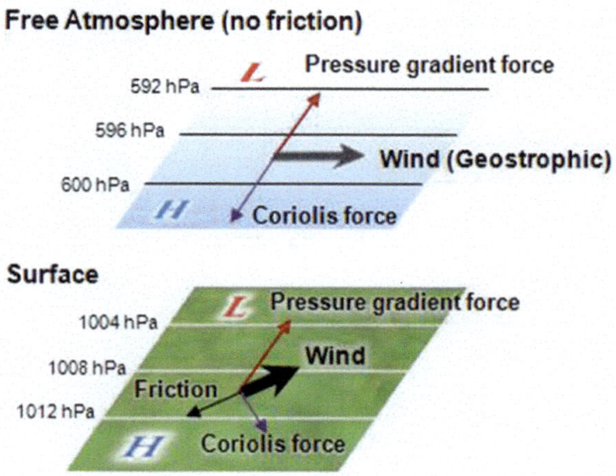

Fig. 3.4 Upper-level and surface airflow associated with straight isohypses or isobars. *Source* UCAR MetEd

Now that airflow associated with straight (upper-level) isohypes or (surface) isobars is apparent, what about flow around an enclosed area of surface low pressure (i.e., a **cyclone**)? There, air flows slightly inward across the isobars—converges toward the low-pressure center from all directions, in accordance with the direction that the pressure gradient force is oriented. The inward motion causes it to pile up, rise, cool, and if it cools to its *dew point temperature* or *frost point temperature*, condense or deposit. In such places, the low pressure is associated with cloudiness and *precipitation*. But the circulation is not solely inward; because the Coriolis effect turns the air to the right (in the Northern Hemisphere) as it moves inward, circulation around surface cyclones is counter-clockwise and inward in the Northern Hemisphere.

Because the Coriolis effect turns air to the left in the Southern Hemisphere as it moves inward, circulation around surface cyclones is clockwise and inward in the Southern Hemisphere. The inward flow for cyclones "fills them in," eventually killing them by bringing in higher pressure from outside, across the isobars, to increase the lower pressure in the middle until its pressure is no longer any lower than that of the surrounding air. This spells the end of the cyclone.

Enclosed areas of high pressure (**anticyclones**, indicated by the blue "H" in Fig. 3.5) are associated with air sinking (which exerts extra pressure on the surface), whereupon the air diverging out away from the high-pressure center in all directions, in accordance with the direction the pressure gradient force is always oriented. The sinking motion over the surface anticyclone restricts upward vertical cloud development, making skies generally fair or clear. But once the air begins to move out from the high-pressure center, the Coriolis effect moves the air to the right (in the Northern Hemisphere), but not all the way to the right until it is parallel to the isobars. Thus, anticyclones near the surface in the Northern Hemisphere have clockwise and outward flow. Anticyclonic flow in the Southern Hemisphere is counter-clockwise and outward. Nature "spreads out" the "extra" air near the anticyclonic center by

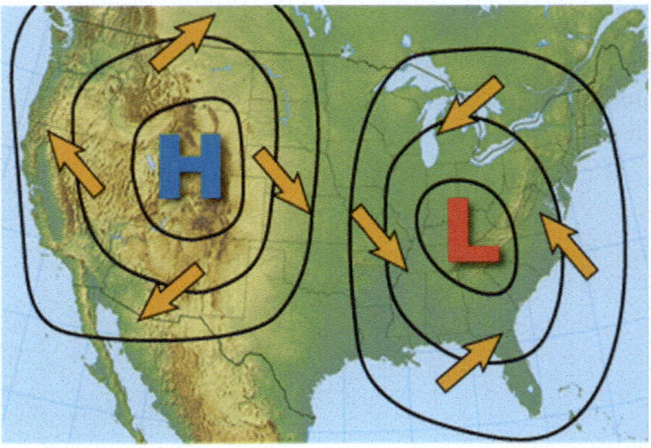

Fig. 3.5 Hypothetical surface anticyclone and cyclone positioned over the U.S., with isobars drawn and airflow indicated. *Source* NWS Teaching Resources

moving the accumulated air outward to all directions, thereby eventually eliminating the higher pressure in the middle to eventually kill them.

Primary Circulation: Introduction

If air or water movement across the Earth would be monitored over a long period of time, certain broad-scale tendencies of that circulation would be noticed to be most common and influential. The *primary circulation* or **general circulation** refers to this prevailing, steering movement of air or water at broad spatial and long temporal scales. For example, the term **mid-latitude westerlies** is used to describe the primary circulation in the mid-latitudes. Recall that winds are always named for the direction from which they blow, so the mid-latitude westerlies blow generally from west to east. However, the convention in oceanography is the opposite—ocean currents are named for the direction to which they flow, so for example, a north current flows to the north, but a north wind blows from the north.

However, the primary circulation is not necessarily apparent every day, at every spatial scale. Winds in the mid-latitudes are not always from west to east. Instead, smaller-scale winds at the *synoptic scale*, and even at the **regional scale** (or *meso scale*) and *micro scale*, all of which are the subjects of subsequent chapters, often mask or suppress the prevailing pattern. The net influence of atmospheric and oceanic circulation at all scales gives character to the natural environment of a place.

Nevertheless, it is important to understand the prevailing patterns because they set the stage for the movement of energy, matter, and momentum throughout the Earth-atmosphere-ocean system. This simplifies the descriptions of the atmospheric circulation, by sorting out events according to size. At the same time, it is important to remember that atmospheric circulation is complex. Imagine it as a huge set of rivers in three dimensions extending continuously around the Earth's atmospheric environment, all influencing and being influenced by each other, and each with a smaller spin-off circulation known as an **eddy**—and more, smaller eddies embedded within larger eddies within even larger eddies. The purpose of the circulation—primarily the atmosphere's general circulation—is to try to redistribute energy that is not evenly received from the Sun.

Our knowledge of global winds comes from both empirical pressure and wind data and theoretical studies of fluid motion in the 1930s called **dishpan experiments**. These involved a circular, colored-dye-in-water-filled dishpan that was heated on its periphery (representing the equatorial part of Earth), cooled in its center (representing the polar part of Earth), and spun on an axis in a counter-clockwise direction when viewed from above (representing Earth's rotation), all in direct proportion to the Earth's *temperature* gradient and speed of rotation.

The dishpan experiments revealed that the dye moved in a relatively predictable way. The fluid moved in an attempt to equalize the heat differences between the center and the periphery. These experiments gave hope that the Earth's general circulation in the atmosphere could be understood because its motion follows certain predictable patterns. But the modeling of that predictable motion would have to wait a few decades for computational advancements.

Tropical Flow Associated with the Hadley Cells

Near the periphery of the dishpan, it was observed that the water was exchanged more or less directly between the periphery and the center, but this would not continue all the way to the center. In the "real world," because the area near the Equator generally experiences greater Q* than the rest of the world, that hot air rises *adiabatically*—without gaining or losing any heat from surrounding air. This rising air cools, and if it cools to the dew point temperature, it will *condense* into clouds. As was described in Chap. 2, the condensation process releases *latent heat* to keep the air parcel buoyant.

After the air rises, it moves poleward. However, by the time it reaches about 30° N and 30° S, it will have chilled aloft to such an extent that its density would increase enough, supplemented by effects of Earth's rotation, to cause the air to sink. After the air sinks, the air near the surface moves toward the Equator, causing the circulation to complete a cell, as shown in Fig. 3.6. One complete cell is predominantly in the Northern Hemisphere and the other is predominantly in the Southern Hemisphere. Each is known as a **Hadley cell** (Fig. 3.6). The part of the Hadley cells that flow poleward aloft play an important role in transporting excess Q* poleward (Fig. 3.7).

The area where the air is rising, shown in Fig. 3.6 as the low-pressure belt (designated by the "L's") along the equatorial edge of each Hadley cell, is called the **intertropical convergence zone** (ITCZ), also known as the **equatorial trough**. These long but descriptive names remind us that the ITCZ generally remains between the Northern and Southern Hemisphere tropics and supports convergence of air as a thermally-induced low-pressure belt near the Equator draws air converging toward the belt, and then reinforcing the rising motion as it is forced upward (see again Fig. 3.7). A **trough** is simply an elongated zone of low pressure—the opposite of a **ridge**—which is an elongated zone of high pressure. The ITCZ is characterized

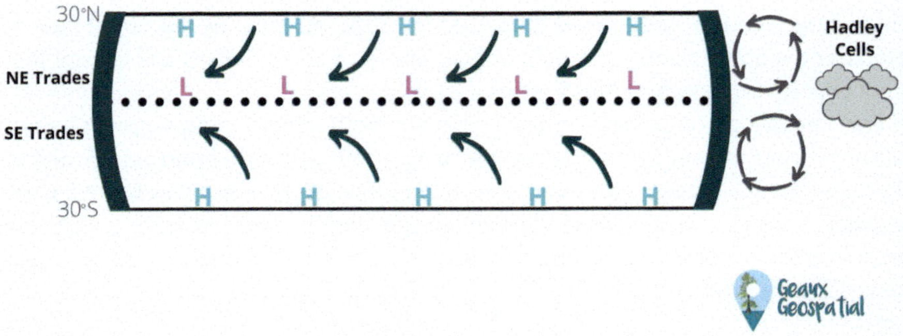

Fig. 3.6 Tropical component of the primary circulation in map view (large diagram) and profile (side), with vertical extent greatly exaggerated. *Source* Geaux Geospatial

Fig. 3.7 Hadley cells in profile view, with vertical extent greatly exaggerated, depicting the poleward heat transport. *Source* UCAR. https://scied.ucar.edu/docs/why-monsoons-happen

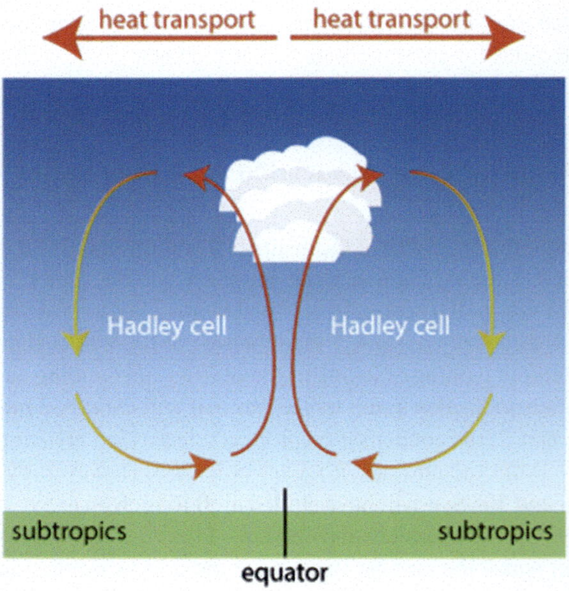

by cloudy and rainy climates, as any low-pressure area with rising air and abundant humidity would support much rainfall.

Winds move toward this rising air and low surface pressure near the Equator (Fig. 3.8), but because of the Coriolis effect, which is weak at these latitudes, they are deflected somewhat to the right in the Northern Hemisphere and to the left in the Southern Hemisphere. The result is a very consistent belt of winds that blow from northeast to southwest in the Northern Hemisphere, known as the **northeast trades** or **northeast trade winds**, and a very consistent belt of winds that blow from southeast to northwest in the Southern Hemisphere, known as the **southeast trades** or **southeast trade winds**.

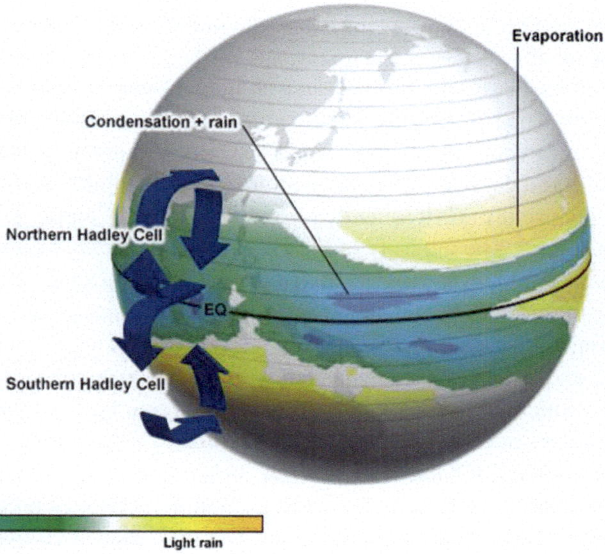

Fig. 3.8 Precipitation zone in association with the rising air, with vertical extent greatly exaggerated, in the inter-tropical convergence zone. *Source* UCAR MetEd

The extensive ocean coverage along the middle part of the Hadley cells where the trade winds exist allows the trade winds to keep surfaces cooler than would otherwise be expected at these latitudes. The cooling effect weakens vertically but reducess the vertical gradient of cooling temperatures with height.

This vertical profile of having a small amount of cooling with increasing height—*environmental lapse rate*—is associated with *atmospheric stability*, as was shown in Chap. 2. In some cases, the chilling from the ocean may be so intense that it actually makes temperatures increase with increasing height in this trade wind zone. Such a **trade wind inversion**—characterized by a negative environmental lapse rate—often exists where the trade winds blow over cool or even cold ocean currents.

Trade wind inversions have implications for the climates where they are common. First, trade wind inversions suppress precipitation. Remember that an increase of temperature with height is called a *temperature inversion*, and air in such areas is unlikely to rise spontaneously, since the coldest air is on the bottom and the warmest air is on top. This is the "cap" on vertical cloud development. The trade wind inversions do allow evaporation of abundant water vapor over the lengthy treks of the trade winds across the tropical oceans. Thus, air in trade wind belts tends to have large *specific humidity* values; in other words, there is abundant water vapor in the air. In places away from the trade wind inversioin where there is a means of lifting the parcels of air, copious amounts of precipitation result. Another implication of the trade wind inversions is that the abundant humidity is trapped beneath the inversion, which can often produce **stratus** clouds, which are low, thick, wide, blanket-like, and they also produce frequent fog as the humid air is chilled from beneath by the cold ocean current.

Within the ITCZ, cloud vertical development is unhindered except by water vapor availability. The ITCZ is discontinuous, with large temporal and spatial variability in the cloud clusters. *Convection* along the ITCZ is strongest over land in summer, which heats substantially to promote more atmospheric instability and vertical lifting. Such conditions occur most frequently in South America, Central Africa, and especially Southeast Asia. The intensity of the ITCZ over Southeast Asia is driven by the heat accumulated both over the terrestrial areas and over the adjacent warm water that the northeast trades pool up as they pull very warm surface water westward across the equatorial Pacific Ocean. By contrast, the ITCZ tends to be weaker over the eastern Atlantic and eastern Pacific where cooler pools of water exist. Off the Pacific coast of South America, the ocean current is so cold (coming from the Antarctic) that the ITCZ remains north of the Equator in that area.

Winds of the ITCZ itself are generally calm or only very light, because of the even pressure gradient and strong vertical ascent of air. This area is called the **doldrums**, as the original meaning of the term was "foolish," which described sailors who tried to move through this area where sailing vessels would stall for long periods of time.

At the poleward edges of the trade wind belt within each Hadley cell, near 30° N and 30° S, the air is sinking semi-permanently. The associated semi-permanent chain of enclosed high pressure—anticyclones—are called the **Subtropical High-pressure Belt**. This name refers to the high pressure exerted by the sinking air near 30° N and 30° S. One of the most persistent areas of quasi-stationary subtropical high pressure is over the North Atlantic Ocean—the **Bermuda-Azores High**, as shown by the global mean surface pressure patterns for January and July in Fig. 3.9a, b. Because high pressure prevails in these areas, convection, cloud cover, and precipitation are suppressed. But in addition, the Sun is nearly directly overhead for an extended period of time, with little moisture or cloud cover to absorb, scatter, or reflect the *shortwave radiation* ($K\downarrow$). The result is that the hottest temperatures on Earth are near the 30° N and 30° S latitudes, as are many of the world's deserts. These belts are known as the **horse latitudes** (Fig. 3.10) because the winds were so calm under these weak pressure gradients that sailors often were forced to throw their horses overboard to conserve food and water during the long trips during the Age of Exploration.

The semi-permanent high pressure associated with the horse latitudes is not uniformly intense across the belt of 30° N and 30° S latitude. Generally, the pressure is strongest over water in summer, because water is cooler than land in summer and will chill the air overlying it, to a greater extent than air at the same latitude that is over the hotter land. This chilling of the air, at least to a greater extent than the air at adjacent latitudes and longitudes, brings a tendency for sinking motion that exerts additional pressure on the surface. Thus, in summer, the oceanic subtropical areas tend to support semi-permanent, cool, dense, sinking air (i.e., high-pressure systems) to a much greater extent than subtropical inland areas.

This semi-permanent anticyclone situated in the subtropical North Atlantic Ocean is known by different names in different parts of the world. Americans refer to it as the "Bermuda High" and Europeans tend to prefer the "Azores High" designation, so here

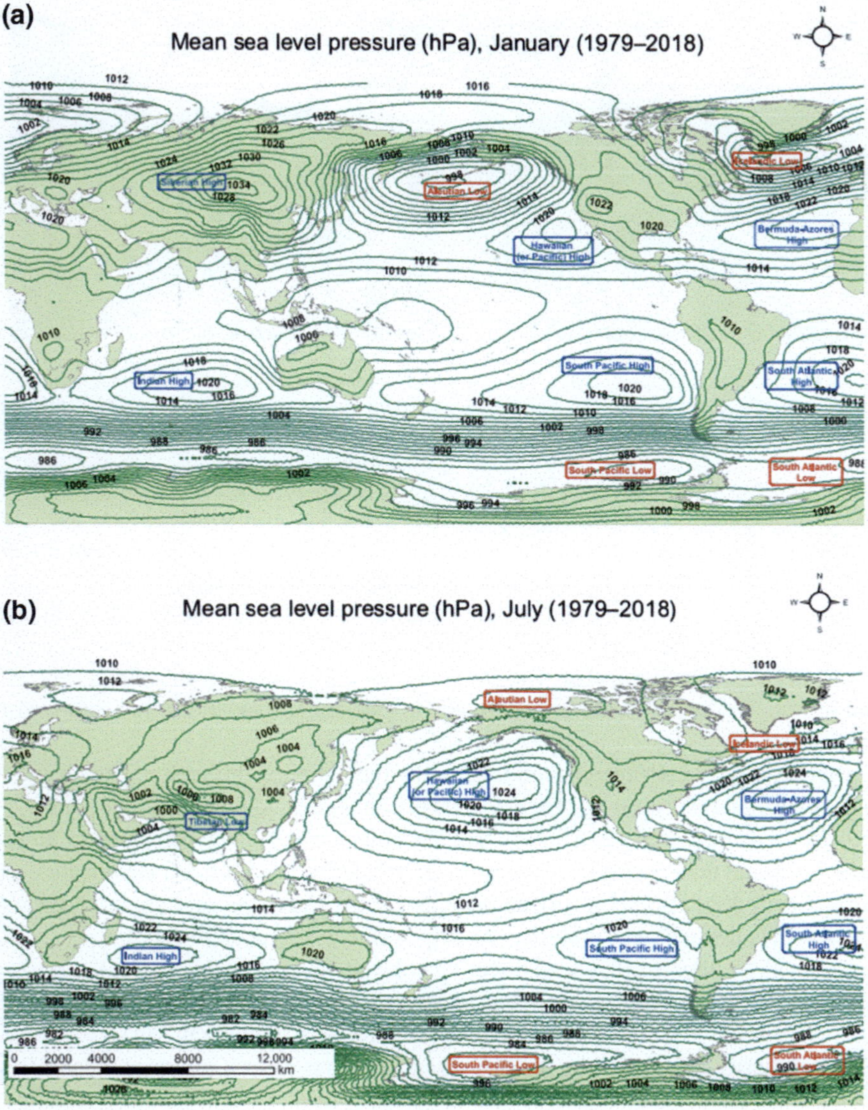

Fig. 3.9 Global mean sea-level pressures for **a** January and **b** July. *Source* Elsevier. Bushra and Rohli (2021). NCEP/DOE 2 Reanalysis data provided by the NOAA/OAR/ESRL. PSD, Boulder, Colorado, USA

the term *Bermuda-Azores High* is adopted. There is also a semi-permanent Southern Hemisphere subtropical anticyclone in the Atlantic Ocean, the **South Atlantic High**, that will be covered in Chap. 19.

In the subtropical North Pacific Ocean is the **Hawaiian High**, also known as the **Pacific High**. Americans tend to prefer the term, Hawaiian High, and East Asians

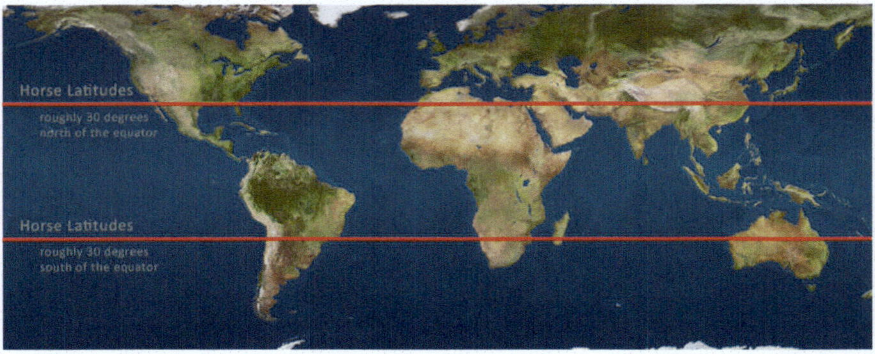

Fig. 3.10 The horse latitudes. *Source* NOAA. https://oceanservice.noaa.gov/facts/horse-latitudes.html

tend to refer to it as the Pacific High. Here the term **Hawaiian-Pacific High** is used, but again, beware, especially in keyword searches, because some authors use one or another of the names. Likewise, there is also a Southern Hemisphere subtropical anticyclone in the Pacific Ocean known as the **South Pacific High**.

The Hadley cells and the circulation patterns within them—the ITCZ, northeast trades, southeast trades, and subtropical anticyclones in each hemisphere—migrate seasonally throughout the year. The slow migration follows the Sun's direct rays, so that in June they are shifted northward, and in December they are shifted southward (Fig. 3.11).

But the Hadley cells respond only partially to the motion of the Sun's direct rays, as the ITCZ shared by the two Hadley cells may only extend poleward to around 12° N to 15° N in June, and 12° S to 15° S in December. The Hadley cells tend to move more over land than over ocean. As more land exists in the Northern Hemisphere to support the initial afternoon surface heating to drive the rising motion at the ITCZ, the entire Hadley cell pair is shifted slightly northward, even at the *equinoxes*.

Tropical weather is influenced directly by the position of the Hadley cells. Wherever the ITCZ is located at a given time, the rising air cools, condenses, forms clouds, and precipitates, leaving conditions cloudy and rainy. As shown in Fig. 3.12, some places on Earth are affected by the ITCZ year-round, even as it migrates seasonally. Such places include the Amazon basin and parts of West Africa. Rain is not falling all the time, but it does fall on most afternoons, when the surface heating drives the convection.

When the subtropical anticyclones are shifted over a location, the weather is clear, dry, and often hot. Hadley cell-induced *subsidence* is stronger on the winter Hadley cell than the summer side. This is because a bigger pressure gradient force exists on the winter side, where much colder air is adjacent to the cell. Places around 25–30° are situated under the subtropical anticyclones year-round. Many famous deserts are in this zone. Some examples in the Northern Hemisphere are the Sahara of North Africa, The Arabian, the Mojave of California, Nevada, and Arizona, and the Sonoran and Chihuahuan of the southwestern U.S. and northern Mexico. In the

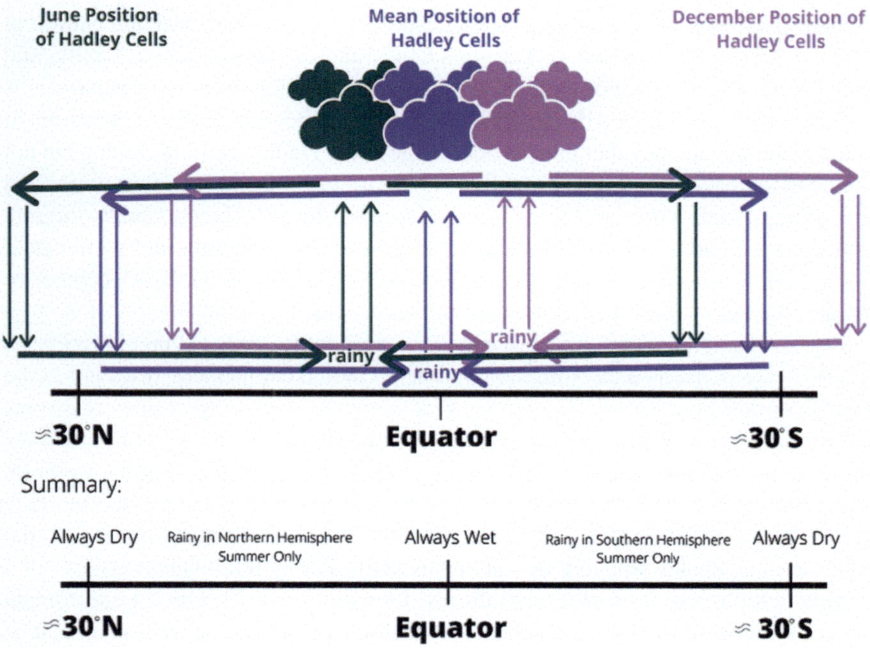

Fig. 3.11 Schematic depicting seasonal shift of the Hadley cell system from June (black) to March/September (purple) to December (pink). *Source* Geaux Geospatial

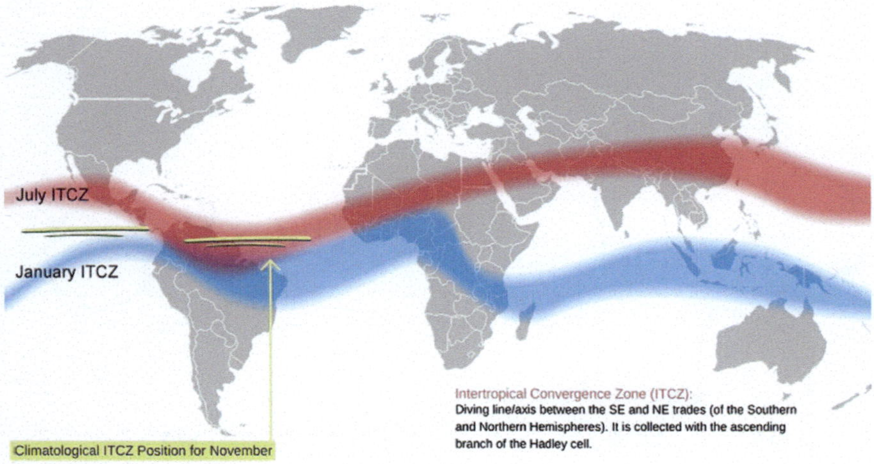

Fig. 3.12 Typical position of the inter-tropical convergence zone in July and January. *Source* NOAA/ESRL Physical Division

Southern Hemisphere, examples of such deserts include the Kalahari and Namib of southern Africa, the Atacama of Peru and Chile, and the Gibson, Great Sandy, and Simpson in Australia. All of these are in the horse latitudes described earlier.

Other places are affected by the ITCZ during part of the year and by the subtropical anticyclones during the other part, as both of the entire Hadley cells, including but not limited to the ITCZ, migrate seasonally (Fig. 3.13). In June and July, the subtropical anticyclones, trade winds, and ITCZ are displaced northward. They gradually migrate southward as Northern Hemisphere summer gives way to autumn and winter (and as Southern Hemisphere winter gives way to spring and summer). By December or January, the entire Hadley cells are as southward as they will be.

Many tropical locations experience sharp seasonal differences in precipitation, as the ITCZ passes overhead as summer approaches in that hemisphere, dries out as the ITCZ is farther poleward during the peak of summer, experiences rainy conditions again as the ITCZ again passes overhead as autumn approaches, and then dries out again as the subtropical anticyclone is overhead in winter in that hemisphere. Such places have a bimodal precipitation peak throughout the year. In fact, to many people around the world, including people who live in this part of the world, the word "season" evokes imagery of wet versus dry rather than hot versus cold.

Even though the near-surface air initially flows equatorward from the subtropical anticyclone toward the ITCZ, and the flow aloft moves poleward from overhead of the ITCZ toward the area overhead the subtropical anticyclone, the Coriolis effect deflects these flow patterns. A deflection to the right in the Northern Hemisphere and to the left in the Southern Hemisphere of this initially equatorward-moving surface circulation results in the northeast trades and southeast trades. These trade winds

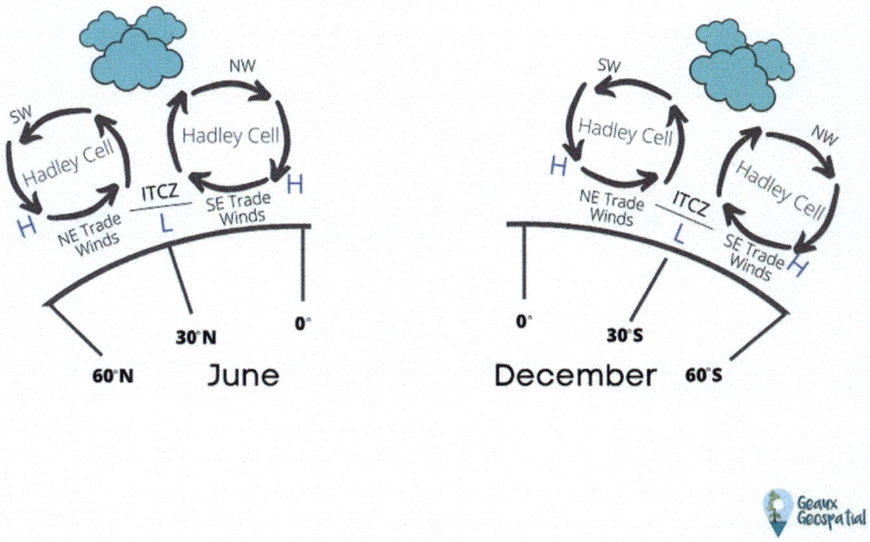

Fig. 3.13 Side view of the major circulation cells from North Pole to South Pole, with vertical extent greatly exaggerated. *Source* Geaux Geospatial

provided the sail power to explore the New World. Christopher Columbus landed in the Caribbean and Central America because the northeast trades blew his ship southwestward from Spain.

Above the Earth's surface, the surface northeasterlies in the Northern Hemisphere and southeasterlies in the Southern Hemisphere give way to southwesterlies in the Northern Hemisphere and northwesterlies in the Southern Hemisphere, because of this Coriolis deflection. But it is important to remember that the Coriolis effect is weak in the low latitudes of the Tropics and is non-existent at the Equator. As a result, geostrophic winds do not really exist in the Tropics. Instead, tropical climatologists and meteorologists need to analyze weather maps in a different way than those studying the mid-latitudes.

The Polar Cells

In the "dishpan" experiment, near the center of the pan, the water would move from the periphery to the center, but again, only to a certain latitude (about 60°). In the "real world," the poles are so cold that there is a high-pressure zone situated there. Cold, dense air at the poles produces sinking and a high-pressure system over the poles called the **polar highs**, with the North Polar High in the Northern Hemisphere and the South Polar High in the Southern Hemisphere (see again Fig. 3.13).

Surface winds at the polar highs blow away from the poles until the air reaches near 60° N and 60° S, whereupon the air warms to some extent, which, when combined with dynamic forces, gives the air a tendency to rise. This rising motion relieves the surface of some air pressure. Thus, a surface low-pressure zone is located in a belt near 60° N and 60° S. Enclosed areas of particularly low pressure (i.e., semi-stationary cyclones) along this belt are known as the **subpolar lows** or **polar front lows**. The low surface air pressure at the subpolar lows is reinforced as the air streams flowing equatorward out of the polar high and poleward out of the subtropical anticyclones converge and rise. The subpolar lows are shown in green in Fig. 3.14.

The subpolar lows tend to be strongest near 60° of latitude in the Northern and Southern hemispheres. Furthermore, because oceans are usually warmer than land at adjacent latitudes in winter, the oceans are generally more supportive of rising air associated with subpolar lows during winter. The two dominant, semi-permanent subpolar lows in the Northern Hemisphere are the **Icelandic Low** in the Atlantic Ocean and the **Aleutian Low** in the Pacific. Notice the areas of relatively low average pressure in January in the North Atlantic (Icelandic Low) and North Pacific (Aleutian Low) in Fig. 3.9a. The circulation cell completes itself as air flowing aloft after it rises at the subpolar lows moves poleward. This produces the permanent **polar cells** (one in the Northern Hemisphere and another in the Southern Hemisphere). But, again, remember that the rotation of the Earth produces a Coriolis effect, and these winds are deflected to the right (in the Northern Hemisphere) and to the left in the Southern Hemisphere. The high latitudes give a strong deflection, which results in surface air in the polar cells, in both hemispheres, flowing from east to west as the

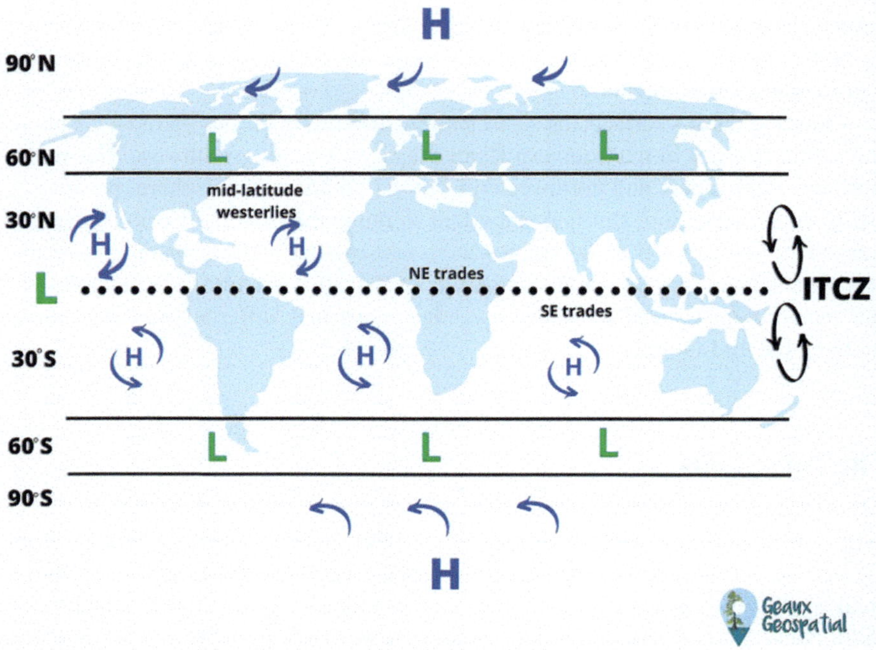

Fig. 3.14 Approximate location of subtropical anticyclones, subpolar lows, and polar highs. *Source* Geaux Geospatial

polar easterlies. Aloft, the Coriolis deflections cause air in the polar cells to flow from west to east—the **polar westerlies**. The boundary between the cold polar cell and the warmer mid-latitude "cell" to be discussed in the next section, embedded within the westerlies itself, is known as the **polar front**. Embedded within the polar front is the very important *polar front jet stream*—a particularly fast-moving circulation—that affects weather in the mid-latitudes directly. It, along with the other features of the general circulation, migrates seasonally, with the direct rays of the Sun (Fig. 3.15a, b).

One major exception to the subpolar pressure systems occurs in Northern Hemisphere winter. The Eurasian landmass—the largest land area on the planet—cools so intensely at the subpolar latitudes that the subpolar belt of low pressure is interrupted in winter as the cold air sinks to create high pressure in a part of the world that would otherwise have low pressure, according to the dishpan experiments. This **Siberian High** is characterized by frigid conditions—even as the air sinking at its core warms adiabatically, with clear skies and anticyclonic (clockwise and outward in the Northern Hemisphere) circulation near the surface that extends across much of Asia and eastern Europe. By contrast, the absence of large land masses in the Southern Hemisphere subpolar latitudes allows a simple linear series of subpolar lows encircling that hemisphere.

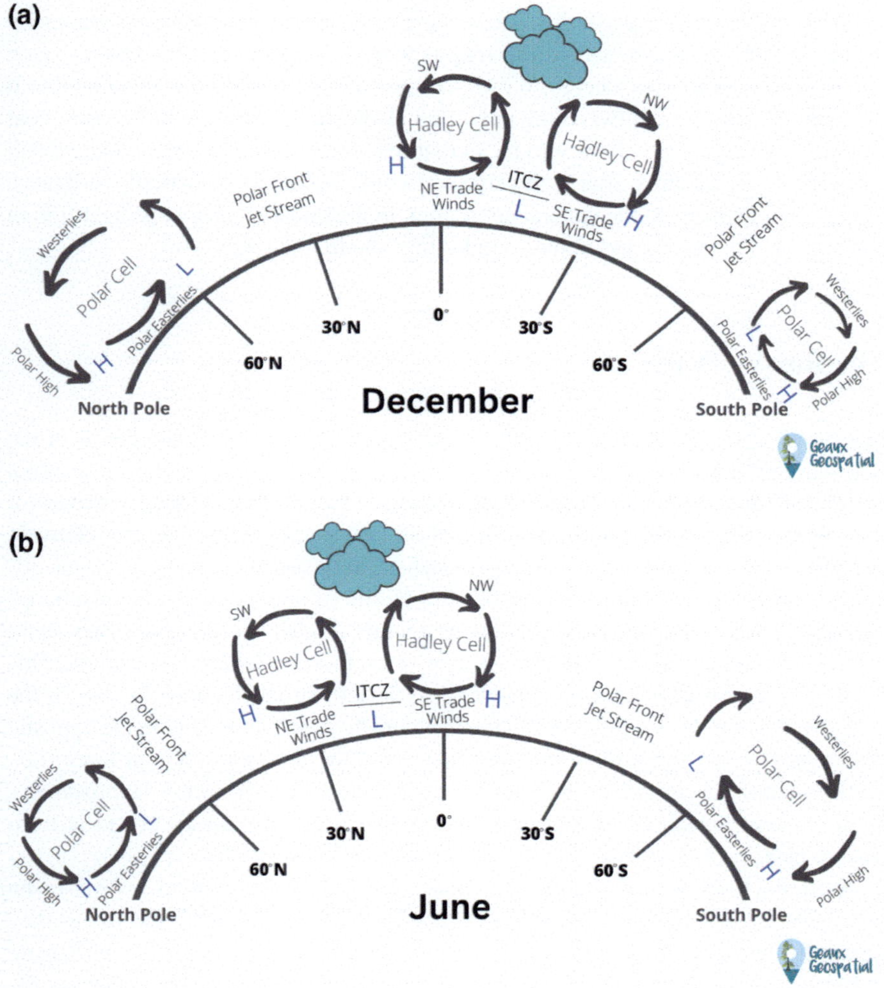

Fig. 3.15 Typical position of circulation cells and the polar front jet stream, with vertical extent greatly exaggerated: **a** December; **b** June. *Source* Geaux Geospatial

Mid-Latitude Surface Circulation

The two circulation cells in each hemisphere covered in the previous sections—the Hadley and polar cells in each hemisphere—are termed **thermally direct**, because the air is rising where conditions are theoretically warmest in the cell, and it is sinking where the air is theoretically coldest in each cell. However, the middle latitudes do not have a cell-based circulation. While the other two cells have easterlies at the surface (in the form of the northeast trades and southeast trades in the Hadley cells) and westerlies aloft, the middle latitudes are different.

In the mid-latitudes, westerlies occur at the surface, with even stronger westerlies in the middle and upper part of the **troposphere**—the atmosphere's lowest layer, where nearly all weather occurs. Why? To begin to understand why, imagine that if the third cell (the one between the polar and Hadley cells) did exist, then the shear from the two adjacent cells would mean that the coldest air in this hypothetical cell would be rising and the warmest air in the cell would be sinking (**thermally indirect**). This cannot happen, so the general circulation in the mid-latitudes must be different from that in the tropics and polar latitudes in some fundamental way.

The Rossby Waves

Instead of having a circulation cell, the prevailing, generalized circulation in the mid-latitudes is for westerlies (i.e., flow from west to east) to exist at the surface (but with local circulations often differing across space and time) and **Rossby waves** (or **long waves**) embedded within westerlies aloft. The surface westerlies are relatively straightforward to understand. Because surface high pressure exists at about 30° and low pressure at about 60° in both the Northern and Southern Hemispheres, the geostrophic wind rule (and air flow around the highs and lows) says that there should be westerly winds between 30° and 60°—the mid-latitude westerlies. This would be the case regardless of hemisphere, as the leftward deflection of winds by the Coriolis effect in the Southern Hemisphere would be supplemented with the opposite orientation of the high-to-low-pressure flow from the pressure gradient force (i.e., north to south, as compared to south-to-north flow in the Northern Hemisphere). Most surface prevailing circulation in the mid-latitudes around the world is from west to east, even though the day to day winds vary substantially.

Aloft, the Rossby waves, depicted in black in Fig. 3.16, undulate like waves on an ocean, except that they are invisible and that instead of an up-down motion as they propagate laterally, they are north–south undulations as they generally propagate from west to east, in both the Northern and Southern hemispheres. The ridges are the parts of the undulations that extend farthest poleward and the troughs are the parts that extend farthest equatorward. Because the poleward side is north in the Northern Hemisphere and south in the Southern Hemisphere, ridges are oriented oppositely in the hemispheres—in the shape of upside-down "U's" in the Northern Hemisphere and "U's" in the Southern Hemisphere. Likewise, troughs appear as "U's" in the Northern Hemisphere and upside-down "U's" in the Southern Hemisphere.

Rossby wave-types of undulations, with the characteristic ridges and troughs of shapes influenced by the rotation and spherical shape of the planet, are characteristic of the flow that would occur in a fluid that has a gradient of heat, in accordance with basic principles of fluid dynamics. The degree of amplification (i.e., **amplitude**) and width (i.e., *wavelength*) of the individual wave and the forward propagation speed of the waves are all related to the strength of the heat gradients. But in addition, of course, the Coriolis effect comes into play when the undulations are in a fluid that is part of the Earth-ocean–atmosphere system. But because the Rossby wave flow is

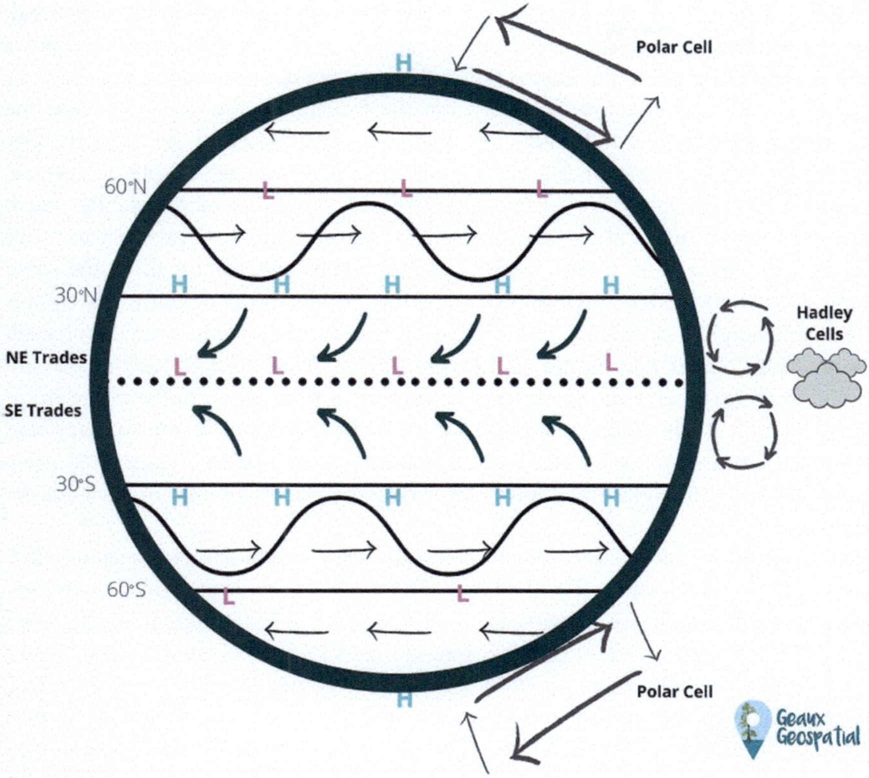

Fig. 3.16 General circulation of the atmosphere, with vertical extent greatly exaggerated, including the mid-latitude Rossby waves. *Source* Geaux Geospatial

aloft, friction is generally not a very important factor in Rossby wave flow, except where the waves pass over high mountains.

Like the Hadley cells and polar cells, the Rossby waves function to redistribute surplus heat poleward in the Earth-ocean-atmosphere system. The poleward-moving parts of the undulations (i.e. the trough-to-ridge sides of the Rossby waves) transport surplus heat poleward. In turn, *advection* of colder air equatorward by the ridge-to-trough sides of the undulations mixes this less energetic air with warmer local, ambient air.

The Rossby waves are embedded within the mid-latitudes and flow generally west-to-east aloft in both hemispheres, but with jogs poleward and equatorward at the ridges and troughs, respectively. The approximate west-to-east overall flow occurs at all times, not just on a "general" basis as at the surface. But why are the upper-level winds in the mid-latitudes always from this general direction?

The **Ideal Gas Law** states that: $P = \rho R_d T$, where P, ρ, R_d, and T represent pressure, density, the dry gas constant, and *Kelvin temperature*, respectively. So, at a given pressure, if T increases, ρ decreases. This equation explains the familiar notion that hot air is less dense than cold air. The Ideal Gas Law also implies that in general, **isobaric surfaces** (i.e., constant pressure levels) slope from the Equator to the poles with increasing height in the atmosphere. This concept is illustrated in Fig. 3.17.

The pressure gradients aloft slope from the Equator to the poles, because the tropics are warmer than the poles. The warmer surface tropical air rises, thereby "stretching" out the atmospheric mass upward at each successive isobaric surface, compared to the polar areas. At the 850-mb level, this difference is subtle, but it causes the "extra" mass of air from the tropics having 850 mb of pressure to move poleward, in an attempt to even out this pressure imbalance, as the air at the same height as the 850-mb level in the tropics has less pressure than the air at the tropics at the same height (see again Fig. 3.17). This difference gets compounded with each increment up higher and higher in the atmosphere, from the 850- to the 700-mb level, and so forth, generating an increasing need for the air to move poleward at a given height. At the highest heights of the weather layer of the atmosphere, near 300 mb of atmospheric pressure, the difference in the pressure surface is so great that the pressure gradient force moves air very swiftly in an attempt to even out the imbalance.

However, upon "initial" movement poleward, the Coriolis effect curves the flow to the right (in the Northern Hemisphere). The negligibility of friction at such high altitudes means that flow is geostrophic, and thus flows at a 90° angle to the pressure gradient, from west to east, or into the page in Fig. 3.17, the mid-latitude westerlies.

This type of circulation is sometimes called the thermal wind because the circulation is set up by differences in thicknesses of isobaric surfaces, which are in turn set up by differences in temperature. The "tilting" pattern of the thermal wind, as

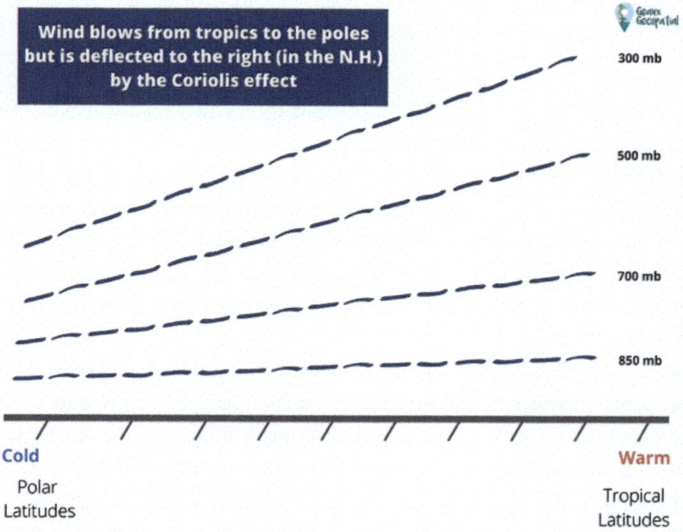

Fig. 3.17 Greatly exaggerated cross-section of the height at which various atmospheric pressure values occur, from the tropics to the poles. *Source* Geaux Geospatial

shown in Fig. 3.17, is much stronger in winter because the temperature gradient between the Equator and the poles is much greater than in summer. Therefore, the slope of the isobaric surfaces is steeper. Because of large temperature contrasts in the mid-latitudes, very fast flow occurs at the boundaries between warm and cold air masses.

The Polar Front Jet Stream

This narrow belt of extremely fast sustained winds embedded within the core of the general westerly flow—at the polar front, where upper-level pressure gradients are strongest—is the polar front jet stream. These narrow ribbons of high winds, with widths of perhaps 100–500 km (60–300 mi) meander around the hemisphere through ridges and troughs in the upper part of the atmosphere's weather layer. They are usually strongest at between 7.5 and 12.2 km (4.7–7.5 mi) above the surface—higher and nearer to the poles in summer and lower and equatorward in winter (Fig. 3.18), with the 300-mb pressure height level a good approximation—and are only a few kilometers thick. The height of the polar front jet extends to the top of the weather layer of the atmosphere—the **tropopause**. Although the polar front jet stream is discontinuous over the hemisphere, it can remain intact for thousands of kilometers before slowing down and then speeding up again.

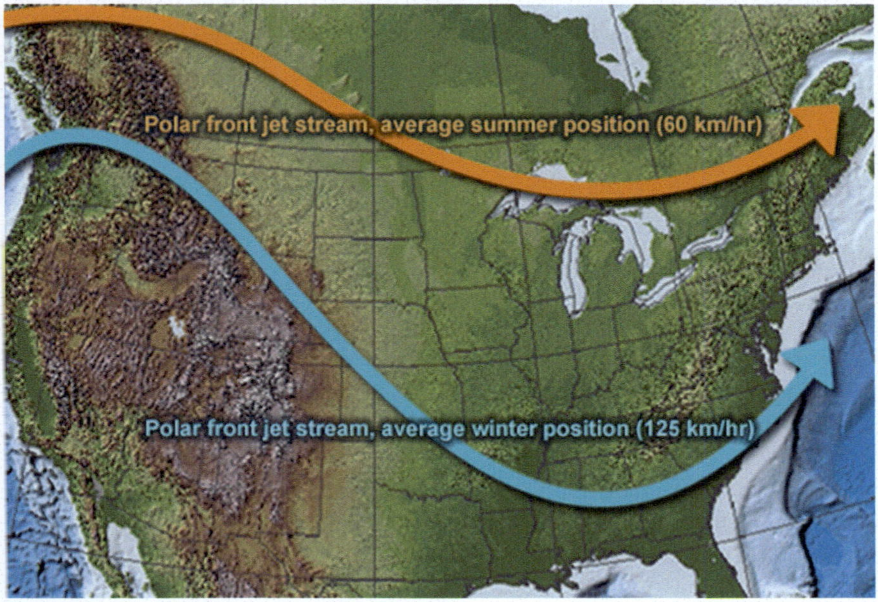

Fig. 3.18 Idealized position and speed of the polar front jet stream over North America. *Source* UCAR MetEd

Wind speeds in the polar front jet stream average 65 km h^{-1} (40 mph) in summer and 130 km h^{-1} (80 mph) in winter. This difference is attributed to the stronger temperature gradient and contrast in winter. However, the winds frequently exceed 190 km h^{-1} (120 mph), especially in the winter hemisphere. On rare occasions, the speed may even exceed 390 km h^{-1} (240 mph). For obvious reasons, airplane pilots prefer to fly in the polar front jet stream when traveling from west to east and avoid the jet stream when flowing from east to west. The polar front jet stream was first noticed in the 1930s by early pilots attempting long-distance travel. However, it was only fully documented during World War II, as the explanation for the unexpectedly poor fuel economy of planes flying to Japan from the U.S.A., but good fuel economy when flying in the opposite direction.

Regardless of the part of the world where it occurs, the fastest part of a jet stream—its core—is called a **jet streak**. Jet streaks form as air accelerates, typically west-to-east or southwest-to-northeast in a jet stream, from the confined region of a trough axis toward the downstream ridge. Upstream from the trough axis/jet streak entrance area, air streams are converging on the trough-to-ridge side of the Rossby wave. This pattern initiates upper-level sinking motions that support surface high-pressure cells. As the air moves through the entrance, **confluence**—lateral motion that merges from two sources—restricts the airflow, initiating a speed increase immediately downstream of the *trough* axis. **Diffluence**—the opposite of confluence—occurs toward the ridge, leading to *divergence* aloft. The result is a jet streak (Fig. 3.19). The upper-level divergence near the jet streak exit region (i.e. the shaded region in Fig. 3.19) creates a momentary upper-level vacuum that supports rising air in the low-level atmosphere, inducing cyclogenesis. Near the trough, the **positive vorticity advection**—or increasingly cyclonic spin—occurs throughout the vertical body of the atmosphere. Such conditions lead to the development of precipitation and potentially severe storm systems, particularly over areas where cold and warm conditions are adjacent to each other, such as those east of the Asian coast.

The Circumpolar Vortex

The part of the middle latitudes in which the coldest air and warmest air meet, overlapping with the polar front jet stream, is where the most action in the mid-latitudes takes place. The collection of densely-packed isohypses at upper-levels and geostrophic winds that circumnavigates the poles at upper levels is the **circumpolar vortex**.

In Fig. 3.20, tracing the complicated pattern of the Northern Hemisphere's circumpolar vortex by following a single isohypse in the middle of densely-packed isohypses around the hemisphere reveals the area and shape of the circumpolar vortex and the "waviness" or circularity of the associated Rossby waves. The warm air is generally beneath the ridges, where air warms *adiabatically* as it sinks under the high pressure, and cooler areas are in the troughs and inside the circumpolar vortex. Areas of surface high pressure have clearer skies than the areas of low pressure and are

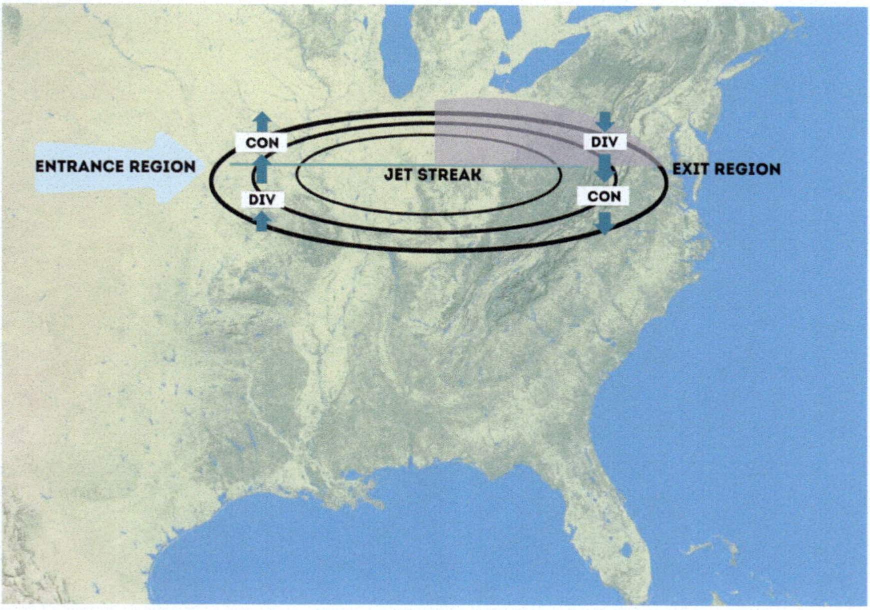

Fig. 3.19 Hypothetical jet streak position and most likely severe weather areas. *Source* Geaux Geospatial

commonly beneath the ridge-to-trough sides of the Rossby waves, while the surface cyclones are associated with cloudy conditions and tend to be located beneath the trough-to-ridge sides of the Rossby waves.

The circumpolar vortex, with the polar front jet at its leading edge, expands equatorward as winter approaches and retreats poleward with the onset of summer. Although the primary upper-level flow is from west to east in a merry-go-round motion around the pole, the circumpolar vortex circularity undergoes changes across space and time. Moreover, its centroid is often displaced from the pole some distance, especially toward the Pacific Ocean sector. A distinct north–south component to the flow is also often present at least in some parts of the circumpolar vortex at some times.

Primary Circulation in the Mid-Latitudes: Putting It All Together

The terms **meridional** and **zonal** flow are often used to characterize the circumpolar vortex, the individual Rossby waves in it, and the associated polar front jet stream. Zonal and meridional flow in the circumpolar vortex are shown in Fig. 3.21a, b. There may be periods of a week or more when the flow is zonal, as in Fig. 3.21a.

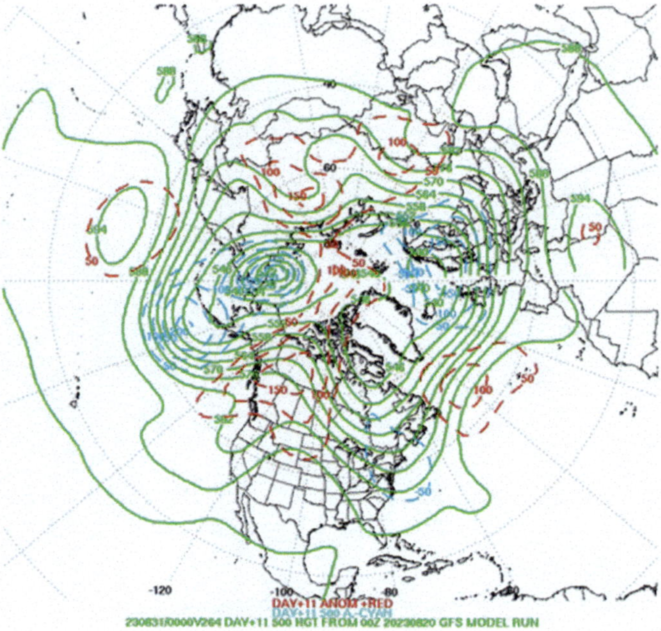

Fig. 3.20 Eleven-day 500-mb geopotential height forecast model output, for 31 August 2023. Circumpolar vortex can be delineated by the locations where the green solid lines are closest, such as south of Alaska and over Lake Superior. Red/blue dashed lines show the anomalies of 500-hPa geopotential heights—where higher/lower than normal. *Source* NOAA. cpc_NHEM_f264wbg.gif (1024×768) (noaa.gov)

In such cases, storms are generally weak or non-existent, largely because the cold air and warm air do not interact intensely. Storms that do exist during a zonal flow regime tend to be pushed forward quickly in the mid-latitude westerlies, so that they are generally less of a menace to any particular location. Periods of zonal flow allow frigid polar air within the circumpolar vortex to accumulate, without being pulled equatorward to mix with warmer air, and restrict the poleward movement of warmer, tropical air.

Eventually the zonal flow regime must break, as the Equator-to-pole heat gradient eventually becomes too substantial and the redistribution of the surplus heat must occur. Thus, the flow aloft begins to meander into meridional flow (Fig. 3.21b), with cold air advected equatorward and warm air pushed poleward. In such regimes, storms tend to be more violent, as the mixing of warm and cold air breeds storminess. Dynamic conditions in a meridional flow regime favor storminess as well, as will be shown shortly. Eventually, the energy is redistributed adequately for the steering flow to return to a zonal regime.

In winter, there are usually only three or four sets (i.e., ridge-trough pairs) of Rossby waves in the circumpolar vortex, while in summer there can be as many as six. Thus, Rossby wavelengths are between 4000 and 6500 km (~2500–4000

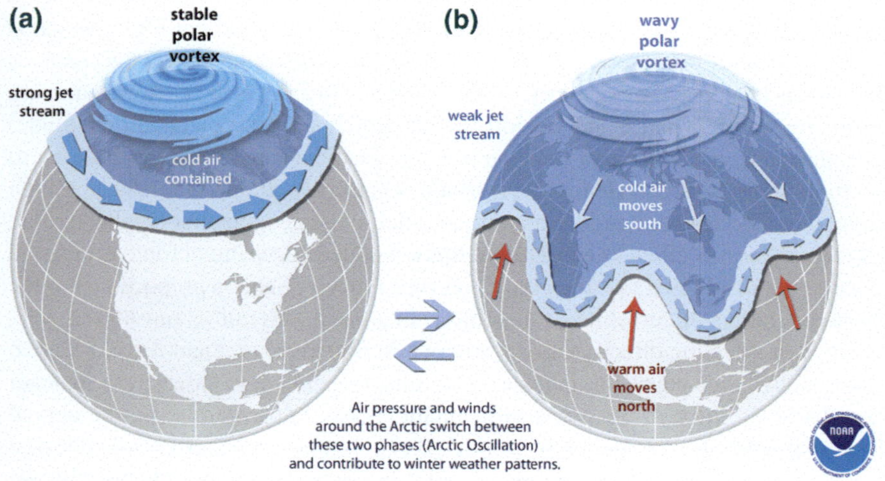

Fig. 3.21 Schematic of the variability in area and circularity or "waviness" of the circumpolar vortex (dark blue), with the polar front and associated polar front jet shown at its leading edge. **a** zonal flow. **b** meridional flow. *Source* NOAA. https://www.noaa.gov/sites/default/files/styles/no_style/public/thumbnails/image/GRAPHIC%20-%20polarvortex_explained_updated%20012919%20-%204034x2912.png?itok=4YZfZ0tg

mi). The difference in number and amplitude of Rossby waves between winter and summer relates to the fact that stronger transfer of energy is needed to even out the greater energy gradients in winter, compared to summer, but that only three or four waves are dominant enough to "carry the load" at the expense of other waves that might develop. In winter, the Rossby waves have somewhat more tendency to be meridional than in summer, but with exceptions to this generalization. To understand why the energy gradient is stronger in winter than summer, compare Louisiana to Minnesota temperature. In summer, Minnesota may be 10 Fahrenheit degrees colder than Louisiana, but in winter, Minnesota might be 40 Fahrenheit degrees colder than Louisiana. This difference is because in summer, there are more hours of daylight in Minnesota than Louisiana, which can partially offset the weaker intensity of sunlight in Minnesota. But in winter, Minnesota has both a smaller number of daylight hours and a weaker intensity of sunlight than Louisiana.

A close relationship exists between the location of the polar front jet stream and Rossby wave ridges and troughs and surface conditions. When the jet is equatorward of a given location, then weather at that location will generally be colder than average for the time of year at that place, because the place sits within the circumpolar vortex. When the jet is poleward of a location, the weather will generally be warmer than normal at that time of year, as the place lies outside of the circumpolar vortex.

When the upper-air wind becomes curved, as it moves through the Rossby wave troughs and ridges, it is no longer geostrophic. However, the curved flow, known as a **gradient wind**, remains parallel to the isobar/geopotential height pattern as the forces remain in balance with the additional support of another force. In purely geostrophic balance, the Coriolis effect is proportional to the pressure gradient force, which is directed toward lower pressure/geopotential heights toward the pole. This creates a flow that is in balance between two opposing forces. In a ridge, the airflow is moving poleward, which increases the Coriolis effect (which is in part dictated by latitude) such that it becomes stronger than the pressure gradient force. This turns the air to the right in the Northern Hemisphere, which is generally equatorward once past the ridge axis. The gradient airflow remains parallel to the geopotential heights despite a stronger Coriolis effect as **centrifugal force** is added. Centrifugal force is initiated toward the outside of a moving object's curved trajectory and is proportional to the speed and radius of curvature of the object. In a ridge, this pull is toward the pole, which is in the same direction as the now-weaker pressure gradient force. The two forces align on the poleward (outside) side of the ridge curve to balance the stronger Coriolis effect through the curve to maintain a flow that is parallel to the curved geopotential heights. However, the general direction becomes equatorward once past the ridge axis. The air flow continues toward the Equator and straightens with the balance of forces existing between the Coriolis effect and the pressure gradient force (or purely geostrophic). As the Coriolis effect weakens with decreasing latitude, the pressure gradient force dominates to turn the air toward the pole. This creates a curved trough in the flow pattern. The centrifugal force then aligns on the equatorward side (outside) of the curve with the combined centrifugal force and Coriolis effect, maintaining flow parallel to the curved geopotential height pattern through the trough. The flow then straightens into geostrophic balance but at that point in a generally poleward direction. With such a trajectory, the Coriolis effect increases with latitude enough to eventually initiate the development of another ridge, and the sequence repeats.

Gradient flow for curved isohypses is slightly slower than geostrophic winds for straight-line isohypses—**subgeostrophic**—at troughs. Likewise, the air flows slightly faster than geostrophic winds would predict for straight-line isobars or isohypses—**supergeostrophic**—around ridges. This is because the flow curves to the right in order to remain parallel to the isohypses and retain the balance of forces, as it moves through ridges. Turning to the right is in accord with the direction that Coriolis effect is turning it. Therefore, the air can move a little faster as it is being pulled around the ridges. But around the trough, the air must turn to the left in order to remain parallel to the isohypses and retain the balance of forces. The air slows down in such a case, to minimize the Coriolis effect that is turning it in the wrong direction for maintaining the balance of forces. Said another way, the flow to the left around the trough slows down the flow for that pressure gradient, because it works against the Coriolis in order to turn to the left, but yet the flow in the downstream ridge is turning in the direction that aligns with the Coriolis effect, so it can flow faster. This creates a "spreading out" or divergence between a trough and the downstream ridge, in a manner analogous to cars on a highway with the cars going slower than the ones in front of them. Since there is divergence in this area, air rises up to replace the air that "spread out." This scenario leads to low-level convergence beneath the trough-to-ridge side of the wave, rising air, and a surface cyclone, and possible precipitation, between the trough and downstream ridge.

The same logic can be used to establish that a surface anticyclone will be situated between a ridge and downstream trough. As the air flows around the ridge supergeostrophically, it "catches up with" the air ahead of it that is flowing subgeostrophically around the downstream trough. This scenario creates a pile-up, or convergence, of air, just as if cars would be moving faster than the cars ahead of them on the highway. The air that has converged cannot go even higher because gravity pulls it down. The balance of forces will not allow it to go laterally. The only place for the converged air to go is downward. This is coincident with higher pressure at the surface—an anticyclone—with little tendency for cloud cover and precipitation. The ridge-to-trough side of the Rossby wave is therefore the "fair weather side" of the wave.

In the Southern Hemisphere, the same patterns still hold because, as described earlier, even though Coriolis bends flow to the left in the Southern Hemisphere, what we call troughs in the Northern Hemisphere appear opposite of troughs in the Southern Hemisphere (where elongated zones of low pressure from the low pressure at 60° S are bulging northward). Similarly, ridges bulge northward in the Northern Hemisphere but southward in the Southern Hemisphere, although in the Southern Hemisphere the flow tends to remain much more zonal.

Even though the Rossby waves propagate across the mid-latitudes from west to east in each hemisphere, when ridges or troughs propagate to a "preferred" region, the wave propagation tends to stall there, usually for an extra day, but occasionally for much longer. Ridges tend to be anchored over mountain ranges for longer periods than average. Troughs are prone to be anchored over land/sea boundaries for a longer time than average.

The Subtropical Jet Stream

Another important jet stream forms where the air moving aloft from a position above the ITCZ poleward, in each hemisphere's Hadley cell, experiences an increase in Coriolis effect (as the Coriolis effect increases with latitude). This causes this so-called **subtropical jet stream** to circumnavigate the hemisphere roughly from west to east at about 30° N and 30° S. Typical positions and zonality/meridionality of the subtropical and polar front jets are depicted in Fig. 3.22.

In contrast to the polar front jet stream, the subtropical jet is unrelated to and largely unaffected by differences in air mass properties. Instead, it forms primarily for dynamic reasons (i.e., air acquiring enough Coriolis effect) rather than the *thermodynamic* reasons (i.e., air at different temperatures adjacent to each other). In each hemisphere, the subtropical jet is at about the 200-mb level. It is even higher than the polar front jet stream because in the subtropics the warmer air causes the weather layer of the atmosphere to extend upward higher than at higher latitudes where the air is colder. It is likely that the subtropical jet has a more important role in producing weather, especially over subtropical locations, than is currently understood. Much

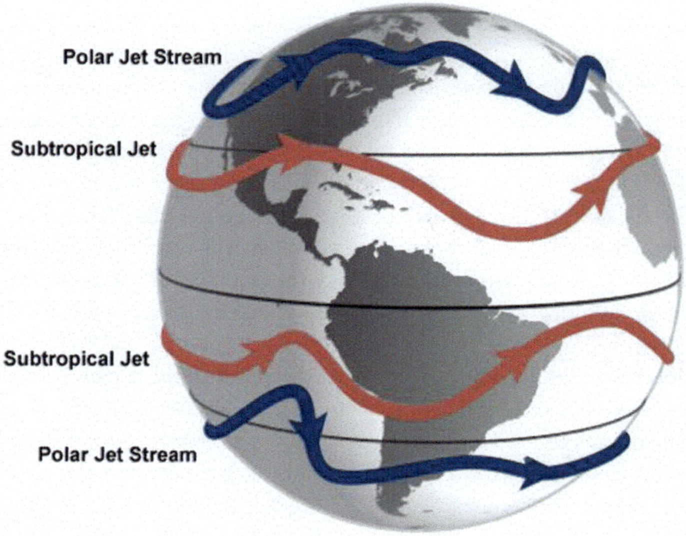

Fig. 3.22 Typical positions and zonality/meridionality of the polar front jet stream and subtropical jet stream, although both vary across time and space. *Source* UCAR CoMet. Permission from Tara Torres

more is known about the polar front jet, because scientists at the major meteorological research institutions were more interested in the jet stream that affected them more directly.

The Mid-Latitude Circulation that Wasn't: The Ferrel Cells

Many textbooks and other information sources show a theoretical but non-existent mid-latitude circulation "cell" in the mid-latitudes of each hemisphere. This theoretical cell is called the **Ferrel cell**, and it would be one of three in each hemisphere—Hadley and Polar, with Ferrel between the two. Even though this theorized cell does not really exist, it does appear as an apparent circulation when the westerly wind waves are averaged in the longitudinal direction.

The explanation for why the Ferrel cell is unimportant follows the earlier explanation for why the mid-latitude westerlies overlie surface westerlies in the mid-latitudes. In theory, the Ferrel cell would be generated because the sinking air at the poleward edge of the Hadley cell in each hemisphere, near 30° N and 30° S latitude, but migrating poleward or equatorward seasonally, would induce a shear that promotes sinking air at the equatorward edge of the theoretical Ferrel cell. Likewise, at the equatorward edge of the polar cell in each hemisphere the rising motion around 60° N and 60° S, but migrating poleward or equatorward seasonally, would induce a shear that promotes rising air at the poleward edge of the theoretical Ferrell cell. The air that accumulated near the surface near 30° N or 30° S would then move to fill in the gap that the near-surface air near 60° N or 60° S vacated as it rose. Aloft, the accumulated air near 60° N and 60° S that rose from the surface would then flow laterally equatorward, to fill the gap vacated by the air near 30° N and 30° S that sank. The result would be a circulation cell in each hemisphere with sinking air near 30° N and 30° S, poleward near-surface airflow to about 60° N or 60° S, rising motion near 60° N or 60° S, and then equatorward flow aloft to about 30° N or 30° S.

But the Ferrel cell does not exist, primarily because in order for such a cell to exist, the warmest air in the cell (near 30° N and 30° S) would need to be rising. And likewise, the coldest air in the cell (near 60° S and 60° S) would need to be sinking—thermally indirect. The dynamic tendency caused by the shearing from the adjacent Hadley and polar cells is offset by the thermodynamic tendency caused by density differences of the air within the "cell." Instead of a "cell," the mid-latitudes of each hemisphere have westerlies at the surface with faster westerlies aloft where friction is reduced.

Overall Assessment

The main features of the general circulation have been covered in this chapter, as a product of the balance of forces—pressure gradient force, Coriolis effect, and friction—amid a rotating set of two fluids—the atmosphere and the oceans—that interface with each other and are in a constant state of change—daily, seasonally, annually, and interannually. They also affect and are affected by other components of the Earth-ocean-atmosphere system, including terrain and land cover in the natural and built environment, and the *cryosphere*, all at a range of spatial and temporal scales. The main features of this circulation are shown in Fig. 3.23. It is an oversimplification to say that we fully understand the drivers of the *climate system* and how/why they change.

The next chapter will zoom in on circulation features at medium scales rather than the hemispheric to global scale. Rossby waves will be covered in more detail, with more attention to their role in supporting or suppressing surface weather systems.

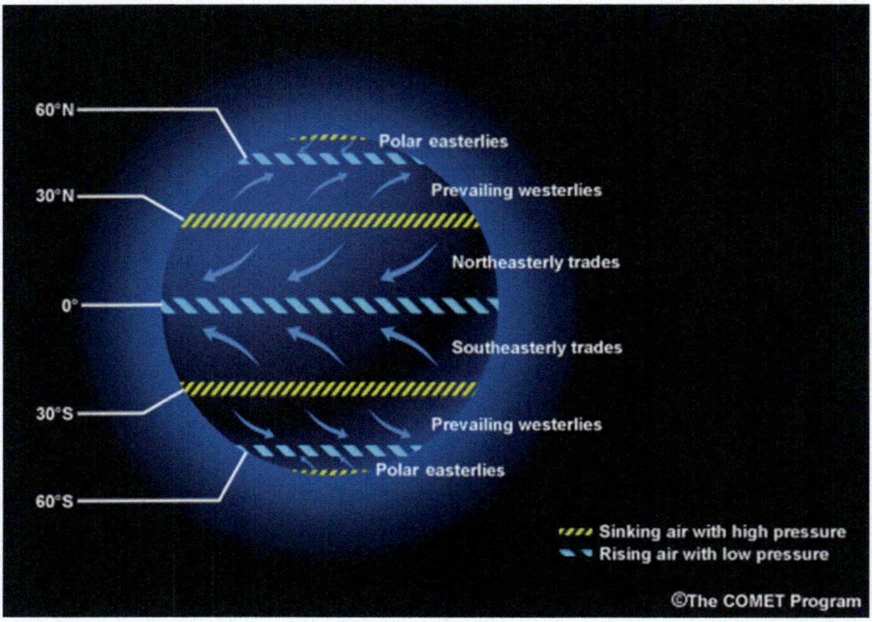

Fig. 3.23 Summary of major features of the general circulation of the atmosphere. *Source* UCAR CoMet. Permission from Tara Torres

Further Reading

Bushra, N., & Rohli, R. V. (2021) *Annotated atlas of coastal and marine winds* (157 pp). Elsevier Limited. 978-012-820061-2.

Knoben, W. J., Woods, R. A., & Freer, J. E. (2019). Global bimodal precipitation seasonality: A systematic overview. *International Journal of Climatology, 39*(1), 558–567. https://doi.org/10.1002/joc.5786

Nguyen, H., Evans, A., Lucas, C., Smith, I., & Timbal, B. (2013). The Hadley circulation in reanalyses: Climatology, variability, and change. *Journal of Climate, 26*(10), 3357–3376. https://doi.org/10.1175/JCLI-D-12-00224.1

Pena-Ortiz, C., Gallego, D., Ribero, P., Ordonez, P., & Alvarez-Castro, M. D. (2013). Observed trends in the global jet stream characteristics during the second half of the 20th century. *Journal of Geophysical Research Letters—Atmospheres, 118*(7), 2702–2713. https://doi.org/10.1002/jgrd.50305

Rossby, C.-G. (1939). Relation between variations in the intensity of the zonal circulation of the atmosphere and the displacements of the semi-permanent centers of action. *Journal of Marine Research, 2*, 38–55. https://doi.org/10.1357/002224039806649023

Waugh, D. W., Sobel, A. H., & Polvani, L. M. (2017). What is the polar vortex and how does it influence weather? *Monthly Weather Review, 98*(1), 37–44. https://doi.org/10.1175/BAMS-D-15-00212.1

Chapter 4
Secondary Circulations in the Atmosphere

Abstract Weather systems comprising the secondary circulation of the atmosphere are steered by the primary circulation and include air masses and transitory cyclones and anticyclones. Air masses are broad expanses of relatively homogeneous thermal and humidity properties that move within the general circulation to create weather patterns. In places where air masses with greatly different properties interact, usually in the mid-latitudes, fronts form and are pushed along by the Rossby waves and polar front jet stream embedded within the mid-latitude westerlies. These fronts often trail from a counter-clockwise- and-inward-rotating (in the Northern Hemisphere), surface-based, low-pressure system known as a mid-latitude wave cyclone, which can form for numerous reasons, including localized "hot spots," vertical stretching of air as it descends a mountain slope, and position relative to Rossby wave ridges and troughs. Anomalous steering flow in the mid-latitudes can cause cutoff lows or blocking patterns that make weather patterns persist for days to weeks. In the tropics, an important secondary circulation system is the easterly wave, which is a kink in the trade wind flow. The eastern side of the easterly wave trough axis supports storm intensification into tropical depressions, tropical storms, and hurricanes/typhoons, but the western side suppresses storm intensification.

Keywords Air mass source region · Arctic air mass (A or cA) · Blocking pattern · Cold versus warm versus occluded fronts · Conservation of potential vorticity · Continental polar (cP) air mass · Continental tropical (cT) air mass · Cutoff low · Cyclogenesis and cyclolysis · Easterly wave · Equatorial (E or mE) air mass · Maritime polar (mP) air mass · Maritime tropical (mT) air mass · Mid-latitude wave cyclone

General

Daily weather conditions are dictated by traveling air masses, *cyclones*, and *anticyclones*. Because these are embedded in and steered by the primary circulation (also known as the *general circulation*), this topic builds on understanding of the general circulation from Chap. 3. The topic is important because these secondary circulations

R. V. Rohli et al., *Atmospheric and Oceanic Circulation*,
https://doi.org/10.1007/978-3-031-66142-6_4

govern daily weather, which when integrated across time, contribute to climate. Just as water in a stream can circulate in different directions even while it is embedded in the primary flow downslope out to sea, the air also has secondary circulations embedded within the steering flow.

Air Masses

It is typical for weather conditions to remain generally similar across distances of hundreds to even thousands of kilometers. But if you traveled across some imaginary boundary, the conditions would become drastically different across a relatively short distance. And then, when you continue traveling for hundreds or even thousands of kilometers, the conditions remain largely similar again. The weather map shown in Fig. 4.1 shows an example of this kind of situation, at least when the "similar conditions" refer to *temperature*. Notice how the temperatures (in red) are mostly similar for large distances south of the blue curved line with triangles across the U.S. East and Southeast, but those temperatures are much different than the ones north of that line. The same is true for *dew point temperatures* (in green).

The situation described above that you have likely experienced can be interpreted easily in a meteorological sense. A parcel of air with uniform weather characteristics over large areas (temperature, stability, humidity, etc.) across a wide region is known

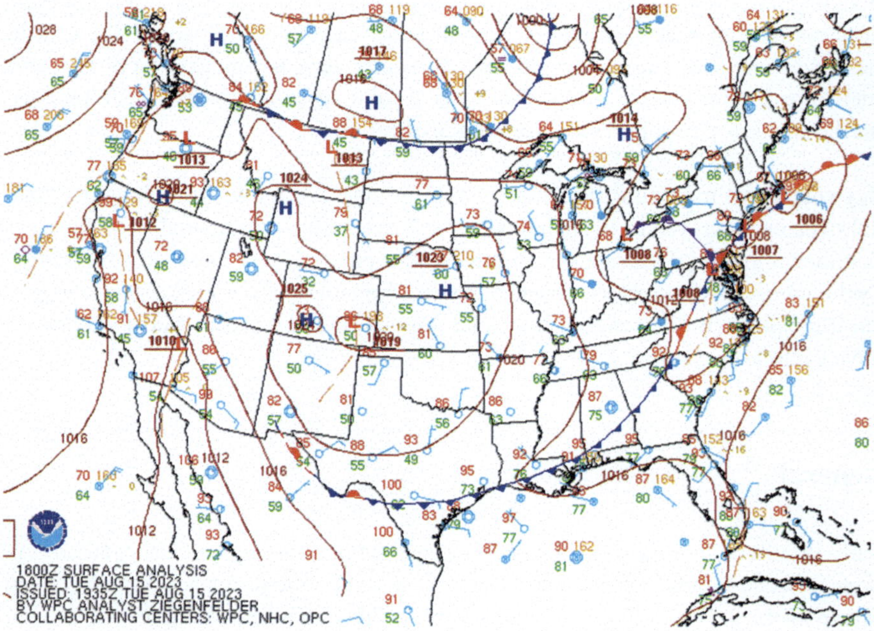

Fig. 4.1 Surface weather map for the U.S., at noon Eastern Daylight Time on 15 August 2023. Temperatures (°F) appear in red and dew point temperatures are in green. *Source* NWS

as an **air mass**. Air masses attain their characteristics through the transfer of energy (represented by temperature) and matter (represented by humidity), upward from the surface beneath them, via *convection*, and from the transfer of *momentum* (in the form of wind speed) downward from the faster upper-level winds, especially if beneath the *polar front jet stream* or *subtropical jet stream*.

The place where an air mass forms and obtains its characteristics is called its **source region**. Typically, source regions are flat and homogeneous across a large area, so that they can allow air masses to develop. So, for example, tropical and subtropical oceans, or flat plains in Arctic areas would be likely source regions, as they would transfer similar amounts of energy and matter upward, and momentum downward, across a wide area.

Meteorologists have developed a standardized, simple manner of classifying air mass types. The system uses a two-letter abbreviation. The first letter represents the nature of the source region. It is either "c" for "continental" or "m" for "maritime." "c" is dry, and likely has an inland source region, and "m" is humid, likely because it comes from an oceanic source region. The second letter is either "A" for "Arctic" or "Antarctic," "P" for "polar," "T" for "tropical," or "E" for "equatorial." At first glance, "A" and "P" appear similar, but "P" air masses are not as bitterly cold as "A" air masses. Likewise, "E" air masses are more extreme versions of the "T" air masses. "Tropical" has a little more seasonality, with cooler conditions and not quite as much *sensible heat flux* for the "cooler" months than "equatorial" air masses.

The air mass type is defined as the combination of the two letters. Notice that only the second letter is capitalized. Therefore, the six possible types are **Arctic** (A or cA), **continental polar** (cP), **maritime polar** (mP), **continental tropical** (cT), **maritime tropical** (mT), and **equatorial** (E or mE) **air masses**. Because little *continentality* exists near the Equator and little unfrozen water at the poles, cE and mA do not appear on Earth. The part of the Earth that is in the equatorial latitudes is therefore inherently maritime. Likewise, the part of the Earth that is in the Arctic or Antarctic latitudes is inherently continental.

The general domains of the various air masses in North America are shown in Fig. 4.2, and those for the rest of the world are depicted in Fig. 4.3. The locations in the maps are simply generalized interpretations. Also, because air masses are modified as they move over different surfaces, they experience changes in the amount of energy and mass that they attain from their surfaces, all the while steered by the general circulation. The exact locations where they cease to have legitimate characteristics of that air mass are ambiguous.

Rossby Waves and Secondary Circulations

The concept of the atmospheric *Rossby wave* was covered in Chap. 3 in general terms. The Rossby waves deliver the secondary circulation features, including air masses, at least in the mid-latitudes and the adjacent parts of the other major circulation systems. This region includes the equatorward edge of the *polar cell* and poleward edge of the *Hadley cell* in each hemisphere. In the mid-latitudes, the westerly winds are moving

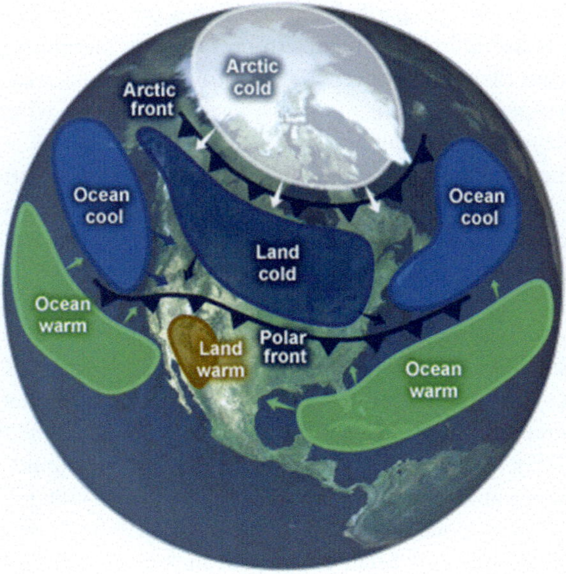

Fig. 4.2 Typical locations of air masses influencing North America. *Source* UCAR CoMet. Permission from Tara Torres

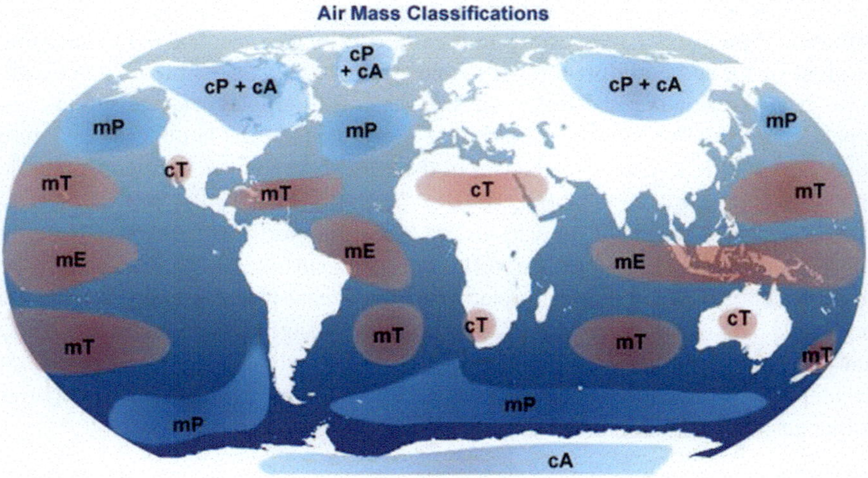

Fig. 4.3 Typical locations of air masses around the world. *Source* UCAR CoMet. Permission from Tara Torres

through the hemisphere in Rossby waves, primarily because of differential heating. These waves follow the areas where the warm and cold air masses come into contact which sets up a thermally-induced pressure gradient from equatorward (where it is warmer) to poleward (where it is colder) side. But then the *Coriolis effect* bends the flow toward the right in the Northern Hemisphere, or left in the Southern Hemisphere, resulting in primarily west-to-east flow, in each hemisphere. Because the Earth is not perfectly heated latitudinally, at some longitudes the Equator-to-pole temperature gradient occurs farther poleward or farther equatorward than the "mean" location. This displacement sets up the "wave" part of the *mid-latitude westerlies*—the Rossby waves.

As was explained in Chap. 3, sometimes the flow is overwhelmingly west-to-east around the hemisphere, with few large poleward (ridges) or equatorward (i.e., troughs) dips. This kind of flow is termed *zonal* (Fig. 4.4a). At other times, the Rossby waves have amplified ridges and troughs which take the flow poleward and equatorward on their general west-to-east trajectory. This type of flow is called *meridional* (Fig. 4.4b). But there is no absolute criteria for what constitutes zonal or meridional flow.

In 1938, meteorologist Carl-Gustav Rossby, introduced in Chapter 1, hypothesized that air in the upper part of the weather layer of the atmosphere—the *troposphere*

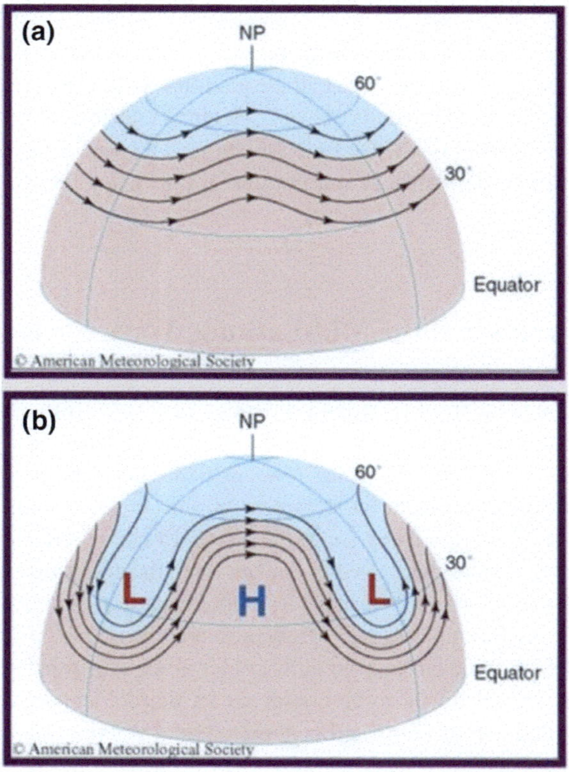

Fig. 4.4 Rossby waves in the Northern Hemisphere. **a** zonal flow. **b** Meridional flow. *Source* UCAR CoMet. https://www.meted.ucar.edu/ams/wim_2014/9a.html. Permission from Tara Torres

(without friction) flows in north–south and west-to-east-oriented waves that circumnavigate the Northern (and also the Southern) Hemisphere, called "planetary waves." Rossby's ideas were formulated without the benefit of upper-air observations from weather balloons or satellite imagery to inform his hypothesis or facilitate testing it. But of course, technology eventually caught up with him and proved him correct, so it was only fair that these waves would be named after him. These Rossby waves have ridges and troughs and are produced because Earth is spherical and rotates around its polar axis. This means that the speed and shape of Rossby waves are influenced by the Coriolis effect, the strength of which is related to latitude.

Rossby defined β as the change in Coriolis parameter (f), which as shown earlier has units of s^{-1}, as north–south distance (in m) changes.

$$\beta = \frac{\Delta f}{\Delta y}$$

And he developed an equation that describes a Rossby wave's rate of propagation (called "c"—the wave's speed of advancing, like the speed that an ocean wave moves onshore, in m s^{-1}):

$$c = U - \left(\frac{\beta L^2}{4\pi^2}\right)$$

Be careful to avoid confusing the "c" in the above equation with the "continental" category of air masses! In the equation above, U represents the west-to-east, or zonal, flow speed (with units of m s^{-1}), L represents the *wavelength*—the west–east distance (in m) from one ridge to the next one, and 4 and π do not have units. This equation is important because it represents the way that the Rossby waves behave in the real world.

Fronts and the Mid-Latitude Wave Cyclone

The polar front jet stream separates very cold from very warm air and tends to steer enclosed mid-latitude surface low-pressure areas along with it in its path. These do not include the *subpolar lows*, which tend to be locked in place too far poleward for the polar front jet stream to be able to move them much. The cyclonic circulation (counter-clockwise in the Northern Hemisphere and clockwise in the Southern Hemisphere) tends to pull cold air from the poleward side of the polar front jet stream equatorward and warm air from the equatorward side of the polar front jet poleward.

As the enclosed area of low pressure forms, cyclonic circulation develops. Because the cyclonic circulation pulls colder air equatorward into warmer air, the warmer air is pushed up over the colder air as the colder air advances equatorward. This is called a **cold front** and is depicted on a weather map as shown in Fig. 4.5a, with a cross-sectional diagram shown in Fig. 4.5b.

On the eastern side of the cyclone, because the cyclonic circulation pulls warmer air poleward into colder air, the warmer air again is pushed up over the colder air, but this time, the warmer air mass is advancing, rather than the colder air. This is called a **warm front** and is depicted on a weather map as shown in Fig. 4.6a and in cross section in Fig. 4.6b.

This entire system of a mid-latitude enclosed area of relatively low pressure (i.e., cyclone), with its associated cold fronts that move equatorward and warm fronts the move poleward, while the whole system is pulled generally from west to east, is known as a *mid-latitude wave cyclone*. This type of cyclone forms as a tiny area of low pressure in the process of **cyclogenesis**.

The low pressure can occur for a variety of reasons. Perhaps a small area of slightly reduced surface pressure occurs because of local-scale *convergence* near the surface due to being a little warmer than the air around it. The low could also be caused by a rougher surface that slows air crossing it, or by two air streams that more or less collide near the surface.

The forward speed of fronts differs between cold and warm fronts, as the whole mid-latitude wave cyclone system is pushed generally from west to east in the westerlies. The cold front is relatively steep compared to the warm front, because the advancing colder air is so dense that it tends to resist forward motion and piles up into a dome shape, with the warmer air forced up over it as the colder air advances. The warmer air does not move "backward" against the steering flow though.

The warm front is relatively gentle compared to the cold front, because the advancing warmer air is so light that it tends to be pushed along aloft more easily. The result is that high, steep clouds form at the cold front, but the warm front is

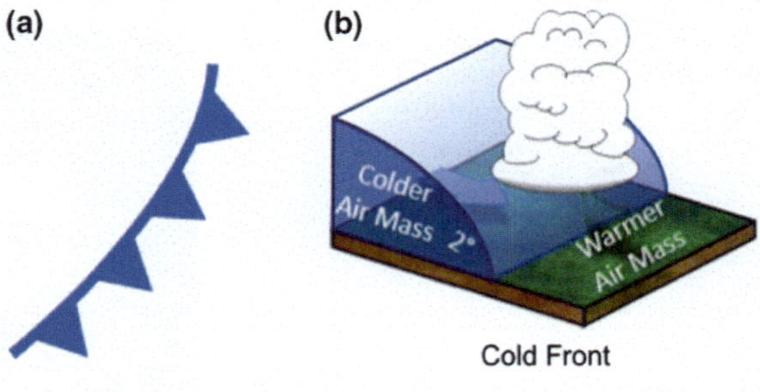

Fig. 4.5 Cold front. **a** Map symbol; **b** cross-sectional depiction of airflow. *Source* Authors. Rohli and Li (2021)—in that source, RVR had revised from NWS Teaching Resources

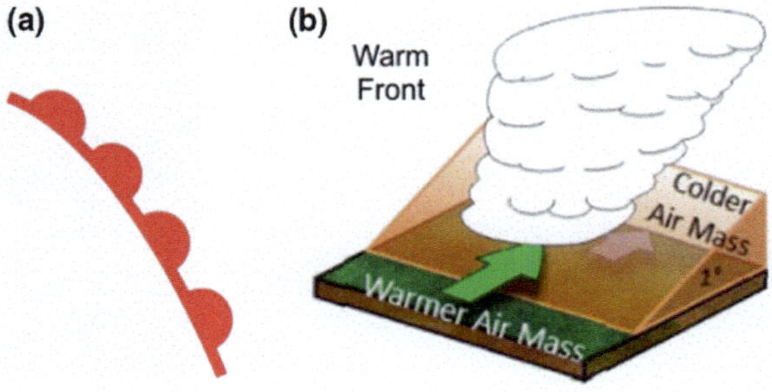

Fig. 4.6 Warm front. **a** Map symbol; **b** cross-sectional depiction of airflow. *Source* Authors. Rohli and Li (2021)—in that source, RVR had revised from NWS Teaching Resources

typically associated with a thicker band of lower, more horizontally-oriented *stratus* clouds (Fig. 4.7). This difference in cloud cover gives the mid-latitude wave cyclone its classic comma shape (Fig. 4.8).

However, the cold front itself advances faster than the warm front, because it is easier for colder air to displace less dense warmer air (i.e., at the cold front) than it is for warmer air to displace denser cold air (i.e., at the warm front). Eventually, the cold front "catches up with" the warm front and lifts the warmer air completely up off the surface. This is called **occlusion**, or an **occluded front**, and is illustrated in Fig. 4.9a. Occlusion is the process that ends the mid-latitude wave cyclone's existence, because

Fig. 4.7 Stratus cloud deck, as is typical at a warm front. *Source* NASA https://scool.larc.nasa.gov/GLOBE/stratus.html

Fig. 4.8 Satellite image of a mid-latitude wave cyclone. *Source* NASA. https://earthobservatory.nasa.gov/images/52297/mid-latitude-cyclone-over-the-united-states

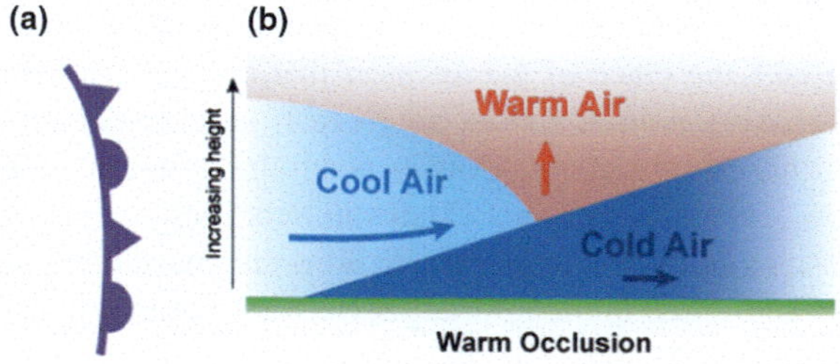

Fig. 4.9 Occluded front. **a** Map symbol; **b** cross-sectional depiction of airflow. *Source* Authors. Rohli and Li (2021)—in that source, RVR had revised from NWS Teaching Resources

once the warmer air is lifted, it cools. This weakens the temperature difference in the three sectors shown in Fig. 4.9b. Furthermore, by the time the mid-latitude wave cyclone reaches the occluded stage, the cold air has migrated far from its polar, Arctic, or Antarctic source region and gained *sensible heat* and *latent heat* from the warmer surfaces along its track. This surface-air exchange causes the cold air to lose its intensity, contributing further to the weakened temperature differences in each sector.

Four conditions favor cyclogenesis. First, weak static stability (i.e., *unstable atmosphere*) contributes strongly to cyclogenesis. This is a major mechanism because it allows the air to rise spontaneously to create the surface low pressure. Second, a nearby source of humidity (especially warm humid air) is also an important contributor to the possibility for rising motion. As was explained in Chap. 2, humid air contributes to saturation occurring at a lower height. At and above this height in the atmosphere, the *adiabatically*-moving air parcel will cool at a lesser rate than the *unsaturated adiabatic lapse rate* because the release of *latent heat of condensation* will partially offset the cooling of the parcel by expansion. This reduced cooling will enhance the possibility that the adiabatically-moving air parcel is warmer than its surrounding environment at a given height in the atmosphere, thereby making it more likely that the parcel will continue rising to greater heights, thereby contributing to an intensifying low-pressure center. These first two conditions described above are *thermodynamic* conditions because they rely primarily on the spontaneous rising associated with atmospheric instability, which is determined by the temperature and humidity of the air parcel.

By contrast, the last two conditions that favor cyclogenesis can be considered as dynamic reasons for the air to rise. The first of these is a strong low-level horizontal pressure gradient, which would cause the air near the surface to move faster before encountering a high-*friction* surface cover. Just as a faster-moving car would be more likely to be forced upward when it hits a pothole, faster wind speeds enhance cyclogenesis by inducing extra surface friction that causes enhancement of rising motion. It is true that the horizontal pressure gradient may originate from temperature differences, but the temperature difference is not the primary mechanism leading to rising.

The fourth condition to enhance cyclogenesis is favorable upper-airflow. As was shown in Chap. 3, air moves faster through ridges and slower through troughs, assuming a constant pressure gradient. Therefore, upper-level *divergence* exists from trough to the next downstream ridge. This upper-level divergence causes air to rise to take the place of the air aloft that diverged, to avoid a vacuum. At low levels, the rising motion causes air to rush in to replace the air that ascended, inciting further rising motion. Thus, beneath the trough-to-ridge side of the Rossby wave is low-level convergence, which supports cyclone formation. This causes climates on eastern sides of continents and adjacent coastal and marine areas (western sides of ocean basins) in the mid-latitudes, where trough-to-ridge sides of the Rossby wave are preferred, to be wet and stormy.

Geographically, the best areas, on a continental scale, for mid-latitude cyclogenesis are on the eastern sides of continents and adjacent offshore waters, especially in winter. This tendency is because on the eastern sides of continents, there is often a cold, cP or cA air mass pushed eastward by the mid-latitude westerlies adjacent to much warmer oceanic air, overlying warm ocean currents.

On a regional scale, **leeward**—or downwind—side of mountain ranges are also favorable areas for cyclogenesis. On the leeward side, a column of air can stretch as it moves downhill, giving it a deeper extent. When the air column becomes thinner and taller, it spins faster cyclonically, which is also known as gaining **positive vorticity**. This process is related to a principle called the **conservation of potential vorticity**. A common experience relating to conservation of potential vorticity occurs when a figure skater starts to spin faster by moving her arms closer to the trunk of her body.

Mid-latitude wave cyclones also die out (called **cyclolysis**) in preferred geographic locations, especially near the semi-permanent low-pressure areas. Prominent areas of cyclolysis are upwind of the *Icelandic Low* and *Aleutian Low* in the Northern Hemisphere, and in a belt of low pressure at subpolar latitudes in the Southern Hemisphere. The semi-permanent low-pressure areas can also revive storms, creating new ones. Of course, semi-permanent high-pressure areas are not areas of cyclolysis, as the cyclones avoid these areas altogether.

In some circumstances, a strong anticyclone may exist to the northeast of the mid-latitude wave cyclone. The clockwise (in the Northern Hemisphere) and outward flow from the high pressure cell may produce easterly winds ahead of the storm system, slowing or stalling the system and producing **stationary fronts**. In rarer circumstances, the anticyclone persists to such great heights that it can even counteract the steering circulation to such an extent that the mid-latitude wave cyclone may actually move from east to west, against the westerlies. This type of atmospheric flow is called a **backdoor front**. In a backdoor frontal situation, the cold front would approach from the east or northeast, but would still cause similar general types of weather conditions as those that approach from the west or northwest. In Southern Hemisphere backdoor fronts, the counter-clockwise flow around the anticyclone located to the southeast would force a backdoor cold frontal passage from the east or southeast instead of the normal approach from the west or southwest.

Mid-Latitude Anticyclones

Mid-latitude anticyclones are distinguished from the large, semi-permanent *subtropical anticyclones*, as they are regional-scale traveling highs that usually appear only at or near the surface, behind cold fronts. They are pushed along by the mid-latitude westerlies and circulate air clockwise and outward in the Northern Hemisphere, and counter-clockwise and outward in the Southern Hemisphere. They can be important circulation features for several reasons. They support sinking air, which leads to fair weather, and if they remain intact under weak circulation, can lead to pollution episodes and even short-term drought. The *advection* around their periphery moves colder air equatorward, such as after a cold frontal passage, and warmer air poleward, such as in the area ahead of the cold front, with a higher-pressure area off to the east of the warm air advection.

Mid-Latitude Westerlies and Flow Anomalies

The west-to-east (i.e., westerly) flow of the polar front jet stream steers mid-latitude wave cyclones and anticyclones. All winds in upper levels from poles to tropics are generally westerly—in both the Northern Hemisphere and Southern Hemisphere. But in the tropics, it is inaccurate to say that the winds are *geostrophic*, because the Coriolis effect is too weak to deflect winds substantially.

Sometimes persistent mid-latitude wave cyclones (i.e., with closed circulation) can become isolated from the main Rossby waves in the westerly wind flow (Fig. 4.10), similar to a spinning *eddy* that is cut off from the main current in a stream meander. These **cutoff lows** can keep weather gloomy, and perhaps with precipitation, for many days, until they are finally picked up by the main flow and moved eastward within the mid-latitude westerlies, or until they dissipate.

Likewise, anticyclones can also become cut off from Rossby waves. These **blocking patterns** (Fig. 4.11) are most common in summer, because in summer the general circulation is usually weaker, so the pressure gradient is often too weak to steer strong and large highs in the mid-latitude westerlies. This leaves them "stuck" and can cause the areas beneath where they are "stuck" to have stagnant air quality, as the sinking beneath the high may persist for days or even weeks. If local topography reinforces the impacts of this broader-scale circulation pattern, such as by being situated in a narrow valley where air cannot move laterally to replace the stagnating air, dangerous air quality can result. Historical examples include the Denora, Pennsylvania, pollution episode of 1948, which killed 20 people, exacerbated by coal burning, the "Great Smog of London" in 1952, and also Mexico City in more contemporary times.

An extreme version of a blocking high pattern, but sandwiched by two cutoff lows, is called an **omega block**, because of its resemblance to the Greek capital letter "omega" (Ω; Fig. 4.12). The effects of heat waves beneath the high as depicted

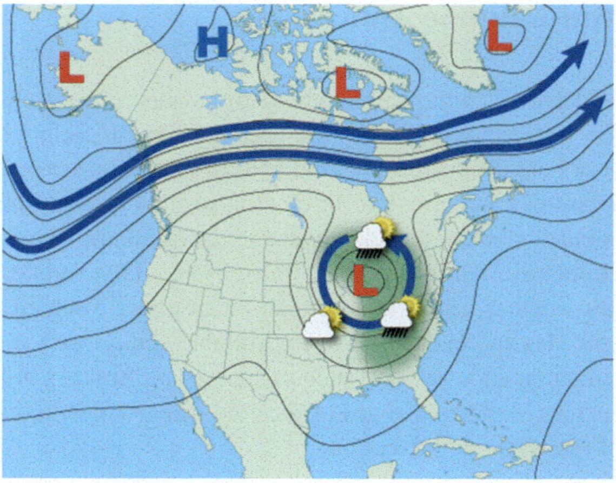

Fig. 4.10 A cutoff low. *Source* NWS. https://www.weather.gov/jetstream/basic

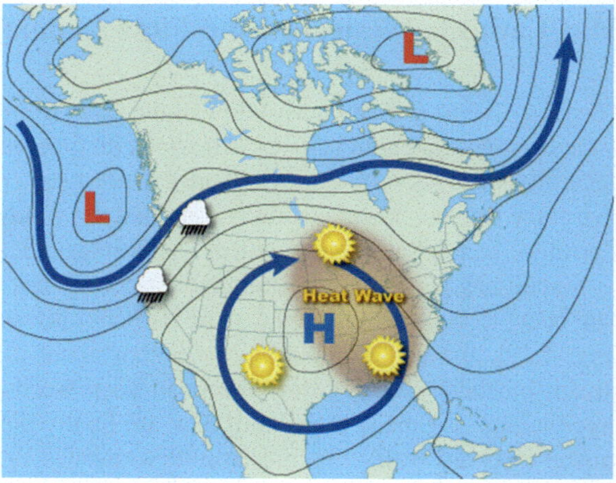

Fig. 4.11 A blocking anticyclone. *Source* NWS. https://www.weather.gov/jetstream/basic

in the image, equatorward of the polar front jet, can be substantial. But in addition, the places that are on the trough-to-ridge sides, such as over the western U.S.A. and the mid-Atlantic coast in the Fig. 4.12 example will be "stuck" in a wet weather pattern because of the persistence of the system.

Occasionally an anticyclone, perhaps from a blocking high or omega block, can be "stuck" poleward of the cyclone. This will leave both stuck, with the clockwise (in the Northern Hemisphere) flow around the high feeding into the counter-clockwise flow around the low, as shown in Fig. 4.13. Both features persist until one of the systems strengthens or weakens, allowing the flow to "correct itself" in the mid-latitude westerlies. These flow patterns are called **Rex blocks** (named for the meteorologist who studied them) and can be associated with odd temperature patterns. The high pressure beneath the ridge often supports clear skies which can lead to warm conditions, while the cyclone equatorward often has cloudy skies that may suppress daytime

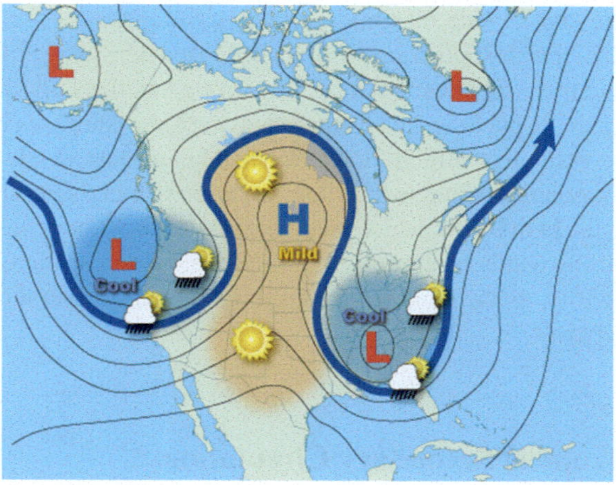

Fig. 4.12 An omega block. *Source* NWS. https://www.weather.gov/jetstream/basic

temperatures, leading to warmer conditions poleward and colder conditions equatorward of the Rex block. Eventually, another pressure system may steer either the high or the low forward in the westerlies more than the other feature, dissipating the Rex block. Subsiding air in the stationary high pressure region heats adiabatically and can create a **heat dome** that can persist for an extended period of time. These heat domes have been responsible for several record-breaking heat waves in recent years.

Regardless of whether a type of secondary flow anomaly exists in the mid-latitude westerlies represented by the Rossby waves at a given place and time, the mid-latitude westerlies represent a vital cog in the primary circulation. They maintain the balance of heat energy between the tropics and the poles. But many other, mostly minor, features of the mid-latitude flow can be important locally at certain times. For example, a **wake low** can form at *mesoscales* when air is cooled by *precipitation* on

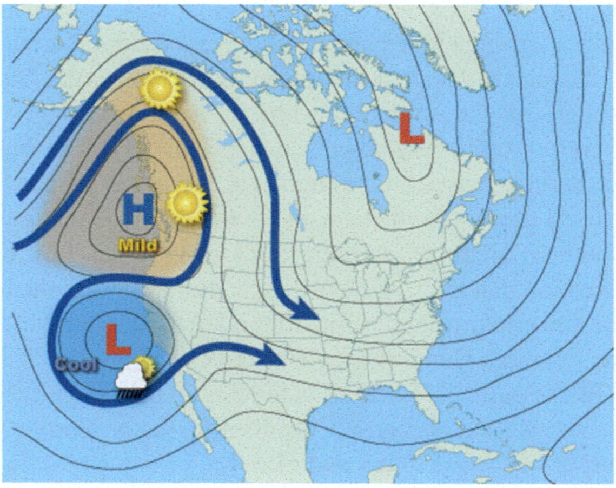

Fig. 4.13 A Rex block. *Source* NWS. https://www.weather.gov/jetstream/basic

the "back" side of a mid-latitude wave cyclone's precipitation shield. In the wake low, the air sinks because of its increased density, warms *adiabatically* on its decent, and then rises nearby, creating a strong localized pressure gradient between the low, the main cyclone, and the incipient high-pressure area between them. Wake lows can be extremely difficult to forecast and can cause deaths and injuries, particularly for those on the water, where they are more vulnerable to strong winds. A similar feature is the **heat burst**, which involves sinking air without the intense pressure gradient and strong winds. Some wake lows and heat bursts can increase the dew point temperature swiftly, as the sinking air is often much drier than the local near-surface air that it replaces.

Tropical Secondary Circulations

The primary secondary circulation system that is embedded in the general circulation in the low latitudes is the **tropical cyclone**. These storms, which can grow from weak **tropical depressions** into larger systems including **tropical storms** and even more powerful **hurricanes** (known as **typhoons** in East Asia), form as the flow direction in the *northeast trades* or *southeast trades* shifts slightly in a wavelike-shape. This feature is known as an **easterly wave** and is detectable on a weather map of **streamlines**, which are arrows representing the flow. Streamlines replace *isohypses*, as geostrophic flow describes flow inadequately in weak Coriolis effect environments of the low latitudes.

The red curved streamlines in Fig. 4.14 depict an easterly wave in the Caribbean Sea. The poleward bulge in the streamlines suggests a poleward extension of the ITCZ, and therefore elongated area of low pressure—a trough. Tropical troughs differ from mid-latitude troughs by bulging poleward rather than equatorward. Likewise, tropical ridges bulge equatorward rather than poleward, as they represent an equatorward protrusion of the high pressure characteristic of the subtropics. For this reason, the tropical troughs are sometimes called "inverted troughs" and tropical ridges are sometimes known as "inverted ridges" to mid-latitude meteorologists.

Tropical cyclone development occurs along the "right" side of the easterly wave trough axis (shown by the white curved lines in Fig. 4.14), where convergence, and therefore uplift, is maximized (labeled "CON" in Fig. 4.14). Convergence occurs in this zone east of the trough axis, regardless of hemisphere, because the air flows around the Northern Hemisphere (inverted) ridge turns to the right (and around the Southern Hemisphere ridge turns to the left), in a direction that aligns with the direction that the Coriolis effect is pulling, while the air flowing ahead of it around the (inverted) trough, turns to the left in the Northern Hemisphere (to the right in the Southern Hemisphere)—the direction opposite that of the direction that Coriolis effect is pulling. When air is moving in the favored direction of the Coriolis effect, it can go faster, because Coriolis effect is proportional to wind speed. Thus, near-surface air slows down and piles up, or converges, in this "traffic jam" east of the trough axis of the easterly wave, regardless of hemisphere. This air is forced upward,

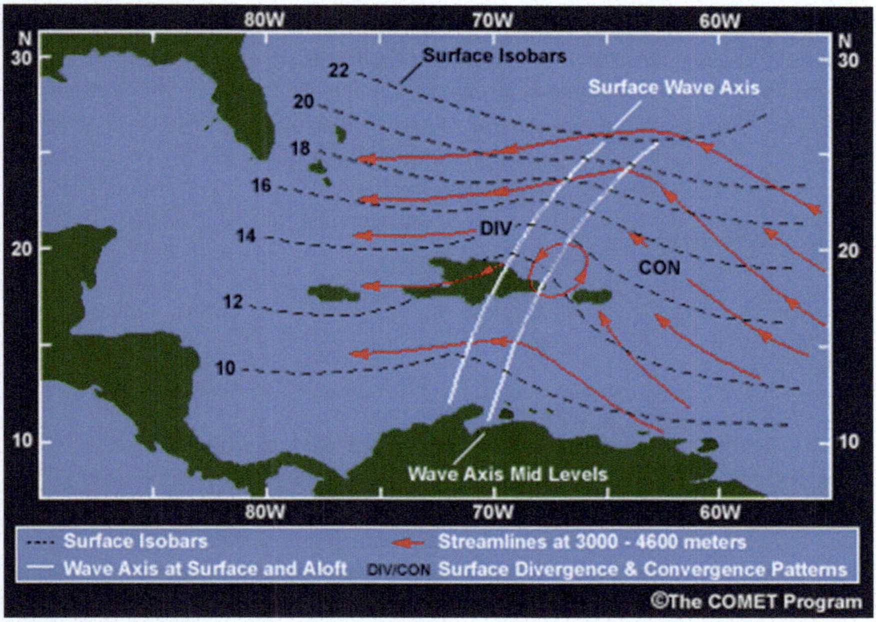

Fig. 4.14 Typical easterly wave axis and associated surface convergence and divergence zones in the Caribbean. *Source* UCAR CoMet. Permission from Tara Torres

contributing to vertical cloud development and surface low atmospheric pressure, and supporting tropical cyclone development on the inverted-ridge-to-trough (i.e., the southeast to northwest) side of the easterly wave.

West of the trough axis in the easterly wave, however, tropical cyclogenesis is not supported. As air flows (leftward in the Northern Hemisphere and rightward in the Southern Hemisphere) around and just past the trough axis, the Coriolis effect is weakened and the flow is slowed. But around the downstream ridge axis, the air is again flowing faster. Thus, to the west of the trough axis, this spreading out of near-surface air causes surface divergence on the trough-to-ridge (i.e., the northeast to southwest) side of the easterly wave, regardless of hemisphere, which causes air from above to sink to replace the surface air that diverged. Tropical cyclone strengthening and vertical cloud growth is discouraged by this sinking motion. In reality, the situation is more complicated because the weakened role of the Coriolis effect in the tropics makes other factors also important. More specific tropical cyclone coverage in relation to the atmospheric and oceanic circulation, including nuances about each of the world's ocean basins, is included in the second half of the book.

The temperature and precipitation patterns experienced at specific locations around the globe are largely a result of secondary circulation features. Air masses bring the temperature and humidity characteristics of their source regions, while transitory anticyclones and wave cyclones, with their associated fronts in the mid-latitudes, create clear skies and storms, respectively. The seasonality, frequency, and geographic locations of these secondary features are controlled by the larger primary circulation features, and together they create the Earth's climate regions. A similar relationship of circulation scales is also present in the ocean which will be explained in the next chapter.

Further Reading

Barriopedro, D., Garcia-Herrera, R., Lupo, A. R., & Hernandez, E. (2006). A climatology of Northern Hemisphere blocking. *Journal of Climate, 19*(6), 1042–1063. https://doi.org/10.1175/JCLI3678.1

Bjerknes, J. (1919). On the structure of moving cyclones. *Monthly Weather Review, 47*(2), 95–99. https://doi.org/10.1175/1520-0493(1919)47<95:OTSOMC>2.0.CO

Gray, W. M. (1968). Global view of origin of tropical disturbances and storms. *Monthly Weather Review, 96*(10), 669–700. https://doi.org/10.1175/1520-0493(1968)096<0669:GVOTOO>2.0.CO

Milrad, S. M., Gyakum, J. R., & Atallah, E. H. (2015). A meteorological analysis of the 2013 Alberta flood: Antecedent large-scale flow pattern and synoptic–dynamic characteristics. *Monthly Weather Review, 143*(7), 2817–2841. https://doi.org/10.1175/MWR-D-14-00236.1

Namias, J., Byers, H. R., Carpenter, A. B., Millar, F. G., Fish, G. V., Brooks, C. F., Willett, H. C., & Stone, R. G. (1936). An introduction to the study of air mass analysis. *Bulletin of the American Meteorological Society, 17*(6–7), 159–242. http://www.jstor.org/stable/26263606

Chapter 5
Ocean Circulation

Abstract Ocean currents help to maintain the Earth's heat balance by advecting heat from the tropics to polar latitudes. The trade winds drive the North and South Equatorial Currents in the Northern and Southern Hemisphere, respectively, flowing from east to west and forming the backbone of surface ocean circulation. The Coriolis effect bends the winds, which drags the surface ocean currents and produces anticyclonic gyres—large, permanent, ocean basin-wide circulations steered around the subtropical anticyclones in the mid-latitude ocean basins. Other spin-offs of the gyres and non-gyre-related regional surface currents are important components of the oceanic circulation as well. The Equatorial Counter-currents move water from west to east. Mesoscale ocean eddies are secondary rather than general circulation features, akin to mid-latitude wave cyclones in the atmosphere, as they are embedded within broader-scale circulation. The Ekman spiral causes ecologically and climatologically important upwelling in many coastal locations where the surface water is pushed oceanward. All of these circulations are connected, from the surface to the depths, by a prevailing three-dimensional conveyor belt of oceanic general circulation known as the global thermohaline circulation that moves in a predictable pattern. The oceans have many other effects on the atmospheric circulation.

Keywords Atlantic meridional overturning circulation (AMOC) · California Current · East and West Wind Drift · Western-boundary intensification · Ekman spiral · El Niño · Equatorial Currents and Counter-currents · Gulf Stream · Gyre · Humboldt Current (Peru Current) · Kuroshio Current · Mesoscale ocean eddy · South Pacific convergence zone · Thermohaline circulation

Purpose

Ocean currents play an important role in maintaining the Earth's heat balance by transferring heat from the tropics to polar latitudes. Ocean transport accounts for about 25% of the job of energy redistribution, compared to the 60% handled by atmospheric *advection* in the form of winds and the remaining 15% that is transported

R. V. Rohli et al., *Atmospheric and Oceanic Circulation*,
https://doi.org/10.1007/978-3-031-66142-6_5

by *latent heating*. The latter involves the evaporation of water—which absorbs solar *radiant energy*—in the part of the Earth that has a surplus of solar radiant energy and abundant oceans for supplying the water, and then the movement of that humid air poleward, where it cools, with its water vapor condensing, falling as *precipitation*, and releasing the stored latent heat via *condensation* and *deposition*.

Surface Currents

Because the source of virtually all of the energy in the Earth-atmosphere-system is the Sun, energy at the atmosphere–ocean interface is usually passed from the air to the water. Sometimes, however, energy passes from the ocean to the atmosphere, too—such as with *El Niño*, which will be explained more completely later in the chapter and book. But in general, the drag exerted by the wind causes the surface water to move. Therefore, the wind is the major driving force of surface ocean currents.

But what are these surface ocean currents? Starting in the low latitudes, the trade winds drive the **North Equatorial Current** in the Northern Hemisphere and the **South Equatorial Current** in the Southern Hemisphere. These flow from east to west in their respective hemisphere and form the backbone of the system of ocean currents (Fig. 5.1).

Because of the *Coriolis effect*, these currents are deflected poleward slightly, but the low-latitude location leaves the Coriolis effect very weak. Nevertheless, the slight poleward curvature is enough to form the equatorward edge of the circulation around the *subtropical anticyclones* in each hemisphere. This produces anticyclonic **gyres**—large, permanent, ocean basin-wide anticyclonic circulations that undergo spatial and temporal changes in intensity—in the mid-latitudes of each ocean basin (Fig. 5.2).

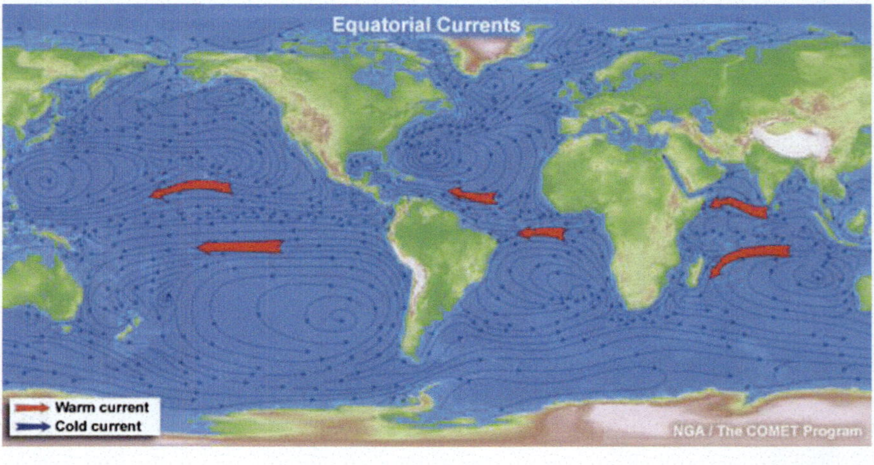

Fig. 5.1 Equatorial ocean currents. *Source* UCAR CoMet. Permission from Tara Torres

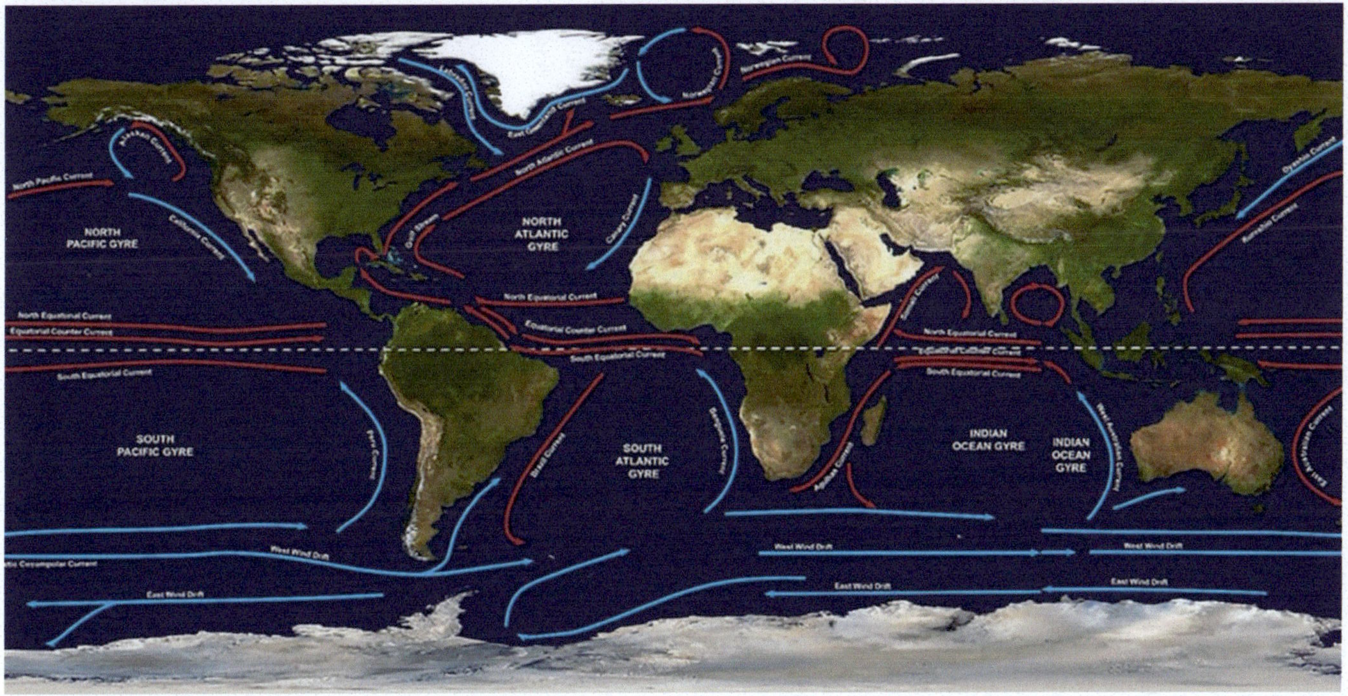

Fig. 5.2 Major surface ocean currents depicted in idealized annual average positions; note that the direction of monsoon currents, such as the Somali Current, changes substantially from summer to winter. *Source* NOAA. https://oceanservice.noaa.gov/facts/gyre.html

As a result, warm currents are found along east coasts of continents (western sides of ocean basins) and, depending on basin geography, transport warmer water across higher latitudes. Cold currents are found along west coasts of continents (eastern sides of ocean basins). For example, the **North Atlantic Drift** (Fig. 5.2) slips eastward north of the cold current near the west coast of Europe, keeping Europe much warmer than would otherwise be expected for these latitudes.

On the other hand, cold currents enhance stability and desert-like conditions by chilling the column of air from beneath, which stabilizes the overlying atmosphere. The Atacama Desert in Peru and Chile—the driest desert on Earth—is the classic example of cold-current-influenced stability. The Namib Desert in the southwestern African country of Namibia, is another example. Both of these examples are partly caused by the offshore current of cold water which makes the *environmental lapse rate* less steep and may even cause a *temperature inversion*. Despite the lack of precipitation, though, the cooler *temperature* and abundant *humidity* in the air often produce foggy and damp conditions. Thus, not every desert has predominantly clear skies, high temperatures, and little humidity.

Other surface features are important too. For example, the eastward-flowing **North Equatorial Counter-current** and **South Equatorial Counter-current** exist immediately along the lowest latitudes in their respective hemispheres (Fig. 5.2), with each flowing on the equatorward sides of the equatorial currents. The Coriolis effect is negligible for the equatorial counter-currents at these low latitudes. The equatorial counter-currents persist to approximately 100–150 m (300–500 ft) beneath the surface.

The other spin-offs of the gyres and non-gyre-related regional surface currents are important components of the oceanic circulation as well. These have distinctive names that usually indicate the geographical area where they are most dominant. They and their environmental impacts will be introduced in the regional chapters of their strongest influence.

Mesoscale ocean eddies are another circulation feature common in the oceans. One of the most fundamental principles in fluid dynamics is that in *turbulent* flow, as would happen when there is a temperature, density, or pressure gradient, spinning flow called an *eddy* will result, which will shed smaller eddies from it as it spins, with the main eddy sometimes persisting for days or months. These mesoscale ocean eddies typically have diameters of a few kilometers up to hundreds of kilometers, and the longest lasting are given names, akin to *hurricanes*. It is likely that they are even more common than currently thought, because those trapped beneath ice are difficult to detect even with today's remote sensing technology.

Mesoscale ocean eddies are secondary rather than general circulation features, akin to *mid-latitude wave cyclones* in the atmosphere, as they are embedded within broader-scale circulation features. Their impacts are the oceanic equivalent of storms as they cause *convection* and redistribute energy, matter, and *momentum*, particularly in places where the gradients of energy, matter, and momentum are sharpest. One major difference between mesoscale ocean eddies and storms is that the mesoscale ocean eddies form in a fluid that is naturally more *stable*, with the warmest part of the fluid—heated by the Sun—at the top. By contrast, storms form in environments in which the warmest, least dense part of the fluid—the air—is generally at the bottom of the fluid, with a steep environmental lapse rate indicative of atmospheric instability and facilitating vigorous *updrafts*. Thus, it is not surprising that atmospheric storms are generally much more violent than mesoscale ocean eddies.

Other forms of general circulation in the near-surface ocean are also important. However, because they act in tandem with vertical motion to produce truly three-dimensional flow, they will be explained after the next section. Aspects of the oceanic general circulation that interact directly with the atmosphere and *cryosphere* are also covered in Chap. 9.

Vertical Motion

The rotation of the Earth induces the Coriolis effect, which causes an apparent turning of winds with height upward from the surface, and of ocean currents with depth downward from the surface. The surface water is pushed by the wind (which itself has a direction that is influenced by the Coriolis effect) in the direction that the wind is blowing. Once that water starts accelerating, it is turned to the right (in the Northern Hemisphere) by the Coriolis effect, resulting in a near-surface water that is deflected slightly to the right of the surface wind. That water then imparts a shear on water beneath it, causing it to flow to the right and then the Coriolis effect pushes it farther to the right. Successively deeper layers of water in the Northern Hemisphere are pushed to the right of the direction that the higher layer of water is being pushed. But at the same time, *friction* dissipates momentum in successively deeper waters, so the speed decreases as does the deflection. This helix-shaped pattern of ocean current vectors is known as the *Ekman spiral* (Fig. 5.3).

The Ekman spiral occurs in the near-surface atmosphere just as in the near-surface ocean. Like the atmospheric counterpart, the ocean's Ekman spiral is predictable mathematically. But also similar to the atmosphere, surface waves and eddies in the ocean can disrupt the predicted motion.

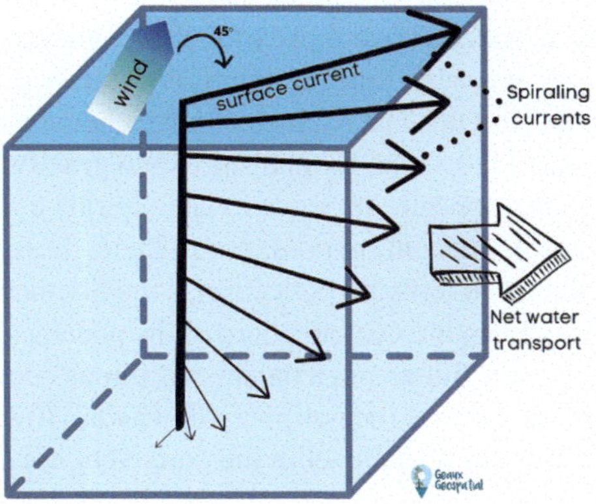

Fig. 5.3 The Ekman spiral. *Source* Geaux Geospatial

The net transport of water in the Ekman spiral will be at an angle of 45° anticyclonically (i.e., clockwise with increasing depth) from the surface wind direction. Thus, the near-surface water in the Ekman spiral is pushed to the right of the surface airflow in the Northern Hemisphere and to the left of the airflow in the Southern Hemisphere. In successively deeper layers the movement will be deflected farther and farther (anticyclonically) from the wind direction, and the speed will decrease.

The depth at which the vector first points 180° from the wind vector is called the **depth of frictional influence**, or **depth of frictional resistance**. At this depth the speed is $e^{-\pi}$ (or 0.0432) times that at the surface. So even a strong surface wind of 20 m s^{-1} (45 mph) will only cause a current of less than 1 m s^{-1} (2 mph) at the depth of frictional influence.

The layer from the surface to the depth of frictional influence is called the **layer of frictional influence**, or the **depth of the oceanic mixed layer**. This is analogous to the height of the **planetary boundary layer** in the atmosphere—the layer immediately adjacent to the surface, where friction is important. If the velocity vectors from the surface to the depth of frictional influence are integrated, the resultant vertically integrated motion is 90° anticyclonically from the wind direction.

Boundary Currents

As surface waters move equatorward on the eastern side of a gyre, the Coriolis deflection and Ekman spiral push the water toward the central ocean region. This deflection toward the center continues as the North Equatorial Current or South Equatorial Current has its waters deflected inward toward the middle of the ocean basin, regardless of the hemisphere. The western side of the basin also has a similar effect, with the poleward-moving water being pulled rightward in the Northern Hemisphere or leftward of the flow in the Southern Hemisphere, but in both cases, the displacement is toward the middle of the basin. The net effect of the Coriolis and Ekman deflections is a mound of water near the middle of the northern and southern ocean basins beneath the atmospheric subtropical anticyclones. Sea levels in these mounds are elevated above those of the surrounding areas.

At the same time, gravity is constantly tearing down the mounds as they continue to be built up. The balance of forces between the Coriolis effect and centrifugal force manifested as the Ekman spiral and gravity generates **inertial flow**.

The opposite occurs as water spirals cyclonically around the water beneath the semi-permanent *subpolar lows*. There, the rightward deflection in the Northern Hemisphere and leftward deflection in the Southern Hemisphere cause water displacement away from the cyclonic core of the oceanic circulation and therefore a depression of sea level. But again, a balance of forces occurs, with gravitational forces pulling the higher water at the periphery downward toward the depression. This force would fill in the core if sea level in the core were not simultaneously being lowered.

One complication, however, is that because the effects of the Coriolis deflection and Ekman spiral increase with increasing latitude, the shape of the gyres is asymmetrical. Currents moving toward the poles have increasing amounts of water pulled toward the center of the gyre while equatorward-moving currents have decreasing amounts of water pulled toward the center. So at the poleward margins of the gyres, the water is pulled strongly toward the middle of the ocean basin by the Coriolis effect and Ekman spiral from the west, east, and poleward directions. But at the equatorward margins, the water is pulled only weakly from the west, east, and equatorward directions. The net effect in both hemispheres is a pile-up of water offset to the west of the gyre center.

This pile-up creates a strong gradient of mass, with a narrow "valley" of water immediately adjacent to the most prominent peak in sea level. The balance of forces requires this substantial imbalance to be removed, so water initially moves from the peak on the western side of the centroid toward the west, in both hemispheres. However, the Coriolis effect changes its direction to become a poleward-moving current that accelerates along with the poleward increase in Coriolis effect, as the two are directly proportional. The water piles toward the western boundary and constricts the flow, which induces a speed increase of the western gyre current. This produces the **western-boundary intensification**—particularly narrow, deep, fast-moving currents in the subtropical to middle latitudes of the western sides of ocean basins (Fig. 5.4). The most prominent examples of western-broundary intensification are the **Kuroshio Current** of the western North Pacific basin and *Gulf Stream* of the western North Atlantic basin, but western-boundary currents also play important roles in the coastal and marine environments near eastern Australia, Brazil, and eastern Republic of South Africa.

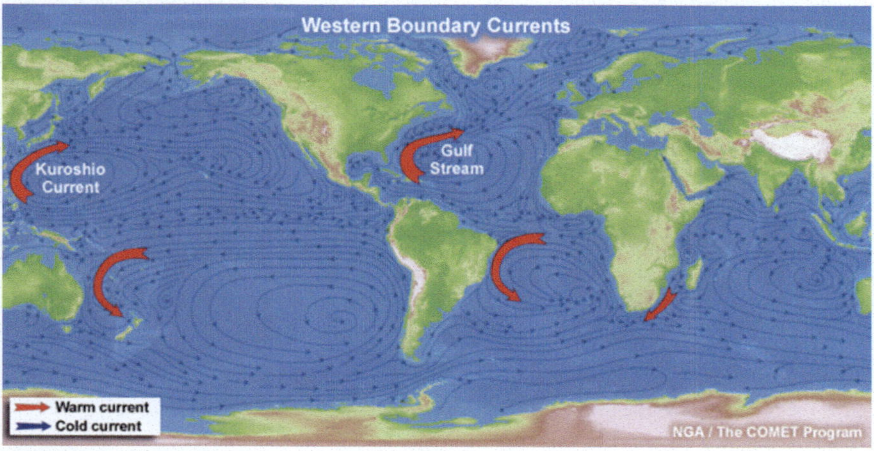

Fig. 5.4 Major western-boundary intensifications. *Source* UCAR CoMet. Permission from Tara Torres

The western-boundary intensification can also be explained by the difference in net Ekman transport between the western and eastern sides of the subtropical anticyclones in each ocean basin. Winds that drive the clockwise subtropical circulation (in the Northern Hemisphere) are the westerly winds (i.e., blowing from west to east) on the poleward side and the easterly winds (i.e., blowing from east to west) on the equatorward side. Assuming that these two winds have the same strength, the higher latitude of the westerly winds gives a larger Coriolis parameter and an enhanced Ekman transport, known as the **Sverdrup transport**, as the water turns south on the eastern side of the gyre. By contrast, the net northward Ekman transport due to the easterly winds is weaker on the western side of the ocean current because of the small Coriolis parameter associated with the easterly winds. For steady clockwise circulation to be maintained, the northward flow on the western side of the gyre must be concentrated in a narrow frictional layer along the west coast—the western-boundary intensification.

By contrast, the eastern sides of the gyres support wide, shallow, slower-moving currents in the subtropical to middle latitudes of the eastern sides of ocean basins. The most prominent examples of these instances of **eastern-boundary currents** have specific names that will be presented in later chapters, but their areas of influence are shown in Fig. 5.5.

Subpolar gyres exist as well. These are steered by the subpolar lows in each hemisphere. However, they are more prominent in the Northern Hemisphere (Fig. 5.6), as the uninterrupted nature of the Southern Hemispheric ocean basins favors more straight flow rather than gyres. Figure 5.6 shows an example of such straight flow, as the **Antarctic Coastal Current**, also known as the **East Wind Drift**. As air flows seaward downslope across the steep Antarctic continent, the very strong Coriolis effect at these latitudes deflects the flow strongly to the left. The shear from these surface winds drags the surface ocean water from east to west in the Antarctic Coastal Current, which parallels the shore.

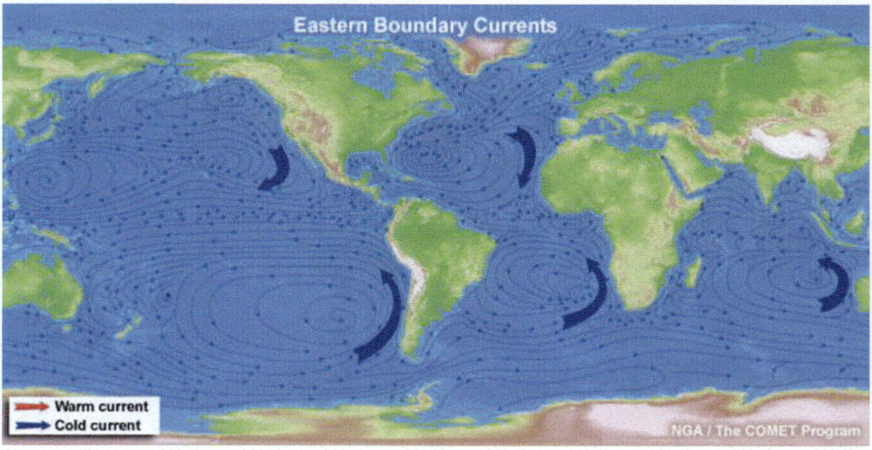

Fig. 5.5 Major eastern-boundary currents. *Source* UCAR CoMet. Permission from Tara Torres

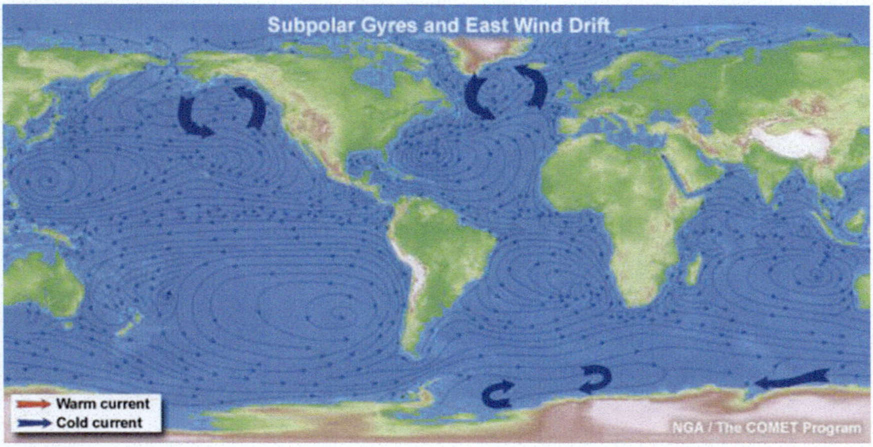

Fig. 5.6 Subpolar gyres and the Antarctic Coastal Current, or East Wind Drift. *Source* UCAR CoMet. Permission from Tara Torres

Upwelling

Some parts of ocean currents, especially near coasts, support strong upward vertical movement, known as *upwelling*. A prominent example is the cold eastern-boundary current known as the **California Current**, which flows from north to south, just off the U.S. Pacific coast (Fig. 5.7).

Upwelling occurs as the Ekman spiral strips water away from the coast, with water from deeper layers moving upward to replace the surface water. Regardless of whether the Ekman spiral pulls air to the right in the Northern Hemisphere or to the

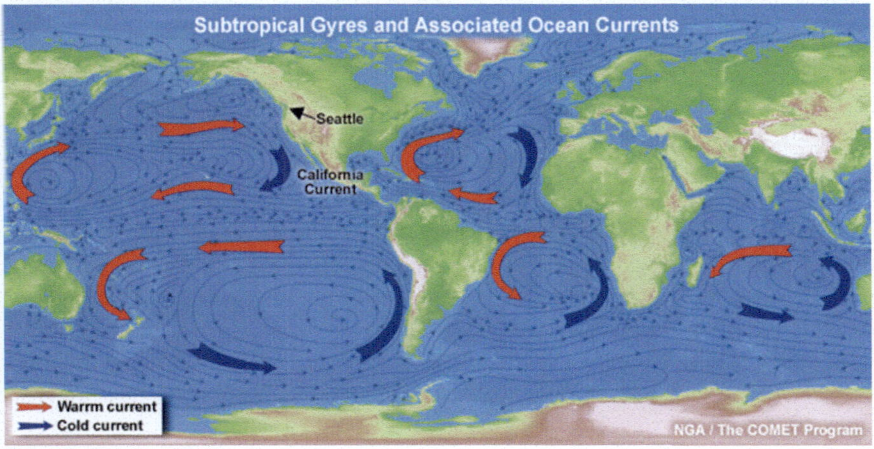

Fig. 5.7 The California Current and subtropical gyres. *Source* UCAR CoMet. Permission from Tara Torres

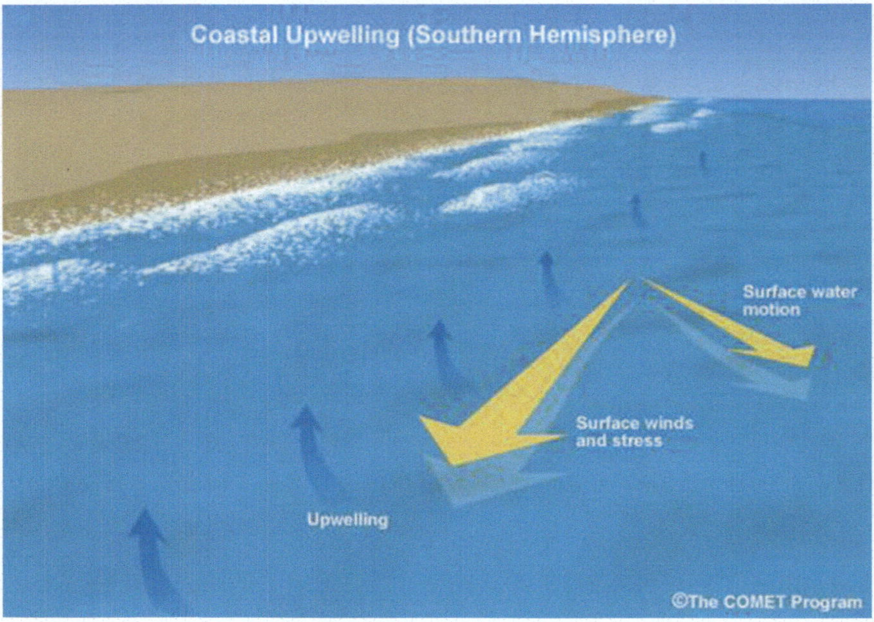

Fig. 5.8 Schematic of the coastal upwelling pattern in the Southern Hemisphere. *Source* UCAR CoMet. Permission from Tara Torres

left in the Southern Hemisphere, the effect is to pull near-surface water seaward on the eastern sides of the ocean basins, at eastern-boundary intensifications. A Southern Hemisphere example is shown in Fig. 5.8; note that the deflection of the surface wind is to the left of the flow, not necessarily to the left as it is being viewed in a diagram or map.

Upwelling occurs slowly, at a rate of approximately 5–10 m (16–33 ft) per day, with a vertical range from about 300 m (1000 ft) to the surface. The uplifted water is much colder than the original surface water and creates a characteristic zone of lower sea surface temperature near the shoreline. For example, in August, sea surface temperatures along the U.S. Atlantic coast is 21 °C (70°F) or above, while those along the central California coast is 15 °C (60°F) or below.

In addition to the colder water, coastal upwelling also brings high concentrations of dissolved nutrients, such as nitrates and phosphates. These nutrient-enriched waters promote plankton growth, which in turn, support large fish populations, which in turn often support large bird populations. More information about oceanic effects—specifically El Niño—appears in the next chapter.

Upwelling can also occur away from the eastern-boundary current, under different circumstances. For example, two localized wind systems pushing surface water in streams that diverge away from each other will generate a "shortage" of water in the widening downwind area between them. Water will move upward—upwelling—to replace the surface streams of water that were pulled apart as they were being forced downwind. In this case, the localized nature of the circulations will make the Ekman spiral and Coriolis effect negligible, so it is simply the lateral motion of the wind-stressed surface water and the vertical motion of the below-surface water that would produce the upwelling.

Thermohaline Circulation

The surface currents, mesoscale ocean eddies, vertical motion, and upwelling do not occur in isolation. Instead, these and other marine circulation features are connected, from the surface to the depths, by a prevailing general circulation of the oceans. This global **thermohaline circulation** is a continuous three-dimensional conveyor belt of water that moves in a predictable pattern. It has been estimated that the thermohaline circulation would disperse a drop of dye evenly throughout the global ocean in about 180 years. Of course, this also means that oil spills, microplastics, and other forms of pollution are also dispersed globally, eventually.

The thermohaline circulation is ultimately driven by the Sun, which produces the energy differences that drive the general circulation of the atmosphere covered in Chap. 3. These winds steer the gyres, which produce upwelling and other motions. But aside from the external features that drive the thermohaline circulation, the broader-scale upward and downward movement of water as it migrates globally is determined by its density. Two factors determine the density of ocean water, which in turn guide the thermohaline circulation: temperature (the "thermo" in "thermohaline") and salinity (the "haline").

Fluids generally become denser as temperature decreases. This density increase is because the individual molecules move closer together as they have less and less energy available to spread apart. This leads to the axiom that colder fluids are more dense than warmer fluids.

In water, including the oceans, density also increases with cooling, but only until it cools to 4 °C (39 °F). Unlike other substances, water begins to decrease in density as it cools from 4 °C (39 °F) to freezing. This property is because of the repelling force between hydrogen atoms in adjacent water molecules. More specifically, at temperatures below 4 °C (39 °F), the repelling force starts to exceed the disruptive random motions of the less energetic molecules, and the molecules begin to organize themselves into a more regular arrangement that occupies more volume and decreases density.

But then, when the temperature drops below freezing, the density of water decreases sharply and suddenly, as the crystalline structure dominates, with adjacent molecules about 9% farther apart than they were when the temperature was 4 °C (39 °F). This is why ice floats on water. As the temperature falls below freezing, density again increases as random molecular motions continue to decrease, as the molecular kinetic energy content decreases more. The relationship between water's density as its temperature is depicted in Fig. 5.9.

The fact that water's density peaks near 4 °C (39 °F) has tremendous implications for life on Earth. If water were like most other fluids, its density would increase as it

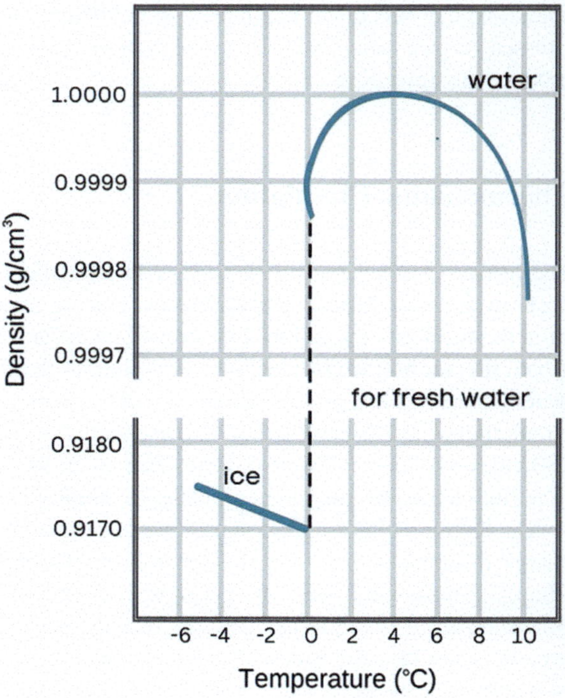

Fig. 5.9 Density of pure water as a function of temperature. *Source* Geaux Geospatial

cooled, thereby causing the deepest parts of the oceans—the parts with the densest water—to be frozen. The abundance and diversity of marine life would be stunted, and the ice block in the deepest parts of the oceans would be unlikely to receive enough warmth from the Sun to ever melt. The oceans would freeze from the floor upward, eventually including nearly all of the oceans except the top-most layer that can be warmed by the Sun—the **epipelagic zone** or **photic zone**. Instead, in the "real world," water that begins to be cooled by *Arctic air masses* or *polar air masses* begins to sink, but only until it reaches 4 °C (39 °F), whereupon it then rises in the ocean column. The oceans remain largely unfrozen. But notice that the temperature vs. density graph in Fig. 5.9 assumes that there is no salt in the water.

A second cause of density differences in the world's oceans is variations in salinity. The more saline the water is, the denser it is, for a given temperature. In parts of the ocean basins where local precipitation, freshwater runoff into the oceans, and/or advection of fresher water from elsewhere is inadequate to offset the local evaporation, waters are more saline. This often happens beneath the subtropical anticyclones, where precipitation and runoff are sparse, or in enclosed or semi-enclosed basins like the Mediterranean Sea and Red Sea, as shown in Fig. 5.10.

The presence of ice also influences salinity, which in turn influences temperature of the ocean water. Ice forms from fresh water, so places that have ice formation tend to have more saline waters. Likewise, when ice melts, the meltwater freshens the ocean water. The net result of the constant balancing between rising water (as it

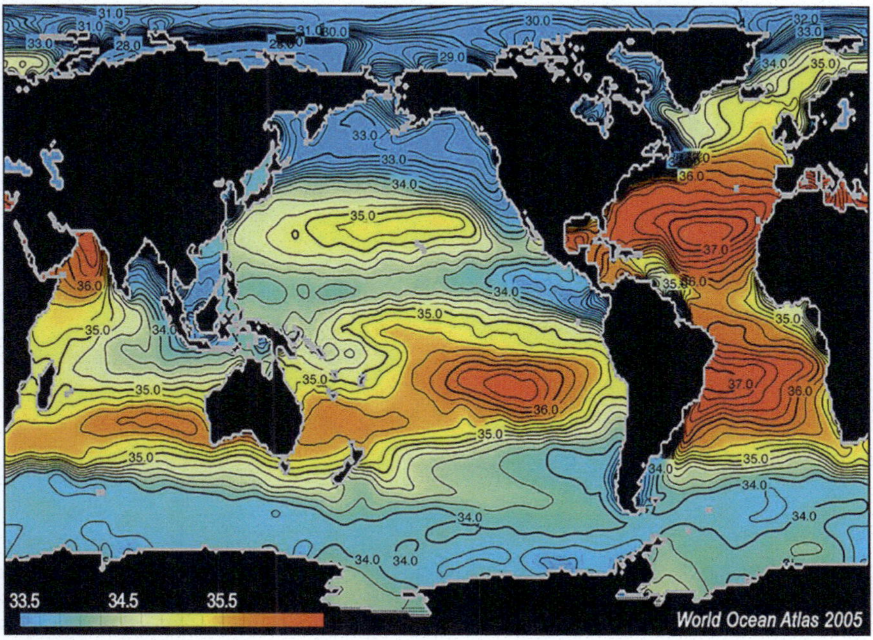

Fig. 5.10 Salinity of the world's oceans, with values shown in parts per thousand. *Source* NOAA/NWS. https://www.noaa.gov/jetstream/ocean/sea-water

becomes fresher and/or warmer above 4 °C (39 °F)) and sinking water (as it becomes more saline and/or colder but still above 4 °C (39 °F)) on its trek from basin to basin, steered by the wind, is the thermohaline circulation shown in Fig. 5.11.

One particularly important component of the thermohaline circulation is the **Atlantic meridional overturning circulation** (AMOC), which has been found to be weakening in recent decades. This trend is of concern to the overall global climate, as a slowing or weakening of the AMOC beyond a yet-unknown "tipping point" could plunge parts of the world into suddenly colder conditions. Debate is ongoing about the role of humans in the weakening of the AMOC versus the natural long-term variability in this important marine circulation. Other regional components of the circulation and specific impacts in each part of the conveyor belt will be described in the appropriate regional chapter.

Sudden (at least on geologic time scales) changes to the thermohaline circulation can occur as *ice sheets* melt or freeze, which can drastically change the temperature and/or salinity of the water, which would then stall or alter the thermohaline circulation. The **Younger Dryas** was a period of abrupt global or nearly-global cooling, from about 13,000 years ago to about 11,000 years ago and is believed to have occurred because ice sheets started to melt suddenly as the world warmed. There is concern that this could be happening again as ice sheets melt.

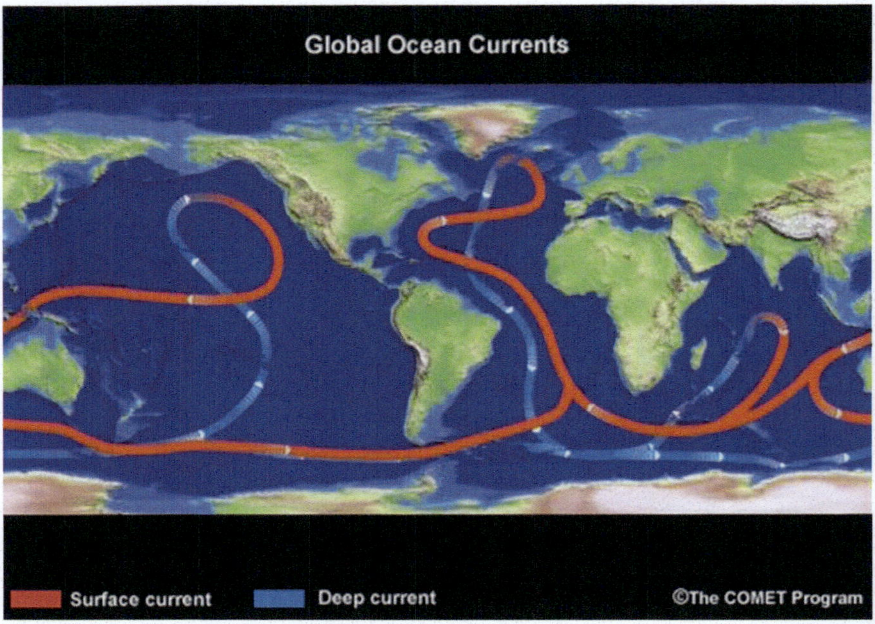

Fig. 5.11 Generalized and idealized view of the thermohaline circulation. *Source* UCAR COMET Program. Permission from Tara Torres

The thermohaline circulation and other more localized forms of ocean circulation discussed here offer opportunities for exchanges of energy, matter, and momentum between the air and the oceans. Part III of the book explores these kinds of interactions between "spheres" in the climate system, beginning with the next chapter that describes interaction between the atmosphere and oceans.

Effect of the Ocean on Global Circulation

In the tropical oceans, well-defined areas of water that are warmer than surrounding waters, such as the **Pacific Warm Pool**, generally enhance convection and precipitation. While the most prominent example of tropical convection zone is the ITCZ, it is not the only important belt of warm-pool-induced convection. A second warm-pool-induced convergence zone extends as an arm protruding diagonally from northwest to southeast across the tropical and subtropical South Pacific Ocean—the **South Pacific convergence zone** (**SPCZ**). The influence of the SPCZ on Oceania in the form of robust thunderstorms protruding from areas known as mature cloud clusters will be covered more completely in Chap. 21. In general, rainfall totals of 250–500 cm (100–200 in) per year are common over the oceanic warm pools.

Warm pools also play an important role in strengthening *easterly waves* of motion within the *northeast trades* and *southeast trades* in the atmosphere. This enhancement is important because easterly waves can grow into tropical disturbances. The dynamics of the easterly waves and how they support or suppress the growth of tropical disturbances into *tropical depressions*, *tropical storms*, *hurricanes*, or *typhoons* was described in Chap. 4. Over warm water, easterly waves can strengthen and cross the Atlantic from Africa to the Gulf of Mexico. In a typical year, more than 100 easterly waves occur, with more than 25 of these becoming tropical depressions. This proportion exhibits considerable interannual variability, and the extent to which the proportion is likely to change in the future is unclear.

In the extratropics, the ocean-influenced circulation affects both temperature and precipitation patterns. Ocean temperatures are relatively uniform by latitude, especially in the Southern Hemisphere, except in areas of cold currents. Notice how the colder colors protrude northward off the western coast of South America and the southwestern coast of Africa—these are because of cold currents (Fig. 5.12).

Oceans and their circulation also affect broad-scale atmospheric circulation through their influence on atmospheric sea level pressure patterns. In winter, the ocean is warmer than land, which leads to lower sea-level pressures. This pressure pattern deepens the *Aleutian Low* and *Icelandic Low*, giving them even lower pressure. The intensified subpolar lows would then have increased capacity to pull mid-latitude wave cyclones in toward them, making these important areas of mid-latitude wave cyclone demise (i.e., *cyclolysis*).

In summer, the ocean is generally cooler than land, which supports greater sea level pressure over the oceans. Such a situation is conducive to large, strong semi-permanent high-pressure zones such as the *Hawaiian-Pacific High* and *Bermuda-Azores High*. The result is generally clear skies over the subtropical oceans. One exception occurs when and where a cold current chills the warmer, overlying air

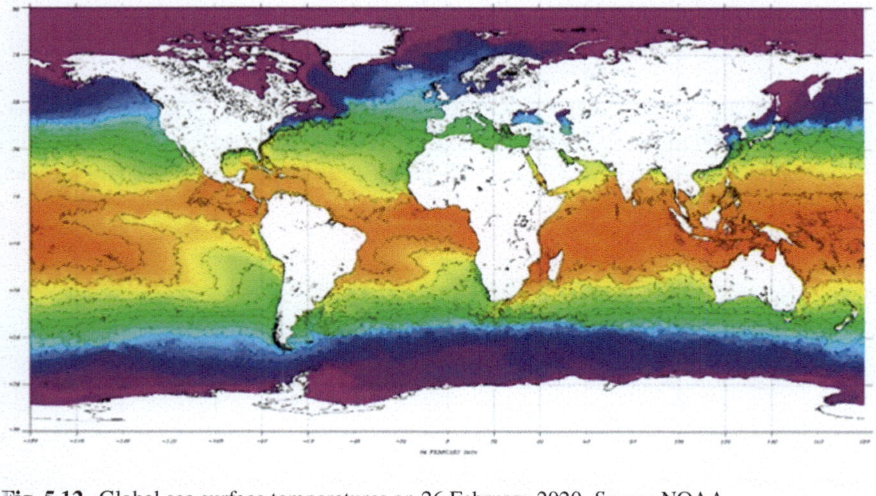

Fig. 5.12 Global sea surface temperatures on 26 February 2020. *Source* NOAA

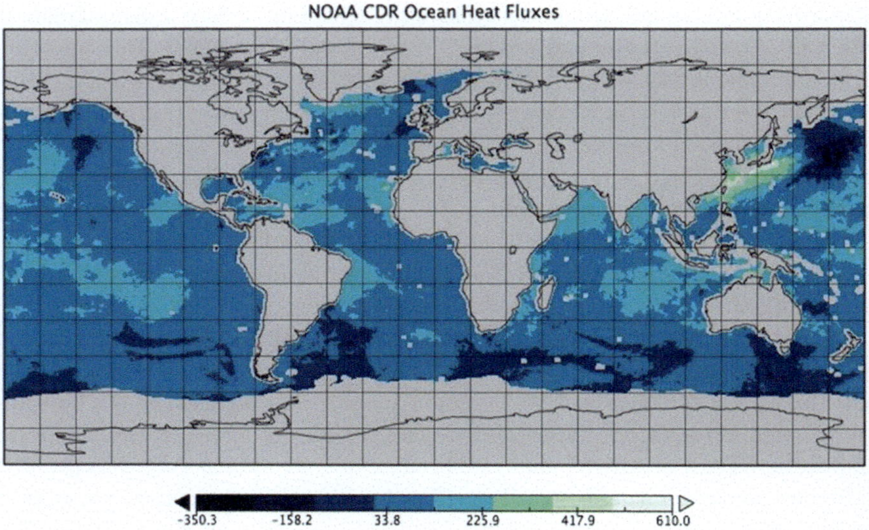

Fig. 5.13 Ocean-air heat fluxes (in W m^{-2}), with ocean-to-air fluxes shown in green and air-to-ocean fluxes in blue. *Source* NOAA

from underneath, causing **advection fog** and low *stratus* cloud decks, like in San Francisco. Similarly, advection fog also occurs when winds move warmer air over a colder surface regardless of whether a cold current is involved. Nevertheless, the effect is accentuated when particularly colder water from cold currents is present.

As a primarily water-covered planet, the Earth's atmospheric circulation is strongly influenced by the hydrosphere both in its interactions with oceanic circulation and through interactions with the cryosphere. The oceans, cryosphere, lakes, mountains, and vegetated surfaces are overlapping forcing mechanisms that influence atmospheric circulation patterns at all spatial scales. These patterns, characterized in the next chapter, define the structure of the Earth's climate patterns that are covered in Parts V, VI, and VII of the book.

The ocean-air heat flux magnitudes are depicted in Fig. 5.13. By definition, areas where the ocean supplies energy in the form of sensible and latent heat fluxes to the air above it have positive fluxes. Places where the atmosphere provides energy to the ocean have negative values.

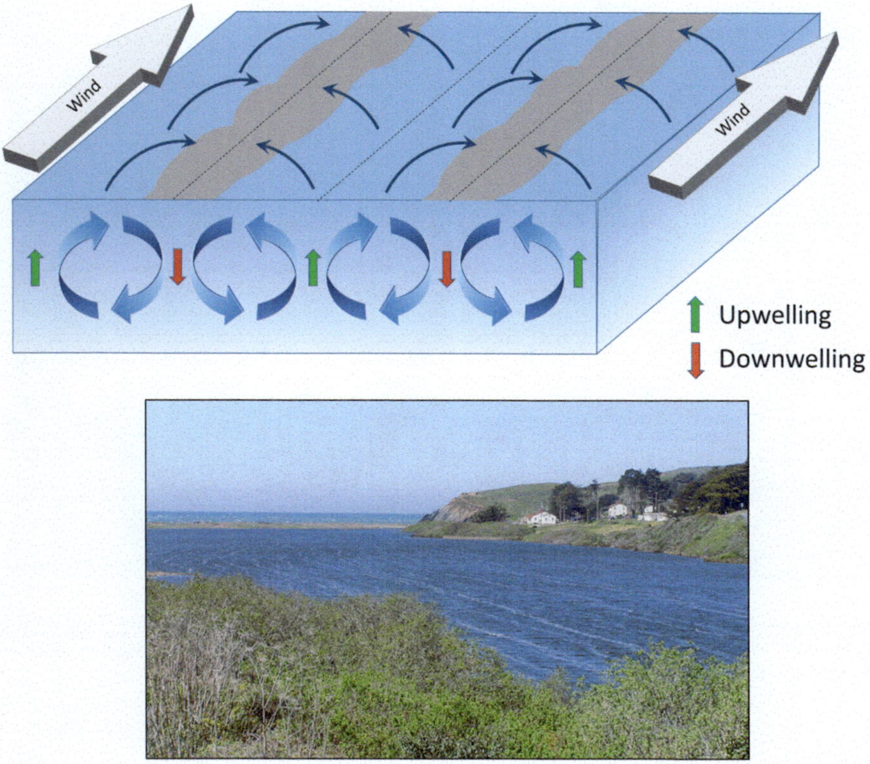

Fig. 5.14 Schematic and photograph of a Langmuir circulation system. *Source* Roger Williams University. https://rwu.pressbooks.pub/webboceanography/chapter/9-7-langmuir-circulation/

The corkscrew-shaped circulations that move upward and downward in the mixed layer of the oceans, parallel to and pushed along by a steady, prevailing wind, as shown in Fig. 5.14, are called **Langmuir circulations**. Notice how an upward-thrusted mass of water is next to a downward-thrusted mass of water in the Langmuir circulation as the upward-moving water mixes with the air. The circulations have important ecological roles in both concentrating and mixing the near-surface water and the nutrients, pollutants, and organisms in it. The analogous pattern in the atmosphere is called a **cloud street**, a series of linear clouds, as in the impressive satellite imagery shown in Fig. 5.15.

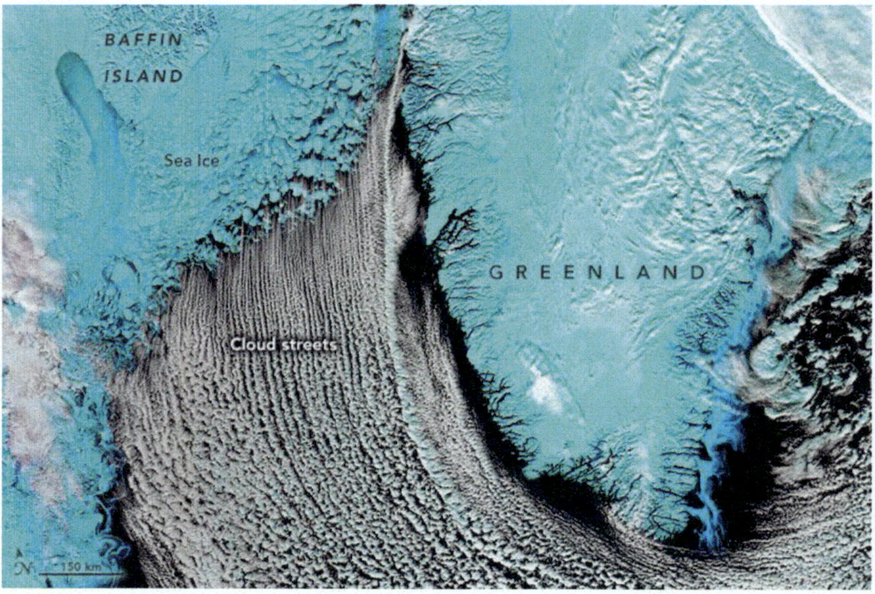

Fig. 5.15 Cloud streets as shown from NASA Visible Infrared Imaging Radiometer Suite on the Suomi NPP satellite on 2 March 2020. *Source* NASA. https://earthobservatory.nasa.gov/images/146387/cloud-streets-over-the-labrador-sea

Further Reading

Anderson, D. (1997). Younger Dryas research and its implications for understanding abrupt climate change. *Progress in Physical Geography, 21*(2), 230–249. https://doi.org/10.1177/030913339702100203

Barstow, S. F. (1983). The ecology of Langmuir circulation: A review. *Marine Environmental Research, 9*(4), 211–236. https://doi.org/10.1016/0141-1136(83)90040-5

Buckley, M. W., & Marshall, J. (2016). Observations, inferences, and mechanisms of the atlantic meridional overturning circulation: A review. *Reviews of Geophysics, 54*(1), 5–63. https://doi.org/10.1002/2015RG000493

Chavez, F. P., & Messie, M. (2009). A comparison of eastern boundary upwelling ecosystems. *Progress in Oceanography, 83*(1–4), 80–96. https://doi.org/10.1016/j.pocean.2009.07.032

Clark, P. U., Pisias, N. G., Stocker, T. F., & Weaver, A. J. (2002). The role of the thermohaline circulation in abrupt climate change. *Nature, 415*, 863–869. https://doi.org/10.1038/415863a

Griffies, S. M., Winton, M., Anderson, W. G., Benson, R., Delworth, T. L., Dufour, C. O., Dunne, J. P., Goddard, P., Morrison, A. K., Rosati, A., Wittenberg, A. T., Yin, J. J., & Zhang, R. (2015). Impacts on ocean heat from transient mesoscale eddies in a hierarchy of climate models. *Journal of Climate, 28*(3), 952–977. https://doi.org/10.1175/JCLI-D-14-00353.1

Hu, D. X., Wu, L. X., Cai, W. J., Sen Gupta, A., Ganachaud, A., Qui, B., Gordon, A. L., Lin, X. P., Chen, Z. H., Hu, S. J., Wang, G. J., Wang, Q. Y., Sprintall, J., Qu, T. D., Kashino, Y., Wang, F., & Kessler, W. S. (2015). Pacific western boundary currents and their roles in climate. *Nature, 522*, 299–308. https://doi.org/10.1038/nature14504

Part III
Interactions Between the Atmosphere and Other Spheres

Chapter 6
Ocean-Air Teleconnections

Abstract Teleconnections, or "seesaw" correlations of energy, matter, or momentum between distant places, have regional to hemispheric-scale atmosphere-ocean impacts and interactions. The El Niño-Southern Oscillation (ENSO) teleconnection is a periodic anomalous warming/cooling of tropical equatorial Pacific waters associated with drastic environmental consequences globally. The Pacific Decadal Oscillation (PDO) is similar to ENSO but with extreme modes that dominate for 20-to-30-year periods rather than six to 18 months as in ENSO. In the Indian Ocean Dipole (IOD), sea surface temperature anomalies seesaw between the western and eastern Indian Ocean. The Pacific-North American (PNA) pattern has simultaneous amplified (meridional) atmospheric ridging over northwestern North America and troughing in southeastern North America vs. de-amplified (zonal) flow across these areas. The North Atlantic Oscillation (NAO) is a pressure/geopotential height seesaw between the Bermuda-Azores High and the Icelandic Low, which are simultaneously anomalously strong or weak, causing either zonal or meridional winds, respectively, across the North Atlantic. The Arctic Oscillation (AO), or Northern Annular Mode, is a hemispheric-scale expression of the NAO, with a seesaw in simultaneous strength of the North Polar High and Northern Hemisphere, mid-latitude, mid-level geopotential heights. These and other patterns contribute to climate variability far beyond the seesaw's location.

Keywords Antarctic Oscillation (AAO) (Southern Annular Mode) · Arctic Oscillation (AO) (Northern Annular Mode) · Atlantic Multidecadal Oscillation (AMO) · Atmospheric river · El Niño-Southern Oscillation (ENSO) · Indian Ocean Dipole (IOD) · Labrador Current · Maritime Continent · ENSO Modoki · Niño 3.4 index · North Atlantic Oscillation (NAO) · Pacific Decadal Oscillation (PDO) · Pacific-North American (PNA) pattern · Pacific Warm Pool · Pineapple Express · Teleconnection · Walker circulation

R. V. Rohli et al., *Atmospheric and Oceanic Circulation*,
https://doi.org/10.1007/978-3-031-66142-6_6

The Concept of Teleconnections

Energy, matter, and momentum are mostly conserved in the Earth-ocean-atmosphere system, so we do not expect to find a sudden increase or decrease in the total amount of these entities in the system. But the distribution of these entities within the system changes over space and time. For instance, as was shown in Chap. 3 regarding the semi-permanent atmospheric pressure features, some specific places have more atmospheric mass in the form of higher *geopotential heights* or *pressure* than others.

Certain, specific sets of places on Earth tend to work as a seesaw, with one side of the seesaw experiencing greater than normal (for that place) energy (in the form of temperature), matter (in the form of sea level pressure or geopotential heights above the surface), or momentum (in the form of stronger winds) at the same time that the same other place on Earth is experiencing less than normal of the same entity. A **teleconnection** is a correlation between energy, matter, or momentum across a large part of the Earth's surface.

Sea-level pressure or geopotential height teleconnections are often of particular interest because they govern variability in broad-scale flow patterns. Extreme phases of these teleconnections, which would leave one side of the seesaw with much higher pressure or geopotential height and the other with much lower pressure or geopotential height, can strengthen or weaken and persist predominantly in that mode over many years.

Teleconnections represent a source of regional to hemispheric-scale atmosphere-ocean impacts and interactions. They are quasi-predictable because scientists can observe the seesaw starting to change and understand the general features associated with each phase of the seesaw. Nevertheless, each seesaw episode is a little different from the others, so prediction skill remains imperfect.

El Niño-Southern Oscillation

The *Southern Oscillation* is the atmospheric component of the oceanic *El Niño* and *La Niña* phenomenon. The entire phenomenon, known as **ENSO**, is a periodic altering of the surface ocean currents in the Pacific Ocean, which can have drastic consequences both in the area and around the globe. ENSO is the biggest producer of atmospheric variability in the world, other than the seasons themselves. ENSO has been known for centuries, but was only considered a local phenomena until the 1960s. Peruvian fishermen noticed abrupt changes in their catch; others around the Pacific noticed changes in weather patterns. But today we know that El Niño impacts the global circulation pattern, both in the atmosphere and ocean. But how?

Under "normal" conditions, the surface air over the western tropical Pacific Ocean (near Indonesia—sometimes called the **Maritime Continent**) has particularly low atmospheric pressure. This is in part because of the warm pool of water that is pushed westward by the trade winds to form the *Pacific Warm Pool*, which warms the

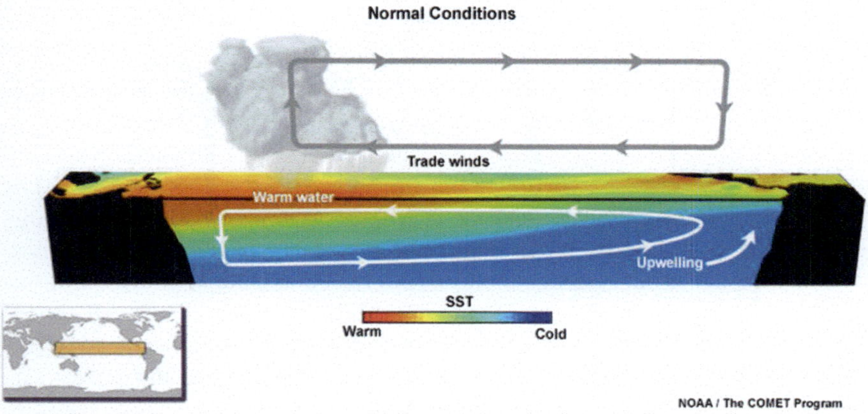

Fig. 6.1 "Normal" circulation in the equatorial Pacific Ocean. *Source* UCAR CoMet. Permission from Tara Torres

atmosphere immediately above the surface, increasing the *environmental lapse rate*, thereby causing the atmosphere to be *unstable* frequently. The rising air associated with instability causes the area to have low pressure even compared to other areas along the *inter-tropical convergence zone* (ITCZ), as air rises above the surface and creates high, tall, frequent cloud cover that produces abundant *precipitation* (Fig. 6.1).

Meanwhile, the eastern tropical Pacific Ocean (near South America) usually has much higher atmospheric pressure. This is in part because of the *upwelling* along the South American coast, which brings colder water to the surface, chilling the air near the surface, making the air more dense, and giving the near-surface atmosphere a tendency to sink. As air pushes down harder on the surface, the near-surface, east-to-west air pressure gradient (from higher to lower pressure) initiates a *pressure gradient force*, which is manifested as the *trade winds* that blow air from east to west, piling up the warm tropical water in the western equatorial Pacific Ocean as the Pacific Warm Pool (see again Fig. 6.1). Aloft, the air that rose over the Maritime Continent typically moves back eastward to replace the air that sank over the higher pressure of the eastern tropical Pacific, near South America. Meanwhile, water also escapes the pile-up over the Maritime Continent via the **Indonesian Throughflow** into the Indian Ocean, where it participates in the global *thermohaline circulation.*

This entire circulation system, with rising air over the Maritime Continent, eastward- flowing air aloft toward equatorial South America, sinking over the eastern Pacific Ocean, and trade wind flow back toward the Maritime Continent, is known as the **Walker circulation** (Fig. 6.2). This pattern is the "normal" circulation (also termed neutral, or, in older literature, La Nada) that is observed in the tropical Pacific Ocean.

For some still unknown reason, however, every three to seven years, the "normal" Walker circulation weakens or breaks down. One component of this breakdown is the oceanic phenomenon of El Niño. The onset of the El Niño phase involves unusually high atmospheric pressure developing in the western Pacific and lower pressure than

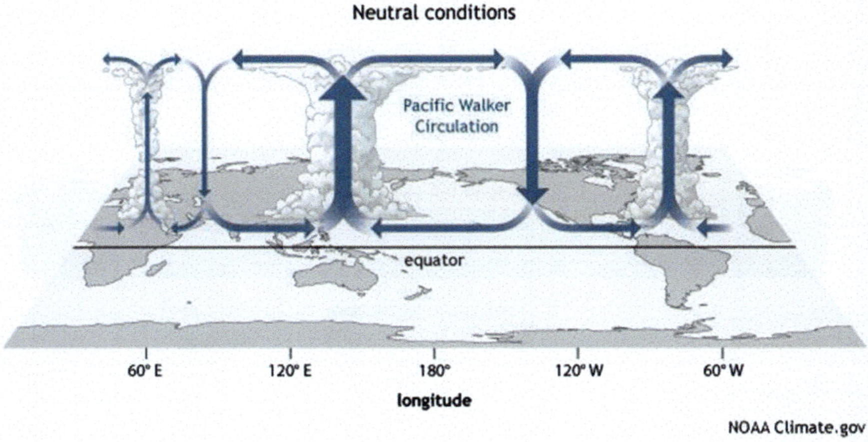

Fig. 6.2 The Walker circulation. *Source* NOAA Climate.gov. https://www.climate.gov/news-features/blogs/enso/walker-circulation-ensos-atmospheric-buddy

normal developing in the eastern Pacific. This pressure anomaly often reaches its peak around Christmas, hence the name El Niño which is Spanish for "the baby". With the weakening, or even reversal, of pressure patterns during El Niño, trade winds weaken and can completely reverse directions, and the entire Walker circulation weakens. Thus, the easterly trade winds which usually pile up water in the western Pacific—the Pacific Warm Pool—break down and force water to slosh back eastward (Fig. 6.3).

Meteorologists recognize two major types of El Niño events. The "conventional" El Niño event is one in which the atmospheric circulation anomaly appears in Fig. 6.3—with rising motion near where sinking motion would normally be, and vice versa. The second type of El Niño event is known as an ENSO **Modoki**. In a

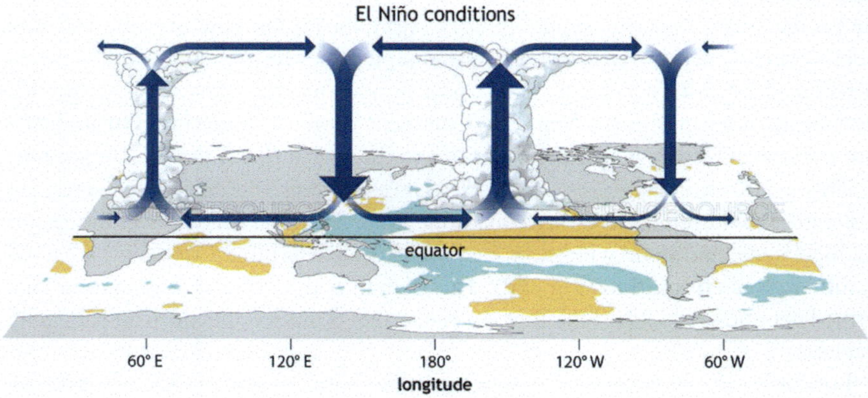

Fig. 6.3 Atmospheric circulation and sea surface temperature anomalies (orange is warmer than normal and blue is colder than normal) during the El Niño phase of the Southern Oscillation. *Source* NOAA Climate.gov. https://www.climate.gov/news-features/blogs/enso/walker-circulation-ensos-atmospheric-buddy

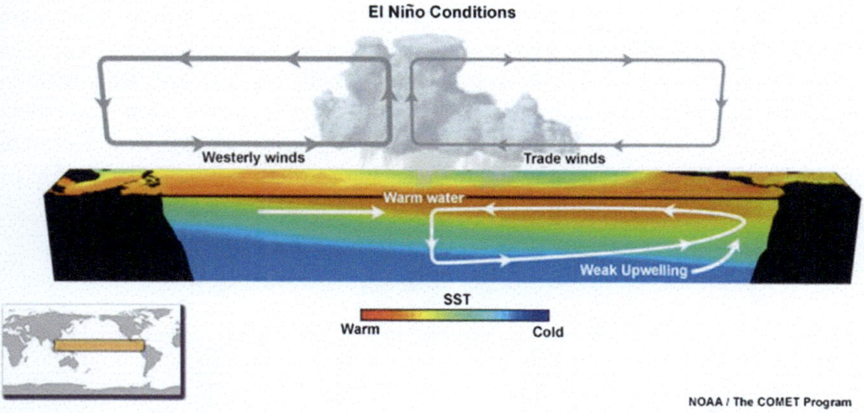

Fig. 6.4 Oceanic temperature profile during a Modoki type of El Niño phase of the Southern Oscillation, with the shifting of the Walker circulation shown. *Source* UCAR CoMet. Permission from Tara Torres

Modoki, the anomalously high sea surface temperatures do not extend all the way to the eastern side of the Pacific basin adjacent to the Americas. Instead, the anomalously warm sea surface is concentrated in the central Pacific Ocean, as suggested in Fig. 6.4. Thus, the Modoki acts almost as a hybrid between the Walker circulation and a conventional El Niño.

Regardless of the type of El Niño, the eastward movement of the warm water causes sea surface temperatures near South America to increase above normal. This anomalous warmth is exacerbated by the suppression of the normal upwelling of cold, nutrient-rich water along that coast because of the extra pile-up of water near that coast. This explains why the El Niño phase of the Southern Oscillation is sometimes known as a "warm event." The "warm" is a culturally derived term from an American perspective, as the warmth is closest to the Americas, while at the same time, anomalous cold sea surface temperatures are near Asia, the Maritime Continent, and northeastern Australia.

The name "Southern Oscillation" also reflects a bit of cultural bias; a better term would have been tropical (or equatorial) oscillation. However, the discoverers were British, and to them, it was "Southern." To Southern Hemisphere dwellers, "Northern Oscillation" might seem like a more appropriate name! The "oscillation" part of the "Southern Oscillation" name is reminiscent of the shifting of atmospheric pressure and wind patterns across the Pacific Ocean.

However, when pressure is lower than normal over the Maritime Continent and higher than normal over western South America, the tropical easterlies (east to west flow) strengthen to become even more intense than during the Walker circulation. In the ocean, this is part of the *La Niña* phenomenon, whereby the sea surface is even colder than its "normal" low temperature in the eastern tropical Pacific Ocean near South America (Fig. 6.5). La Niña occurs when the "normal" pattern intensifies too much and is sometimes known as a “cold event,” with the terminology again conforming to the conditions in the ocean near the Americas.

The atmospheric signal of La Niña is for the normally low pressure in the western tropical Pacific, near Indonesia—the Maritime Continent, to be even lower than normal. And likewise, the normally higher atmospheric pressure in the eastern tropical Pacific, near South America, is even higher than normal. The profile of atmospheric and oceanic conditions during La Niña is shown in Fig. 6.6.

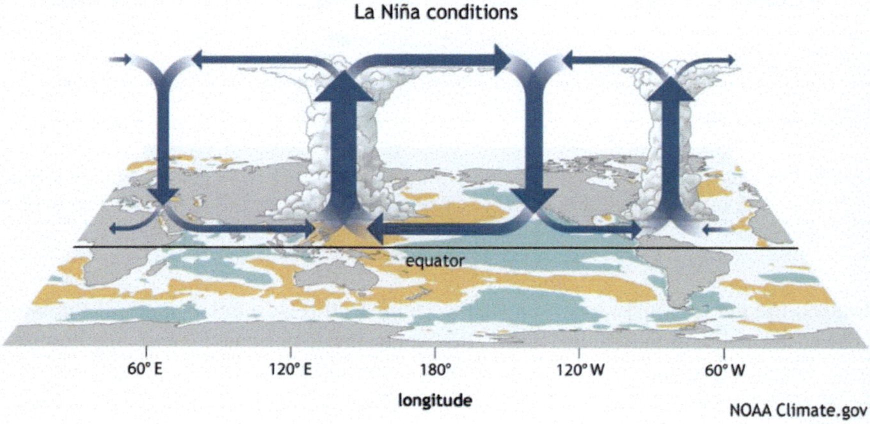

Fig. 6.5 Atmospheric circulation and sea surface temperature anomalies (orange is warmer than normal and blue is colder than normal) during the La Niña phase of the Southern Oscillation. *Source* NOAA Climate.gov. https://www.climate.gov/news-features/blogs/enso/walker-circulation-ensos-atmospheric-buddy

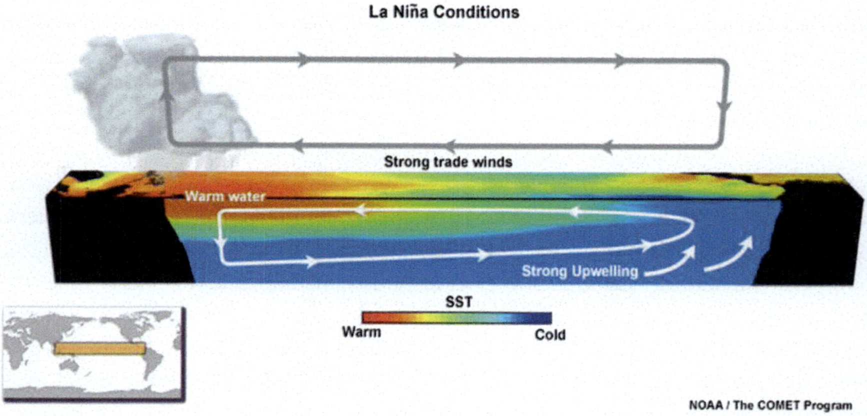

Fig. 6.6 Oceanic temperature profile during the La Niña phase of the Southern Oscillation, with the intensification of the Walker circulation shown. *Source* UCAR CoMet. Permission from Tara Torres

The effects of the extreme phases of ENSO—the La Niña and especially the El Niño phase—are felt worldwide. Among the most impactful properties of El Niño is the tendency for the Asian monsoon systems to weaken. Monsoons will be covered in much greater detail in the East Asia (Chap. 16) and South Asia (Chap. 17) chapters, but a simplified explanation is that the summer vs. winter monsoonal shifts in the steering circulation caused by reversals of pressure patterns are altered during El Niño.

The weakening or reversals of the "normal" monsoonal circulation patterns are important for people, because they may bring unexpected extended weather conditions. Because they are so important in Asia and important to a lesser degree in some other parts of the world, where there are many mouths to feed, their weakening can have enormous human impacts.

The monsoons weaken during El Niño because the normal rising motion over the low pressure near Indonesia is not as vigorous, and this causes the westward and eastward outflow aloft over that rising zone to be weakened. With weaker westward outflow, the weakened *Coriolis effect* is less effective at deflecting that humid air onto Asia, so the phase of the monsoon that brings wetter conditions tends not to be so wet. This presents a major problem when water is needed to serve (including feeding) large populations.

El Niño events also have other impacts. They tend to suppress Atlantic *tropical cyclone* activity because they cause abnormally strong upper-level westerly winds over the Caribbean (in opposition to the trade winds). This westerly flow shears off the tops of many or most developing tropical cyclones, which include storms of tropical origin with intensities ranging from weak tropical disturbances to the strongest *hurricanes* and *typhoons.*

El Niño events are often characterized by a bifurcation, or splitting, of the *polar front jet stream* and *subtropical jet stream* over North America. This happens because the abnormal warm pool over the eastern tropical Pacific can get "stuck" near the shore and migrate northward, allowing an elongated area of warm conditions, pushing the sharpest temperature gradient northward, and amplifying the *ridge* over western North America. This "abnormal" warm water zone is not to be confused with the "normal" Pacific Warm Pool in the western tropical Pacific Ocean. The northward displacement of the polar front jet stream—which exists near the poleward boundary of the warm water—brings warmer than normal winters in the northern U.S. and Canada. Yet at the same time, the need to transfer energy, matter, and momentum westward persists, creating the subtropical jet stream. The abnormally warm water in the central tropical Pacific also enhances the *Hadley cell* in the vicinity of the *Hawaiian-Pacific High*, activating a wave train including the *Aleutian Low*, which can sometimes force meridional flow into North America.

The El Niño event that persisted for an unusually long time throughout the early 1990s displaced the polar and subtropical jet streams across North America to such an extent that Texas received double the normal precipitation in December 1991. The strong subtropical jet, which was fed by more rising conditions and warmer sea surface temperatures than normal, had steered these storms repeatedly toward Texas. A similar narrow, persistent "conveyor belt" of abundant, persistent moisture in the air, known as an **atmospheric river**, tends to occur over California during El Niño

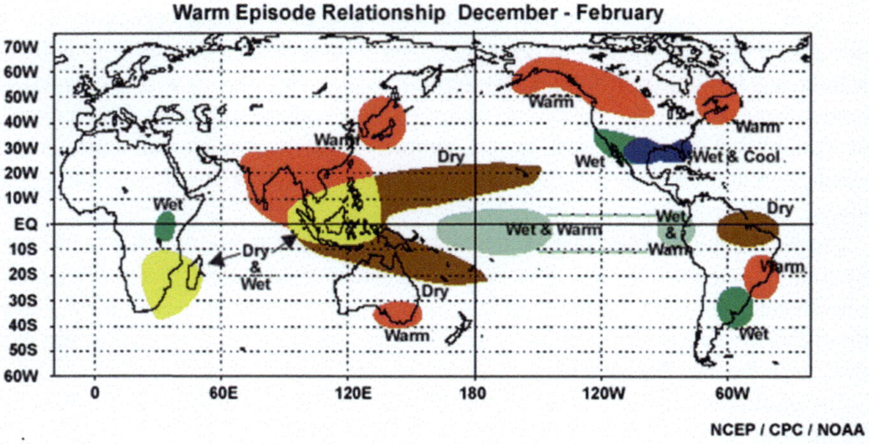

Fig. 6.7 Impacts of El Niño phase of the Southern Oscillation in December through February. *Source* UCAR CoMet. UCAR CoMET cites NCEP/CPC/NOAA—all U.S. government sources. Permission from Tara Torres

events, where the resulting parade of storms is known as the **Pineapple Express**. The winter of 2023 may be the best recent example of a Pineapple Express, even though that winter was characterized by overall “normal” rather than El Niño conditions.

Many other impacts of the El Niño phase are felt in wide-ranging parts of the world (Fig. 6.7). For example, El Niño tends to bring wet conditions to normally-arid parts of Peru and Ecuador and the eastern United States. El Niño also causes drought in Indonesia, Australia, Philippines, near locations where the normal Pacific Warm Pool weakens or vacates.

La Niña also has impacts, though not as severe as those from the El Niño phase. For instance, La Niña brings drought to many of the same places that tend to be wetter than normal during El Niño, including California and the U.S. Gulf of Mexico coast.

The impacts of El Niño and La Niña events are variable, depending on the strength of the episode and, in the case of El Niño, whether it is a conventional or Modoki event. In general, conventional El Niño events tend to have stronger and more wide-ranging impacts. Some evidence indicates that even though El Niño events are projected to occur more frequently in the changing climate, a greater proportion of those events may be the less impactful Modoki. During El Niño, an area may experience flooding one time and drought the next.

Several indices have been developed to represent the presence and intensity of ENSO events. One of these is the **Niño 3.4 index**, which is a statistically-standardized representation of the sea surface temperature in the so-called Regions 3 and 4 of the central equatorial Pacific Ocean (Fig. 6.8). The Niño 3.4 has proven to be among the most useful indicators of ENSO because the geographical extent of sea surface temperatures upon which it is based—the region bounded by 5° N, 5° S, 170° E, and 120° W (i.e., the stippled area in Fig. 6.8, which overlaps Region 4 in green and Region 3 in blue)—aligns very well with the area that is abnormally warm with *convection* during both conventional and Modoki El Niño events but cool and non-convective during La Niña. Positive values of the Niño 3.4 index mean that the sea surface temperatures in the stippled area of Fig. 6.8 are warmer than normal, which corresponds well with the El Niño phase (see again Figs. 6.4 and 6.5). A very strong signature of El Niño appears in the central and eastern equatorial Pacific Ocean if a statistical correlation is mapped between the Niño 3.4 index and sea surface temperature. Figure 6.9 shows that warmer than normal sea surface temperatures occur in the central and eastern equatorial Pacific and northward and southward of this area along the North and South American coast, with cooler than normal sea surface temperatures in the western Pacific Ocean, when the Niño 3.4 index is positive—an El Niño event (Fig. 6.9).

The opposite sea surface temperature anomalies would occur when the Niño 3.4 index is negative—a La Niña event. Several other indices of the intensity of El Niño or La Niña have been developed, such as the **Southern Oscillation index**, which is the difference between sea level pressure between Tahiti, in Polynesia, and Darwin, Australia.

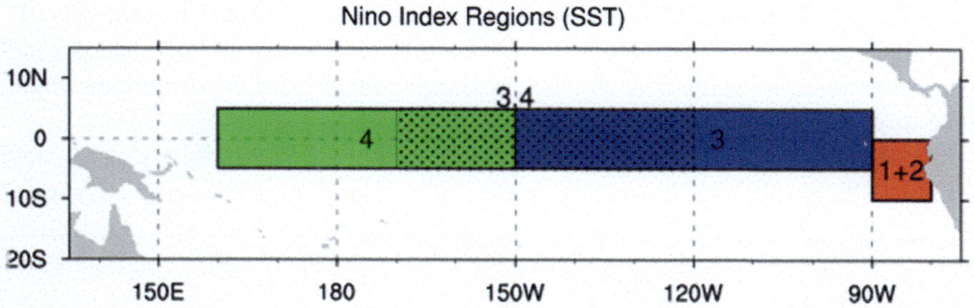

Fig. 6.8 Sea surface temperature regions in the tropical Pacific Ocean. *Source* NCAR climate data guide. https://docs.google.com/document/d/1KtG9gvPc8YqXfd7u73rr8hkm8a64YEg7iso1PylTV_s/edit

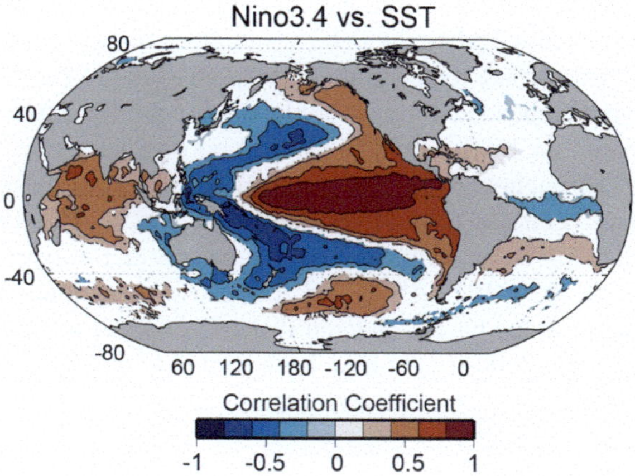

Fig. 6.9 Statistical correlation coefficient between the Niño 3.4 index and sea surface temperature. *Source* Frontiers in Earth Science. Rohli et al. (2022)

Pacific Decadal Oscillation (PDO)

Another important Pacific Ocean teleconnection is similar to ENSO but operates on multidecadal time scales, with most, but not all, times within the period of 20-to-30 years conforming to one of the two dominant patterns, in contrast with the normally six to 18-month period of dominance of an El Niño or La Niña within ENSO. This **Pacific Decadal Oscillation** (PDO) has become known as the **Pacific Multidecadal Variability** (PDV) in recent years, as it has become uncertain to some whether the pattern really represents a back-and-forth oscillation of modes, or whether one mode will be proven to persist indefinitely into the future.

A second major difference between ENSO and the PDO is that the PDO's sea surface temperature signature is strongly apparent in the extratropical North Pacific Ocean in addition to the tropical Pacific, while ENSO's primary oceanic signature is confined to the tropical Pacific. In the PDO's positive or "warm" phase, warmer-than-normal sea surface temperatures occur in the tropical equatorial Pacific and up the western coast of North America, while colder-than-normal sea surface temperatures are in the northwestern Pacific near Japan and the Aleutian Islands and in a triangular region of the South and Central Pacific Ocean (Fig. 6.10). In its "cold" phase, the opposite sea surface temperature anomalies persist—warmer than normal in the northwestern Pacific and in the same triangular region, and colder than normal in the tropical equatorial Pacific and near North America.

Just as with ENSO, the "warm phase" and "cold phase" terminology associated with the PDO is culturally biased, named to make sense from an American perspective. This is because the PDO was discovered and named by fisheries scientists working in northwestern North America in 1998, who noticed a different magnitude and type of fish catch off the North American coast through the different temperature regimes. To someone from Asia, the warm or cold phase would likely evoke images of the opposite pattern of anomalies. To avoid confusion, "positive mode" is becoming more common in usage to refer to the warm phase, and "negative mode" is often used in lieu of "cold phase".

Recent research continues to confirm that the effects of the El Niño phase tend to be magnified when the PDO is simultaneously in a warm phase, as the two patterns of sea surface temperature overlap and reinforce each other. Similarly, a La Niña event embedded within a cold phase of the PDO will likely have stronger impacts than if the ENSO and PDO are in opposite phases.

Throughout the period of instrumental records, timelines of major PDO regimes have been identified. Specifically, the 1925–1946, 1977–1999, and 2015–present intervals were dominated by the warm (or positive) PDO regime. Likewise, the 1890–1924, 1947–1976, and 2000–2014 were dominated by the cold (or negative) phase. Some have wondered if the "hiatus" in warming global air temperatures in the early 2000s might have been related to the dampening effect of the PDO's cold phase.

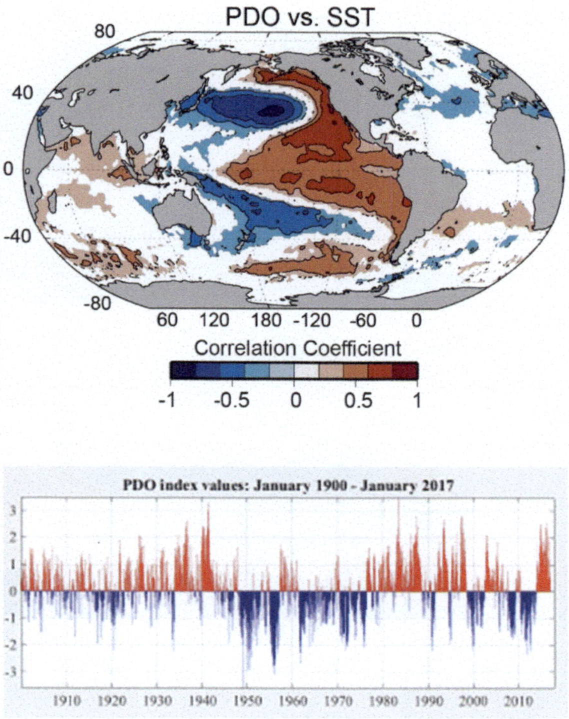

Fig. 6.10 Top: Statistical correlation coefficient between the Pacific Decadal Oscillation index and sea surface temperature. Bottom: PDO time series. *Source* Frontiers in Earth Science. Rohli et al. (2022)

Indian Ocean Dipole (IOD)

A similar pattern of oscillation in sea surface temperature anomalies occurs in the Indian Ocean. This **Indian Ocean Dipole** (IOD) refers to anomalies of sea surface temperature between the western and eastern Indian Ocean. The IOD is sometimes called the "Indian ENSO."

During the so-called "positive" IOD phase (Fig. 6.11a), sea surface temperatures are abnormally warm in the western Indian Ocean, from southeastern African coast northward near Somalia in what is known as "the Horn" of East Africa. Positive IOD events are linked to much wetter-than-normal conditions in this normally very dry coastal area. At the same time, sea surface temperatures are abnormally cold in the eastern Indian Ocean, near Indonesia and Australia, where drought conditions prevail in these normally wet environments.

Positive IOD events are often, but not always, coincident with the El Niño phase of ENSO and tend to accentuate the impacts of El Niño in this part of the world. This support is because shear from the anomalous sinking air associated with anomalously high atmospheric pressure over and near Indonesia (i.e., the Maritime Continent) during El Niño causes sinking air adjacent to the anomalously sinking side of the Walker circulation on its west. Likewise, because atmospheric pressure is much higher than normal near the surface of the Maritime Continent during El Niño, the

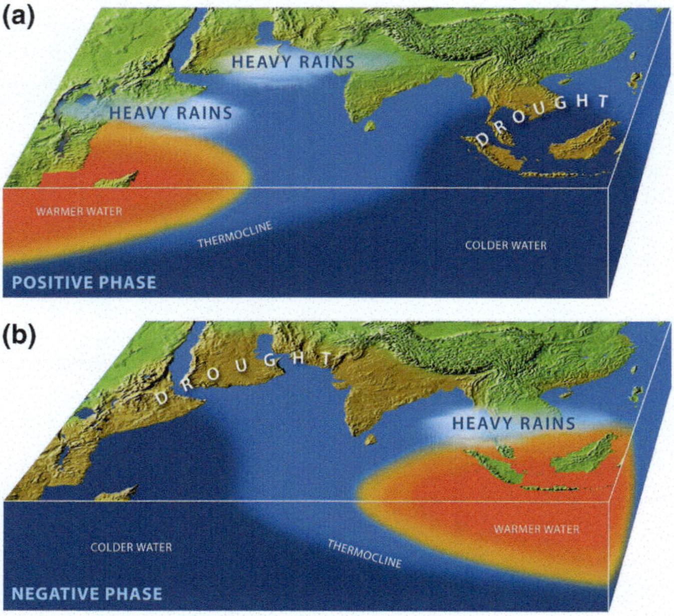

Fig. 6.11 Sea surface temperature anomalies associated with the **a** positive and **b** negative phases of the Indian Ocean dipole. *Source* Woods Hole Oceanographic Institute

surface air spreads out laterally away from the high and pushes near-surface air from east to west over the Indian Ocean—the positive IOD signature. At the same time, it also pushes near-surface Pacific air from west to east, allowing the Pacific surface water to slosh eastward toward South America—the El Niño signature. Therefore, the Indian Ocean has enhanced trade winds during positive IOD events and El Niño events, unlike the Pacific Ocean, which has weakened trade winds during El Niño events. This pile-up of warm water in the western tropical Indian Ocean destabilizes the atmosphere in concert with the anomalously rising air, to create wet conditions.

The "negative" phase of the IOD (Fig. 6.11b) brings opposite conditions. The normally dry areas are even drier than usual, and the normally wet areas are even wetter than usual. Specifically, drought persists and intensifies in East Africa, coincident with colder-than-normal sea surface temperatures and associated enhanced stable atmospheric conditions in the western Indian Ocean. Simultaneously, the Maritime Continent is even rainier than usual, coincident with warmer-than-normal sea surface temperatures in the eastern Indian Ocean. Negative IOD events are often, but not always, coincident with the La Niña phase of ENSO.

These negative IOD patterns occur because the enhanced rising motion near the Maritime Continent associated with the Walker circulation causes a shear that induces rising motion adjacent to it in the eastern Indian Ocean, reinforcing the even-lower-than-normal atmospheric pressure over the Maritime Continent. The reduced air pressure over the Maritime Continent causes a pressure gradient, with the pressure gradient force initiating an anomalously strong west-to-east near-surface airflow (weakening or reversing the trade wind flow) over the tropical Indian Ocean in an attempt to fill the pressure gradient. This near-surface airflow drags surface water with it, with very little Coriolis effect at these low latitudes. The eastward movement of surface water, in turn, causes upwelling and colder-than-normal sea surface temperatures near the coast of tropical East Africa and the Horn of Africa, reinforcing the stable atmosphere there.

Atlantic Multidecadal Oscillation

Like the Pacific and Indian Oceans, the Atlantic Ocean is also the site of ocean-atmosphere teleconnections. One of the most important is a multidecadal oscillation with a period of about 70 years, known as the **Atlantic Multidecadal Oscillation** (AMO). Like the PDO, because of expressions of uncertainty that the AMO is really an oscillation and not simply a long-term trend toward one phase, in more recent years the AMO has also been called the **Atlantic Multidecadal Variability** (AMV).

The AMO is expressed as sea surface temperature anomalies in the North Atlantic Ocean. The so-named "warm phase" is characterized by above-normal temperatures in most of the North Atlantic basin, and the "cold phase" has below-normal temperatures in the same basin.

Figure 6.12 shows the areas with the strongest relationship between the AMO (or AMV) and sea surface temperatures. Areas in red are warmer in the positive AMO mode and colder in the negative AMO mode. Locations in blue are colder in the positive AMO mode and warmer in the negative AMO mode. Notice from Fig. 6.12 that a few areas near the coastline of North America show little or no relationship to the temperature patterns in the rest of the North Atlantic. The nearshore waters are associated primarily with the continental shelf which receives a significant influx of generally colder continental fresh water from rivers. Also, the circulation around the **North Atlantic Gyre** pushes the *Gulf Stream* moves away from the North American coast with increasing latitude, keeping the anomalously warm water associated with the AMO to the east. The Gulf Stream flow eventually becomes the *North Atlantic Drift* and extends to Europe. Along the North American coast, the cold **Labrador Current** flows equatorward to fill the location between the Gulf Stream and the coastline.

Figure 6.13 shows the variation in AMO phase over time. The warm phase—higher than average sea surface temperatures in most of the North Atlantic Ocean—dominated from approximately 1930 (before official data collection began) to 1965 and from 1995 to the present. The cold phase—lower than average sea surface temperatures in most of the North Atlantic Ocean—dominated from 1900 to 1930 (not shown), and from 1965 to 1995.

The AMO phase has been shown to be related to temperature, precipitation, drought, hurricane activity, and glacial growth and shrinkage, mostly in the Northern Hemisphere, with the impacts varying in different places. However, the extent of

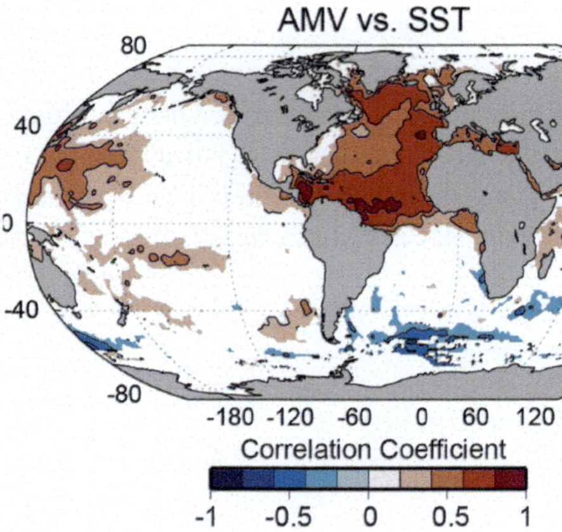

Fig. 6.12 Statistical correlation coefficient between the Atlantic Multidecadal Oscillation index and sea surface temperature. *Source* Frontiers in Earth Science. Rohli et al. (2022)

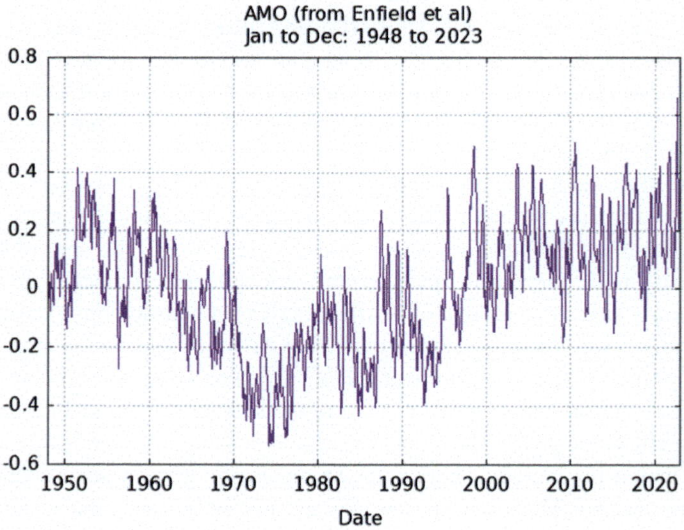

Fig. 6.13 Atlantic Multidecadal Oscillation index, 1948 through July 2023. *Source* NOAA. https://psl.noaa.gov/data/climateindices/

the cause-effect relationship between AMO phase and anthropogenic warming remains controversial.

All of the teleconnections described thus far (ENSO, PDO, IOD, AMO) have their main signature in the ocean in the form of sea surface temperature variability, with atmospheric responses to these sea surface patterns. The next teleconnections to be covered have their main signature in the atmosphere in the form of airflow variability, with oceanic responses to these atmospheric patterns. They might be considered as air-ocean teleconnections rather than ocean-air teleconnections.

Pacific—North American (PNA) Pattern

The first of the air-ocean teleconnections characterizes the long-wave flow over North America in the middle atmosphere, such as at the 500-mb level, at a given time. The positive mode of this **Pacific-North American** (PNA) **pattern** has amplified ridging over western North America and simultaneous amplified troughing over eastern North America (Fig. 6.14).

During a positive PNA regime, the warm water from the northern tropical Pacific Ocean extends northward near North America. This drives the sharpest gradient between warm and cold air northward, reinforcing or setting up the ridge over western North America. This phenomenon is related to El Niño, as the warm water pool off the South American coast migrates poleward on the eastern side of the Pacific Ocean during El Niño.

This amplified ridge-trough configuration is one form of a *meridional* (i.e., north-to-south or south-to-north motion) circulation pattern. Some argue that variability is greater in a meridional regime since a slight eastward or westward displacement of the ridge-trough configuration would shift storm tracks, leading to a major change in conditions. But others argue that the opposite is true because the systems are more locked in place in meridional regimes.

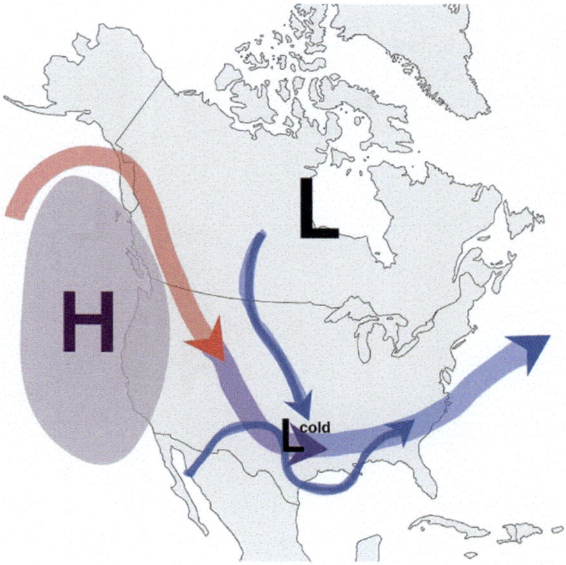

Fig. 6.14 Schematic of the positive mode of the Pacific-North American (PNA) pattern. *Source* Geaux Geospatial

The neutral mode of the PNA pattern has only weak ridging over western North America and simultaneous weak troughing over eastern North America. The weak ridging results from dynamic conditions in the atmosphere that tend to anchor *Rossby wave* ridges in place over high north-south-running mountain ranges.

The negative mode of the PNA pattern has very weak ridging or even troughing over western North America and simultaneous very weak troughing or even ridging over eastern North America (Fig. 6.15). This pattern is one form of *zonal* circulation, which is said to occur when the *mid-latitude westerlies* are simply flattened, with the flow largely west-to-east, with little north-to-south or south-to-north component to them. At such times, the periphery of the *circumpolar vortex*, which occurs along these Rossby waves, is nearly circular in shape over North America.

Figure 6.16 shows the areas in which the 500-mb geopotential heights correlate best with the mode of the PNA pattern. When the PNA pattern is in its positive mode, 500-mb geopotential heights are higher than normal over the tropical central North Pacific Ocean and Rocky Mountain cordillera of western North America, indicative of ridging, and below normal over the extratropical northeastern Pacific Ocean and the southeastern U.S., representing troughing. As described above the negative mode of the PNA pattern shows ridging and troughing in the opposite locations—ridging over the extratropical northeastern Pacific Ocean and the southeastern U.S. and troughing over the tropical central North Pacific Ocean and Rocky Mountains.

The PNA pattern is a source of regional impacts. It is well-established that the PNA-related variability is expressed as temperature and/or precipitation variability in much of the U.S., such as colder than normal conditions in the U.S. Southeast during the positive mode. The PNA pattern is generally considered a source of weather/climate variability, but like all other teleconnection patterns, if global climate

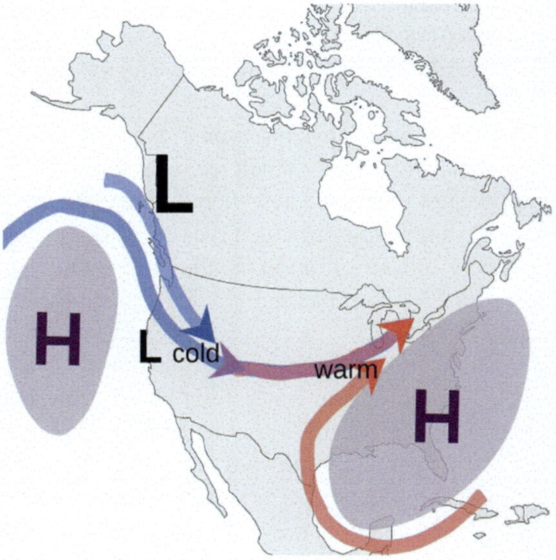

Fig. 6.15 Schematic of the negative mode of the Pacific-North American (PNA) pattern. *Source* Geaux Geospatial

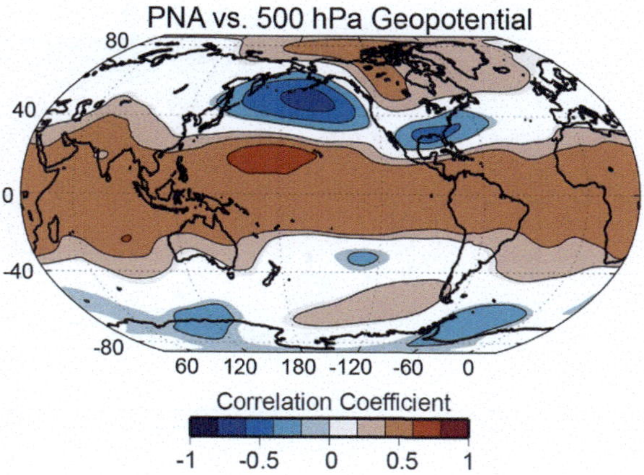

Fig. 6.16 Statistical correlation coefficient between the Pacific-North American (PNA) pattern index and 500-mb geopotential height. *Source* Frontiers in Earth Science. Rohli et al. (2022)

change were to make a particular mode more prominent and/or permanent, it could be considered an element of climate change.

North Atlantic Oscillation (NAO)

Another atmosphere-focused teleconnection is most strongly manifested as a seesaw in sea-level pressure throughout much of the *troposphere*, rather than solely near the 500-mb geopotential height, between the *Bermuda-Azores High* and *Icelandic Low*. The positive phase of the seesaw of this **North Atlantic Oscillation** (NAO) occurs when the *Bermuda-Azores High* has higher pressure than normal (i.e., is stronger than normal) and the Icelandic Low simultaneously has lower pressure than normal. This atmospheric regime would be depicted by a dense set of *isobars* or *isohypses* between these two pressure features. The correlation field between the sea-level pressure and the NAO mode is shown in Fig. 6.17, with the seesaw evident in the North Atlantic Ocean. Figure 6.17 confirms that in the positive PNA pattern, higher sea-level pressure than normal occurs in the subtropical Atlantic and lower sea-level pressure than normal occurs in the subpolar Atlantic. During the negative PNA pattern, the opposite occurs.

Under such circumstances, the tightening of the north–south pressure gradient in the mid-latitude Atlantic Ocean causes stronger-than-normal west-to-east winds blowing toward Europe—more west-to-east (i.e., zonal) flow. This pressure gradient persists through a major part of the weather layer of the atmosphere, from the surface upward. Thus, surface storms are steered rapidly to Europe, but because the flow is largely zonal, there is little upper-level support for such systems. Many other impacts of a positive NAO are apparent as well, including the confinement of bitter cold air masses poleward of the zonal flow, leaving much of Europe warmer than normal (Fig. 6.18).

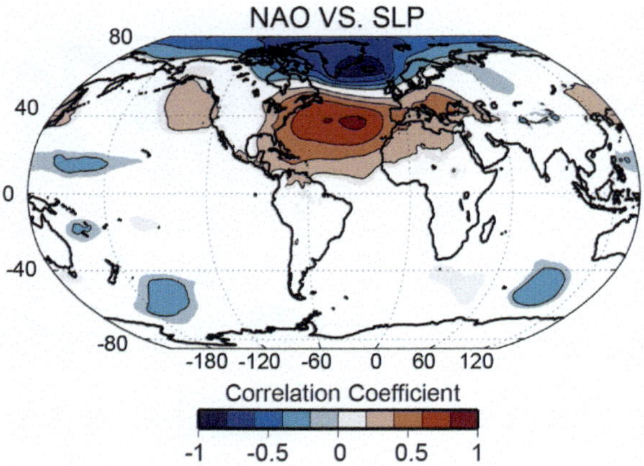

Fig. 6.17 Statistical correlation coefficient between the NAO phase and sea level pressure. *Source* Frontiers in Earth Science. Rohli et al. (2022)

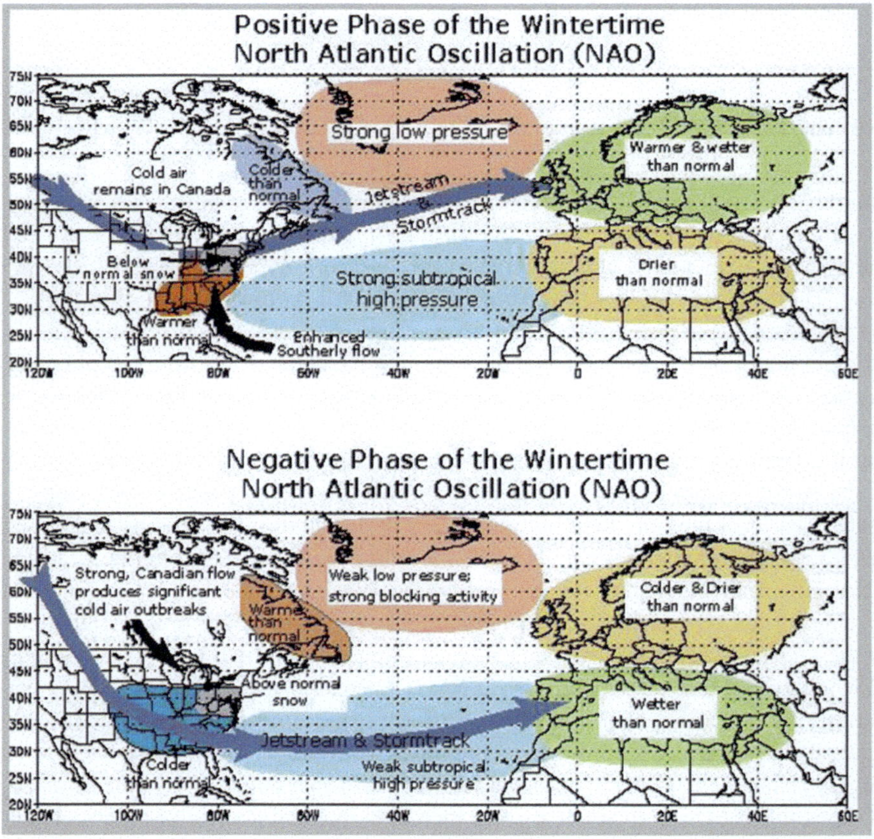

Fig. 6.18 Impacts of the positive and negative phases of the North Atlantic Oscillation. *Source* NWS. https://www.weather.gov/mfl/winteroutlookforsouthflorida

In its negative mode, both pressure systems are less intense than usual; this pattern would be depicted by fewer isobars or isohypses around the Bermuda-Azores High and the Icelandic Low. Under such circumstances, the weakening of the north–south pressure gradient in the mid-latitude Atlantic Ocean causes weaker-than-normal west-to-east winds blowing toward Europe and a meridional flow pattern. Again, this pressure gradient persists through a major part of the weather layer of the atmosphere.

The meridional nature of the flow in the negative NAO regime amplifies storm systems beneath the trough-to-ridge side of the wave. This type of flow can also be locked in place with the ridges and troughs in stationary positions for days or even weeks at a time. Such a *blocking pattern* can have important impacts, such as producing drought beneath the ridges and steering storm systems to the northern or southern parts of the major zone of the Rossby waves.

Also, instead of creating a blocking pattern, the amplified Rossby wave can also propagate elsewhere. Thus, in a negative NAO regime, anomalously amplified troughs can occur in places such as the southeastern U.S., as would also occur during a positive PNA pattern. Places poleward of these amplified troughs would be much colder than average. Unusual snows in subtropical locations such as the southeastern U.S. would be more likely to occur during a negative NAO regime.

Arctic Oscillation (AO; Aka: Northern Annular Mode)

A final air-ocean teleconnection is strongly related to the NAO, but with a signature of variability in its structure and impacts occurring on longer time and space scales. This **Arctic Oscillation** (AO), also known as the **Northern Annular Mode**, involves a seesaw between the strength of the *North Polar High* and mid-level geopotential heights, such as at the 500-mb level, in the Northern Hemisphere mid-latitudes. An analogous **Antarctic Oscillation** (AAO), or **Southern Annular Mode**, exists in the Southern Hemisphere, but less is known about it and its impacts.

While the NAO-related variability occurs largely in and near the North Atlantic Ocean, the AO has strength and impacts across the entire extratropical Northern Hemisphere. The correlation field between the AO and 500-mb geopotential height reveals these two seesaw areas, as shown in Fig. 6.19. Debate continues about whether the AO is the hemispheric-scale expression of the NAO or whether the two are less intricately intertwined.

In the "warm" or "positive" phase of the AO, the North Polar High has lower sea-level pressure and 500-mb geopotential heights than usual, even though they are still higher than elsewhere. At the same time, higher-than-normal pressure and geopotential heights exist over the Northern Hemisphere mid-latitudes. When these patterns occur, the Northern Hemisphere's circumpolar vortex is retreated poleward, as suggested in Fig. 6.20. During such times, warmer-than-normal conditions can penetrate farther northward in the Northern Hemisphere. The mid-latitude westerlies are displaced northward too, along with the main storm tracks that the westerlies steer.

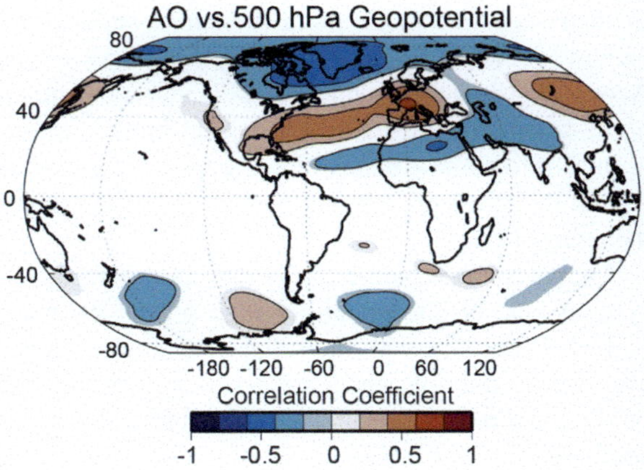

Fig. 6.19 Statistical correlation field between the Arctic Oscillation index and 500-mb geopotential height. *Source* Frontiers in Earth Science. Rohli et al. (2022)

The *northeast trades* are also stronger and more extensive, as they have a wider range of latitudes in which to dominate the atmospheric circulation.

In the "cold" or "negative" phase of the AO, the North Polar High has higher pressure than usual and there is lower pressure than normal over the Northern Hemisphere mid-latitudes. Under these circumstances, the Northern Hemisphere's circumpolar vortex is expanded equatorward and colder-than-normal conditions can penetrate farther equatorward. Snow in the southeastern U.S. is more common in a cold phase of the AO, just as during a cold-phase NAO, in winter. At the same time, the mid-latitude westerlies are displaced equatorward too, along with the main storm tracks that the westerlies steer. The northeast trades are also weaker and less extensive than normal, as their domain is "squeezed" into a smaller-than-normal range of latitudes. One implication of the cold-phase AO on world history was that the Nazis assumed it to be impossible for three consecutive brutal Russian winters to occur, but the cold phase of the AO dominated for several years in a row in the 1940s.

The mean values of the AO from 1950 through 2022 are shown in Fig. 6.21. The 1960s and 1970s were dominated by the cold (negative) phase, corresponding to the larger circumpolar vortex in the Northern Hemisphere and generally cold winters in the U.S. during those decades. Since the 1980s, the AO has been in the warm phase more often than not, corresponding to the warming global temperatures over the last several decades. The frequency and intensity of the modes of each of these teleconnections can help to explain why climates and ocean circulation regimes are the way they are.

Current research suggests that many other ocean/atmosphere relationships similar to ENSO and the other modes of variability described in this chapter exist in other ocean basins. While none of these are as strong as the ENSO of the tropical Pacific and may not affect climates across the Earth, they do have similar effects in their local regions. Future research is likely to uncover even more of these connections between the atmosphere, the ocean, and climates.

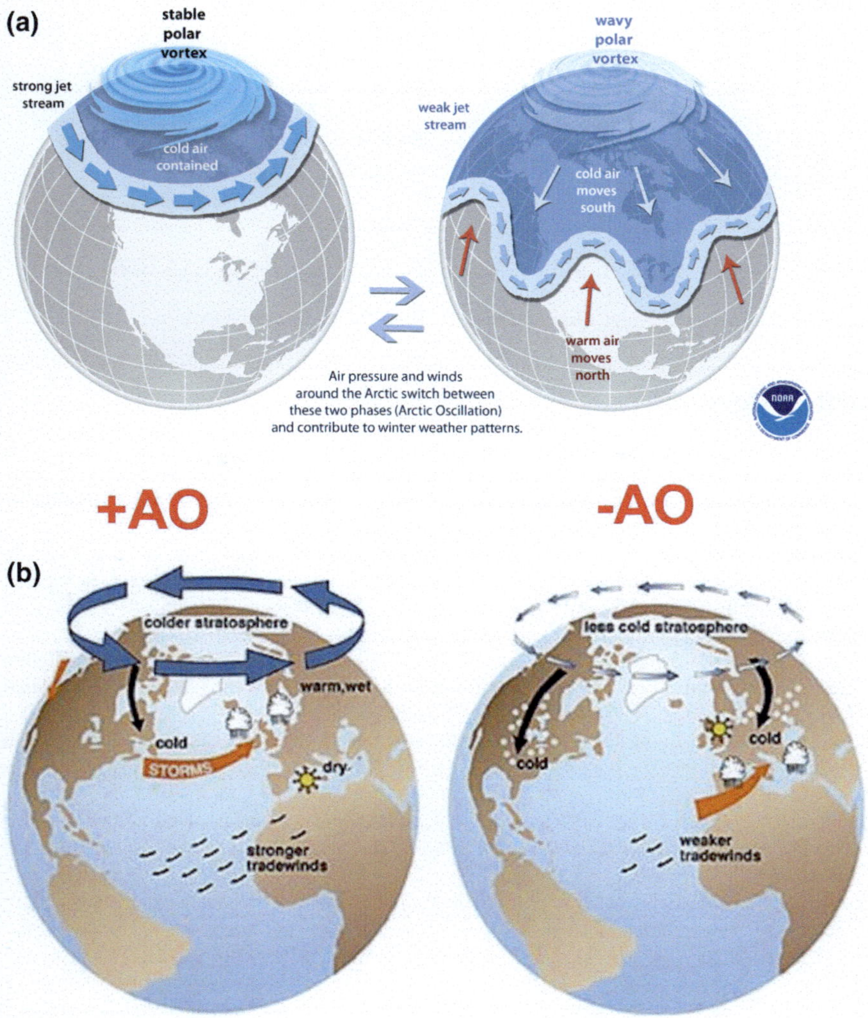

Fig. 6.20 Contrast between the circulation characteristic of the positive and negative modes of the Arctic Oscillation regarding **a** circumpolar vortex and Rossby waves. *Source* NOAA. https://csl.noaa.gov/news/2021/301_0128.html, and **b** impacts. *Source* NOAA/NWS. https://www.weather.gov/ffc/climate_faq

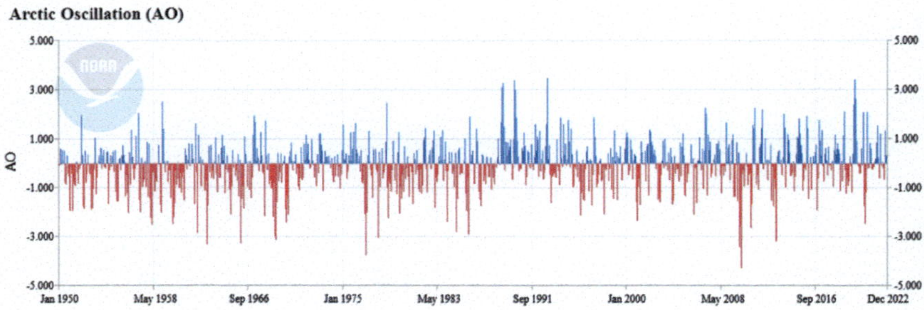

Fig. 6.21 Time series of Arctic Oscillation index, 1950–2022. *Source* NOAA

Further Reading

Barnston, A. G., & Livezey, R. E. (1987). Classification, seasonality and persistence of low-frequency atmospheric circulation patterns. *Monthly Weather Review, 115*(6), 1083–1126. https://doi.org/10.1175/1520-0493(1987)115%3c1083:CSAPOL%3e2.0.CO;2

Capotondi, A., Wittenberg, A. T., Newman, M., Di Lorenzo, E., Yu, J. Y., Braconnot, P., Cole, J., Dewitte, B., Giese, B., Guilyardi, E., Jin, F. F., Karnauskas, K., Kirtman, B., Lee, T., Schneider, N., Xue, Y., & Yeh, S. W. (2015). Understanding ENSO diversity. *Bulletin of the American Meteorological Society, 96*(6), 921–938. https://doi.org/10.1175/BAMS-D-13-00117.1

Enfield, D. B., Mestas-Nunez, A. M., & Trimble, P. J. (2001). The Atlantic multidecadal oscillation and its relationship to rainfall and river flows in the continental U.S. *Geophysical Research Letters, 28*(10): 2077–2080. https://doi.org/10.1029/2000GL012745

Hurrell, J. W., & van Loon, H. (1997). Decadal variations in climate associated with the North Atlantic Oscillation. *Climatic Change, 36*(3–4), 301–326. https://doi.org/10.1023/A:1005314315270

Knight, J. R., Folland, C. K., & Scaife, A. A. (2006). Climate impacts of the Atlantic multidecadal oscillation. *Geophysical Research Letters, 33*(17), L17706. https://doi.org/10.1029/2006GL026242

Leathers, D. J., & Palecki, M. A. (1992). The Pacific North American teleconnection pattern and United States climate. Part II: Temporal characteristics and index specification. *Journal of Climate, 5*(7): 707–716. https://doi.org/10.1175/1520-0442(1992)005<0707:TPATPA>2.0.CO;2

Mantua, N. J., & Hare, S. R. (2002). The Pacific Decadal Oscillation. *Journal of Oceanography, 58*(1), 35–44. https://doi.org/10.1023/A:1015820616384

McPhaden, M. J., Zebiak, S. E., & Glantz, M. H. (2006). ENSO as an integrating concept in earth science. *Science, 314*, 1740–1745. https://doi.org/10.1126/science.1132588

Rohli, R. V., Snedden, G., Martin, E., & DeLong, K. L. (2022). Impacts of atmosphere-ocean teleconnections on the south-central United States. *Frontiers in Earth Science, 10* (Art. No. 934654). https://doi.org/10.3389/feart.2022.934654

Saji, N. H., Goswami, B. N., Vinayachandran, P. N., & Yamagata, T. (1999). A dipole mode in the Tropical Indian Ocean. *Nature, 401*, 360–363. https://doi.org/10.1038/43855

Thompson, D. W. J., & Wallace, J. M. (1998). The Arctic Oscillation signature in the wintertime geopotential height and temperature fields. *Geophysical Research Letters, 25*(9), 1297–1300. https://doi.org/10.1029/98GL00950

Wallace, J. M., & Gutzler, D. S. (1981). Teleconnections in the geopotential height field during the Northern Hemisphere winter. *Monthly Weather Review, 109*(4), 784–812. https://doi.org/10.1175/1520-0493(1981)109%3c0784:TITGHF%3e2.0.CO;2

Chapter 7
Effects of Large Lakes

Abstract Large lakes and other inland bodies of water can impact regional-scale atmospheric circulation and properties substantially. The change in net radiation (Q*) from a deficit in winter to a surplus in summer sets the stage for the intra-annual cycles associated with the storage and release of heat within the lake, with the cycle of air temperature near the lake lagging behind the typical warmest and coldest times of year. Likewise, seasonal tendencies for atmospheric stability and instability, and the stable stratification within the lake, near large mid- and high-latitude lakes causes lags in seasonal preferences for cyclogenesis and anticyclone activity and have implications for lake pollution dispersion. Moisture added to the air by large, unfrozen lakes initiates fog and/or lake effect snowfall when air temperatures differ substantially from the underlying lake temperatures. The lake versus land breeze occurs diurnally when air switches direction from offshore to onshore as a result of diurnal temperature oscillations between the lake and the adjacent land surfaces. Horizontal convergence caused by increased friction when air blows onshore promotes cloud cover and precipitation near lake shores. When persistent winds along the major axis of a lake end suddenly, waves known as seiches can be initiated.

Keywords Advection fog · Anticyclonic shear · Cold-air funnel · Cyclonic shear · Double lake breeze · Fetch · Lake breeze · Lake effect snow · Land breeze · Radiation fog · Seiche · Stable and unstable season · Stable stratification · Thermal bar

General Weather and Climate Effects

Just as oceans have a huge impact on the broad-scale, steering circulation of the overlying air, large lakes and other inland bodies of water can impact and be impacted by their surrounding area on a regional scale. Differences in *friction*, *specific heat* between land and water, and, for mid-to-high-latitude lakes, the phase of water between summer and winter, are all important contributors to the regional atmospheric and water circulation associated with large lakes. Throughout the year, the

R. V. Rohli et al., *Atmospheric and Oceanic Circulation*,
https://doi.org/10.1007/978-3-031-66142-6_7

lakes create regionally milder conditions as they store abundant energy throughout summer and early autumn that is released as winter approaches, when they become warmer than the overlying air. These thermal features offset what would otherwise be extreme continentality for landlocked large lakes.

For large lakes in the mid-latitudes, such as the Great Lakes of the U.S. and Canada and Lake Tahoe and the Great Salt Lake of the U.S., the change in *net radiation* (Q*) from a deficit in winter to a surplus in summer sets the stage for interesting climatological effects. The storage and release of heat continues beyond the typical warmest and coldest time of year, especially if the lake is deep enough to continue storing or releasing energy even after the time of year when Sun reaches its highest or lowest point in the sky. As long as Q* is positive, shortwave *radiation* (K↓) will continue to be stored in the lake, and as long as Q* is negative, longwave radiant energy will continue to be released upward by the lake. Thus, the onset of the seasons over lakes and especially over oceans, is delayed compared to land areas.

During summer, large lakes typically aid in producing a generally *stable atmosphere* as the cooler lakes are often overlain by relatively warmer air, especially during daytime hours outside the tropics. This *temperature inversion*, with a negative *environmental lapse rate*, is an extreme case of atmospheric stability. Under such conditions, there is typically little cloudiness and *precipitation* over and on the downwind side of the lake.

The opposite is often true in late fall and early winter, when mid- and high-latitude lakes are often warmer than the air overlying them. In these cases, the environmental lapse rate is larger than the *unsaturated adiabatic lapse rate*, and the air parcel moving *adiabatically* would continue to rise spontaneously, creating an *unstable atmosphere*. In these conditions, the accumulation of *humidity* over an unfrozen lake would make it likely that the rising air would cool even faster than usual as it encounters a cold *air mass*, reaching its *dew point temperature* relatively soon on its ascent, and create thin, wispy pillars of rising fog. This phenomenon is known as **steam fog** or **Arctic sea smoke**.

Regardless of location and season, cloud cover and precipitation are augmented over and downwind of the lake. Outside of the tropics, winters can be gloomy because of the combination of low *temperature* and extensive cloud cover. Thunderstorms may be active during the warm months, with higher frequencies on the equatorward side than on the poleward side. The thunderstorms are driven by *convection* associated with surface heating.

Cyclones and Anticyclones Over Large Lakes

The seasonal tendencies for stability and instability of large mid- and high-latitude lakes causes preferences for *cyclogenesis* and *anticyclone* activity at certain times of the year. In autumn, cyclogenesis is more likely in most cases, because the rising motion associated with the unstable atmosphere created by warm air with much colder air overhead draws near-surface air from all directions to replace the air that ascended.

If the lakes are large enough to generate strong regional flow in response, the *Coriolis effect* will be strong enough to generate counter-clockwise (in the Northern Hemisphere) and somewhat inward flow that characterizes a cyclone. If upper-level support in the form of a trough-to-ridge side of the *Rossby wave* lies high overhead, the rising motion will be sustained, particularly in the presence of humid air from the lake, and a cyclone forms. For the same reason, anticyclone frequency is at a minimum in autumn, as the unstable conditions are not conducive to formation of surface high pressure. Likewise, cyclones that already formed upwind from the lake tend to strengthen over the lake, while anticyclones tend to bifurcate around large lakes, or weaken if the steering flow forces them over the lake. If the lake is deep enough to retain enough heat, the pattern continues into the winter months. The world's deepest lake, Lake Baikal in eastern Russia, stretches to a maximum depth of more than 1.6 km (1 mi), is larger than the U.S. state of Maryland and contains more than twice as much water as all of the North American Great Lakes combined, but its high latitude limits the amount of energy storage.

As winter continues, the *polar front jet stream* may slip equatorward of the mid- and high-latitude lakes, positioning them on the **cyclonic shear** side of the jet, which encourages cyclonic spin in the upper atmosphere. The faster motion of the polar front jet equatorward of a lake, or any other area, makes air spin cyclonically (i.e., counter-clockwise in the Northern Hemisphere and clockwise in the Southern Hemisphere), which promotes cyclogenesis. Smaller-scale weather systems are likely to strengthen over such areas, assuming that the other ingredients for development are in place. If the polar front jet slips equatorward of a location early enough in winter, the seasonal instability of autumn may compound the chance of intensification. For example, for the Great Lakes, maximum cyclone frequencies occur during December and January. Figure 7.1 depicts the process of upper-level shearing by the polar front jet stream graphically, with the cyclonic shear side in this case over the Caspian Sea – the world's largest lake (despite the "Sea" in its name) that is almost as large as the U.S. state of Montana.

Fewer cyclones and more anticyclones occur during summer over mid-latitude lakes and other mid-latitude areas, as the polar front jet stream retreats poleward. High-pressure systems are more frequent at that time because the lakes are often cooler than the overlying air and because of the position beneath the **anticyclonic shear** side of the polar front jet at these times (see again Fig. 7.1).

Severe Weather Over and Near Lakes

Large lakes increase nocturnal thunderstorm activity because of the delayed warming and cooling cycles. The distribution of the severe weather is uneven because the distance over which the air flows over a uniform surface (the lake in this case), known as the **fetch**, differs strongly from the upwind to the downwind side. For example, the amount of thunderstorm precipitation is reduced on the eastern (i.e., downwind) side of the Great Lakes than on the western (i.e., upwind) side, as air

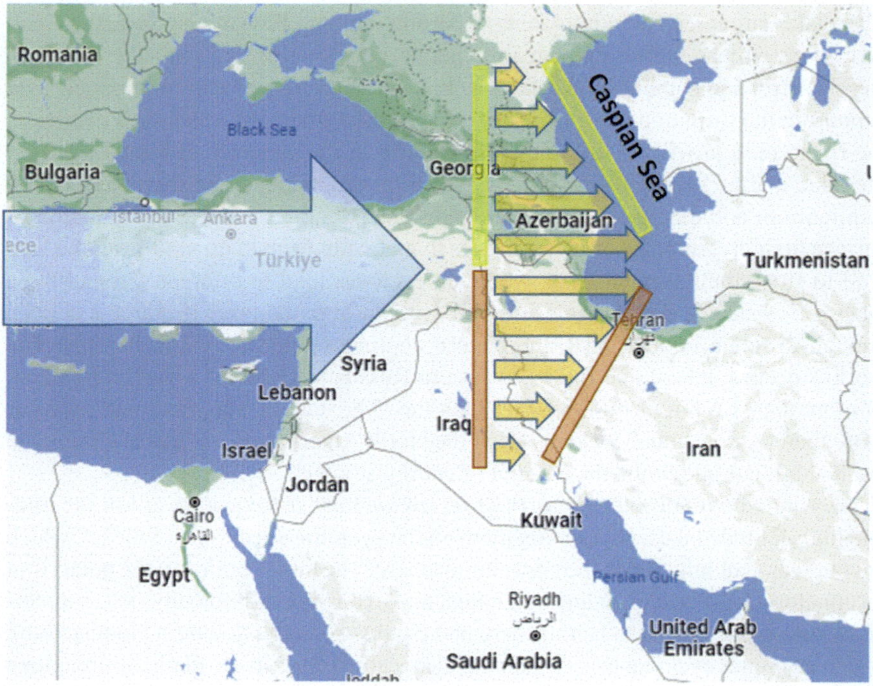

Fig. 7.1 Depiction of the cyclonic and anticyclonic shear side of the polar front jet stream, poleward and equatorward of the polar front jet, respectively. *Source* Authors

mass contrast is greater in the west because of increased continentality on the western side.

During summer, large lakes may cause a decrease in the number of hail days by 10 to 60 percent. However, when the lake becomes warmer than the overlying air in autumn, the lake may contribute to an increase in the number of hail days as the surface warmth destabilizes the overlying atmosphere.

Cold-air funnels are an unusual phenomenon that may occur over large lakes, but little is known about their genesis. The funnels form beneath showers or weak thunderstorms when the air aloft is especially cold. They are usually harmless, but can develop into weak tornadoes and cause damage.

Lake Effect Snow

The addition of moisture to the atmosphere by large unfrozen lakes also promotes snowfall when and where temperatures are below freezing and cold air is pushed over warmer lakes. **Lake effect snow** may occur in situations when the surface weather map indicates no apparent cause for a snowstorm. Often the surprising culprit is a departing rather than an approaching *mid-latitude wave cyclone*, in the area behind (i.e., west of) the cold front with an approaching anticyclone. Thus, lake effect snows are often generated with increasing atmospheric pressure, which is unusual because as most storms approach, atmospheric pressure is decreasing with the arrival of the low-pressure cyclonic core. When air encounters a mountain slope, its upslope movement—**orographic** (or mountain-associated) **lifting**—can enhance lake effect snow in some areas, as the air cools to its dew point temperature. Whether or not a migratory anticyclone or some other feature such as a short wave in the Rossby wave flow is to blame, *mesoscale* features such as these are uncommon causes of immense snows in most areas of the world, but they are a prominent feature of the U.S. and Canadian Great Lakes and other large lakes.

The greater the difference between lake and air temperature, the greater the chance for the development of lake effect snow, assuming air temperatures are below freezing. Even as relatively warm lakes in autumn may induce bifurcation of anticyclones around the lakes, the associated clockwise (in the Northern Hemisphere) and outward circulation induces lake effect snows. These events occur by *advection* of very cold and dry air over the warm surface of the lakes. Evaporation from the surface quickly saturates the air and induces clouds and heavy snowfall on the lake's *leeward* side, which is often on the eastern shore because of the *mid-latitude westerlies* (Fig. 7.2). In the case of the Great Lakes, leeward locations may experience nearly continuous broad, low-level, precipitating cloud cover known as **nimbostratus** clouds and frequent snowfall through long periods extending from autumn through spring as the lakes inject copious amounts of water vapor into relatively cold air. The lake effect "snow machine" is suspended once a lake freezes, thereupon preventing water vapor transfer to the atmosphere.

Fetch is an important consideration when analyzing the likelihood of lake effect snow. A larger fetch increases the likelihood of lake effect snow as water vapor is added along the fetch. Figure 7.3 gives an example of the longest fetch in the case of Lake Erie. The numerals in Fig. 7.3 refer to compass angles, with 270 degrees being

Fig. 7.2 Lake effect snow areas of the Great Lakes region. *Source* UCAR CoMet. Permission from Tara Torres

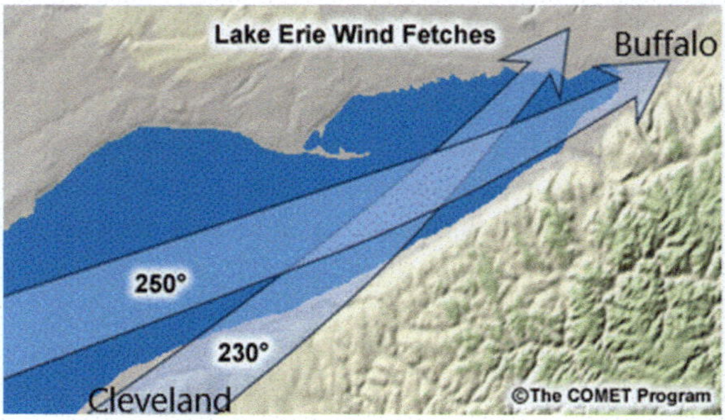

Fig. 7.3 Converging air streams conducive to enhanced lake effect snows. *Source* Authors. Modified from UCAR MetEd

"from the west" and 180 degrees being "from the south." The converging air streams maximize water vapor content and enhance the resulting snow totals leeward of the lake.

Linearly-oriented lake effect **snow bands** often form when the winds are aligned parallel to the major axis—the longest axis—of a lake. As a general rule, the heaviest snows fall in narrow snow bands that move along the major axis. If the flow is perpendicular to the major axis of the lake, snow bands are less likely to form because the fetch is shorter. Instead, a more continuous cloud deck is more likely.

Lake effect snows are most common in early winter, before ice clogging (especially in Lake Erie) occurs. Distance that the snow penetrates inland depends on the wind speed and the local terrain. When winds are lighter, such as in the core of the winter in many places, the snow bands remain closer to the shores, leaving heavier snow totals there.

Lakes modify regional climates downwind. For example, fruit orchards exist along lee shores at latitudes that otherwise would not support the crops. The coldness of a lake in spring compared to the land prevents premature budding of the fruit. The primary hazard of fruit at such latitudes is premature budding as early budding may result in death from a late freeze. Conversely, the warmth of the lake in late summer and autumn compared to the land extends the growing (harvest) season, ensuring crop harvest before the first killing frost.

Physical Limnology

The scientific study of lakes, or **limnology**, includes physical, geological, biological, and chemical features and processes. The most impactful property of lakes for the atmospheric and oceanic circulation is temperature. Lakes have a similar impact on nearby air temperatures as oceans, but on a smaller scale and to a smaller degree. The difference between thermal features of water versus land surfaces, as described in Chap. 2, drive many relevant processes, including lake effect snows. With a specific heat of about 4180 Joules kg^{-1} K^{-1}, a kg of water requires about five times more energy (or work) to be heated by one Kelvin (or 1 °C or 1.8 °F) than the solids that make up the Earth's crust, such as dry sand (830 J kg^{-1} K^{-1}) or wet mud (2510 J kg^{-1} K^{-1}). Thus, a large amount of energy must be added to water bodies to increase temperature even by a small amount. The reverse is also true—a large amount of energy must be extracted from water bodies to cause a temperature decrease of even a modest amount. Land surfaces, by contrast, heat and cool quickly with far smaller energy exchanges.

Besides having a large specific heat, water is somewhat transparent to incoming solar energy. The penetration of shortwave radiation (K↓) is largely dependent on water clarity, but on average may extend down to approximately 100 m (330 ft). Water is almost completely opaque to *longwave radiation*, however, which means that *counter-radiation* (L↓) remains within only a few millimeters of the water surface. Nevertheless, the downward transfer of shortwave radiant energy makes the surface cooler during times of positive net radiation (Q*) compared to opaque solid surfaces.

Lakes also affect surface evaporation. Evaporation increases when the near-surface Q* is positive, especially as the temperature increases and as the vertical vapor gradient increases—with much drier air overlying more humid near-surface air. These conditions are best met in daytime hours. Lakes enhance the likelihood that the surface air will be near saturation, thereby increasing the vertical humidity gradient, if all other factors are equal.

Lakes and other water bodies also have vertical and horizontal currents that redistribute energy throughout the water body in accordance with the *Second Law of Thermodynamics*. The rate of penetration of heat from the surface into a substance can be expressed in terms of how many Watts (or Joules per second) of power are needed to change the temperature one meter away by one Kelvin. This quantity is termed **thermal conductivity**, and, like specific heat, it differs significantly between land and water surfaces. Specifically, the liquid nature of water allows heat to be transferred much more easily when the water is flowing turbulently, such as in a wind-stressed lake or ocean, than on a land surface. For solid surfaces, the heat transfer can only occur by molecular processes (or *conduction*) and is about 0.20 W m^{-1} K^{-1} for dry sand and 0.25 W m^{-1} K^{-1} for wet sand. The thermal conductivity of water varies greatly with its temperature and amount of turbulence (or convection) present, but even for absolutely still water, its thermal conductivity is approximately 0.6 W m^{-1} K^{-1}. In turbulent flow, the thermal conductivity of water is many times larger because of eddy motion.

For these reasons, water bodies tend to maintain a fairly consistent temperature through the seasons. Tropical ocean surfaces may hold temperature ranges of less than 5 °C (9 °F), but the seasonal range depends on latitude. Continental locations have much wider *annual temperature ranges*.

Seasonal Lake Processes

In high-latitude lakes, the warmth of spring causes the surface water temperature to increase from near freezing to 4 °C (39 °F). At that temperature, the density reaches its maximum, causing it to sink, which allows colder, near-freezing waters from depth to rise to the surface. The process repeats until the entire water column warms to 4 °C (39 °F). The result is a slow warming of the lake compared to the surrounding land surfaces, especially if the lake is deep.

By the time that the entire column has a temperature exceeding 4 °C (39 °F), the high *Sun angle* and abundant Q* (particularly shortwave radiation) allow for a more rapid warm-up of the water body. The lake then warms from the top downward, assisted by convective transfer of the radiant energy. The warmed surface water then literally floats on top of the colder, denser water beneath it—a condition known as **stable stratification**.

During the process of warm-up, shallower water columns, particularly near the shore, will warm throughout their depth faster than the deeper water columns, such as near the middle of the lake. Thus, stratification occurs near the shores earliest in summer. Areas of intense mixing known as **thermal bars** mark the boundary between the stratified shallow waters and the still-overturning deeper waters. Pollutants may become trapped and concentrated near the shore, where the waters are stably stratified and further vertical mixing is hampered, and also where people are most likely to interact with the lake by swimming, fishing, boating, and other recreational activities.

By mid-August or early September in Northern Hemisphere lakes, the lakes begin to cool as solar energy receipt is reduced and cold air masses overlie them. As surface waters chill from their warm summer temperatures, their density increases, so they sink and are replaced by the still-warmer waters from the depths. These, in turn, are cooled, sink, and replaced by other water from beneath, with the process repeating until the entire water column cools to 4 °C (39 °F)—the maximum density. The exact time within the cold season when the lake reaches maximum density depends on latitude, lake depth, air mass temperature, and interannual temperature variability.

Once the lake has cooled to maximum density, with the time depending on the various factors, the surface layers may cool further. At that point, a reverse stratification may develop with the colder surface waters floating at the slightly warmer (at 4 °C (39 °F)) and denser deep waters. In high-latitude and/or high-elevation lakes, ice may then form. The ice formation will generally begin in the shallow near-shore water columns where the reverse stratification process is completed first, except where inflow to the lake from rivers keeps the water flowing. In such places, turbulent mixing prevents the water from freezing as quickly. Regardless of the extent to which the lake ices over in winter, when the Q* values of spring increase enough to warm the surface above 0 °C (32 °F), the ice melts from the top downward and the process begins anew.

The period when the lake is cooler than the land is known as the **stable season**. During the stable season, winds crossing the lake are chilled and cause cooler temperatures on the downwind side. Evaporation rates (and therefore the *latent heat flux*) are small (remember that positive latent heat flux is positive in the upward direction) and the transfer of water from the lake to the air is minimized. The cooling of the air causes localized high-pressure (and *temperature inversions*) to form over each lake.

Cyclones that approach the area during the stable season tend to weaken as the lakes fail to supply heat to the overlying atmosphere. Energy transfer is from the air to the water body. Anticyclones increase in strength in this situation.

The period when the lake is warmer than the land is known as the **unstable season**. The reverse process occurs in the unstable season as energy is transferred from the warmer lake surface into the atmosphere. Cyclones and storms crossing the area increase in strength as low-level energy is added to the atmosphere. Anticyclones, by contrast, tend to weaken over lakes during this period.

The onset of the stable and unstable seasons differs for different lakes, based on their size, depth, and latitudinal extent. For example, Lake Michigan's stable season extends from March to August while the unstable season occurs from August to March. The influence of the lake is generally strongest at the peaks of the stable and unstable seasons, when the temperature differences with the atmosphere are steepest.

Effect of Lakes on Local Winds—The Lake Breeze

A lake can alter local atmospheric circulation, primarily when steep air pressure gradients occur across small areas due to the differential heating and cooling of the lakes and the adjacent land surfaces. These pressure gradients are most likely to occur because of different rates of heating of air over the lake compared to the air over the adjacent land masses.

During generally warm periods, as on a sunny afternoon, air adjacent to the land near a lake will usually heat more substantially than air over the lake at the same height, for all the reasons described previously. The relatively high temperatures induce rising motion and relatively low pressures over land, and the relatively low temperatures over the middle of the lake tend to produce higher pressure there. Near-surface air will then flow from the higher pressure over the lake to the lower-pressure air over land. Because winds are always named for the directions from which they blow, this atmospheric circulation is termed a **lake breeze**. During these times, the cooler onshore lake breeze can provide natural air-conditioning for near-shore residents. In the lake itself, warm surface water is piled near the shore, as the heat stored in the shallow water column cannot easily disperse laterally to the middle of the lake. A **double lake breeze** can form over narrow strips of land sandwiched between lakes, such as on parts of the upper peninsula of Michigan, creating strong surface *convergence* and additional updrafts.

On the other hand, during generally cold periods, particularly at night, the air immediately above the lake is usually warmer than the air immediately above the adjacent land surfaces. This is because the radiant energy stored during daytime hours can then move back upward to the lake surface, while the stored radiant energy over land, concentrated at the surface, is lost to the air above it via longwave radiation much more quickly. The relatively warm air overlying the lake would have a tendency to rise, inducing relatively low air pressure to form over the lake while higher air pressures occur over the adjacent land. The near-surface flow of air would then occur from the land to the lake, a **land breeze**. During these times, the movement of surface water to the middle of the lake allows colder water from the deeper part of the lake to move upward in the process of *upwelling* near the shore. Cities on the shore of a large lake, such as Chicago, would often find nighttime temperatures on the lakefront to be much colder than those a few kilometers inland.

The lake breeze-land breeze phenomenon actually happens cyclically, with the strongest lake breeze happening at the time of day when the lake is coolest compared to the land and the strongest land breeze happening at the time of night when the lake is warmest relative to the land. The lake breeze would gradually weaken as the Sun sets, with no circulation at some point in the early night, and then the circulation reversing, slowly at first and then strengthening as the land becomes much colder than the lake.

The lake-land breeze circulation would only be important in the absence of larger, stronger *synoptic-scale* pressure gradients and associated storms, which would easily obliterate the local to regional lake-land breeze. Depending on the individual situation, lake breezes may occur during 30 to 50% of the summer days, extending inland between a few kilometers to as much as 40 km (25 mi). The lake breeze may supply enough cool air to the shoreline areas in summer to suppress any spontaneous rising of convective thermals near the coast. But even though the coast may have clear skies, clouds are likely to form much more easily a short distance inland.

Other Weather Effects of Lakes

Lakes and other large water bodies affect the prevailing wind speed. Their relatively uniform and smooth surfaces compared to the surrounding land areas cause reductions in friction, allowing wind speeds to increase over them. When winds move from the rough land to the smooth lake as in the case of a land breeze or other broader-scale wind motion, they increase in velocity. When they move from the lake to the land, as in a lake breeze or circulation associated with a broader-scale weather system, they slow. As they slow, near-surface air piles up along the shore, causing horizontal convergence, which then initiates rising motion, as shown in Fig. 7.4. This rising motion can cause cloudiness and perhaps precipitation. For this reason, shorelines and coastlines are more likely to have precipitation than adjacent areas.

Wind direction changes over a large lake as the wind speed changes. An accelerating wind over a lake picks up additional rightward (in the Northern Hemisphere) deflection as the Coriolis effect increases with increasing velocity. When the winds are slowed by the rougher terrain on the leeward side of the lake, their rightward deflection weakens, giving the appearance of a slight leftward motion as the Coriolis effect decreases. Small lakes and other water bodies have no noticeable shifts in wind speed or direction.

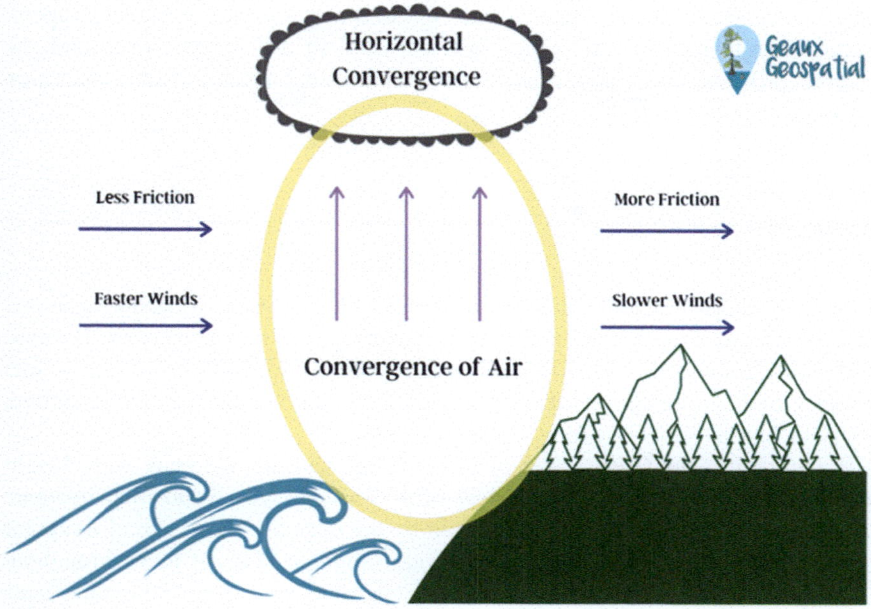

Fig. 7.4 Horizontal convergence along the margins of a lake. *Source* Geaux Geospatial

Lakes also increase instances of fog. In general, fog tends to occur at the height of the stable season when lakes are much cooler than the land and the overlying air. When a cold water body or current chills warmer air that passes over it, the overlying air is chilled from underneath. If the temperature of the passing air cools to the dew point temperature, *advection fog* results. Advection fog associated with lakes tends to form in late spring to early autumn, when the lakes are colder than the ambient air. In many cases, the air may only need to be chilled by a few degrees to reach *saturation* and form fog, because evaporation from the lake surface is adding vapor to the air, raising the dew point temperature simultaneously as the air temperature decreases with the cooling from beneath. The advection fog is blown to the downshore side of the lake, but generally dissipates quickly at that point, as the air warms quickly over the heated land surface, thereby increasing the *water vapor capacity*.

Other types of fog are also produced by lakes. As described earlier, steam fog occurs when lakes are warmer than the land, and colder air masses move over the relatively warm waters. The evaporation process is accelerated in this unstable atmospheric environment, allowing turbulence to aid the transfer of water vapor into the overlying air. The added water vapor quickly brings the cold air to the saturation point to form a steam fog. Unlike advection fog which can be thick, steam fogs appear as wispy pillars of rising clouds from the water surface.

Radiation fog forms primarily during the driest time of year for the lake, during clear nights with little wind, and with evaporated water from earlier in the day increasing the nighttime humidity. Under such conditions, longwave radiation escapes from the land and water surfaces out to space, allowing the local temperature over the lake to fall. After this vapor-enriched air drifts slowly over land, the radiation processes that cool the land surface rapidly chill the humid air to saturation to produce a radiation fog.

Waves in Large Lakes

Wave heights are related to three main factors: wind velocity, wind duration, and fetch. Maximum wave heights are achieved when all three factors increase. Because air pressure gradients are usually stronger in winter, these conditions tend to be met best in winter. An exception, of course, is the case of a violent summer or autumn storm, such as a *tropical cyclone.*

In the case of the Great Lakes of the U.S. and Canada, the peak wave height month is November, which is at the peak of the unstable season and when cyclonic storms intensify, often as they are overlain by the cyclonic shear side of the polar front jet stream. Significant wave heights may continue to occur there throughout winter and spring with the passage of strong migratory cyclones. Maximum waves of about 20 m (70 ft) are not uncommon in Lake Superior during winter.

Lake shores are prone to erosion with large waves associated with cyclones. However, during the later winter months, ice forms a protective rim around some extratropical lakes, including the Great Lakes, sparing them from more severe erosion.

A **seiche** is a free oscillation in lake waters, formed by a storm or sometimes even in the absence of a strong forcing mechanism such as a storm. Seiches are especially common in extensive, shallow lakes such as Lake Erie and are produced by sharp pressure gradients as persistent strong winds from one direction move surface waters in the same direction for a period of many hours. When wind velocity decreases abruptly, the seiche begins, with waters piled on the leeward side of the lake sloshing back and forth between the opposite sides of the lake. Seiches are particularly well developed when aligned along the major axis of the lake and may last for many hours, leading to accelerated erosion of the associated shores.

Lakes and other large inland bodies of water can have important impacts on the atmospheric and oceanic circulation at local to regional scales. But several other terrain features also have important impacts. The next chapter explores the major impacts of other terrain features.

Further Reading

Adrian, R., O'Reilly, C. M., Zagarese, H., Baines, S. B., Hessen, D. O., Keller, W., Livingstone, D. M., Sommaruga, R., Straile, D., van Donk, E., Weyhenmeyer, G. A., & Winder, M. (2009). Lakes as sentinels of climate change. *Limnology and Oceanography, 54*(6), 2283–2297. https://doi.org/10.4319/lo.2009.54.6_part_2.2283

Angel, J. R., & Isard, S. A. (1997). An observational study of the influence of the Great Lakes on the speed and intensity of passing cyclones. *Monthly Weather Review, 125*(9), 2228–2237. https://doi.org/10.1175/1520-0493(1997)125%3c2228:AOSOTI%3e2.0.CO;2

Kirillin, G., Lepparanta, M., Terzhevik, A., Granin, N., Bernhardt, J., Engelhardt, C., Efremova, T., Golosov, S., Palshin, N., Sherstyankin, P., Zdorovennova, G., & Zdorovennov, R. (2012). Physics of seasonally ice-covered lakes: A review. *Aquatic Sciences, 74*(4), 659–682. https://doi.org/10.1007/s00027-012-0279-y

Niziol, T. A., Snyder, W. R., & Waldstreicher, J. S. (1995). Winter weather forecasting throughout the eastern United States. Part IV: Lake effect snow. *Weather and Forecasting, 10*(1), 61–77. https://doi.org/10.1175/1520-0434(1995)010<0061:WWFTTE>2.0.CO;2

Sills, D. M. L., Brook, J. R., Levy, I., Makar, P. A., Zhang, J., & Taylor, P. A. (2011). Lake breezes in the Southern Great Lakes region and their influence during BAQS-Met 2007. *Atmospheric Chemistry and Physics, 11*(15), 7955–7973. https://doi.org/10.5194/acp-11-7955-2011

The Engineering Toolbox. (2023). *Solids, liquids, and gases: Thermal conductivities*. https://www.engineeringtoolbox.com/thermal-conductivity-d_429.html. Last accessed Aug 27, 2023.

Chapter 8
Interactions with Landscapes

Abstract While global and regional circulations were considered in earlier chapters, this chapter emphasizes several major relationships between circulation and local landscapes. Rugged terrain can enhance precipitation on the windward (i.e., upwind) side and suppress it on the leeward, or rain shadow, side. The slope angle and the compass point direction that a slope faces—the slope aspect—affect net radiation (Q*), which in turn affects temperature. In general, north of the Tropic of Cancer at 23.5° N—the Northern Hemisphere extratropics, south-facing slopes—south aspects—have greater Q* than north-facing slopes. Topography also impacts local winds, as colder air drains downhill, particularly at night in the form of mountain breezes or katabatic winds, and warmer air ascends upslope, particularly during daytime hours in the form of valley breezes, or anabatic winds. Coastal areas often experience a shift in wind direction between day and night, from sea breezes to land breezes. Desert, grassland, and forest land covers have characteristic energy balances that are affected by different albedo (K↑) and evapotranspiration rates. Urban areas also impact the thermal distribution, in the form of the urban heat island, along with precipitation, humidity, and wind regimes in their local areas.

Keywords Albedo · Anabatic wind · Chinook wind · Condensation nuclei · Foehn wind · Fog or frost hollow · Katabatic wind · Land breeze front · Mountain breeze · Orographic lifting · Pollution front · Rain shadow · Santa Ana wind · Sea breeze · Slope aspect · Understory · Urban canyon · Urban heat island · Urban precipitation island · Valley breeze · Valley inversion

Introduction

In Chap. 1, "*site*" was described as the characteristic landscape features at that place, with emphasis on those features that influence the local environment substantially. Likewise, "*situation*" corresponds to the broader-scale phenomena that have implications on the atmospheric and oceanic environments at a place. The topic of this chapter, which includes topography, the nature of the slope in the local area, and

R. V. Rohli et al., *Atmospheric and Oceanic Circulation*,
https://doi.org/10.1007/978-3-031-66142-6_8

the coastal and land cover features, are components of "site." They influence and are influenced by the general circulation features that were covered in Chaps. 3 and 5, for the atmosphere and ocean, respectively. Chapter 7 then explored the impacts on one particular type of terrain feature—the lake and other large bodies of inland water. This chapter zooms in on several other "site" components that are typically important factors for the atmospheric and oceanic circulation in local areas.

Topographic Features of Local Climates: The Orographic Effect

The flow of air over mountains produces drastic differences in climates across short distances. As air ascends a mountain slope—known as *orographic lifting*—it cools *adiabatically* as it rises. The cooling rate proceeds at the *unsaturated adiabatic lapse rate*, as long as it does not reach *saturation*, which occurs when the air cools to its *dew point temperature*. If at any point on its ascent, the air cools to its dew point temperature, it reaches saturation and will begin to cool at the lesser *saturated adiabatic lapse rate* beginning at that level. This side of the mountain facing upwind is known as the *windward* side (Fig. 8.1).

But as air descends on the downwind (*leeward*, or lee) side, it warms at the *unsaturated adiabatic lapse rate* on its entire descent, because once it descends even a little, it warms a little and therefore increases its *water vapor capacity* and is no longer saturated. Thus, on the leeward side, the *temperature* increases at a faster rate than it cools on the windward side, with the difference between the windward and leeward side temperatures (assuming that they are at the same elevation) exaggerated as the level at which saturation is reached on the windward side is lowered.

Thus, the adiabatic warming on the leeward side exceeds the adiabatic cooling on the windward side, as long as the air parcel reaches saturation on the windward side.

Fig. 8.1 Windward (left, in this image) and leeward (right) sides of a mountain slope. *Source* UCAR CoMet. Permission from Tara Torres

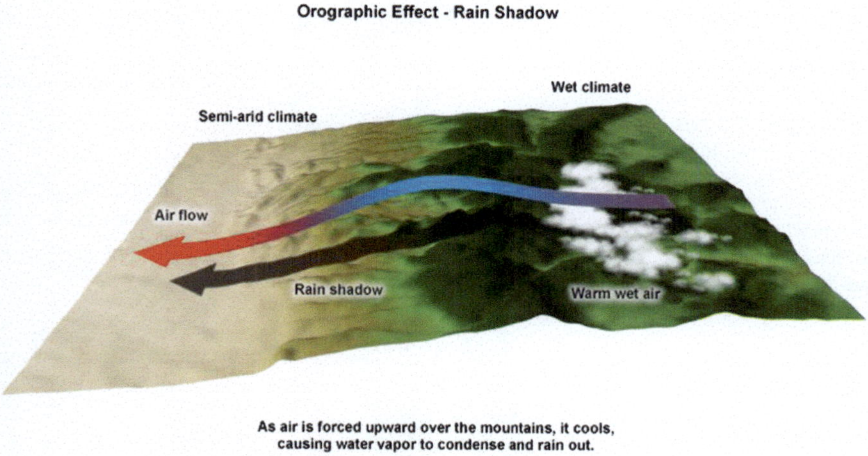

Fig. 8.2 Rain shadow phenomenon and associated climates. *Source* UCAR CoMet. Permission from Tara Torres

This difference causes the **rain shadow** side on the leeward side of the mountain to be hotter and much drier than the windward side, assuming that the elevation is similar from one side of the mountain to the other (Fig. 8.2).

Areas affected most directly by the orographic effect are those in which the spines of major mountain chains run perpendicular to prevailing wind directions. The Cascade Mountains of North America and the Andes Mountains of South America are excellent examples, as these mountain ranges lie perpendicular to the *mid-latitude westerlies*, which pick up abundant moisture from the Pacific Ocean before the air ascends. This arrangement means that the air is nearly saturated before it even begins the ascent, which accentuates the differences between temperature and moisture on the windward versus the leeward sides. The U.S. Pacific Northwest (Fig. 8.3), coastal Chile, and other places have great differences in landscapes between the western and eastern sides of the slopes. The windward and leeward sides, along with the impacts of topography on the environment, should be apparent in Fig. 8.3. Notice that there are actually two, and in some places more than two, linear north–south-oriented mountain ranges.

Topographic Features of Local Climates: Slope Aspect and Angle

Mountain slopes create surface differences other than the orographic effect, such as in the *net radiation* (Q*)—the amount of *radiant energy* gained (i.e., absorbed, not reflected) by the surface from the Sun as *shortwave radiation* (K↓) and from the atmosphere as *longwave radiation*, minus the amount of energy lost from the

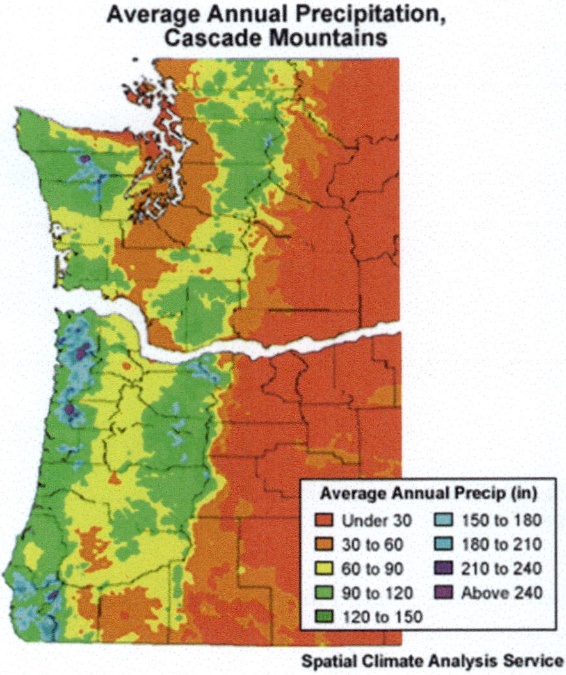

Fig. 8.3 Average annual precipitation in the U.S. states of Washington and Oregon. *Source* UCAR CoMet/Spatial Climate Analysis Service. In Rohli and Li (2021), we have the same graphic and it's referenced only as UCAR CoMet. Permission from Tara Torres

surface to the atmosphere as longwave radiation. If the solar energy reflected from the surface (i.e., *albedo*) were included, it would need to be subtracted back out of the equation.

The compass point direction that a slope faces—the **slope aspect**—affects Q*. North of the *Tropic of Cancer* at 23.5° N—the Northern Hemisphere extratropics, south-facing slopes—south aspects—have greater Q* than north-facing slopes. This is primarily because of the extra solar radiation received on the south-facing slope as it strikes the surface from a more perpendicular angle, as the Sun is always in the southern sky in the extratropical Northern Hemisphere, except near dawn and dusk in summer in the high latitudes. By contrast, north-facing slopes in the Northern Hemisphere are much more often shadowed by the mountain itself. Even when not in a shadow, the angle at which the Sun's radiation strikes the north-facing surface is far from perpendicular. South of the *Tropic of Capricorn* at 23.5° S—the extratropical Southern Hemisphere, north-facing slopes—north aspects—have greater Q* than south aspects, because the Sun is situated in the northern sky at the vast majority of the time.

In the low latitudes, or tropics (23.5° N to 23.5° S), the Sun generally passes overhead twice each year, so the slope aspect—north-facing or south-facing—that experiences more Q* depends on the time of the year. But regardless, in the tropics, Q* between north and south slopes would vary little, since the Sun is nearly overhead at mid-day and goes back and forth between being north of the location and south of the location during the year.

The amount of difference in Q* between the north-facing and south-facing slopes depends on the latitude and the slope angle. In the mid-latitudes and high latitudes in the Northern Hemisphere, north slope aspects have less flow of moist energy upward in the form of evaporated water (known as the *latent heat flux* (Q_E)) than south slopes, because there is less solar radiation to force evaporation. Therefore, the north side is usually cold and damp, and is often tree-covered. In the Northern Hemisphere middle and high latitudes, the south slope aspects have more solar radiation, more Q*, and more Q_E when water is available. Therefore, these places are hotter and drier and are often covered with shrubs rather than trees, because the shrubs can exist in a hotter, drier environment.

Regarding slope angle, steeper sloping surfaces generally have a different daily and seasonal range in Q* than flatter surfaces. In low-latitude terrain, the difference in Q* between the time of maximum and minimum values for the day, or through the year, will generally be less for sloping than for flatter surfaces. This difference is because the peak radiation intensity occurs at a time when a steeper slope angle can cause the Sun's rays to arrive at the surface at a more oblique angle than on a flat surface, where they would strike the surface at an angle closer to being perpendicular. And likewise, the minimum solar radiation occurs at a time (i.e., near sunrise and sunset) when a steeper slope angle would be increasingly likely to place the location in a shadow near either (but not both) sunrise or sunset, and increasingly likely to give the surface a more perpendicular angle of incidence of solar radiation at either sunrise or sunset. In high-latitude locations, the opposite is true. An increasing slope would be more likely to enhance the range of Q* by providing a more perpendicular angle of Q* receipt in the high-Sun part of the day, with little receipt near sunrise or sunset whether the surface is sloped or not.

Topographic Features of Local Climates: Valley and Mountain Breezes

Mountain slopes also create characteristic local-to-regional-scale wind systems, especially in the absence of much broader-scale *pressure gradients* induced by *mid-latitude wave cyclones* and their associated *fronts*, or *tropical cyclones*. During the daylight hours, local winds are likely to move upslope. This pattern is because warm air in the valley tends to rise upslope. Air between peaks is cooler than air near the surface at the same height, so it sinks into the valley, pushing the warm air already in the valley upslope. These upslope winds near the surface originate in the valleys, so these are called **valley breezes** (Fig. 8.4).

The naming convention follows that used for all winds. Specifically, as shown in Chap. 3, winds are named by the direction from which they blow—from the valley to the mountaintop, giving these winds the name "valley breeze." These winds are also called **anabatic winds** and are usually weak but apparent. If hiking up a mountain,

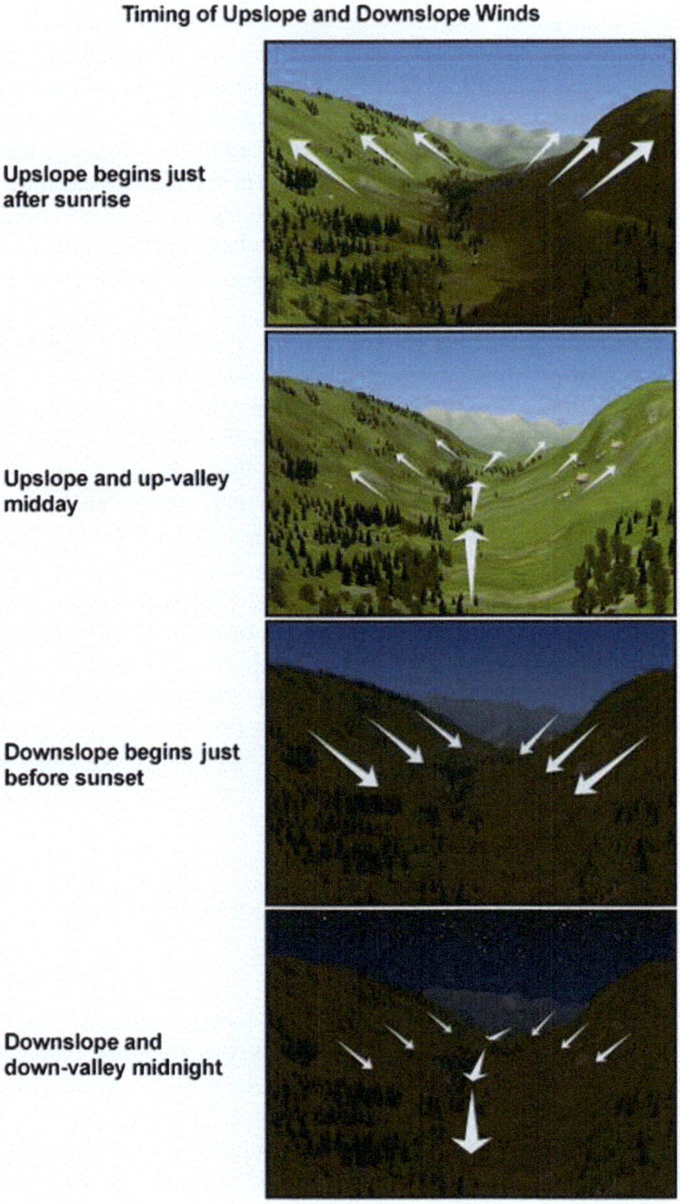

Fig. 8.4 Schematic of idealized anabatic and katabatic winds. *Source* UCAR CoMet. Permission from Tara Torres

the anabatic wind will be a help on the ascent but a hindrance on the descent, if the hike takes place during daylight hours.

At night, however, air adjacent to the surface at the peak is cooler than the air at the same height away from the surface. Thus, this cooler surface air at the summit moves downslope into the local valley (see again Fig. 8.4). Thus, these winds are called **mountain breezes** or **katabatic winds**. Mountain climbers who persist on their descent into the nighttime hours will have the wind at their backs, despite other obvious disadvantages of hiking on slopes at night. However, the fact that different types of land covers heat and cool at different rates does complicate the "anabatic by day, katabatic by night" generalization, particularly for snow-capped mountains.

Phenomena related to the anabatic and katabatic winds can be important components of local and regional atmospheric circulation. For instance, **valley inversions**, in which cold, *stable* air is stuck in the valley bottoms by katabatic winds from two or more surrounding slopes, are common at night, even though the katabatic winds warm adiabatically on descent. One notorious example of the impacts of a valley inversion occurred in 1948 in the industrial town of Denora, Pennsylvania. Because of the steepness of the valley in which Denora is situated, the air pollution could not escape, especially with the very stable atmospheric conditions persisting for several days.

A related example of circulation features associated with anabatic and katabatic winds is the **pollution front** that crosses into and out of valley towns on a daily basis. The diurnal reversal of wind directions on slopes brings air pollution upslope by day and downslope by night. A given town midway up a slope might experience semi-daily spikes in air pollution.

On an even more localized scale, **frost hollows** or **fog hollows** can make for circulation complications. These are simply local depressions that accumulate frost or fog because of cold air drainage into them at night. The citrus trees in Florida that sit in local depressions are more vulnerable to cold-weather damage and flood damage.

Anabatic and katabatic winds are also common on broader scales. As inland areas near the coast can have much higher elevation than nearby coastal areas, anabatic winds can bring air onshore by day, for the same reasons described above. Likewise, katabatic winds can bring air offshore by night (Fig. 8.5).

Depending on the local prevailing winds, one or the other of these can persist beyond the typical daylight or non-daylight hours. For example, when persistent, quasi-stationary *anticyclones* can force circulations downslope under fair-weather conditions, regardless of the time of day or night. Such winds are very warm and dry and are known by different localized names in different areas. They are called **Chinooks** in the western U.S., particularly in Washington and Oregon, **Santa Ana** winds in southern California, and **Foehn** winds in the European Alps. These downslope winds tend to be best developed in spring, when continents are cool with relatively large high-pressure areas.

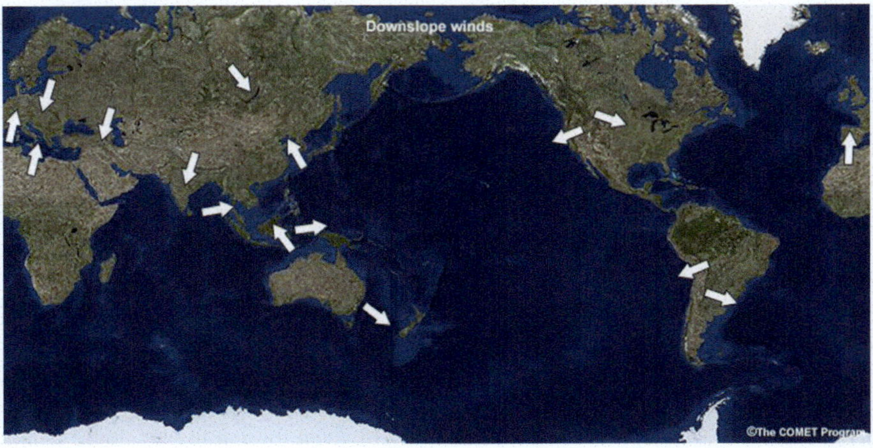

Fig. 8.5 Locations where broad-scale katabatic winds are common. *Source* UCAR CoMet. Permission from Tara Torres

Coastal Features of Local Climates: Sea and Land Breezes

In coastal areas, the difference in thermal properties of land and water can set up interesting and important local-to-*mesoscale* circulation shifts diurnally. By day, land heats up faster (and more) than water, setting up a thermally-induced pressure gradient, and a local to mesoscale circulation cell. The hotter air over the inland area rises, generating a localized zone of relatively low atmospheric pressure over land. The cooler air over the water body sinks, generating a localized zone of relatively high atmospheric pressure over water. The *pressure gradient force* always works to move air from a place with higher pressure to a place with lower pressure. In this case, this means that the near-surface wind will move onshore—from the water body to the land. This phenomenon is called the **sea breeze** (Fig. 8.6).

At the scale of the sea breeze, the *Coriolis effect* is generally negligible so it has little effect on the wind direction, so the winds are approximately perpendicular onshore.

The sea breeze generally blows at 6–20 km h^{-1} (4–12 mi h^{-1}) from the coast to approximately 32 km (20 mi) inland. However, in some cases, the sea breeze can persist to twice or three times that distance.

The leading edge of the sea breeze over land is sometimes called the **sea breeze front**. This is not a "front" in the sense of those associated with a mid-latitude wave cyclone; the sea breeze front is much smaller, with little or no *air mass* contrast across the frontal boundary. Nevertheless, rain or snow showers or thunderstorms are possible along the sea breeze front, especially where changes in the coastline orientation can cause multiple advancing sea breeze fronts that converge. Because the water is cooler than the land during daytime hours, the onshore sea breeze generally feels pleasant on a hot day. The sea breeze is one attraction for going to the beach

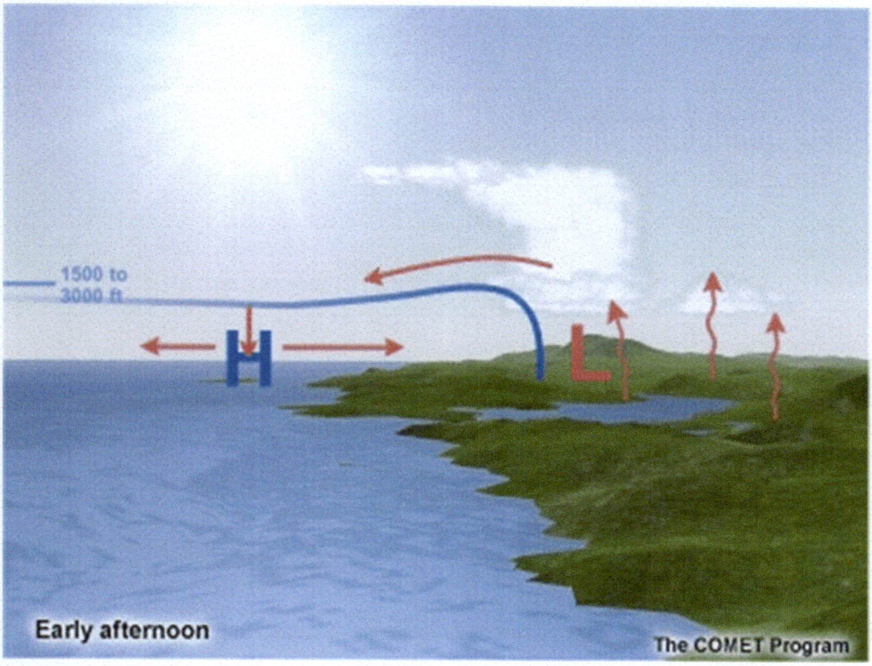

Fig. 8.6 The sea breeze circulation. *Source* UCAR CoMet. Permission from Tara Torres

on a summer afternoon, when the land is so hot that the low pressure is lowest and the pressure gradient force is strongest.

By night, land gets cooler than water; again, there is a thermally-induced pressure gradient, but this time it is directed in the opposite direction. This is called the **land breeze,** which is shown in Fig. 8.7.

The land breeze is generally only about 3–6 km h^{-1} (2–4 mi h^{-1}). It typically extends to a few kilometers seaward. It is weaker and smaller in scale than the sea breeze because the thermal differences, and therefore the pressure gradients, are weaker by night than by day. Daytime *unstable atmospheres* produce *turbulence*, which causes vertically-oriented eddies to transfer the momentum from faster upper-level winds to the surface. Such turbulence is more likely to be damped at night, when the atmosphere is stable. The result is stronger surface winds by day than by night, in general, including a stronger sea breeze than land breeze.

In cases of relatively narrow water bodies with land on each side, two or more land breezes can each create a **land breeze front**, and the fronts can converge. Again, these mesoscale "fronts" have little air mass contrast across the frontal boundary. However, the *convergence* associated with two or more land breeze fronts can induce sudden thunderstorms, particularly because abundant moisture is likely to be present over the water body (Fig. 8.8). Convergence on island shores has similar effects (Fig. 8.9).

Fig. 8.7 The land breeze circulation. *Source* UCAR CoMet. Permission from Tara Torres

Land Cover Features of Local Climates: Deserts

Specific terrain features and atmospheric and oceanic circulations support desert environments. Because these features tend to occur in particular regions of Earth, it is relatively easy to guess where on Earth deserts are found. Regarding terrain features, locations on the leeward sides of mountains, where air sinks and warms adiabatically, tend to have the dry conditions characteristic of the rain shadow. Likewise, distance from major bodies of water, especially oceans, is another important terrain feature that supports desert environments. For example, Kazakhstan in central Asia is many thousands of kilometers from a major source of moisture. Nebraska in the U.S.A. is not a desert, but its distance from the oceans means that it receives far less *precipitation* than would otherwise be expected for a mid-latitude location frequently near the polar front, the trough-to-ridge side of the *Rossby waves*, and the path of many mid-latitude wave cyclone tracks. Parts of both Nebraska and Kazakhstan are also affected by positions in the rain shadow side of major mountain chains.

Atmospheric and oceanic circulation features that support deserts are relatively simple and distinctive. Areas beneath a subtropical anticyclone, especially near the *eastern-boundary current*, are usually arid. Because subtropical anticyclones are situated at the core of *gyres*, the "desert" is actually over the ocean. Huge swaths of oceanic subtropical areas have a "desert" climate type.

Deserts are also often near cold ocean currents, which are often found near the eastern-boundary current. There the cold water produces surface inversions, where

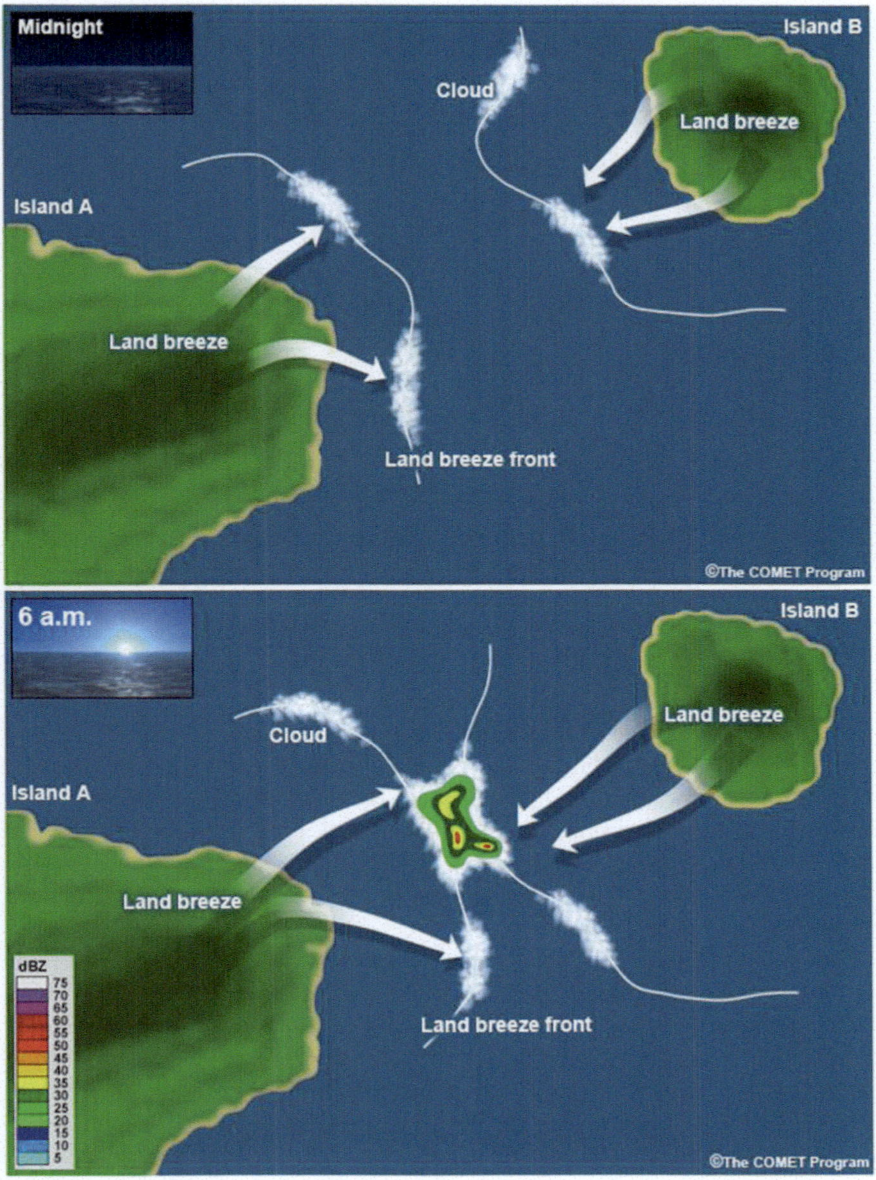

Fig. 8.8 Schematic of converging land breeze fronts. *Source* UCAR CoMet. Permission from Tara Torres

Fig. 8.9 Afternoon cloud development along the shore of a Caribbean island. *Source* UCAR CoMet/NASA. Permission from Tara Torres

air above the cold current is even warmer than the lower air immediately adjacent to the cold current. The stable conditions make it difficult for humid air to rise and to create cloud growth and precipitation, because the colder, denser air lies below the warmer air in a stable atmosphere. Turbulent eddies will generally be elongated horizontally, not vertically, in those situations. The horizontal elongation is a sign that it is difficult for air to rise sufficiently high to reach saturation, condense, and precipitate. Thus, the area is arid.

Energy budgets of deserts are distinctive. Deserts have a large Q*, because there are few clouds to reflect and absorb insolation (K↓). Q* is dominated by heating that can be sensed—*sensible heating* because there is little water to be evaporated for going into *latent heating*. But the relatively large *albedo* (which, as described in Chap. 2, is K↑) from the often light, sandy surfaces causes a fairly high percentage of K↓ to be reflected rather than absorbed by the surface. The lack of clouds and water molecules to emit longwave radiation down to the surface means that the K↓ is much greater than the L↓ emitted by the atmosphere downward to the surface. Heat accumulated on the surface from Q* on the desert surface is poorly conducted downward and (especially) upward to air. Therefore, desert surfaces tend to heat quickly. However, because the longwave radiation from these hot surfaces is unlikely to be absorbed by the atmosphere with so little water in the air, most of the longwave energy emitted goes to space and relatively little is re-emitted back down to the surface. These conditions result in very hot days but much cooler nights.

Regarding stability, at the broadest scale, deserts are in very stable atmospheres. However, at local scales, there is a layer of great instability near the surface, caused by great surface heating, which increases the environmental lapse rate substantially. Most of the time, though, there is too little *specific humidity* to produce precipitation from the unstable atmospheres.

Land Cover Features of Local Climates: Grasslands

Grasslands have radiation budgets that are characterized by lower K↑ (albedo) than in deserts, but with wide seasonal variation. When grasses are higher, the albedo decreases as more *absorption* of the *shortwave radiation* that was reflected can occur within the grass canopy.

Grasslands typically support more *evapotranspiration* than in deserts. This feature is simply because of the presence of vegetation, which can use its roots to pull water from the better-developed soils, where it is stored between precipitation events, rather than simply running off the surface as in deserts. During times of moisture stress, plant leaf apertures through which evapotranspiration occurs known as **stomates** close, decreasing evapotranspiration rates. This mechanism conserves water for the plant and is partially responsible for the seasonal and daily variations in evapotranspiration. In grasslands, Q* is first spent on Q_E, so Q_H sometimes remains somewhat low, though Q_E may be small in drier grassland environments (which would increase Q_H).

Land Cover Features of Local Climates: Forests

Most forests have two or more climate layers – the **canopy** (upper layer) and the **understory** (floor). The floor experiences very little direct K↓, but the K↓ that penetrates the canopy and the longwave radiation that is emitted from the surface are both more likely to be absorbed within the canopy than in open terrain. Nevertheless, within the forest Q* is small and Q_E generally exceeds Q_H. The understory has less wind, greater relative humidity, lower temperatures, and less sunlight than the canopy. A medium density of trees contributes most to increasing *friction* of the overlying air, as a closed canopy smoothens the flow over it. A closed canopy creates a microclimate within the forest that is largely separated from that in the outside world.

Urban Influences on Local Climates

Urban areas cause the local water balance to be altered from what would occur naturally. This change occurs in two primary ways. First, evapotranspiration rates (and therefore Q_E) generally decrease because of the lack of vegetation. Second, **runoff** of water from the surface toward local streams is increased. The abundance of concrete prevents precipitation from infiltrating into the soil and instead causes a greater percentage of it to run off the surface. Furthermore, artificial drainage is built, such as gutters on buildings, storm sewers, gullies, canals, pipelines, crowns along the middle of roadways, and other features designed to get water out of the area as quickly as possible to minimize flooding. Such water tends to run off before there is time for it to evaporate.

These two factors contribute to the **urban heat island**—the "bullseye" of higher temperatures centered on cities compared with their surrounding rural hinterlands (Fig. 8.10). Temperatures increase by 1–2 °C (2–4 °F) even in moderately-sized cities, and the effect in the world's megacities is much greater. Without as much evaporative cooling or standing water to evaporate water during intervals between precipitation events, a larger percentage of the Q* goes into sensible heating rather than latent heating. This difference causes a larger *Bowen ratio* in urban areas.

Aside from these two factors, many other factors also contribute to the urban heat island. For instance, urban areas generally have less *outgoing longwave radiation*—the longwave radiation emitted from the surface all the way out to space without being absorbed, reflected, or scattered—which means that more longwave radiation remains confined near the surface. This effect is because of longwave radiation absorption by pollution, with additional subsequent *counter-radiation* emitted by the pollutants back downward, further warming the surface. Also, the linear set

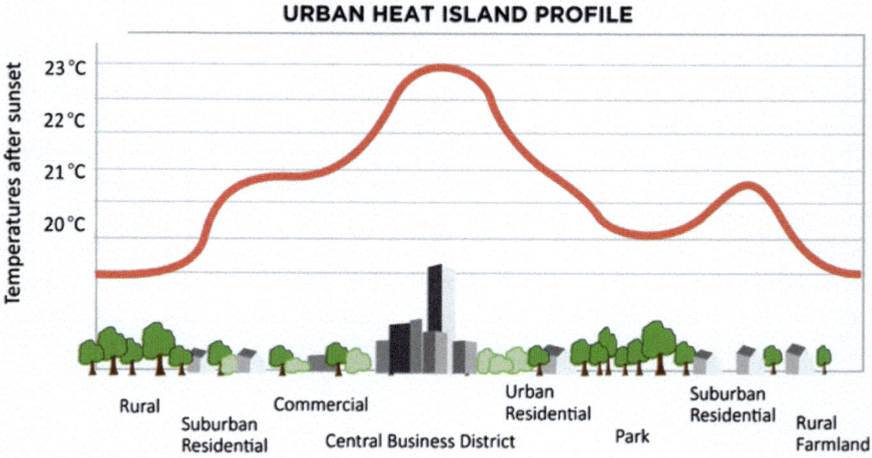

Fig. 8.10 The urban heat island effect on temperatures. *Source* UK Met Office. https://blog.metoffice.gov.uk/2023/01/10/heat-in-cities-the-health-impacts-of-a-changing-climate/

of vertical walls formed by the string of buildings on each side of city streets—known as the **urban canyon**—absorb longwave radiation from the surface, thereby hindering its ability to escape upward out of the urban area either from *reflection* off the building surface or from direct loss from the surface to space. However, some of this additional absorption is offset by the artificial shadows created by the same building walls, which hinders *insolation* from reaching the surface. Another factor that causes urban areas to be artificially hot is the additional absorption of shortwave radiation by pavement and buildings over that by most natural land covers, despite the shadows. In addition, human activities generate waste heat that is much more concentrated in cities, such as by streetlights, fireplaces, any type of combustion via motors, burning of trash, and other activities.

Aside from the differences in temperature, urban areas also have other distinctive atmospheric conditions compared to rural areas. First, insolation is reduced by 5–30% due to absorption and reflection by the enhanced air pollutants, and a decrease in surface area that can absorb insolation directly from space.

On a related note, cloudiness is increased, both because pollutants act as **condensation nuclei**—microscopic solid particles known as *aerosols* or *particulates*—that encourage the formation of a liquid water droplet or ice crystal by providing a tiny surface around which water vapor can undergo *condensation* or *deposition*. Another reason that urban areas are cloudier is that higher temperature causes greater water vapor capacity, thereby eventually leading to more condensation, even if only after the air rises to great heights to cool to its dew point temperature.

Third, interruption of circulation by urban buildings causes weaker prevailing wind speeds. However, those same structures cause stronger channelized winds between them. The reduced prevailing winds tend to cause the extra heat to remain concentrated locally as it is constantly being produced from the mechanisms described above.

Fourth, the moisture and precipitation regime is different in an urban area too. Relative humidity is reduced by 2–8%, because the increased temperatures cause an increased water vapor capacity. Thus, even if the *specific humidity* is the same as in the rural areas, the *relative humidity* will be reduced. Somewhat paradoxically, precipitation increases by 5–10% in the city, especially on the downwind side of town, largely because of the extra condensation nuclei from pollutants. This phenomenon has come to be known as the **urban precipitation island**. However, there fewer snow days, because the cities are warmer than their surrounding rural environs.

Conclusions

This chapter examined the nuances of land cover as they affect the atmospheric circulation. One conclusion can be drawn is that weather data records in varied topography should be evaluated with caution. Traditionally, most records come from cities in valleys and coastal areas, which is largely inadequate to represent the "full picture" of the Earth's climate. Fortunately, this bias in the data records is being reduced with the arrival of longer-term data derived from satellites, buoys, ship measurements, aircraft, and other sources to supplement the traditional land-based stations. These minimize the effect of local nuances when average values comprising broad-scale features are studied, while still leaving the nuances in place when more localized analysis is needed.

Further Reading

Houze, R. A. (2012). Orographic effects on precipitating clouds. *Reviews of Geophysics, 50*. https://doi.org/10.1029/2011RG000365

Miller, S. T. K., Keim, B. D., Talbot, R. W., & Mao, H. (2003). Sea breeze: Structure, forecasting, and impacts. *Reviews of Geophysics, 41*(3), 1011. https://doi.org/10.1029/2003RG000124

Oke, T. R. (1982). The energetic basis of the urban heat island. *Quarterly Journal of the Royal Meteorological Society, 108*(455), 1–24. https://doi.org/10.1002/qj.49710845502

Shepard, J. M. (2005). A review of current investigations of urban-induced rainfall and recommendations for the future. *Earth Interactions, 9*(12), 1–27. https://doi.org/10.1175/EI156.1

Snyder, P. K., Delire, C., & Foley, J. A. (2004). Evaluating the influence of different vegetation biomes on the global climate. *Climate Dynamics, 23*(3–4), 279–302. https://doi.org/10.1007/s00382-004-0430-0

Stewart, I. D., & Oke, T. R. (2012). Local climate zones for urban temperature studies. *Bulletin of the American Meteorological Society, 93*(12), 1879–1900. https://doi.org/10.1175/BAMS-D-11-00019.1

Zardi, D., & Whiteman, C. D. (2013). Diurnal mountain wind systems. In f. Chow, S. De Wekker, & B. Snyder (eds), *Mountain weather research and forecasting*. Springer Atmospheric Sciences. https://doi.org/10.1007/978-94-007-4098-3.2

Chapter 9
Atmosphere–Cryosphere–Ocean Interactions

Abstract The part of the Earth covered by water in its frozen state (snow or ice) is known as the cryosphere. This includes land ice, seasonal or multi-year sea ice, seasonal snow cover, and permanently frozen soil water known as permafrost. The ice-albedo feedback is a process in which the high albedo, or shortwave energy reflectivity, of snow- or ice-covered surfaces causes the cryosphere to expand relatively rapidly during cold times, but also to melt back relatively rapidly during times of warming. Polar amplification is also an important feature of the cryosphere, as the high latitudes warm at a faster rate than the tropics, thereby reducing the need for tropical to polar transfer of energy. The result is that the general circulation theoretically slows, leaving storm systems outside the cryosphere hovering over the same areas for longer periods of time. This situation makes extreme wet events even wetter, and lengthens the inter-arrival times between precipitation events, thereby lengthening droughts too. Polar cyclogenesis often occurs in baroclinic zones where the cryosphere meets much warmer water. Other interactions between the cryosphere, hydrosphere, and atmosphere are important but often poorly-understood mechanisms in driving and regulating the climate system.

Keywords Ablation · Accumulation · Equilibrium line (Firn line) · Ice cap · Ice sheet · Iceberg · Langmuir circulation · Mountain glacier · Nor'easter · Polar (Arctic) amplification · Positive and negative feedbacks · Riming · Seasonal and multi-year sea ice · Sublimation

The Cryosphere

The part of the Earth covered by water in its frozen state (snow or ice) is known as the *cryosphere*. The four primary components of the cryosphere are the land ice, sea ice, snow cover (which is seasonal), and permanently frozen soil water known as **permafrost**. This chapter will explain how the cryosphere fits into the "bigger picture" that makes the atmospheric and oceanic circulation the way that it is, so that it can provide the niches and habitats for life on Earth as we know it.

R. V. Rohli et al., *Atmospheric and Oceanic Circulation*,
https://doi.org/10.1007/978-3-031-66142-6_9

Effects of Ice on the Radiation Balance

The presence of snow or ice influences the local to regional radiation balance, primarily by increasing the *albedo* (K↑). Fresh snow may have an albedo exceeding 80–85%, while albedo for old snow that has become "dirty" may hover between 50 and 65%. For ice, the albedo is usually 20–40%. These values all exceed those for ice-free surfaces considerably. When ice exists, the surface albedo is unaffected by the medium beneath the ice.

Net radiation (Q*) is generally small in the *cryosphere* and can even be negative. A negative Q* in such a case means that the air supplies the ice- or snow-covered surface with radiant energy, on a net basis, instead of the more common case which is for the surface to supply the air above it with energy. But why would this be the case in the cryosphere?

As shown in Chap. 2, Q* is the sum of the *net shortwave radiation* (K↓ − K↑, or K*) and *net longwave radiation* (L↓ − L↑, or L*) components. This can be written as $Q^* = K^* + L^*$. In the K* component, K↓ is relatively small in the cryosphere because the Sun is so low on the horizon and may even stay below the horizon for many consecutive winter days. Yet K↑ is almost as large as K↓ in the cryosphere, because of the high albedo. In other words, a relatively high percentage of the Sun's incident radiation on the surface is reflected back up from the surface.

In the L* component of Q*, *counter-radiation* (L↓) is generally small in the cryosphere because the atmosphere is too cold to send much L↓ to the surface. L↑ often exceeds L↓ in the cryosphere because the surface has more energy to emit than the overlying atmosphere has to emit downward to the surface. Even though a small amount of *longwave radiation* emitted by the atmosphere down to the surface as L↓, is actually reflected by the surface, emitted longwave radiation should not be confused with reflected *shortwave radiation*. The latter involves much more *radiant energy*.

Therefore, instead of having a positive Q* in which the *sensible heat flux* and *latent heat flux* move upward via *convection* from the surface as the more energetic molecules contain *sensible heat* and *latent heat* from evaporated water, the opposite is often the case in the cryosphere, as Q* is often negative. The sensible and latent heat fluxes are usually negative in the cryosphere, implying that heat and moisture are transferred downward toward the icy/snowy surface. Negative sensible heating means that eventually the air will lose its little bit of heat to the ground below it, causing winter to get brutally cold through the core of winter. Nowhere on Earth is this more evident than over the *ice caps*. Negative latent heating produces **riming**—a *deposition* of water vapor directly onto a solid surface as ice, skipping the liquid phase. The process results in accretion of ice on or around a (usually) ice core or surface. Annual accretion layers form on continental *ice sheets* and these layers may be cored and examined to understand past climatic conditions.

The *temperature* pattern produces a surface *inversion* (a negative *environmental lapse rate*) that may last quasi-permanently in the most extreme parts of the cryosphere. In such places, the atmosphere is very stable, and very sharp temperature gradients exist at the edges of the ice caps. These edges are preferred areas for storm formation, despite the lack of substantially warm air to interact with the frigid polar air.

Land Ice

There are five components of land ice: ice sheets (> 50,000 km^2; 30,000 mi^2), ice caps (somewhat smaller), **ice shelves** which protrude over the ocean water, **mountain (alpine) glaciers**, and **icebergs**. Mountain glaciers are quite common, as they are found in approximately 50 nations, at all latitudes including the tropics, such as in Indonesia, Tanzania, Peru, Venezuela, and Myanmar. But they are generally small. Icebergs are fresh water from the land that floats on the ocean. All land ice is fresh water—not salt water—even the part of the ice shelves that extend over the ocean.

Approximately 15 million km^2 (6 million mi^2) on Earth is covered by land ice. However, the total varies seasonally and in geologic time. During Northern Hemisphere winter, the global land ice area expands because a larger expanse of the more continental Northern Hemisphere undergoes a seasonal free-thaw cycle than the more ocean-covered Southern Hemisphere. Today, land ice is only found at high altitudes and latitudes, but at most other times in geologic time, the area was even smaller, if it existed at all. At a few times in geologic time—in the core of the ice ages—the total planetary coverage would have been much greater.

The thickness of ice on a mountain can be higher on the ocean side than on the land side, because the additional *humidity* from the ocean allows for the production of more snowpack. If the ocean is on the windward side, such as in the northwestern U.S., cloudiness on the windward side would keep daytime and summer temperatures cooler, thereby maintaining land ice for a longer part of the year. Furthermore, in summer (when melting can occur), the oceans are cooler than land, so the ocean side may permit the ice to last longer into the summer season. For these reasons, many mountain glaciers are on the northeastern side of the mountain in the Northern Hemisphere and on the southeastern side in the Southern Hemisphere.

The growth of a glacier is typically assessed in terms of its glacial **mass balance**—the glacier's added mass minus its lost mass. When the balance is positive, snowfall, blowing snow, avalanches, and riming lead to an **accumulation** rate that exceeds the **ablation** rate. Ablation occurs through *melting*, *sublimation* – the transformation from solid ice directly to water vapor—the opposite of deposition, and **calving** in which some of the ice breaks off and falls into the ocean, forming an *iceberg*. The zones of accumulation and ablation are situated at the opposite ends of a glacier, with the **equilibrium line** (or **firn line**) separating the two zones.

Climate variability and change can be measured by changes in the location of the equilibrium line. Glaciers advance and retreat because of the changing mass balance between the accumulation above the equilibrium line minus the ablation below the equilibrium line. The mass balance is tied to climate fluctuations such that glacier movement serves as a proxy for climate change and variability. Ablation is a direct response (responds to temperature) while accumulation is an indirect response (delayed reaction). In total, total changes may be slow, taking 20–1000s of years to see a fundamental response. This timing makes it difficult to analyze some trends.

Glaciers grow and advance if temperature decreases (an immediate reaction) and/or accumulation increases from increased *precipitation* (delayed response). Glaciers retreat if temperatures increase (immediate) and/or precipitation decreases, and therefore accumulation decreases (delayed response). Ice sheets, large glaciers, and glaciers on gentle slopes take longer to react to climate variations than smaller alpine glaciers, simply because of the mass involved.

Glaciers and land ice have important impacts on the overlying atmosphere. For instance, they encourage the formation of very *stable* air with little convection. They also affect regional circulation by forcing air up and over them. They modify the radiation and energy balance locally, by reflecting a higher percentage of incoming shortwave radiation in the form of K↑ (albedo) than a bare or vegetated surface would.

On a broader scale, glaciers and land ice also affect sea level by storing fresh water, thereby preventing it from running to the ocean. Thus, the presence of more glaciers and land ice on Earth coincides with saltier oceans and lower sea levels. Periods in Earth history in which glaciers and land ice are actively melting, as today, are characterized by freshening ocean waters and rising sea levels. The marine fossil record can also yield effective clues regarding Earth temperature and salinity, under the assumption that plants and animals would have existed in the same kinds of temperature and salinity environments in the geological past as they do today.

Sea Ice

The reduced density of water in its frozen state causes *sea ice* to float on the rest of the ocean. The freezing and melting of sea ice actually contributes very little to global sea level changes, because the amount of water involved is small compared to other sources of ice. Nevertheless, a distinct decrease in global sea ice extent has been observed since the late 1970s when modern satellites have been monitoring its extent closely (Fig. 9.1). This is important for several reasons. First, the presence of the sea ice helps to minimize the effects of further warming by covering more of Earth's surface with a high-albedo surface that reflects a large percentage of the *insolation.* This concept will be explored in more detail shortly. Second, the presence of the ice can impact the location and intensity of the *thermohaline circulation*, which in turn drives climate changes globally. And third, the sea ice provides habitats and niches for marine life.

There are two forms of sea ice. The first is **seasonal sea ice**, which melts in summer and is approximately 0.3–1.2 m (1–4 ft) deep. Some scientists classify the seasonal sea ice based on its thickness, using such terms as "thin," "medium," and "thick" categories within this general range of thickness. Some scientists also may use alternative names for seasonal sea ice, such as "new" or "first-year," but these

Fig. 9.1 Arctic sea ice minimum seasonal extent, as of 16 September 2021. *Source* NASA. https://www.nasa.gov/sites/default/files/thumbnails/image/sea_ice_2021_min.jpg

other terms are confusing because they might imply that the distribution of sea ice is expanding over time rather than just over the natural course of summer and winter.

The second type is **multi-year sea ice**, which continues building thicker, year after year, but still not more than 3–4 m (10–13 ft) thick. Again, multi-year sea ice may be described under other names, such as "old ice," but these other terms are more likely to lead to misinterpretation of their features. In winter, multi-year sea ice is snow-covered, giving it an even greater albedo than at other times of year. In summer, water pools can form over the sea ice surface, affecting the radiation/energy balance, albedo, and other environmental features.

Approximately 12% of the world's oceans is covered by sea ice. But its distribution is uneven. The Northern Hemisphere has a much greater amount of multi-year sea ice than the Southern Hemisphere, because the polar Arctic is much more confined by land masses that prevent warm ocean waters from melting sea ice in summer. Warm ocean currents (especially the *Gulf Stream*) affect the size of sea ice. The coastline shapes of Asia and North America limit the northward extent of the warm *Kuroshio Current*, contributing to the relatively far equatorward extent of sea ice, particularly the multi-year type, in the Pacific. In the Northern Hemisphere, the winter/summer ratio of sea ice is approximately 2:1, while the ratio in the Southern Hemisphere is about 8:1. Again, this difference is explained by the position of Antarctic landmass, which allows warm currents to melt back and confine Southern Hemisphere sea ice to only the highest marine latitudes, which themselves are far from the South Pole, because of the position of Antarctica.

Sea ice has significant domino effects on the atmosphere. As discussed previously, an increased albedo initiates a **positive feedback** that encourages even more cooling. This cooling happens as the increased albedo prevents more surface heating from occurring, which allows extra cooling, which allows extra ice formation, which reflects more radiation, which cools the surface air even more, producing even more sea ice. In general, a positive feedback means that something happens which causes something else to occur, which amplifies the effect of the first thing that happens. For example, one student yawning in class initiates a suggestion that causes another person to yawn, which causes everyone to yawn in class. The ice-albedo effect described above is a positive feedback of great importance in understanding the relationship between all components of the cryosphere, not merely sea ice, and the temperature changes that affect and are affected by atmospheric and oceanic circulation.

The ice-albedo positive feedback can also work the other way when sea ice begins to melt. In that case, the increased insolation, or increased warming from other sources, melts a small amount of ice, which allows slightly less insolation to be reflected, which warms the near-surface air even more, which melts a little more sea ice, and so forth. Future climate projections tend to suggest that this ice-albedo feedback is a cause for concern as a producer of accelerated warming in the future.

Moreover, current prevailing thought is that the steering atmospheric circulation is weakening over time. This idea is based on the observations that the polar parts of the Earth tend to be warming at a greater rate than the tropics, as melted snow and ice in the cryosphere is replaced by darker land with less albedo. This reduced pole-to-tropics temperature gradient means that there is less need to re-distribute energy across the latitudes, which reduces the need for the steering circulation. This phenomenon, known as **polar amplification**, or sometimes **Arctic amplification** despite its possibility for occurrence in the Antarctic too, may have impacts on future weather and climate. For example, with a weaker steering circulation, storm systems would stall or move more slowly, and therefore have more time to drop precipitation on a given area. But the weaker circulation would also increase the interarrival time to the next storm system passage, lengthening dry periods and exacerbating drought conditions too. This rationale forms the basis of why the future is expected to have more intense periods of both wet and dry extremes. Of course, the extremes cause the most impacts to life and property, so polar amplification is a cause for concern.

Not surprisingly, the opposite of a positive feedback is a **negative feedback**. In a negative feedback, stimulus A causes response B, which prevents A occurring or at least amplifying. An atmosphere–ocean–cryosphere example of a negative feedback would be that additional warming of Earth causes more evaporation and unstable atmospheres, which drives the evaporated water up higher in the atmosphere, making it more likely to condense and form clouds, which then absorbs and reflects more $K\downarrow$ and perhaps causes more snowfall atop the sea ice, thereby reducing additional warming through the increased *reflection* from additional cloud- and snow-cover. Will this negative feedback offset the tendency for the positive feedback system to accelerate warming? Output from most current climate models suggest that it will not.

Sea ice also reduces interaction between the ocean and atmosphere, which can impact evaporation and the locations where precipitation freshens the ocean water. In turn, these changes can affect the thermohaline circulation and marine ecology. Sea ice is also a factor in the atmospheric energy budget, as it relates to *latent heating.* As was shown in Chap. 2, latent heat is released as ice forms. This release can warm up the atmosphere, delaying winter (ironically!) as the ice forms, by a few weeks to a month.

In spring, however, when seasonal sea ice melts, insolation—expressed as latent heat (energy)—is "used" to melt ice in spring. Thus, summer is delayed (also ironically!) by a few weeks to a few months. One important impact of this process is that a very large temperature gradient is established on the edges of the sea ice. Locations immediately offshore from areas that freeze tend to experience the most extreme cyclones, as the sharp thermal gradient encourages the formation of *fronts*, as was shown in Chap. 4.

Polar lows form in these **baroclinic zones** – places where cold water is immediately adjacent to warm water. Many such baroclinic zones occur near Japan and the eastern U.S. and Canada. As the leading edge of the ice/water interface extends farther oceanward and southward in winter, so do the storms. A good rule of thumb is that *cyclogenesis* may be most common at about 10° of latitude south of the ice margin, where enough of the effect of the warm/cold interface can combine with the more abundant warm, humid air needed to fuel the storm through latent energy transformations, especially with assistance from the *polar front jet stream*, as described in Chap. 3. The resulting storms can affect environmental conditions in islands and the west coasts of continents both in chronic and catastrophic ways.

Snow Cover

Global snow cover displays wide interannual and intra-annual variability. Figure 9.2 shows the average maximum and minimum coverage of terrestrial snow and sea ice. The extent is strongly correlated with temperature, of course.

As the snow line moves equatorward as winter approaches, the temperature gradient at the boundary (i.e., the snow line) moves equatorward too. An especially sharp temperature gradient occurs if the snow cover is near the coast, owing to land/ocean temperature contrasts. In such cases, potentially explosive cyclones can form easily along a sharply defined baroclinic zone produced by the temperature gradient, which can then produce even more snow in a positive feedback. The intense winter storms of the U.S. Atlantic coast and East Asian Pacific coast called **nor'easters** fall into this category.

Circulation can get locked in place for weeks because troughs are common on the eastern halves of continents in the mid-latitudes. Persistent eastern continental troughing means that successive cyclones are likely to form between trough and downstream ridge—near the coast. The counter-clockwise circulation around the cyclone wraps humid air from the Atlantic and Great Lakes around the low.

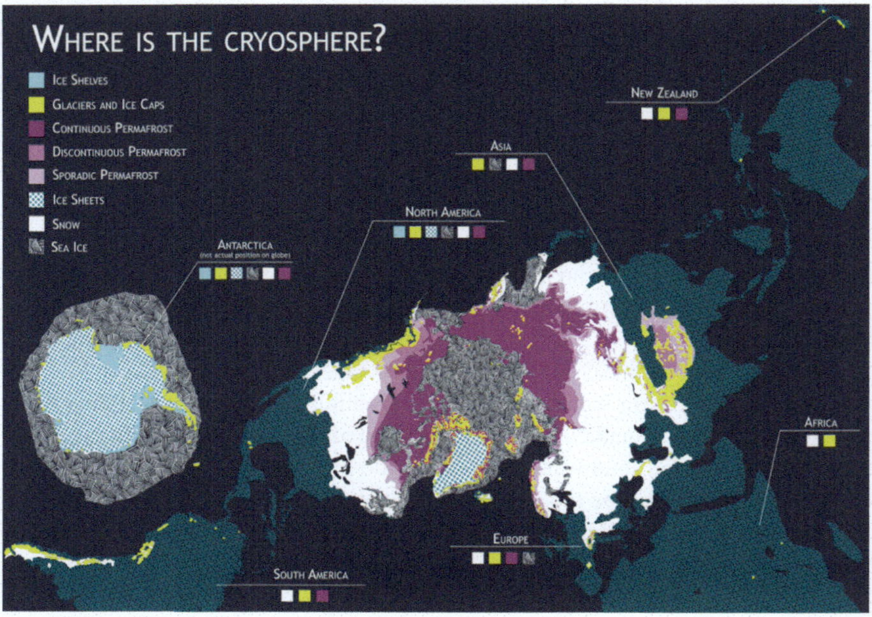

Fig. 9.2 Extent of snow cover and cryosphere. *Source* https://nsidc.org/our-research

Ocean–Atmosphere–Cryosphere Interactions

Across the majority of the Earth, outside of the highest latitudes, the atmosphere is interacting with water in liquid form. The cryosphere adds a whole new dimension to these interactions. The details of how the largest pools of water—the oceans—interact with the atmosphere and cryosphere are some of the most confounding mysteries of how the *climate system* works. Little high-quality, long-term data are available. Nevertheless, these ocean–atmosphere–cryosphere interactions are probably a key hurdle that limits current ability to predict future atmospheric and oceanic circulation, and therefore climates, more successfully.

In general, the components within each "sphere" are related to each other in a chain-reaction series of changes that could ripple through the system. Evaporation leads to precipitation, which alters salinity. The salinity changes, in turn, affect the currents, which alter ocean temperatures in various places, which impact the overlying air temperatures. Changes in air temperatures then affect and are affected by cloud cover and winds, as the *pressure gradient force* is affected.

The ocean is the largest recipient of K↓ and L↑ radiation on Earth's surface. Figure 9.3 shows mean daily L↑ in *Watts* per square meter. Other than very hot land masses in summer, L↑ is generally greatest over tropical oceans because of the relatively large *specific heat* of water. Yet L↑ is also a non-negligible amount of radiative energy even in the exceedingly cold cryosphere as well. Oceanic heat is mixed relatively easily, especially compared to land. Thus, the mixed layer of the

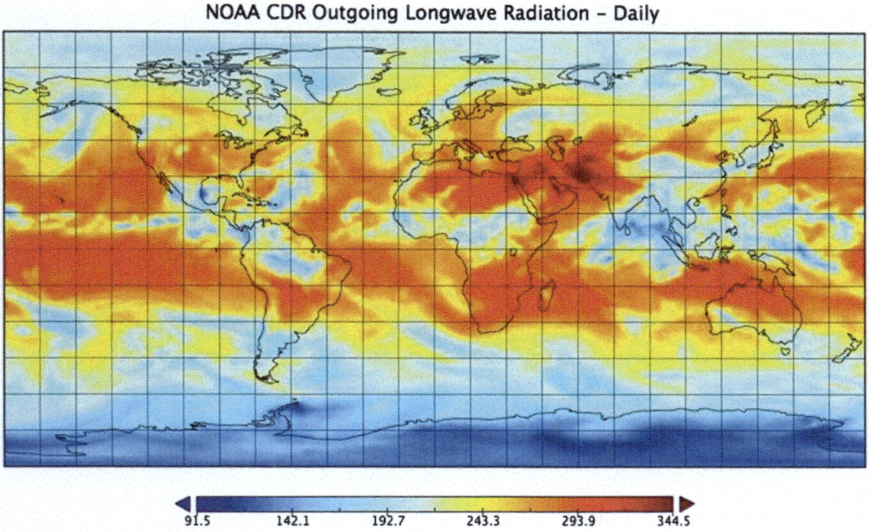

Fig. 9.3 Mean daily outgoing longwave radiation in W m^{-2}. *Source* NOAA

ocean holds more than 30 times as much heat as the atmosphere. This disparity is a major reason that interactions between the hydrosphere, cryosphere, and atmosphere are so important. In areas of a warm ocean current and *unstable atmosphere* overlying it, such interactions are even more vigorous.

The ocean is also important for lateral heat transport. Sea surface temperatures can affect the stability of the air above them. Coastal deserts tend to be adjacent to the coldest ocean sea surface temperatures, while high sea surface temperatures support warm, unstable atmospheres. Therefore, the large-scale circulation is largely determined by pools of ocean water, as they affect the positions of ridges and troughs. We have already seen that ocean-air teleconnections like ENSO, IOD, PDO, and AMO are very important aspects of ocean–atmosphere interactions.

Further Reading

Chapin, F. S., Sturm, M., Serreze, M. C., McFadden, J. P., Key, J. R., Lloyd, A. H., McGuire, A. D., Rupp, T. S., Lynch, A. H., Schimel, J. P., Beringer, J., Chapman, W. L., Epstein, H. E., Euskirchen, E. S., Hinzman, L. D., Jia, G., Ping, C. L., Tape, K. D., Thompson, C. D. C., … Welker, J. M. (2005). Role of land-surface changes in Arctic summer warming. *Science, 310*, 657–660. https://doi.org/10.1126/science.1117368

Cohen, J., & Entekhabi, D. (1999). Eurasian snow cover variability and Northern Hemisphere climate predictability. *Geophysical Research Letters, 26*(3), 345–348. https://doi.org/10.1029/1998GL900321

Deser, C., Tomas, R. A., & Peng, S. L. (2007). The transient atmospheric circulation response to North Atlantic SST and sea ice anomalies. *Journal of Climate, 20*(18), 4751–4767. https://doi.org/10.1175/JCLI4278.1

Koyama, T., Stroeve, J., Cassano, J., & Crawford, A. (2017). Sea ice loss and Arctic cyclone activity from 1979 to 2014. *Journal of Climate, 30*(12), 4735–4754. https://doi.org/10.1175/JCLI-D-16-0542.1

Large, W. G., & Yeager, S. G. (2009). The global climatology of an interannually varying air-sea flux data set. *Climate Dynamics, 33*(2–3), 341–364. https://doi.org/10.1007/s00382-008-0441-3

Serreze, M. C., & Barry, R. G. (2011). Processes and impacts of Arctic amplification: A research synthesis. *Global and Planetary Change, 77*(1–2), 85–96. https://doi.org/10.1016/j.gloplacha.2011.03.004

Vavrus, S. (2007). The role of terrestrial snow cover in the climate system. *Climate Dynamics, 29*(1), 73–88. https://doi.org/10.1007/s00382-007-0226-0

Part IV
Circulation Analysis Based on the Climate System

Chapter 10
Synoptic Analysis

Abstract Over long periods of time, the primary circulation features in the atmosphere and ocean, secondary atmospheric circulations, and mesoscale ocean eddies tend to produce "expected" patterns. Our ability to recognize these patterns and the impacts that they typically produce in the form of weather and climate, along with the temporal changes in those patterns, is important, not only for weather and climate forecasting purposes, but for making our society more resilient to weather and climate extremes. Scientists can identify the patterns and their changes over time through two primary strategies. The modeling approach is used mainly (but not exclusively) by meteorologists and oceanographers. The empirical approach involves examining atmospheric patterns based on the data and trying to explain deductively why they occur as they do. The study of the relationship between atmospheric circulation patterns understood from either the modeling or empirical approaches and the climate at the surface is known as synoptic climatology, which can be done either as an air mass or flow pattern approach. In recent years, "Big Data" approaches to synoptic climatology have become powerful tools toward understanding relationships between atmospheric circulation and the environment.

Keywords Artificial intelligence (AI) · Artificial neural network (ANN) · Automated (objective) classification system · Cluster analysis · Decision tree (DT) · Deep learning (DL) · Eigenvector analysis · Empirical orthogonal function · Kirchofer technique · Machine learning (ML) · Manual classification · Principal components analysis (PCA) · Random forest (RF) · Self-organizing map (SOM) · Synoptic climatology

Explanations of Climate Patterns

Over long periods of time, the same general circulation features in the *atmosphere* and ocean described in Chaps. 3 and 5, and secondary atmospheric circulations described in Chap. 4 and *mesoscale ocean eddies* described in Chap. 5, tend to produce "expected" patterns. Ocean-air teleconnections (Chap. 6) and interactions

R. V. Rohli et al., *Atmospheric and Oceanic Circulation*,
https://doi.org/10.1007/978-3-031-66142-6_10

with landscapes (Chap. 8) and the *cryosphere* (Chap. 9) further refine those expected patterns across space and time.

Our ability to recognize these patterns and the impacts that they typically produce in the form of weather and climate, along with the temporal changes in those patterns, is important, not only for weather and climate forecasting purposes, but for making our society more resilient to weather and climate extremes. For instance, if it can be known with reasonable certainty that the atmospheric circulation patterns will make for a colder-than-normal upcoming spring, city planners can allocate more appropriate budgets for snow removal, school superintendents can incorporate a few extra school days to offset potential closure days, and hospital administrators can plan for additional need for staff during what might become a longer-than-normal flu season.

Scientists can identify the patterns and their changes over time through two primary strategies. The modeling approach is used mainly (but not exclusively) by meteorologists and oceanographers. The strategy involves using the equations of atmospheric physics to mimic atmospheric processes on a computer. In many components of the modeling procedure, it is necessary to use parameterizations—a kind of adjustment factor—to approximate the processes for which the physics cannot be represented using mathematical equations, usually because the process is too poorly understood.

Like most other sciences, the modeling approach is inherently inductive, meaning that it relies on very specific processes and parameterizations to build a general solution of "what is the atmospheric or oceanic circulation going to be like in the future?" *Hypothesis* testing via the *scientific method* uses an inductive (specific to general) approach. Experiments are run on specific cases, holding other variables constant, with testing and re-testing to try to reach an overall general conclusion.

Weather forecasts are based on these types of models. The models themselves are very complicated. Modeling climate is even more challenging than generating models to forecast short-term weather. And even these climate models have poor performance unless they mimic processes and include effective parameterizations from the entire *climate system*—the atmosphere, *hydrosphere*, *lithosphere*, *cryosphere*, and *biosphere*. It is enough to restrict the modeling to the atmosphere for weather forecasting models, but not for climate models.

The second major strategy that scientists can employ to identify the patterns and changes over time is through the empirical approach. This strategy involves looking at atmospheric patterns based on the data and trying to explain why they are the way that they are. The empirical approach uses deductive reasoning (from general to specific) rather than the inductive approach of the modelers. The patterns are the general "answers" and the causes of those patterns are the specifics.

Synoptic Climatology

The study of the relationship between atmospheric circulation patterns and the climate at the surface is known as **synoptic climatology**. "Syn" means "same," as in a "synonym," and "optic" refers to seeing. Synoptic climatology is all about "seeing the flow patterns at the same time"—a snapshot in time that maps the flow patterns. The typical weather map that appears on apps or television weathercasts is a synoptic map. Synoptic climatology involves classifying weather maps into a reasonable number of categories, or types, in a similar manner as taking a deck of cards and putting them into stacks based on their suit, or value, or color, or some other criterion. Then, the frequency of each weather type can be found, to determine which patterns are most common at which times of year, how the frequency is changing over time, and so forth.

Two general approaches exist for map classification in synoptic climatology. One way is to look at the flow patterns. This would include the *cyclones* and *anticyclones*, locations of *mid-latitude wave cyclones*, or, on an upper-level map, where and how amplified the ridges and troughs are. Any day that has a relatively similar pattern of these kinds of features would go into the same category. The other approach is to base the categorization on the positions of the air masses. This categorization is done by collecting *temperature*, *humidity*, and other *thermodynamic* variables at a series of grid points and matching days having similar thermodynamic properties in similar positions.

Flow Pattern Analysis

The use of flow patterns to do a synoptic climatological analysis assumes a relationship between the upper-air circulation and surface climate. For temperatures in the mid-latitudes, where *Rossby waves* represent the flow patterns in the *mid-latitude westerlies*, warmer air moves beneath upper-level ridges. This happens because ridges of high *geopotential heights* are elongated poleward (which is northward in the Northern Hemisphere), meaning that after the air sinks, warming *adiabatically* on the descent and exerting high pressure on the surface, the warmer air generally extends poleward beneath the "bulge" of the ridge. Likewise, colder air generally settles into troughs, which are elongated equatorward, giving the colder air from poleward locations an opportunity to extend equatorward (which is southward in the Northern Hemisphere) as it moves in the mid-latitude westerlies. The shape of the jet stream can often be used to guess temperatures using this principle.

The distinction between *zonal* versus *meridional* circulation can also be used to categorize flow patterns in the mid-latitudes, where Rossby wave flow is important. In zonal flow (Fig. 10.1), east–west-oriented temperature gradients are small. Meridional flow is characterized by strong east–west temperature gradients.

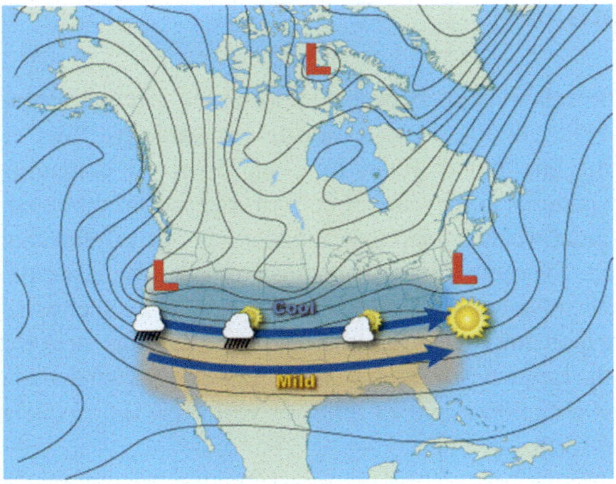

Fig. 10.1 Depiction of zonal flow across the southern United States. *Source* NOAA. https://www.weather.gov/jetstream/basic

Analyzing the positions of ridges and troughs and how they relate to surface conditions is a traditional area of emphasis in synoptic analysis. Ridges and troughs have preferred geographic positions. Ridges are often found over mountain ranges and troughs are favored near land–ocean boundaries. As shown in Chap. 3, *precipitation* and storminess are more likely on the trough-to-ridge side of the upper-level Rossby wave, as shown in Fig. 10.2.

Three steps are taken to creating a synoptic climatology of flow patterns. First, the flow patterns must be classified into a manageable number of categories. The within-group variability for each category should always be minimized compared to the between-group variability. In other words, the flow pattern on days in the same category should be more similar to each other than they are to the flow pattern on days in a different category.

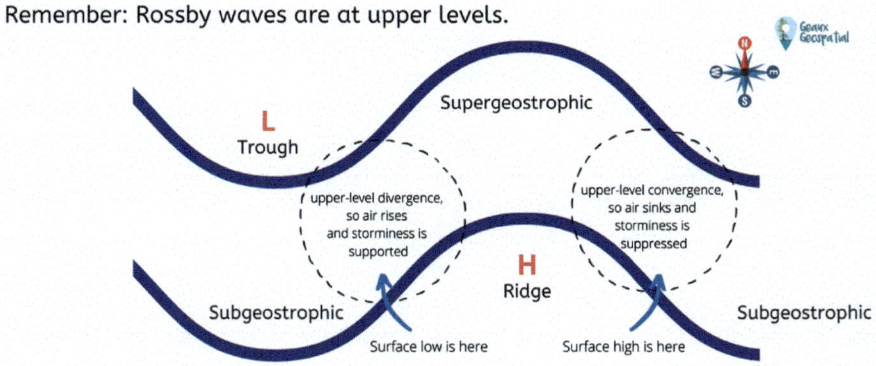

Fig. 10.2 Depiction of three-dimensional air flow associated with Rossby waves. *Source* RVR

The second step in synoptic climatological analysis is to relate the flow patterns to environmental conditions at the surface. The types of environmental conditions depend on the purpose of the classification. Typically, temperature and precipitation are two variables that are often used, but there could be many others. Air pollution variables would be used if the purpose is to understand what kinds of atmospheric flow patterns are most conducive to good and poor air quality, for instance. Wind speed, direction, and the variability of both might be important if analyzing the flow patterns most supportive of harnessing wind energy.

The third step is to understand the frequency and variability of each flow pattern. Another important component of this step is to determine whether the frequency of the pattern is changing over time. This step brings meaning and context to the relationships established in the second step.

Flow pattern analyses are useful because atmospheric flow is linked to so many environmental features. For example, maybe a researcher would want to know why moth migrations to Louisiana vary from year to year. In this case she might hypothesize that different atmospheric flow regimes exist from year to year. Perhaps flow type 1 that she identifies has the most moths and flow type 7 has the fewest. Or perhaps an environmental health scientist is interested in knowing why some Octobers have much worse hay fever cases than others, or why pine pollen is more abundant in some springs than in others. Perhaps a biological oceanographer is interested in testing a hypothesis that harmful algal blooms are related to atmospheric flow variability. The possibilities are limitless.

One way of doing a flow pattern analysis is **manual classification**, in which the weather maps are analyzed by categorizing the types by hand, using a point of reference. Advantages of manual classification are substantial. The manual approach allows the analyst to be fully engaged in the process, therefore giving a strong sense of the data and making use of the classifier's knowledge and expertise. The categories produced are all-inclusive. Furthermore, if it is based on the cyclone model, then similar categories can be set up in other locations in the mid-latitudes, with some modifications.

There are also several major disadvantages of manual classification. The class boundaries are somewhat arbitrary, leading to problems of replicability. Furthermore, the point-based nature of the classification makes transferability of the results, even to nearby locations, inaccurate at least to some extent. Third, the replicability of the analysis is compromised because of the human subjectivity involved in the classification. Not only do different analysts classify differently, but even the same analyst may be unlikely to be able to repeat the classification in the exact same way. Finally, manual classification is time- and labor-intensive.

A second way of doing a flow pattern analysis is to let the computer do the classifying. These **automated (objective) classification systems** range in complexity. On the simpler end, correlation-based methods, such as **Kirchofer techniques**, have been used. Kirchofer approaches would take every grid point's data value (usually pressure or geopotential height) for each day in the data set, and correlate that to the pressure pattern in day 2, then day 3, and so forth, for the entire period of study. The computer would then identify the "key day," which is correlated best with the most other days. The key days become the most common patterns. At that point, the non-key days are put into groups with one of the key days based on its highest correlation using a statistical technique known as **cluster analysis**. The user decides how many categories and whether or not any days in the data set will remain unclassified. Figure 10.3 shows an example of a classification that results in ten surface weather patterns over eastern North America.

Another form of objective, automated classifying can be done by **eigenvector analysis**, of which **principal components analysis** (PCA) or **empirical orthogonal functions** are two common types. These approaches collect geopotential heights or sea-level pressure at a set of gridpoints and use the technique to derive flow patterns (since flow responds to differences in geopotential heights or sea-level pressure) that explain the most variability. Cluster analysis is also used in eigenvector analysis.

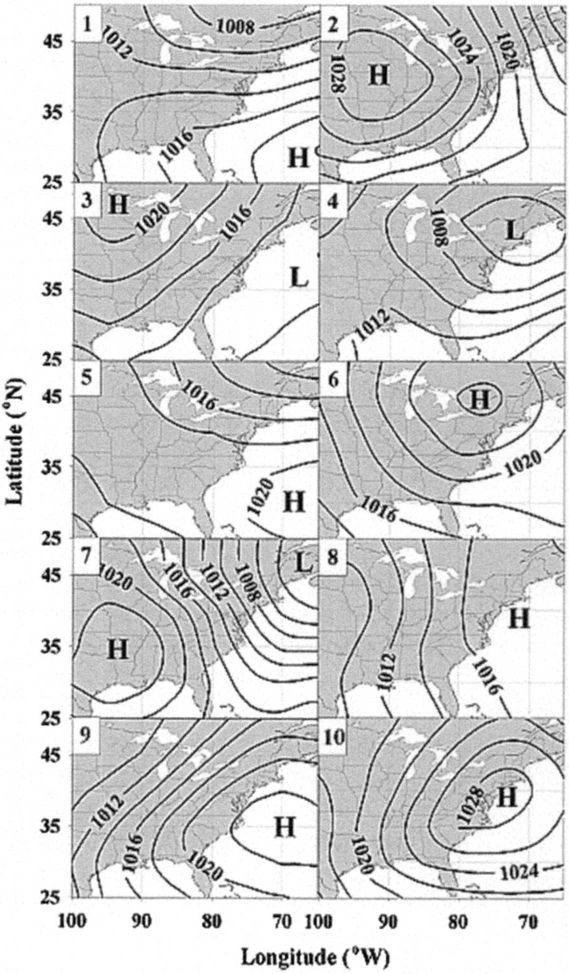

Fig. 10.3 Example of a correlation-based synoptic circulation classification system. *Source* American Geophysical Union, Miller et al. (2006)

Air Mass Synoptic Climatology

In this type of analysis, the classification is based on atmospheric variables rather than on flow patterns. At a single place, the user would take all the weather variables at several levels in the atmosphere, and using a clustering or PCA procedure, identify the X most common air masses over that place. This could, of course, be done for all stations on a given day to produce a map of air mass types over the U.S. for that day.

A natural question might be, "What is better - flow pattern analysis or air mass synoptic typing?" The answer depends on the kind of problem being considered. Flow techniques are better for circulation-based phenomena. This might include research questions like, "Which weather patterns produce floods in Florida?" or "Do moths migrate in pre-established patterns innate to the species, or are the migrations related to the weather patterns at the time?" or "How much acid rain variability from year to year is related to differences in weather patterns versus how much is related to differences in human activities?". Air mass techniques are better for analyzing the variability in atmospheric phenomena at a given place, such as pollution episodes, heat stress, or the *latent heat flux*. Either of these techniques can be used at different spatial scales, including the local, meso, regional, continental, and hemispheric. However, most synoptic climatologies are at the regional or continental scale.

"Big Data" Approaches to Synoptic Climatology

Regardless of whether a flow pattern or air mass-based method is chosen, approaches also exist to classify weather patterns in an automated, "Big Data" style. The use of computers to solve problems by building models that classify or predict future information, such as weather map patterns, is known as **artificial intelligence** (AI). A subclass of AI in which the models "learn" characteristics from an input set of previously observed data, such as maps already categorized into user-defined weather types, is called **machine learning** (ML).

A type of ML model that attempts to mimic the human brain by relaying on/off signals throughout a collection of interconnected elements working together (i.e., neurons) is called an **artificial neural network** (ANN). Some ANNs involve "competitive learning" in the model development in which the elements "compete" for their degree of influence in the model. The type of competitive-learning-based ANN technique that is most useful in synoptic typing is called **self-organizing maps** (SOMs).

Aside from ANNs, another class of ML models involves decision tree (DT) methodologies. Trees are usually used to "bin" phenomena characterized by various types of data, by making rules based on the available parameters. For example, a person who can type more than 100 words per minute and spells words with few errors might be binned as a candidate for a position as a court reporter. One common

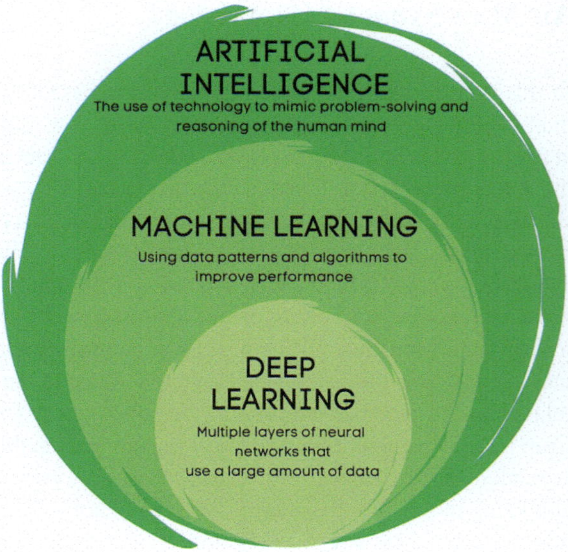

Fig. 10.4 Categories within artificial intelligence

type of DT algorithm is **random forest** (RF). Daily weather maps or **Köppen climate types**—described in the following chapter—are particularly well-suited to DT approaches.

Finally, a subclass of various ML models (most notably those based on ANNs), in which the models are constructed over numerous layers, leading to a more thorough but also more time-consuming analysis, is known as **deep learning** (DL). Synoptic weather typing and synoptic climatology are only two of many different types of applications of AI, ML, SOM-based ANNs, DTs, and DL, but they are particularly well-suited to such approaches. A simple schematic of the most basic categories of AI is shown in Fig. 10.4.

Advantages of objective classification systems include speed, efficiency, and replicability. Disadvantages are that they require more specialized technical training and that subjectivity is still involved. The user still must tell the computer how many patterns there will be, where the grid points will be, the extent to which the classification should focus on generating many categories each with a few maps in the category versus few categories each with many maps, and other organizational information.

Further Reading

Davis, R. E., & Kalkstein, L. S. (1990). Development of an automated spatial synoptic climatological classification. *International Journal of Climatology, 10*(8), 769–794. https://doi.org/10.1002/joc.3370100802

Miller, W. D., Kimmel, D. G., & Harding Jr., L. W. (2006). Predicting spring discharge of the Susquehanna River from a winter synoptic climatology for the eastern United States. *Water Resources Research, 42*(5). Art. No. W05414. https://doi.org/10.1029/2005WR004270

Muller, R. A. (1977). Synoptic climatology for environmental baseline analysis—New Orleans. *Journal of Applied Meteorology, 16*(1), 20–33. https://doi.org/10.1175/1520-0450(1977)016%3c0020:ASCFEB%3e2.0.CO;2

Reichstein, M., Camps-Valls, G., Stevens, B., Jung, M., Denzler, J., Carvalhais, N., & Prabhat. (2019). Deep learning and process understanding for data-driven earth system science. *Nature,* 566, 195–204. https://doi.org/10.1038/s41586-019-0912-1

Sheridan, S. C. (2002). The redevelopment of a weather-type classification scheme for North America. *International Journal of Climatology, 22*(1), 51–68. https://doi.org/10.1002/joc.709

Yarnal, B. (1994). *Synoptic climatology in environmental analysis: A primer* (256 pp). Wiley. 978-0471947967.

Yarnal, B., Comrie, A. C., Frakes, B., & Brown, D. P. (2001). Developments and prospects in synoptic climatology. *International Journal of Climatology, 21*(15), 1923–1950. https://doi.org/10.1002/joc.675

Chapter 11
Climate Classification

Abstract Climate classification is useful for characterizing the atmospheric environment, with the variables included depending on the classification's purpose. Several classification systems have been devised, including Thornthwaite's water-balance-based system, but the Köppen system has been employed frequently and will be used here. The three-letter abbreviation system for Köppen types has a capitalized first letter that describes the annual temperature regime, except for the "B" types described below. "A" (Tropical) climates are warm year-round with no cold season. "C" (or mesothermal or temperate) climates occur in subtropical to mid-latitude areas and have winters that can be cold but not continuously cold. "D" (or microthermal) climates also occur in mid-latitudes but have longer, harsher, uninterrupted winters, with summers that are not continuously hot. "E" (or Arctic or Antarctic) climates are harsher than D climates and are unconducive to tree growth. B (Arid) climates are dry, regardless of the annual temperature. "Dry" is determined by the extent to which potential evapotranspiration (PE)—the amount of evapotranspiration that could have occurred if water were freely available—exceeds its precipitation. The Köppen types are extended over the oceans here and in subsequent chapters to reveal truly global climatic types.

Keywords A (Tropical) climate · B (Arid) climate · Biome · C (mesothermal) climate · D (microthermal) climate · Deficit · E (Arctic/Antarctic) climates · Empirical classification system · Genetic classification system · Köppen climate types · Potential evapotranspiration (PE) · Surplus · Thornthwaite classification system · Water balance diagram

Purpose

All of the processes that shape the natural environment acting in concert, from the unequal arrival of energy from the Sun across the Earth to the pressure patterns that this inequity produces, to the *Coriolis effect*, to the major semi-permanent features, to the secondary circulation features embedded upon the general circulation in both the atmosphere and the oceans, to the energy, matter, and momentum exchanges between

R. V. Rohli et al., *Atmospheric and Oceanic Circulation*,
https://doi.org/10.1007/978-3-031-66142-6_11

the atmosphere and the ocean, and between the atmosphere and the lithosphere, and between the atmosphere and the *cryosphere*, leave us with the world in the condition that we know it. This world, in turn, with all its diversity, supports an amazing diversity of life. How can we even begin to describe the types of natural environments that exist, when it is all so complex and every place is unique? This is quite a chore, and by necessity, it must be an exercise in oversimplification.

Yet climate classification is important as a means of summarizing the weather patterns that emerge over long periods of time. We classify climates to simplify things, such as the range of certain types of environmental impacts that might be expected in a certain place. For purposes such as this book, it would be impossible to study the climate of every place in the world. But yet it is also useless to examine sections of the world at a time that have climates that are too different to study as a homogeneous unit. Therefore, it is important to find the "sweet spot" of having a sufficient number of climatic types to be manageable but yet not so few types that places with much different climates are lumped together.

Classification is a characteristic of the other sciences, and in our everyday lives, too. In everyday speech, we classify a number of different objects as 'chairs' even though they look different as a recliner, executive chair, high chair, and throne. So classification is important for summarization and simplification, and climate classification is no exception.

One gaping hole in the way we classify climates, until recently, has been that the oceans are ignored, an unfortunate reality since over 70 percent of Earth's surface is ocean. Availability of new data sets that make use of amalgamations of various types of weather observations, from weather buoys and ship logs to satellites, and spatial interpolation procedures that transform those observations into gridded format, have now been assembled for so much of the Earth for many decades that is is now possible to identify climatic types over the entire Earth. We will no longer ignore the climatic types over the oceans! One potential growth area is to use ocean climate classification to understand marine biome distributions, from microscopic to cetaceans.

Variables and Thresholds

The decision on which variables to include in a climate classification should depend on the purpose of the classification. For example: As described in Chap. 10, if the purpose of classifying is to study potential sites for wind energy generation, then wind data should form the basis of the classification. Sites have ample mean wind speeds year-round and little variability about that mean would be in the "most desirable" category. Other sites may have mean wind speeds that are adequate, but too much variability to make for economically-viable and convenient energy sources in the current energy landscape. Still others may have adequate speeds and low variability only for some months, and so on. The variables that would be included would be much different if the purpose were for a vegetation cover study, or some other application.

Regardless of the purpose of the classification, data availability, quality, and reliability play an important role in the decisions about the classification. *Temperature* and *precipitation* data are readily available, but not usually sunshine variables, for example. Of course, data from official sources are considered more reliable than other data.

The selection of thresholds for separating climate types also depends on the purpose of the classification. For example, if the purpose is agricultural, it might make sense to separate climatic types at the 0 °C (32 °F) **isotherm** (i.e., a line on a map touching all places that have a given temperature—in this case 0 °C; 32 °F).

Classification Systems: General

There are two types of classification systems for climate. In a **genetic classification system**, the climate types are determined by the major forcing feature that exists. For example, all places that are influenced by the equator-to-pole circulation on the western sides of *subtropical anticyclones*, might be in the same climatic type, while all climates impacted by the *inter-tropical convergence zone* (ITCZ) year-round might be in another category. In these kinds of classifications, it is not the actual temperatures or precipitation totals that need to be similar for all places in the same climatic type. Instead, the processes are the determinant of what climate type exists, with the understanding that those processes will have different impacts on weather, including temperature, precipitation totals, and other weather elements, from place to place.

The second type of classification system is the **empirical classification system**. In empirical systems, places that have similar values of the weather elements—temperature, precipitation, or other variables—in similar seasons would have the same climate type. Table 11.1 summarizes the characteristics of genetic versus empirical classification systems. Two empirical systems are explored in more detail here, for their use in understanding regional and global climate types. However, the genetic mechanisms that produce these characteristic climate types will be explained along with the types, in Parts V and VI.

Thornthwaite Classification System

The **Thornthwaite system** was designed by C. Warren Thornthwaite in the mid-twentieth century. Thornthwaite believed that climates should be classified based on the extent to which water is abundant or limiting on the surface, for *evapotranspiration*, storage between soil particles, runoff to streams, and deep infiltration to replenish groundwater tables. Thus, the variables and thresholds that separate one Thornthwaite type from another are based on the surface **water balance**. More specifically, how much water (in centimeters of precipitation equivalent), if any, is left at the surface, for a given period, after water that moves up to the atmosphere as evapotranspiration is subtracted from the precipitation that fell during that period?

Table 11.1 Comparison of genetic and empirical climate classification systems

Characteristics	Genetic	Empirical
Basis	Based on processes—dynamics—climates that are grouped together are influenced by the same physical mechanisms	Based on observations of climate elements—sites that have similar data are grouped in the same climate type
Examples	ITCZ climates, monsoon climates, east coast mid-latitude climates, etc.	Climates where the average annual temperature is between x and y and the average precipitation > z inches per year
Advantage	Explanation of the causes of climate	"Cookbook" style of classification with consistent application of the rules across space
Problems	Hard to classify climate unless the dynamics affecting the surface climate are understood. (i.e., only reveals that it is a monsoon climate, not how much precipitation falls) Regions often overlap and shift (but is this really a problem?)	Does not address variability and the dynamics of the climate; it is a static classification of a dynamic system Does not give any explanation of why the climate is the way it is
Uses	Not used commonly	Example: Thornthwaite system, Köppen climate types

The calculations made to model the **surplus** of water available to go into runoff to streams, or **deficit** of water—the water that would have been evaporated, restored to the soil storage, or flowed downhill as runoff to streams—in Thornthwaite's classification system are based on the physical interactions between local energy (i.e., temperature) and moisture (i.e., precipitation) availability. The equations are empirical, meaning that they are not based on modeling the physics of the energy availability as it relates to the uptake of water through evapotranspiration, infiltration into the soil, or hydrologic flow. However, others have generated physically-based models that rival those of Thornthwaite. Nevertheless, the Thornthwaite approach has become favored because its empirical nature leaves it less dependent on having accurate data available for a wide variety of variables. This applicability makes it a useful system for agriculture, where accurate weather station-measured data may be distant.

In the Thornthwaite system, each category has a four-letter abbreviation based on the amount and seasonality of abundance of water in the surface water balance, considering precipitation and evapotranspiration, annually and seasonally. The system works best locally and regionally. However, for wider-scale interests, such as global atmospheric and oceanic circulation, it may be useful to identify another system.

The Köppen System

In the late 1800s, the Russian-German botanist Vladimir Köppen (pronounced "KER-pen") sought to design a climatic classification for explaining the natural world patterns of broad-scale vegetation—called **biomes**—based on climate. So he set up the boundaries that separate climatic types to conform to the known, at that time, biomes around the world. Then, because Köppen reasoned that it is largely temperature and precipitation means and extremes that limit vegetative growth, he devised mathematical formulas that characterize the ranges of mean and extremes of monthly temperature and monthly precipitation for each of the climatic types, and therefore biomes. Over the century and a half since its introduction, the system of *Köppen climate types* has undergone several revisions in its mathematical formulas, as imperfections in the overlay of climatic types with biomes became known and as more became known about the biotic realms.

Because the same vegetation type and climate is likely to produce a characteristic soil type, categories of biomes, soil types, and climate types correspond very well to each other in the Köppen system.

Köppen's system has a three-letter abbreviation system. The first letter is always capitalized and, except for the "B" type described below and driven by moisture availability, tells about the annual temperature. An overview of each of the five major types is below.

A climates are tropical, which, by definition, indicates that they are warm year-round; there is no cold season. The A climates occupy far more area over the oceans than over the land area, as relatively little land exists at the equatorial latitudes. A climates should not be confused with *A air masses*. As has already been shown, A (for "Arctic" or "Antarctic") air masses are the coldest, but A climate types are tropical.

C climates in the Köppen system are found in subtropical to mid-latitude areas and are also known as **mesothermal**—medium amount of heat—or **temperate**. Winters in C climates can be cold, but they are not continuously cold for weeks or months at a time. These regions are pleasant for human activities; more people live in C climates than in any other climatic type. C climates are also very common over the oceans. Again, the C climate can be easily confused with the lower-case "c" air mass types. The latter indicate dry air masses, but upper-case "C" climate types indicate mesothermal.

D climates are also found in mid-latitude areas, but these climates have longer, harsher, uninterrupted winters, with summers that are warm to hot but not continuously hot for weeks to months at a time. They are also known as **microthermal**—a small amount of heat. D climates are much less prominent over the ocean than over land because the moderating influences of the ocean restrict the amount of ocean surface that experiences harsh, uninterrupted winters yet warm to hot summers.

E climates are **Arctic** or **Antarctic**. The E climates are distinguished from D climates over land by the **treeline**, which separates the two. In other words, E climates are too cold to support tree growth, while trees can cover terrestrial D climates. E climates are almost non-existent over the ocean because of the moderating influences of these water bodies. Beware that E air masses are the mildest ("equatorial"), but E climate types are the most severe regarding temperature.

B climates are **arid** (or dry), regardless of what the annual temperature is like. According to Köppen and others, "dry" is determined not based merely on precipitation but instead based on whether the place's **potential evapotranspiration**—the amount of evaporation that could have occurred if water were freely available) exceeds its precipitation in most months. For example, Las Vegas receives about as much precipitation as northern Scandinavia, but Las Vegas has a B climate while Scandinavia does not. This difference is because potential evapotranspiration is so low in Scandinavia, where the Sun angle is low and summer is so short; Scandinavia appears to be humid, but Las Vegas appears arid. The B climate occupies more of Earth's terrestrial area than any of the other five major types. In fact, contrary to popular assumptions, the B climates are surprisingly common over the ocean; precipitation does not necessarily fall frequently over an ocean, especially beneath the *gyres* in the subtropical oceans.

The second letter of the Köppen classification has different meanings for the different first-letter types. For A, C, and D climates, the second letter is either "f," "m," "s," or "w." Note that all of these are lower-case because they are adjectives—not nouns, which are capitalized—in German. The "f" represents "wet year-round." So, for example, the **Tropical Rain Forest** (Af) **climate**, is the wettest of all the A climates. The Af climates are far more extensive over the ocean than over land (Fig. 11.1). The Cf and Df families of climates are also wet year-round and occupy large sections of the Earth, especially over land.

The other three possible second letters for A, C, and D climates all designate dryness during some of the year, but not with sufficient aridity to plunge the climate into a B type. The second letter of "m" occurs only in the A type, to represent a tropical monsoon effect in which there is a short dry season. The **Tropical Monsoon** (Am) **climate** is the second-wettest A climate and, like the Af, is far more extensive over oceans than terrestrially (Fig. 11.1). Contrary to what its name might imply, Am climates are usually unrelated to monsoon circulations, but instead the short dry season results because of distance from the ITCZ for a few months. Yet the natural vegetation of Am climates is tropical rain forest because the water balance deficits are sufficiently small so as to allow this lush vegetation to survive the brief dry season. No Cm or Dm climates exist on Earth, as the latitudes that have mesothermal or microthermal climates are either arid for all or much of the year or experience some precipitation year-round from *fronts*.

The second letter of "s" is used to designate dry summers in the D and especially C general climate types. These are not extensive, because enhancement of precipitation through *convection* makes most places wetter in summer than winter. However, exceptions do exist, particularly in the mesothermal, or C, climates, where places having a **Mediterranean** (Cs) **climate** are some of the most desirable tourist destinations. Ds climates exist but are rare.

Finally, a second letter of "w" represents dry winter conditions. In the **Tropical Wet-dry** (Aw) **climate** or **Tropical Savanna** (Aw) **climate**, about half of the year is

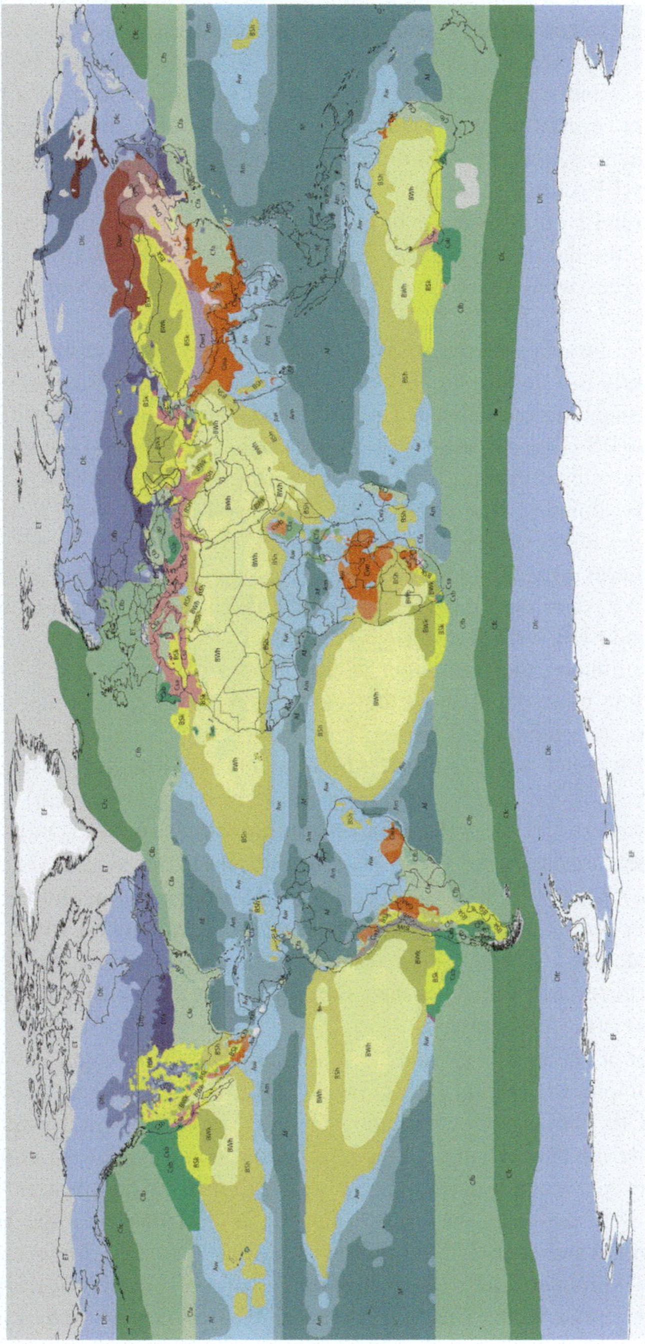

Fig. 11.1 Köppen climatic types around the world. *Source* Modified from Reynolds et al. (2021)

wet and half is dry, as defined by the potential evapotranspiration concept. Winters can also be dry in C or D climates, including the **Mesothermal Winter Dry** (Cw) **climates**, and the **Microthermal Winter Dry** (Dw) **climates**.

For E climates, the second letter is either “T” for **Tundra climate** or “F” for **Ice Cap climate**. Tundra (ET) climates are far less severe, because at least low-to-the-ground tundra vegetation can grow there. These types of vegetation range from bushes down to lichens, which is a combination of mosses and algae stuck to rocks.

In the Ice Cap (EF) climates, however, as not even lichens can grow there, but life does indeed exist in EF climates in the form of bacteria, viruses, and other microbes. Climatologically, the difference between ET and EF is that ET can have a frost in any month of the year, while EF climates have average Temperatures below Freezing in all months.

For B climates, the second letter is “W” for **True Desert** (BW) **climate** and “S” for the less severely dry **Semi-arid** (BS; sometimes known as the Steppe) **climate** type, with the second letter capitalized. The difference is simply in the intensity of the aridity. Typically, BS climates are transitional areas between humid environments and deserts. They often surround the core of the driest areas—the BW climates. Both BS and BW climates exist over the ocean as well as over land. B climates can be further subcategorized with a third letter—“h” for “hot” and “k” for “cold,” with the 54° isotherm separating the two. Thus, four types of B climates exist: **True Desert—hot** (BWh) **climate**, **True Desert—cold** (BWk) **climate**, **Semi-arid—hot** (BSh) **climate**, and **Semi-arid—cold** (BSk) **climate**.

For C and D climates, the third letter is either lower-case “a,” “b,” “c,” or “d,” with each providing more detail about the summer temperature. Note that none of these third letters are capitalized. The “a” represents “hot” summers, such as in the U.S. Southeast, which has a **Humid Subtropical** (Cfa) **climate** type that is hot in summer, wet year-round, and too cold in winter to be an A (or tropical) climate. Most of the Midwestern U.S. has a **Humid Continental—hot summer** (Dfa) **climate** type. There, winters are harsh (i.e., microthermal), and conditions are wet year-round with a hot summer. Much of inland California has a **Mediterranean—hot summer** (Csa) **climate** type, as it is hot and dry in summer there. Parts of inland India, south-central Africa, Mexico, Central America, and Brazil, especially where elevation causes winters to be too cold to be classified as tropical, have a **Mesothermal Winter Dry—hot summer** (Cwa) **climate**.

The third letter of “b” in the “C” and “D” climates is used to designate a “warm” summer. For example, the coastal Pacific northwestern U.S. and much of western Europe have a Cfb climate, which is known as the **Marine West Coast** (Cfb) **climate**. Decoding the abbreviations, this designation indicates that the climate is mesothermal, wet year-round, and having a warm summer. The northern tier of the Midwestern U.S. and adjacent eastern Canada is dominated by the **Humid Continental—warm summer** (Dfb) **climate** type. Much of coastal California has a **Mediterranean—warm summer** (Csb) **climate**; it is mesothermal, and warm (not hot) and dry in summer there. The **Mesothermal Winter Dry—warm summer**

(Cwb) **climate** is rare but occurs in small areas within southeastern Asia, south-central Africa, Central America, and Brazil where elevation prevents summers from being hot.

If the third letter of the C and D climate types is "c," summers are "mild." It is rare for C climates over the continents to have a third letter of "c," since a C climate will almost always have a hot or warm summer. One of the few exceptions is in much of Iceland, which has a **Humid Subtropical—mild summer** (Cfc) **climate**. The warm *North Atlantic Drift* keeps much of Iceland in a C climate despite its high-latitude location, but the oceans moderate summer temperatures too, keeping it cool rather than hot or even warm. The Cfc climatic type is much more common over the open oceans, where winter cold and summer heat are both moderated.

Much more common over land is the **Humid Continental—mild summer** (Dfc) **climate**, which is more popularly known as a **Subarctic** (Dfc) **climate**. Places with strong monsoonal effects are dry in winter, with the Dwc designation used for such locations, which are almost exclusively in northeastern Asia. D climates over land that have a third-letter of "c" include the **taiga** of Siberia or the **boreal forest** of North America. Boreal/taiga forest is the largest type of forest biome—far more extensive than tropical rain forests or mid-latitude forests. However, the **Humid Continental—cool summer** (Dfd) **climate** also known as a Subarctic (Dfd) climate is rare. These are the places with chilly summers, yet winters that are not cold enough to prevent trees from existing. Thus, even Dfd climates are equatorward of the treeline.

The world distribution of Köppen climate types (Fig. 11.1) shows patterns that recur based on latitude and position regarding marine and continental influence, including *air masses*. The major factors that govern weather and climate that were covered in Chap. 2—solar radiation, *humidity*, and atmospheric stability—determine the energy and matter available in the atmosphere, and the general circulation of the atmosphere that was described in Chap. 3 and ocean that was described in Chap. 5 then carry the heat, moisture, and stability features to their destinations. The result is a characteristic set of features for each location around the Earth—unique at each place but also similar across space—that make up the climate of each place. Because places near each other are more likely to have similar features both locally and via the dynamic forcing mechanisms that drive the atmospheric circulation, the patterns of climate types repeat around the world and tend to be similar across some distance.

Figure 11.1 is the first modern, detailed map that includes climatic types over the oceans as well as the continents, calculated according to the revised Köppen system. Now that over 30 years of satellite-sensed temperature data are available, the mapping of marine climatic types is possible, accurate, and important. For example, the ranges of marine species could be mapped and overlain by climatic type, and then climate models can be used to predict the future extents of the climatic types, to assess the prospects for future extinctions, fisheries industries, migrations, and other *primary impacts* and *secondary impacts*.

The major patterns that recur globally can be summarized based on the map. The Af, or Tropical Rain Forest, climate extends in a near-continuous belt along the Equator. The Am, or Tropical Monsoon climate, and the Aw, or Tropical Wet-dry or Tropical Savanna climate, generally surround the Af type. These are the dark, medium, and light shades of blue in Fig. 11.1, respectively.

The B climates tend to be in the subtropical latitudes and/or in the middle of the largest land masses (Fig. 11.1). They also appear prominently off the western coasts of continents in the subtropics, where the influence of the subtropical anticyclones, reinforced by coastal oceanic *upwelling*, supports a *stable atmosphere*, suppresses precipitation and causes *evapotranspiration* to exceed precipitation, thereby generating an arid climate type. The marine area covered by B climates is truly and surprisingly extensive.

The C climates are found in subtropical to middle latitudes, and in the high latitudes that are either over the ocean or influenced strongly by the moderating influence of the ocean. The Humid Subtropical (Cfa) climate is on the eastern sides of continents in the subtropics. Not surprisingly, the Marine West Coast (Cfb) climate are on the western sides of continents, often in the middle to high latitudes poleward of the Cs climates.

The D climates are largely a Northern Hemisphere phenomenon, because little of the Southern Hemisphere supports large landmasses at the latitudes that would allow for harsh winters without being polar winters. E climates are found over Antarctica, Greenland, and at a more local scale, the tops of the highest mountains elsewhere.

In the next series of chapters, the major features of the atmospheric and oceanic circulation, along with their resulting impact on the Köppen types and other features of the environment, will be examined in more detail, one major world region at a time. However, no attempt is made to provide details on every climate around the world. Small sections of nations or ocean basins may have anomalous circulations because of elevation or other local-scale anomalies. Instead, the focus is on broad-scale patterns that contribute to regional climates. Pay careful attention to the forcing mechanisms and the way that they influence different parts of the world in the same way, and similar parts of the world in different ways.

Further Reading

Allen, R. G., Pereira, L. S., Raes, D., & Smith, M. (1998). Crop evapotranspiration—Guidelines for computing crop water requirements. *FAO Irrigation and Drainage Paper 56*. Food and Agriculture Organization.

Köppen, W. (1900). Versuch eine Klassifikation der Klimate vorzugsweise nach ihren Beziehungen zur Pflanzwelt [Attempt of a classification of the climate primarily based on the relationship with the flora]. *Geographische Zeitschrift, 6*, 593–611.

Reynolds, S. J., Rohli, R. V., Johnson, J. K., Waylen, P. R., & Francek, M. A. (2021). *Exploring physical geography* (Third edition, 736 pp). The McGraw-Hill Companies, Inc. 978–1–260–36499–6.

Rohli, R. V., Joyner, T. A., Reynolds, S. J., Shaw, C., & Vázquez, J. R. (2015a). Globally extended Köppen-Geiger climate classification and temporal shifts in terrestrial climatic types. *Physical Geography, 36*(2), 158–175. https://doi.org/10.1080/02723646.2015.1016382

Rohli, R. V., Joyner, T. A., Reynolds, S. J., & Ballinger, T. J. (2015b). Overlap of global Koppen-Geiger climates, biomes, and soil orders. *Physical Geography, 36*(2), 142–157. https://doi.org/10.1080/02723646.2015.1016384

Thornthwaite, C. W. (1948). An approach toward a rational classification of climate. *Geographical Review, 38*(1), 55–94.

Part V
Regional Responses in the Extratropical Northern Hemisphere

Chapter 12
United States, Canada, and Adjacent Oceans (Theory)

Abstract The U.S. and Canada stretch across a wide range of latitudes and climatic types. The multiple north–south-oriented mountain ranges in the West and lower mountains in the East provide opportunities for orographic precipitation, double rain shadow effects, and interaction of strikingly different air masses that produce severe mid-latitude weather, including tornadoes and supercell and multicell thunderstorms. The mid-latitude westerlies, with their embedded Rossby wave ridges and troughs and the polar front jet stream, provide the steering circulation for much of the region. The Bermuda-Azores High drives the North Atlantic Gyre, with the Gulf Stream and spin-off Loop Current forming its western edge. The Bermuda-Azores High brings maritime tropical air masses to the U.S. east of the Rocky Mountains and to southeastern Canada. Preferred mid-latitude wave cyclone tracks include the leeward side of the Rocky Mountains and baroclinic zones associated with winter land-sea temperature differences in the northern Gulf of Mexico and near the North Carolina coast. Ocean–atmosphere teleconnections including the Pacific-North American pattern are important causes of atmospheric variability in this region affected so strongly by both continentality and maritime influence.

Keywords Alaska Current · Cumulonimbus cloud · Double rain shadow · Dry line · Gust front · Gulf of Mexico low-level jet (LLJ) · Loop Current · Mesocyclone · Mid- and South Atlantic Bight · Mid-latitude wave cyclone preferred tracks · Multicell thunderstorm · North Atlantic and North Pacific gyres · Squall line · Storm surge · Supercell · Thermal low · Thunderstorm types · Tornado and Dixie Alley · Vertical wind shear

General Geographic Characteristics

The region comprising the U.S. and Canada contains wide extremes of latitude that support remarkably diverse *temperature* and moisture regimes. The latitudes extend almost from the tropics in southern Florida and Texas to deep within the *Arctic Circle*,

R. V. Rohli et al., *Atmospheric and Oceanic Circulation*,
https://doi.org/10.1007/978-3-031-66142-6_12

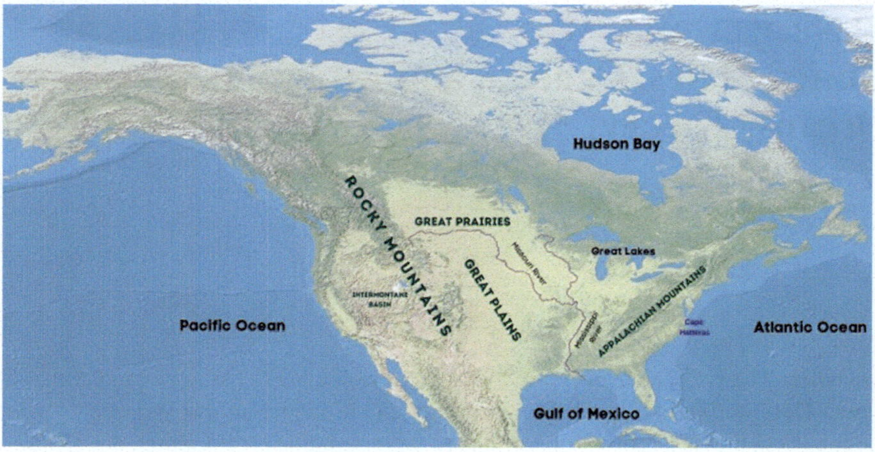

Fig. 12.1 Selected physiographic features of the U.S. and Canada. *Source* Geaux Geospatial

while longitudes span from approximately 180°—near the International Date Line—in Alaska's Aleutian Islands to about 30° W (Fig. 12.1). The world's largest island—Greenland—may be considered with this region because it is similar environmentally to Canada's northeasternmost extent, yet may also be considered part of Europe due to its political and historical ties to Denmark and northern Europe. Nevertheless, here it is included with the Arctic world and covered in Chap. 21. Likewise, the atmospheric and oceanic circulation associated with other islands that are technically part of the U.S. but are distant from the mainland, including Puerto Rico and the Hawaiian Islands, are included in other chapters, particularly Latin America and the Caribbean (Chap. 19), and Australia and the Pacific (Chap. 20), respectively.

Topographic Influences

The topography of North America provides "site" characteristics that influence atmospheric and oceanic circulation, and therefore climatic regimes. The Pacific coastline is mostly straight, with some notable exceptions that create unique climate niches, such as San Francisco Bay in California and Puget Sound in the northwestern U.S. state of Washington. Formidable coastal hills and mountain systems parallel the coastline and form a barrier to the humid air moderated by the Pacific. Thus, only a narrow strip of marine-influenced terrain occurs. A classic *orographic lifting* effect occurs as humid Pacific air moving eastward in the *mid-latitude westerlies* encounters the mountain barriers. The result is that the western sides of the mountain axes are much wetter than the eastern sides. A few hundred kilometers eastward, another, higher north-to-south mountain system serves as an even more formidable barrier to inland penetration of the humid, moderate Pacific air. These mountains are known by different names in different areas, but the most prominent are the Cascades of British Columbia, Washington, and Oregon, and the Sierra Nevadas of California and Nevada.

To the *leeward* side of those higher mountains lies the broad Intermontane Basin. This is a very prominent valley peppered with generally low, disorganized mountain belts, but elevations even in the valley remain high above sea level. The Intermontane Basin includes all of the U.S. state of Nevada and parts of all of the other western U.S. states.

East of the Intermontane Basin but still in the western U.S. and Canada is another north–south-running mountain range—the Rocky Mountains. The Rockies are even higher than the mountains farther west. They are tall enough to initiate *friction* with the steering flow of air passing overhead and disrupt the *Rossby wave* flow in the mid-latitude westerlies. They also create vertical compression of air on their *windward* sides and stretching on the leeward sides that contribute to storm formation.

East of the Rockies are the Great Plains of the U.S.A. and the Great Prairies of south-central Canada. From this region eastward, the Gulf of Mexico is the primary source of moisture for most of the U.S.A. and Canada. The leeward side of the Rockies forms the headwaters of the the vast Mississippi-Missouri River basin, which drains over 41 percent of the U.S.A. and a tiny part of central Canada, with its mouth in one of the world's great, and most rapidly sinking, deltas at the "boot" of Louisiana. The fresh water output from the Mississippi-Missouri River basin is abundant enough to affect the circulation of the Gulf of Mexico, especially when it consists of snowmelt from the northern U.S.A. that mixes with the warm Gulf waters.

The eastern headwaters of the Mississippi-Missouri River basin are on the windward slopes of the Appalachian Mountains, which run in a generally northeast-to-southwest direction. Because of their discontinuous extent and lower elevations, the Appalachians and the local names for the ranges that comprise them, such as the White Mountains of New Hampshire, Green Mountains of Vermont, Berkshires in Massachusetts, Adirondacks and Catskills in New York, Poconos of Pennsylvania, Allegheny in Pennsylvania and Maryland, Blue Ridge from Pennsylvania to Georgia, and Smokies in Tennessee and North Carolina, exert less influence on the regional climates. For instance, *mid-latitude wave cyclones* easily cross these mountains and no pronounced *rain shadow* is produced on their eastern slopes. The Appalachians do cause some local environmental effects, however. For example, the southern slopes of the Appalachians receive more *precipitation* than other areas because of the orographic effect from the *advection* of warm, wet, and *unstable atmosphere* Gulf of Mexico air, and the colder peaks experience more snow. There are also narrow zones, based largely on elevation, where the temperature, orographic effect, and topography all contribute to hazardous freezing rain and sleet during mid-latitude wave cyclone passages.

The orientation of the north–south-running Rocky and Appalachian Mountain chains separated by a vast, uninterrupted, relatively flat Great Plains and Prairies (see again Fig. 12.1) makes it easy for various types of air masses to interact in the middle of North America. The Great Plains and Prairies act as a channel for both tropical and polar *air masses* to stray far poleward and equatorward, respectively, of their *source regions*. This interaction of different air masses can produce severe and extreme weather. Most notably, *cP air masses* or *cA air masses* can move equatorward into a *trough* on the ridge-to-trough side of the Rossby waves, and *mT air masses* can extend poleward, steered by the trough-to-ridge side of the Rossby waves.

The result is frequent interaction between very cold, dry air masses and warm, humid air masses near the mid-section of the continent, creating some of the largest air mass temperature contrasts on Earth. The frequent collision of these very different air masses makes this area between the Rocky and Appalachian Mountains susceptible to *cyclogenesis*, tornadoes, and other severe weather. The area between south-central Canada through the Great Plains to central Texas—known as **Tornado Alley**—experiences the greatest frequency of severe weather on Earth. A secondary area of very high tornado frequency stretches in a swath from Texas eastward to Georgia—a region known as **Dixie Alley**.

The high-latitude part of the U.S. and Canada also has characteristic topographic influences. However, these are dominated by coastal orientations that affect the ocean currents and by local atmospheric features caused by lakes that are relics of past glaciations. *Sea ice* is prominent near Alaska in winter. This sea ice limits heat transfer between the warm ocean and adjacent atmosphere, similar to the influence of land. This ice-induced continental effect aids the development of very low temperatures in Alaska and in Canada's Yukon, Northwest, and Nunavut territories.

The wide array of climates that result from the atmospheric and oceanic circulation features of the U.S. and Canada are partially a result of the effect of *continentality*. Summers are oppressively hot and winters are brutally cold for a large swath of the central North American continent. But unlike other parts of the world where continentality usually comes with dryness, the low elevations between the Gulf of Mexico and the vast flatlands in the middle part of the continent allow mT air masses to penetrate far into the continental interior, bringing moisture inland to places that might otherwise be a desert. This significant moisture *advection* not only enhances precipitation, but it also somewhat mitigates extreme temperatures normally associated with continentality, by allowing cloud cover that absorbs, scatters, and reflects *shortwave radiation* ($K\downarrow$) before it reaches the surface, suppressing summer daytime high temperatures. The moisture also minimizes *outgoing longwave radiation* lost to space, which prevents morning low winter temperatures from dipping even lower. Still, though, the range of conditions, including temperature, brought by the atmospheric and oceanic circulation is so broad that all of the major *Köppen climate types*, from *A climates* (tropical) to *E climates* (polar) are represented in the U.S. and/or Canada.

General Circulation of the Atmosphere

The general circulation of the atmosphere and oceans represents the "situation" as described in Chap. 1. North America and Eurasia are the only two large landmasses in the mid-latitudes, and therefore the mid-latitude weather systems are the dominant "situation" features.

The dominant general circulation feature for much of the U.S. and Canada is the mid-latitude westerlies, which are embedded in the Rossby waves and *circumpolar vortex* associated with the *polar front* and *polar front jet stream*, which in turn steer mid-latitude wave cyclones. Extending between approximately 30° and 60° N, these circulation features occur in a transition zone between the cold and dry polar latitudes and the warm and wet tropics.

As winter approaches, the mid-latitude westerlies extend farther southward with the entire polar front, while maintaining their general west-to-east flow. In some cases, winter upper-level troughs may protrude as far south as the Gulf of Mexico. Likewise, as summer approaches, the mid-latitude westerlies are confined increasingly northward in their general west-to-east flow. Thus, most of the U.S. and Canada experiences weather associated with *cold fronts* and *warm fronts* that trail from the mid-latitude wave cyclone for at least part of the year.

The winter flow pattern is characterized by a prominent *ridge* in the mid-latitude westerlies over western North America in association with the cold landmass and an "anchoring" in place by dynamic conditions associated with the Rocky Mountains. This ridge, paired with a downwind trough over the southeastern U.S., is the signature feature of the positive mode of the *Pacific-North American* (PNA) *pattern*. Notice the western North American ridge in Fig. 12.2, which shows longwave flow for the Northern Hemisphere, moving around the North Pole, which is shown in the center of the images. The circumpolar vortex is easily discernible by the Rossby wave flow.

During the warm season, the circumpolar vortex shrinks as temperatures increase across the high latitudes. The polar front retreats to a position north of the U.S.- Canadian border with embedded mid-latitude wave cyclones primarily affecting Canada and Alaska. The northern tier of U.S. states experiences trailing cold fronts generated by the weaker frontal cyclones that occur during these warmer times.

A second dominant general circulation feature for the U.S. and Canada east of the Rockies, is the *Bermuda-Azores High*. This pressure feature is the semi-permanent North Atlantic *subtropical anticyclone* over the subtropical *North Atlantic Gyre* that drives much of the surface oceanic circulation and water vapor advection into the eastern U.S.A. and Canada. A northwestward migration and/or a substantial increase in size/intensity of the anticyclone blocks migratory mid-latitude wave cyclones moving from west to east in the westerly wind flow across the continent. When the Bermuda-Azores High is displaced toward North America and strong, the interior U.S. may experience abundant precipitation as cyclones stall and dissipate, while the Atlantic Coast experiences *subsidence-* (or sinking-) induced drought from the anticyclone.

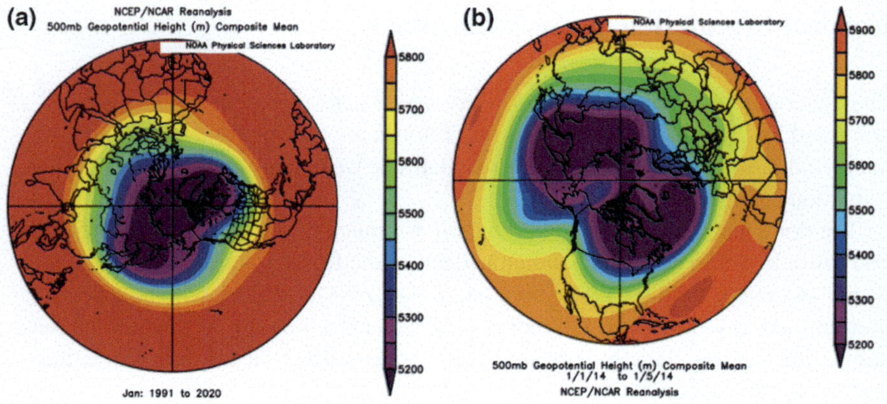

Fig. 12.2 The Northern Hemisphere circumpolar vortex, shown as **a** the height of the 500-mb surface, about halfway up in the atmosphere, averaged for all Januarys over the period 1991–2020. *Source* NOAA. https://psl.noaa.gov/cgi-bin/data/composites/printpage.pl, and **b** the 500-mb height surface averaged for 1–5 January 2014, when much more amplified ridges and troughs existed. *Source* NOAA. https://psl.noaa.gov/data/composites/day/

Both the Bermuda-Azores High and *Hawaiian-Pacific High* in the Pacific Ocean exert their strongest and most extensive influence in summer. Storms are deflected to the north (in the Rossby waves) on the western sides of these clockwise-and-outward-rotating semi-permanent anticyclones, and to the south (in the trade winds) on the eastern sides of these anticyclones.

The subtropical anticyclones support strong sinking all the way to the surface, and therefore more surface *divergence*, on their eastern sides (the western sides of continents) than on the western (eastern continental) sides. Their main influence is tilted vertically downward moving from west to east. This results in the eastern side of the subtropical anticyclone (western side of the continent) inducing sinking motions which suppress cloud development. On the western side of the subtropical anticyclone (eastern side of continent) the influence is minimized, which allows the development of precipitation and thunderstorms. The tilting of the influence of sinking is caused by the stabilizing influence of cold ocean currents on the eastern sides of the subtropical anticyclones (western sides of the continent), and warm currents occupying the surface on the western sides of the anticyclone. The warm currents encourage low-level atmospheric instability which promotes cloud and precipitation processes. The situation occurs not just in North America, but for other ocean basins and adjacent continents as well. On the southward side of the clockwise and outward flow associated with the subtropical anticyclones is generally east-to-west circulation. This is manifested as the *northeast trades*, which affect Hawai'i and the southernmost United States.

In the higher latitudes, around 60° N, the semi-permanent *subpolar lows*—the *Aleutian Low* and the *Icelandic Low*—churn over the Pacific and Atlantic Oceans, respectively. Their low-pressure centers form in response to the relatively warmer waters for the high latitudes of their locations, which reinforces the rising motion.

Thus, the subpolar lows reach peak intensity in winter when the water is usually warmer than the air at comparable latitudes. The relatively warm waters supply energy and water vapor to the atmosphere, creating unstable conditions and resulting storminess.

Northward of the subpolar lows are the surface *polar easterlies*, which have reversed flow—from west to east—aloft. Most of Alaska and all of northern Canada are affected by this atmospheric circulation cell. Thus, surface storms would generally be approaching from the east, but because of the reversal of the direction of the overlying steering circulation, such storms would have their tops sheared off as they moved. This, along with the general lack of unstable conditions and atmospheric moisture in such cold conditions, makes high-latitude storms generally weak.

Additionally, a small, thermally-induced low-pressure area occurs over the southwestern U.S. and adjacent northwestern Mexico during summer. The system, termed a **thermal low**, is a low-pressure area into which water vapor flows, from both the Gulf of Mexico and the Gulf of California, that squeezes between mountain passes. The situation promotes the convergence of humid air into the low over the arid Southwest of the U.S., to humidify the normally dry *cT air masses* in the area. The thunderstorms that result are termed the summer "monsoon." While not a true monsoon as in other parts of the world where prevailing wind direction changes seasonally, miniature monsoon-like conditions with a small boost in summer rainfall result.

Ocean Currents

Climates of the U.S. and Canada are also strongly influenced by the circulation of the surface waters in the nearby ocean basins. Situated between the large North Atlantic and North Pacific basins, the continent experiences contrasting impacts on its coasts. The Pacific Coast is primarily influenced by the *California Current* (see again Fig. 5.2). Moving clockwise around the Hawaiian-Pacific High, the California Current approaches the coast in the vicinity of the U.S. Pacific Northwest, moves southward parallel to California, and gradually moves away westward as it reaches the Mexican Pacific coast state of Baja California. The water originating in the cooler North Pacific Ocean brings lower sea surface temperatures to the California coast. Additionally, the *Coriolis effect* pulls surface water to the right of its path, away from the coast, inducing cold water from deep in the ocean to rise up to take the water's place—a process known as *upwelling*. As a cold current, the California Current encourages subsidence and an overlying *stable atmosphere* on the Pacific coast. Atmospheric stability results from the cold ocean waters chilling the overlying air, leading to subsidence, which stymies cloud and precipitation development. Farther south in Mexico's Baja California, water may slide northward between the California Current and the coast, especially during *El Niño* events.

Surface ocean currents are driven primarily from *friction*-induced energy transfer with overlying wind. Each ocean basin hosts a subtropical anticyclone centered near

30° latitude in both hemispheres. The clockwise and outward circulation of surface anticyclones in the Northern Hemisphere forms cold currents in the eastern sides of the ocean basins while western sectors experience warm currents. The anticyclonic circulation frictionally drags high-latitude (cold) waters toward lower latitudes through the eastern ocean basins, and low-latitude (warm) waters are moved to higher-latitude regions in the western basins.

The North American Atlantic coast experiences the warm *Gulf Stream* as a result. During winter the Gulf Stream creates a steep thermal boundary in contrast to the cold continent. This *baroclinic zone* supplies energy to the atmosphere, enabling mid-latitude wave cyclones to strengthen as they track along the coast. The resulting storms, known as *nor'easters*, may produce heavy snows through the most densely populated regions of the continent, including the large metropolitan areas of New York City, New York; Washington, District of Columbia; Philadelphia, Pennsylvania; Boston, Massachusetts; and Baltimore, Maryland; and the seven most densely populated states in the U.S. (New Jersey, Rhode Island, Massachusetts, Connecticut, Maryland, Delaware, and New York). A transition zone of sleet and freezing rain occurs between the snowfall regions west of the storm core and the rain region to the east in such situations. The system can paralyze transportation and induce electricity outages across a wide swath of heavily populated areas. While intense winter storms that produce destructive winds and/or intense snowfalls in the U.S. and Canada, including but not limited to nor'easters, are sometimes named by the popular media, although no such "official" names are recognized for these systems.

The Gulf Stream is also partially responsible for frequent and heavy fogs along the northeastern U.S. and southeastern Canada. This occurs as the cold *Labrador Current* replaces the departing Gulf Stream. The position of the cold Labrador Current near the coast stabilizes the low-level atmosphere, leading to coastal fog in the Maritime Provinces (New Brunswick, Nova Scotia, and Prince Edward Island in eastern Canada), Atlantic Provinces (Maritime Provinces + Newfoundland and Labrador), and three of the U.S. New England states (Maine, New Hampshire, and Massachusetts). St. John's, Newfoundland, is among the foggiest cities in the world, in addition to being among the cloudiest, wettest, and snowiest in Canada.

In the Gulf of Mexico, the **Loop Current**, an extension of the *North Equatorial Current* (representing the equatorward portion of the North Atlantic Gyre) that moves into the Caribbean Sea as the Caribbean Current and into eastern Gulf of Mexico, forms the Gulf Stream. The Loop Current is responsible for adding energy into the atmosphere off the Gulf coast of Florida. The Loop Current aids in the development and/or strengthening of *tropical cyclones* by providing surface energy from the warm waters. Loop Current interactions also affect cool season cyclogenesis. When the polar front jet stream dips over the Gulf of Mexico, the energy and moisture from the warm surface current combines with the *positive vorticity* of the upper air circulation to produce extremely strong frontal cyclones. The "Blizzard of the Century" of March 1993 is an example. This "meteorological bomb" was a mid-latitude wave cyclone that moved into the Florida panhandle with *hurricane*-force winds and a substantial

storm surge of rising water on landfall. The system migrated northeastward along the Atlantic coast and generated heavy snowfall from central Georgia through Canada's Atlantic Provinces.

Along the eastern coast of North America, the Gulf Stream moves up the Atlantic coast and eventually veers eastward to become the *North Atlantic Drift* as it approaches Iceland. The shape of the continent allows the Gulf Stream to parallel the coast along most of the southeastern U.S. states, but gradually moves farther out to sea north of Virginia. The coastal fogs that are common in Canada's Atlantic Provinces and the U.S. New England states are often the result of the Labrador Current flowing beneath warmer air that becomes chilled from beneath to the *dew point temperature*, whereupon *condensation* produces *advection fog*. The term "advection fog" refers to any such fog produced by the chilling of warmer air moving over colder water.

Along its western coast, Alaska extends to close proximity to Asia with only the 84-km (52-mi-)wide Bering Strait separating the two continents. The land blocks the North Pacific Ocean from protruding northward. Therefore, the North Pacific has no warm current equivalent to the North Atlantic Drift. Instead, the warm-water *Kuroshio Current* which occupies the western North Pacific Ocean does not extend far northward in the Pacific. The current curves back southward as a cold current along the eastern edge of the ocean basin, which is the Pacific coast of North America.

The **Alaska Current** is technically called a "warm current" even though it carries cold water because it is warmer than the surrounding water. It encircles the northern part of the Pacific basin. After the northern part of the **North Pacific Gyre**—the **North Pacific Current**—crashes into the coast near the U.S./Canadian border, the water turns northward as the Alaska Current and southward as the California Current.

Air Masses

Temperature and humidity patterns in the U.S. and Canada are primarily a function of exposure to air masses from different source regions. The main air masses affecting the continent are shown in Fig. 12.3.

All three types of cold air masses affect the U.S. and Canada. Continental Arctic air masses, or *cA air masses* (sometimes written as just "A"), originate over the sea ice and snow-covered land surfaces north of the Arctic Circle, bringing the coldest and driest weather. They usually remain north of the U.S. but can migrate as far south as the Great Plains and northern Midwest in the U.S. in winter. The somewhat milder *continental polar* (cP) *air masses* carry cold, dry air from the Prairie Provinces southeastward into the eastern U.S., especially during periods of high-amplitude troughs in the Rossby waves. Much cooler and more humid *maritime polar* (mP) *air masses* form over the northern Pacific Ocean, spin off the Hawaiian-Pacific High, and affect the coastal areas of the northwestern U.S. and western Canada. Their effects are confined to the coastal zone because they are modified by orographic lifting when they encounter the coastal mountains. A second source region of mP air is located over the Labrador Current in the Northwest Atlantic. While westerly winds usually

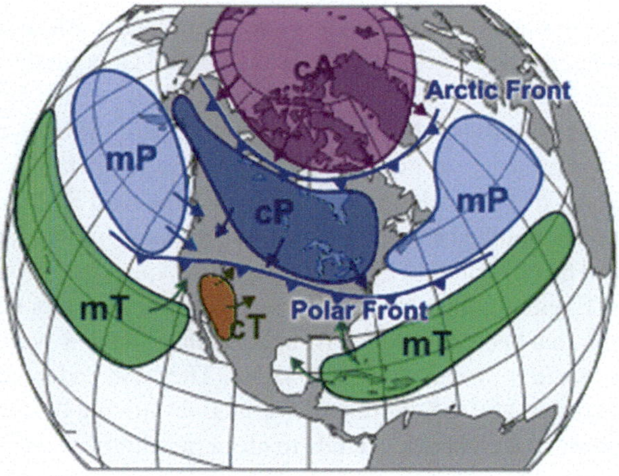

Fig. 12.3 Air mass source regions and paths affecting the U.S. and Canada. *Source* UCAR CoMet

carry this air away from North America, under certain circulation patterns, such as ahead of a strong coastal cyclone, northeasterly winds can move mP air into Quebec, the Atlantic Provinces, and New England.

The U.S. and Canada are also affected by two of the three warm types of air masses, with much wider and more intense influence in summer. Maritime tropical air masses, or *mT air masses*, cover large areas of the southeastern United States. In summer, circulation around the western edge of the Bermuda-Azores High advects this air northward and westward, expanding the region of warm, humid air. On the Pacific Coast, the influence of mT air is minimized by the presence of a cold ocean current and the blocking effect of the mountains. Finally, *continental tropical* (cT) *air masses* form in the southwestern U.S. and adjacent northern Mexico, creating hot, dry conditions across this region.

Effect of the Great Lakes

The Great Lakes of North America are a unique feature in the world because they are large enough to set up regional circulation patterns, both from the effect of the water body and from the adjacent lakes. They are the only place in the world with a series of lakes large enough that the rotation of the Earth comes into play, in the form of the Coriolis effect in determining the lake circulation.

The Great Lakes consist of Lakes Superior, Michigan, Huron, Erie, and Ontario (Fig. 12.4). All are shared between Canada and the U.S., except Lake Michigan which lies solely in the United States. Lake Superior is the largest, deepest, highest in elevation, and northernmost of the Great Lakes. Lake Huron is the second-largest and has the longest coastline. Lake Michigan is the third-largest and the only lake of the five that is longer in the north–south direction than in the east–west direction. These east–west and north–south configurations set up different and sometimes overlapping weather and climate impact zones, as nearby areas are likely to be influenced by thermal, precipitation, and circulation properties associated with multiple lakes. Lake Erie is the shallowest of the lakes, which makes it more likely to freeze in winter despite its southernmost location. Finally, while Lake Ontario is the smallest, its depth ranks second among the Lakes, reducing its likelihood for winter freezing, and its easternmost position makes it the lowest in elevation, as the lakes drain northeastward into Canada's St. Lawrence River and out to the Atlantic Ocean in eastern Canada. The drainage basin of the Great Lakes is small (see again Fig. 12.4), a testament to the unimaginably large and heavy continental *ice sheets* that pushed down the terrain over which they sat as they melted quickly in geologic time as a result of climatic warming, from about 14,000 years ago to about 4000 years ago.

The Lakes' position between about 41° and 48° N latitude (halfway between the North Pole and Equator) makes for substantial climatic seasonality. Wide seasonal variations occur in the *Sun angle* and daily number of daylight hours throughout the year. There are substantial energy reception differences during winter, when the angle of the Sun's inclination is higher and the number of daylight hours is

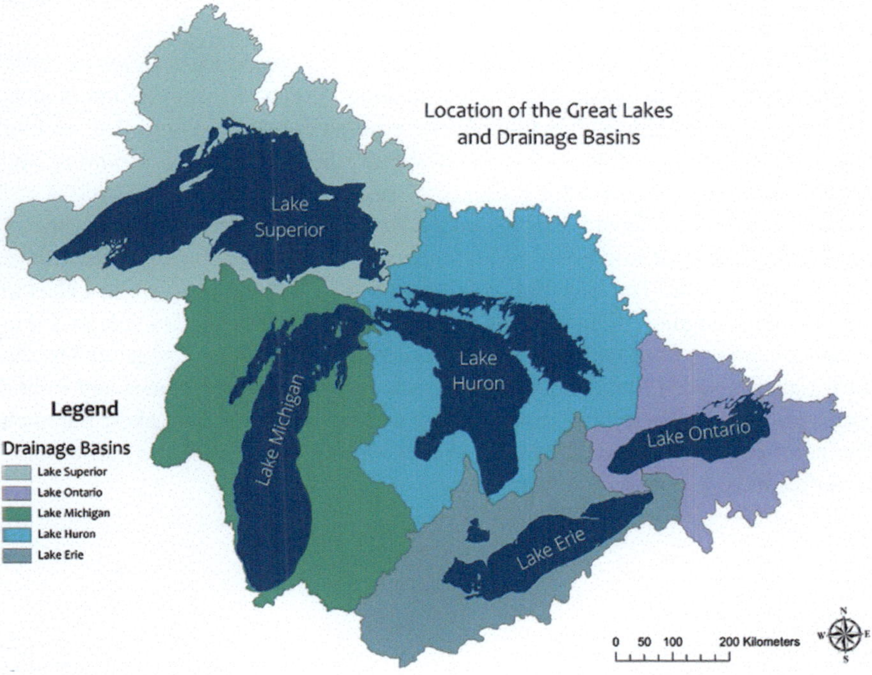

Fig. 12.4 The Great Lakes and the drainage basin of each. *Source* Geaux Geospatial

greater in the southern parts of the basin. Surface anticyclones in winter bring bitter cold temperatures associated with high pressures and sinking air. The conditions support clear nighttime skies that allow radiation cooling of the surface as L $\uparrow$ escapes to space. During summer, however, even though the angle of Sun's inclination angle is greater over the southern lakes, the daylight hours are greater in the northern areas. Smaller variations of energy occur between the northern and southern parts of the Lakes and the temperature gradient is smaller in summer, compared to winter.

The latitude also aligns with the boundary between *surplus* and *deficit* of *radiant energy*, so vigorous energy exchanges between the tropics and poles occur over this area. Thus, active, disturbed weather conditions are common. When the mid-latitude steering circulation over North America is *meridional*, either cP, A (or cA), or mT air masses dominate the Great Lakes region, depending on the mean position of the associated ridge and trough. But regardless, the strongly contrasting air masses are likely to meet over or near the Great Lakes, creating fronts that can produce rapid temperature changes, precipitation, and severe weather.

On the other hand, when the flow is *zonal*, mP air masses of Pacific origin can be important, but the Pacific origin of this air is barely recognizable upon reaching the region. Such a situation may bring mild weather to the area without the temperature extremes, precipitation, or severe weather associated with the cP, A, or mT air masses. The zonal flow may transport air eastward across the Lakes' vast west-to-east extent from 92° W to 76° W, which is the equivalent distance from Detroit to Jacksonville or from Chicago to Dallas.

The Great Lakes basin falls into the *Humid Continental—warm summer* (Dfb) *climate* type in the Köppen climate classification system. Substantial interannual variability is coincident within this climate type as climate is controlled by four major factors: latitude, air masses and atmospheric disturbances, *continentality*, and marine effects resulting from the presence of the Lakes themselves.

The winters of the Great Lakes region tend to be cold and very cloudy because of the high frequency of migrating mid-latitude wave cyclones and associated fronts, and *lake effect snow*. The heat and *humidity* added by the relatively warm Lakes to frigid cP or cA air above it, in the form of the *sensible heat flux* and *latent heat flux* in an *unstable atmosphere*, cause air to rise and condense. Cloudiness generated by the Great Lakes is added to the cloudiness provided by migratory cyclones and fronts. Occasionally, easterly winds can bring fair weather to the Great Lakes, but such flow is infrequent because the Great Lakes lie within the zone of the mid-latitude westerlies in both summer and winter.

Thunderstorms

Precipitation and thunderstorms may result from regional and local features such as convective heating on unstable days. These types of events can be represented by the annual thunderstorm frequencies (i.e., number of days on which thunder is heard), as shown in Fig. 12.5. The summer maximum in thunderstorms occurs in Florida and along the Gulf and Atlantic coasts. There, the summer warmth and humidity often create an unstable atmosphere and associated free convection. Florida has the additional thunderstorm-producing capabilities from convergence associated with the eastward-moving *sea breeze* from the Gulf of Mexico converging with the westward-moving sea breeze from the Atlantic Ocean. A secondary thunderstorm maximum is also found in the southwestern U.S. where *anabatic winds* (i.e., *valley breezes*) cause convergence at mountaintops. In July and August, these storms can be enhanced by the monsoon. The net effect of the Great Lakes is to decrease thunderstorm activity in the southern sections through surface cooling and to increase activity in the northern part of the basin by adding moisture and moderating the colder temperatures. Few thunderstorms occur on the Pacific coast. The strong, sinking (subsidence) side of the Hawaiian-Pacific High, reinforced by the cold California Current, stabilizes the atmosphere across most of the far west.

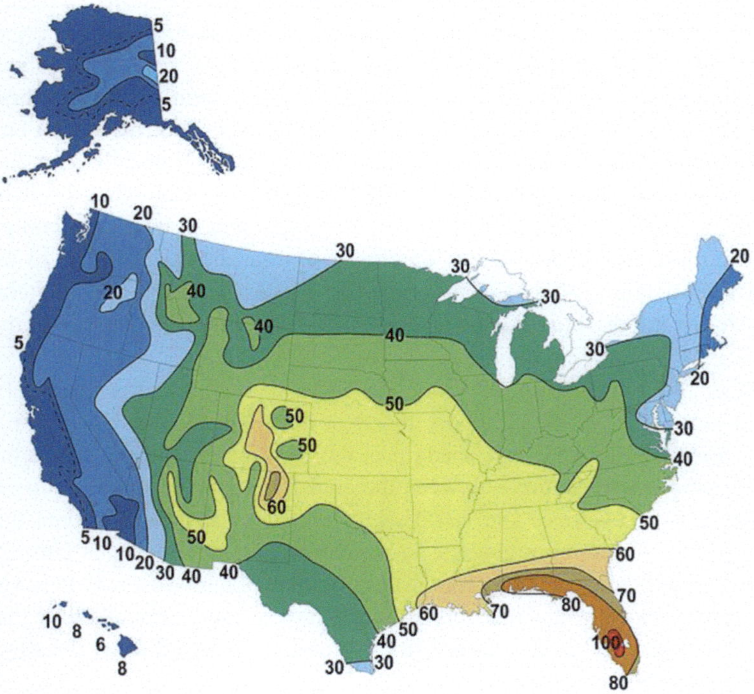

Fig. 12.5 Mean number of thunderstorm days per year for the United States. *Source* NOAA

Severe Mid-Latitude Weather: A United States and Canada Specialty

The extreme air mass contrast causes the U.S. to have some of the most violent weather on Earth. The configuration of the north–south mountains allows very cold, dry air masses to mix regularly with very warm, moist air deep within the continental interior. The strong thermal and moisture gradients can support extremely strong mid-latitude wave cyclones. Embedded *cumulonimbus* thunderstorms produce strong winds, extreme lightning, heavy precipitation, large hail, and tornadoes, in addition to any form of precipitation (rain, snow, sleet, freezing rain, and hail), sometimes with more than one form happening in different parts of the system simultaneously.

Favorable conditions for severe weather in a mid-latitude wave cyclone occur when and where two air masses with strongly contrasting conditions meet. Most commonly in the U.S. and Canada, the air masses involved are a cP converging with a mT air mass. The strong contrasts produce a sharp increase in wind speed with increasing height near the surface—known as **vertical wind shear**—as the warmer and/or more humid air mass is pushed up over the colder and/or drier one. If divergence of winds in the upper atmosphere is also occurring, such as on the trough-to-ridge side of the Rossby wave, conditions are even more favorable for severe weather. The strongest cumulonimbus clouds can occasionally produce a much larger, more severe, and longer-lasting thunderstorm—known as a **supercell**—than typical thunderstorms. Alternatively, a collection of strong thunderstorms that are organized such that the downdraft from one cell displaces upward the air in an adjacent cell, contributing to *updrafts* in a domino fashion, is called a **multicell thunderstorm** system. Particularly hazardous types of multicell systems include **mesoscale convective complexes** (MCC), which may have dozens of thunderstorm cells across an area the size of an entire state. The cells in a MCC all reinforce each other through their organized updrafts and downdrafts that remain separated as the entire complex is moved by the steering circulation. This organization allows the MCC to remain intact for many hours. A cold front-derived type of MCC in which the interacting thunderstorms are oriented in a linear fashion is the **mesoscale convective system** (MCS).

Whether the thunderstorm acts alone as in a supercell, or as a small part of a multicell storm system, or is simply a less severe, typical, unicellular **air mass thunderstorm**, in the U.S. or Canada it is more common near the cold front but still within the **warm sector**—between the cold front and the warm front—of the mid-latitude wave cyclone. The warm sector contains the most abundant thermal energy, greatest water vapor content, most intense low-level unstable atmosphere, and vertical wind shear necessary for the development of these severe thunderstorms. If a cT air mass meets with a maritime tropical air mass roughly parallelling the cold front in the warm sector at a **squall line**, sometimes referred to as the **dry line**, interactions can be particularly severe.

However, severe weather can occur anywhere near the mid-latitude wave cyclone, especially where smaller, less apparent **short waves**, with smaller ridges and troughs, in the Rossby wave flow are embedded within the trough-to-ridge side of the wave (Fig. 12.6). Notice in Fig. 12.6 the kinks on the trough-to-ridge sides of the Rossby waves, where extra uplift has created more humid and likely precipitation conditions over western South Dakota, northeastern Montana, northern Utah, and other places. Notice also that an enclosed area of low heights (and therefore low pressure), the upper-level signature of a surface mid-latitude wave cyclone, exists even as high as the 500-mb level over South Carolina and coastal Georgia. These smaller short waves might seem to be harmless, but they are active baroclinic zones.

Under such favorable conditions, updrafts flowing up the imaginary ramp into the lower pressure of the approaching thunderstorm base rise toward the storm's back side. This is usually the western side, if the storm system is moving from west to east, steered by the mid-latitude westerlies. The updrafting flow is squeezed by gravity as it is trapped over the denser, ambient air cooled by falling precipitation. The updrafting air's velocity then increases as it constricts, much like water flowing out of a half-blocked garden hose. This velocity increase aloft enhances the vertical wind shear and speed divergence, both of which encourage further, stronger updrafts with the speed divergence causing a vacuum.

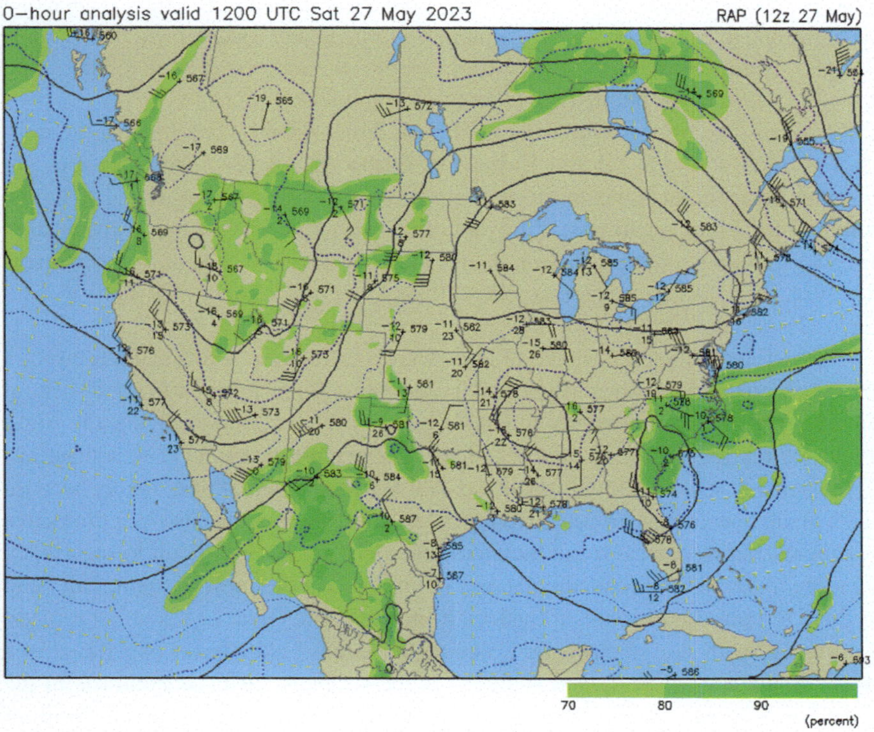

Fig. 12.6 Examples of short waves in the 500-mb flow (shown with isohypses in tens of meters; solid lines), with temperatures (°C; dotted lines) and humidity (green tones) for 1200 GMT on 27 May 2023. *Source* NOAA. https://weather.rap.ucar.edu

Within the precipitating part of the cumulonimbus cloud, usually away from the updraft, the density of the cooled air increases quickly, creating a sudden rush of downdrafting air known as a **gust front**. Despite the term, the gust front is at a much smaller spatial and temporal scale than a cold front or warm front; it is more similar in size, but not duration, as the *sea breeze front*. It is recognizable to anyone outdoors near an approaching thunderstorm as a sudden blast of much cooler air. The breezy sensation is a signal to any amateur meteorologist that the thunderstorm will be approaching within a few minutes, as the regional circulation pushes the thunderstorm in the same direction as it pushes the gust front once it hits the surface.

As the updrafts move into the back (western) side of the thunderstorm, they gain counter-clockwise rotation (in the Northern Hemisphere) from the Coriolis effect, which causes them to break around the cloud, rather than filling in the lower pressure in the middle of the storm system. Friction decreases with the continued updrafting higher above the surface, producing a counter-clockwise-rotating **mesocyclone**, often on the southwestern side of the cloud. The momentary vacuum draws more air upward from beneath the mesocyclone. The rotation may produce a funnel cloud that may protrude downward from the cloud base. A tornado occurs when the funnel cloud makes contact with the surface. The formation of a mesocyclone is a good indicator that a tornado may touch down within a few minutes. Yet the presence of a mesocyclone does not necessarily mean that a tornado will touch down. Fortunately, **Doppler radar** is an effective tool for detecting the early stages of rotation, and therefore warning the public that a tornado is imminent.

Tornadoes are categorized using their peak three-second wind gusts as the indicator of their strength, via the **Enhanced Fujita** (EF) **scale**. The categories are designed to distinguish the ability of various types of construction to withstand the winds. Fortunately, regardless of the part of the world where they occur, the weaker tornadoes are much more common than each successive increase in tornadic intensity (Fig. 12.7), from categories EF0 (105–137 km hr^{-1} or 65–85 mph) to EF5 (323–419 km hr^{-1} or 201 + mph).

The region of greatest tornadic activity, known as Tornado Alley, extends through the U.S./Canadian continental interior from central Texas through to south-central Canada, with the southeastern and midwestern U.S. extensions giving the high-frequency region somewhat of a "C" shape (see again Fig. 12.7). The southeastern U.S. extension (Dixie Alley), has the most tornado-related fatalities, mainly because of the greater percentage of people living in less sturdy homes that also lack protective basements, but also because southeastern tornadoes are often difficult to see as they are more likely to be shrouded in intense rainfall and/or darkness of night. Hilly and forested terrain also obscures tornadoes from view. The Great Lakes affect tornado occurrence, with the southwestern part of the Great Lakes basin experiencing a higher frequency in spring and summer, as the strongest air mass contrasts in the Great Lakes basin exist at that place and time. April is the most common month of tornado occurrence in the southern part of the Lakes, while May is the peak month in the north. The distribution relates to the position of the polar front jet stream, which drifts to higher latitudes as the Northern Hemisphere warms.

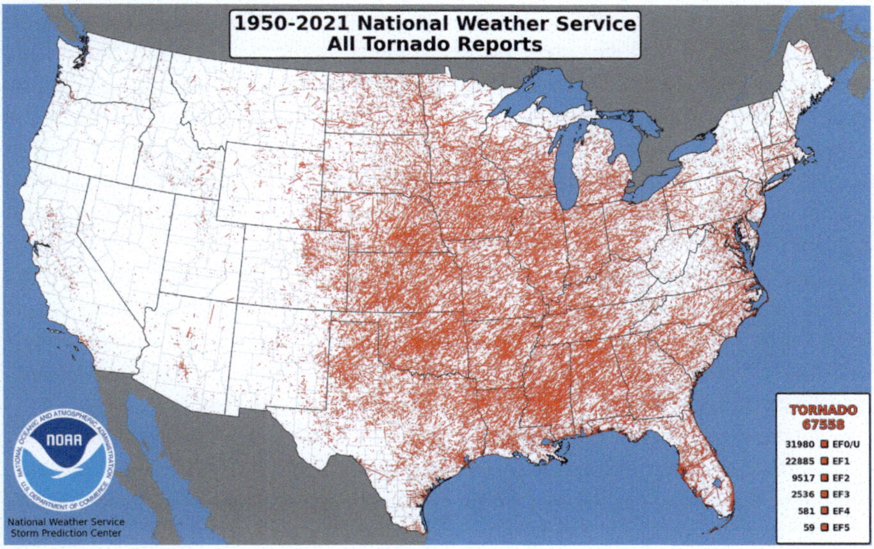

Fig. 12.7 Number and tracks of all reported U.S. tornadoes, by enhanced Fujita category, 1950–2021. *Source* NOAA. https://www.spc.noaa.gov/gis/svrgis/

West of the Rockies, and especially along the Pacific coast, severe weather is uncommon, for a variety of reasons. In the Intermontane Basin, the frequently sparse atmospheric moisture weakens downdrafts that would be important for maintaining an organized conveyor belt-like flow in coordination with the updrafts in other parts of the system, within and below the cumulonimbus cloud. The coastal mountain ranges and rugged parts of the Intermontane Basin induce additional, complicated, local sources of friction that generate local circulations that tend to interfere with the organized rotational flow of a developing tornado. Just as in the Appalachians, updrafts and downdrafts are caused locally by the mountains themselves rather than the tornadic system. Near the Pacific coast, the Hawaiian-Pacific High and its associated subsidence and cold California Current all tend to suppress severe weather. However, severe weather is somewhat more common on the Pacific coast in the cool season when the Hawaiian-Pacific High drifts southward and the polar front jet stream extends to lower latitudes. Under such conditions, mid-latitude wave cyclones can track across the Pacific to the U.S. or Canadian Coast and may be particularly vigorous during cool season times when the ocean waters are anomalously warm. This scenario is more common during El Niño winters.

Of course, tropical cyclones are also among the most severe weather hazards in the United States. The June 1 through November 30 hurricane season typically brings multiple landfalls to the Atlantic or Gulf of Mexico coasts each year, with devastating damage that seems to increase over time as more and more development is concentrated in the coastal zone. However, tropical cyclone frequency is less in the Atlantic-Caribbean-Gulf of Mexico basin than in some other parts of the world.

In the Northern Hemisphere, winds, storm surge, tornadoes, and damage associated with tropical cyclones are typically greatest on the right-forward side of the storm as it moves ashore. This is because the counter-clockwise rotation around the **eye** of the storm brings the right-forward side of the storm the humid wind from the marine area where friction is reduced, which then accumulates the highest wall of water on the same side as landfall occurs. Storm surge is usually the biggest threat to life and property from tropical cyclones, as it can exceed 6 m (20 ft) in major hurricanes.

Occasionally, a mid-latitude wave cyclone merges with a tropical cyclone off the North Atlantic coast to produce an even stronger extratropical system. These storms can be particularly devastating to the coast and near-shore regions as the counter-clockwise circulation drives high surf toward the landmass, if the mid-latitude westerlies are too weak to steer the system away from North America.

A weakening steering circulation associated with *polar amplification* would likely allow not only more cases of mergers of tropical and mid-latitude systems to hit the U.S. Atlantic coast, but it would also mean that weather systems in general, including tropical cyclones, would move slower and have more time to dump precipitation on locations in their path. There would also be a longer interarrival time between precipitation-bearing weather systems, which would lengthen droughts. This is the rationale for future projections of increasing severe weather and extremes of both wet and dry weather. Can you think of recent examples of weather events, in the U.S./Canada or elsewhere, that support or refute this hypothesis?

Common Storm Tracks (Cyclones)

While cyclogenesis and the resulting storm tracks can occur across the U.S. and Canada at any given time, the topography of the continent and the position and amplitude of the Rossby waves favor certain regions. Climatologically, four cyclone types are identifiable. **Alberta Clippers** form year-round on the leeward side of the northern Rockies (Fig. 12.8). The leeward side of mountain ranges is a good place for cyclogenesis because of the dynamics associated with counter-clockwise (in the Northern Hemisphere) spinning in an increasingly "tall" column of air. Furthermore, the acceleration and resulting upper-level *divergence* of the steering flow after it is "squeezed" between the higher mountain elevations and the *tropopause* invites uplift to accompany the counter-clockwise spin.

Sometimes these incipient storms form in the Gulf of Alaska and are called **Gulf of Alaska Lows** before being renamed "Alberta Clippers" when they approach the Canadian Rockies. Because these cyclones are "young" when they start their trek east of the Canadian and (with their trailing fronts) U.S. Rockies, they often have abundant energy associated with them. Thus, they can be severe weather producers if they have enough water vapor to release *latent heat* in the *condensation* process. However, they usually lack enough water vapor to become major severe weather events.

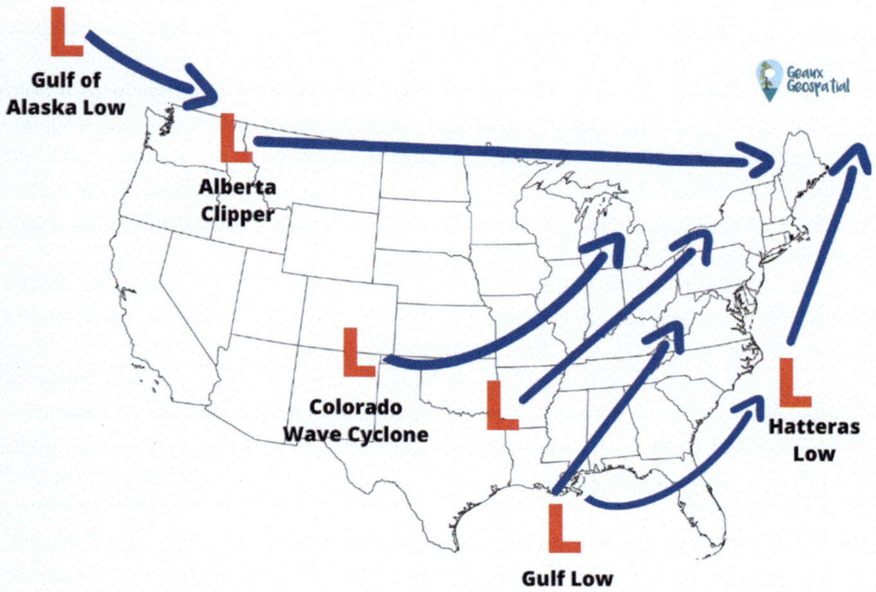

Fig. 12.8 Common mid-latitude wave cyclogenesis regions and generalized trajectories in and near the U.S. and Canada. *Source* Geaux Geospatial

Colorado Wave Cyclones are similar to Alberta Clippers in that they form on the leeward side of the Rockies for dynamic reasons. However, Colorado Wave Cyclones form more often in the cold half of the year, when the cold/warm polar front boundary is farther south. They tend to have more water vapor associated with them than Alberta Clippers because the counter-clockwise circulation around the low-pressure center can pull Gulf of Mexico moisture into them more easily and are thus more likely to produce severe weather. Because the leeward side of the Rocky Mountains is often on the ridge-to-trough side of the Rossby wave, which is the side that is not conducive to storm development, both Alberta Clippers and Colorado Wave Cyclones are more likely to form during periods of *zonal* circulation, which de-amplifies the Rossby wave. Notice that their tracks follow a generally west-to-east trajectory across the continent, which suggests a zonal steering flow (see again Fig. 12.8).

During periods of amplified Rossby waves in the circulation above North America, cold air can push farther south, leading to more equatorward cyclogenesis areas, especially along the Gulf and Atlantic coasts, where the cold and warm air masses often meet. **Gulf Lows** and **Hatteras Lows** form in the area of strong temperature gradient between land and water. This situation makes these regions very strong baroclinic zones in winter, when the land is cold but water is still warm. The Gulf Lows most typically form over the Gulf of Mexico, but they can also form inland, when the Gulf of Mexico coast of the U.S. remains much warmer than the land farther inland, as is represented by the red "L" over southwestern Arkansas in Fig. 12.8.

Nevertheless, the Gulf Lows that form over water are often more likely to produce severe weather because of the abundance of nearby moisture.

In the case of the Hatteras Lows, the warm Gulf Stream is adjacent to a cold continent. Deeper oceanic features near Cape Hatteras also contribute to the frequent cyclogenesis nearby. North of Cape Hatteras is the part of the *continental shelf* known as the **Mid-Atlantic Bight** (Fig. 12.9). A bight is a gradual bend of the shoreline that protects an open bay. Ocean current flow along the Mid-Atlantic Bight is equatorward.

South of Cape Hatteras is the part of the continental shelf known as the **South Atlantic Bight**, where the continental shelf flow is poleward. When the two currents along the shelf meet near Cape Hatteras, there is a strong convergence of water flow, which has the net effect of pushing water laterally out to sea. This seaward flow of relatively warm water during much of the year—because the shelf is so shallow—creates an ocean front where the water temperatures differ strongly across a small lateral distance. These can also create baroclinic zones, where weather patterns and waves are "excited."

All four of these major track types tend to converge over New England, the Maritime Provinces of eastern Canada, and the northeastern United States. That part of the continent thus experiences frequent frontal passages with convection and associated clouds and precipitation. However, the younger, more energetic stages of the

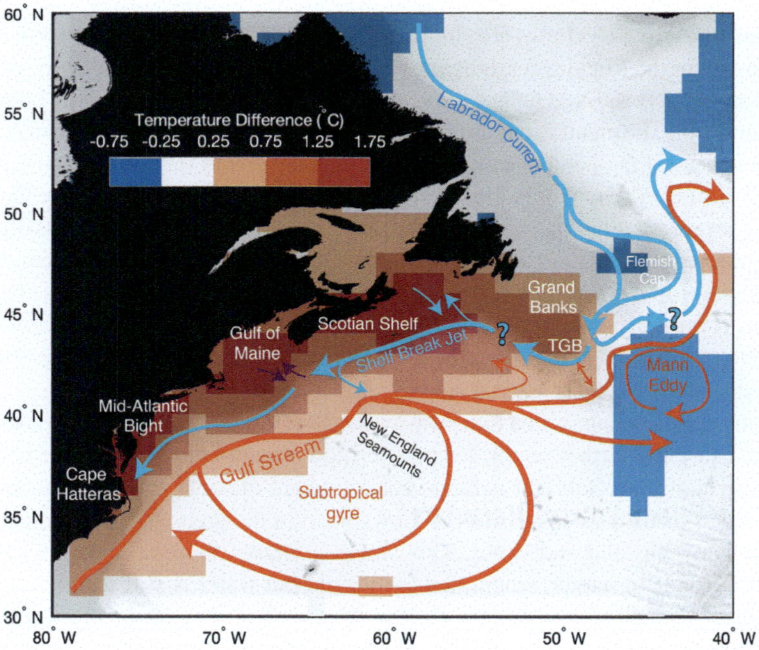

Fig. 12.9 Depiction of the warm water of the Mid-Atlantic Bight along with other temperature effects of ocean currents off the U.S. Atlantic coast. *Source* Nature. https://www.nature.com/articles/s43247-021-00143-5

cyclones are likely to occur over the Great Plains and Midwest, enhancing the likelihood for severe weather in these areas. New England and the Pacific Coast are more likely to experience older *occluded fronts*, resulting in many days of light precipitation in the form of *stratus* clouds. Of course, these descriptions are a simplification and generalization of cyclone activity across North America. Storms do form in other places, and storms within a general type all differ from one another.

Broad-Scale Circulation Variability Effect on Temperature and Precipitation

The PNA pattern involves three areas with strongly-correlated air pressure or *geopotential heights*: the North Pacific (Gulf of Alaska), northwestern North America, and the southeastern United States. More specifically, the North Pacific and southeastern U.S. tend to both have higher 500-hPa geopotential heights (and pressure) than normal or lower geopotential heights (and pressure) than normal, but whatever their anomaly mode at a given time, it is opposite the anomaly mode of northwestern North America and the tropical North Pacific at the same time. The PNA index is used to measure these seesaws and thus circulation modes across North America.

When the PNA pattern is in its positive mode, the normal ridge in the West and the trough in the East are amplified even more than usual. Western areas experience higher geopotential heights (or pressure) than normal while eastern regions experience anomalously low heights/pressure. This scenario brings meridional circulation across the U.S. (see again Figs. 6.14 and 6.15). When the PNA pattern is in its negative mode, then the ridge/trough configuration is suppressed (or even inverted). Zonal circulation occurs in such cases, or in extreme cases a trough/ridge pattern can occur from west to east.

In winter, 20 to 30% of the variability in temperature can be explained just by knowing the height/pressure at the three critical PNA points (**centers of action**). The positive pattern (with its amplified ridge/trough) causes the eastern U.S. to be colder than normal and the western U.S. to be warmer than normal. The negative (zonal) pattern restricts the cold air to the north, causing the southeastern U.S. to be warmer than normal and the northwestern U.S. to be colder than normal.

Precipitation variations are not linked as strongly to the PNA pattern as temperature. While the amplitude of the ridges and troughs control the north/south movement of air masses, and thus temperature, even small changes in the axis of the waves can have important influences on storm tracks, limiting the power of the PNA pattern to explain precipitation. In the positive PNA mode, the Midwest does tend to be somewhat drier than normal, while the Southeast is slightly wetter than normal. The trough over the East in the positive mode tends to favor Gulf Low and Hatteras Low cyclogenesis with coastal tracks over the Southeast. In a negative PNA mode, Alberta Clippers and Colorado Wave Cyclones sweep across the northern and midwestern

U.S. along zonal tracks, making these areas wetter than normal. In extremely negative PNA years an unusual western trough can direct more intense cyclone activity into the central U.S., leading to some of the highest precipitation totals and flooding in that region.

Other forms of atmospheric flow variability are important in the U.S. and Canada as well. For example, during the *El Niño* phase of ENSO, the polar front jet stream bifurcates over North America, with the northern branch preventing most cA air masses from moving southward into the U.S., and the southern branch advecting Pacific Ocean moisture over the southeastern United States. During the positive phase of the *North Atlantic Oscillation* (NAO), the steering circulation tends to be more zonal across the eastern U.S. and Canada, while meridional flow is more likely in a negative NAO regime. Most models suggest that the positive NAO phase will become preferred in the foreseeable future, even though the confidence in that prediction and resulting impacts to temperature and precipitation patterns is only moderate. The other forms of atmospheric flow variability described in Chap. 6 also have a signature atmospheric flow in the U.S. and Canada.

Regional-Scale Circulation: Gulf of Mexico Low-Level Jet (LLJ)

Another major feature of North America's circulation is the **Gulf of Mexico low-level jet** (LLJ), a stream of fast-moving air from the Gulf of Mexico into the southern Great Plains and occasionally as far inland as Nebraska, that occurs mainly at night. Along with the Bermuda-Azores High, the LLJ is the reason that most of the water vapor over the continent east of the Rocky Mountains originates from the Gulf of Mexico. While the Bermuda-Azores High advects moisture into the continental interior most strongly at the surface and has a weakening influence with height, the LLJ is near the top of the *planetary boundary layer* at approximately 1–1.8 km (0.6–1.1 mi) in height, moving at perhaps 40–110 km hr^{-1} (25–70 mi hr^{-1}).

The LLJ forms because of diurnal pressure changes associated with daily heating of the continent, along with the difference in elevation from west to east. As radiative cooling occurs nocturnally in the lower-elevation eastern part of the U.S. west of the Appalachians, the longwave radiation lost from the surface is gained by the atmosphere at approximately the same height as the ground surface of the western United States. At the same time, the loss of outgoing longwave radiation from the surface in the western U.S. is occurring, inducing sinking of cold air over the western U.S., while the relatively warm air at the same height over the eastern U.S. rises. This sets up a pressure gradient from east to west. But as the air begins to move from east to west (against the mid-latitude westerlies), the Coriolis effect turns the air stream to the right of the flow. The result is a southerly flow of air into the southern Great Plains.

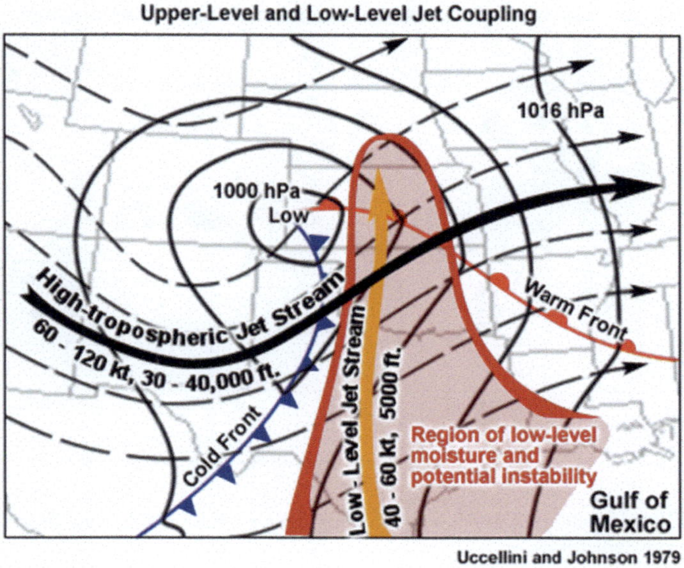

Fig. 12.10 Upper-level and low-level jet coupling for the Gulf of Mexico low-level jet stream. *Source* UCAR CoMet cites Uccelini and Johnson (1979). Permission from Tara Torres

Atmospheric stability and an associated mid-level *temperature inversion* increase wind speeds through the planetary boundary layer by restricting fluid motion through the vertical. The effect occurs at night because daytime shortwave radiation receipt from the Sun reduces the tendency for the surface cooling and sinking air in the western U.S. during daytime hours.

The LLJ can enhance mid-latitude wave cyclone impact (i.e., Colorado Wave Cyclones) by supplying humid air, especially when the polar front jet stream is already supporting that mid-latitude wave cyclone. The LLJ can also occur or be strengthened by the flow around a surface cyclone just northwest of the Gulf of Mexico. In such cases, the counter-clockwise rotation around the cyclone draws humid air northward ahead of the cold front, as shown in Fig. 12.10. The presence of the LLJ enhances the development of severe frontal cyclones and associated tornadoes. Notice in Figs. 12.11 and 12.12 the extent of influence of the Gulf of Mexico LLJ, as compared to similar LLJs around the world.

Influences on Spatial Patterns of Temperature

In the U.S. and Canada, temperature patterns are influenced primarily by latitude and continentality. Average temperatures generally decrease northward as decreasing Sun angles leave less available radiant energy (Q*) to warm the surface. However, the increasing number of daylight hours with increasing latitude in the summer months

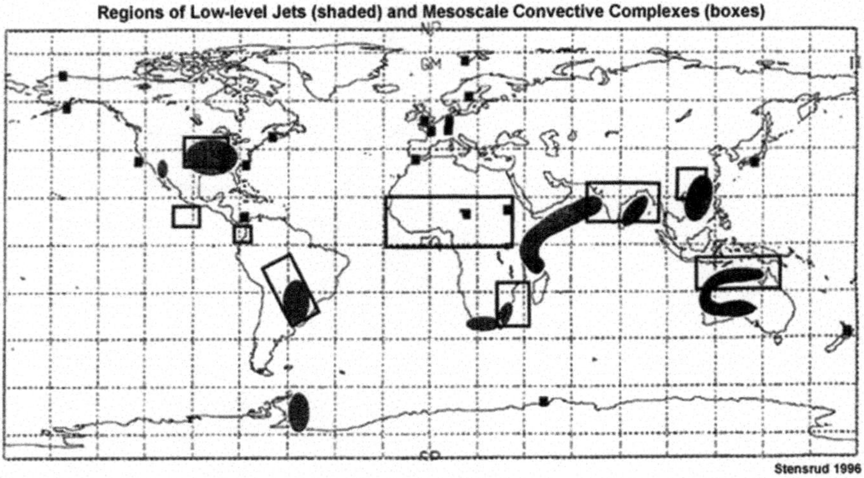

Fig. 12.11 Global locations of low-level jet streams. *Source* UCAR CoMet cites Stensrud (1996). Permission from Tara Torres

helps to offset the effect of lower Sun angle at higher latitudes. Thus, in summer, the temperature difference between the northern part of the U.S. and Canada is less different than in the southern part, compared to the winter temperature differences between the northern versus southern parts of the U.S. and Canada.

Seasonal temperature patterns are also affected by continentality in this large land mass. In winter, the lowest temperatures often occur just to the north of the geographic center of the continent, at longitudes between 90° W and 100° W (see again Fig. 2.16). The western mountains prevent most of the Pacific *humidity* from reaching far inland, so the westerlies have little impact in bringing moisture inland to moderate the temperature of interior North America.

In 1961 the American climatologist Glenn Trewartha developed a continentality index, such that higher numbers represent more continental environments. The index shows that the region centered on Minnesota, the Dakotas, and Manitoba has the strongest continental characteristics. To the east, the Great Lakes moderate the summer and winter temperatures to some extent, resulting in a somewhat smaller continental index. Even the Atlantic coast has somewhat strong continental properties because the mid-latitude westerlies advect continental air to the mid-latitude Atlantic coast. The Pacific coast is the least continental area, as maritime air masses moderate temperatures substantially. The Atlantic region, and to a lesser extent the Gulf of Mexico coast, also display small continentality effects.

North America also experiences considerable interannual variability of temperature. These year-to-year variations are influenced by the position and number of Rossby waves at a given time. The largest area of interannual variability is immediately to the east of the U.S. and Canadian Rockies. This area is near the mean trough/ridge boundary, so slight west–east displacements of the ridge/trough configuration can result in rapidly differing conditions there. The broad-scale circulation there (and

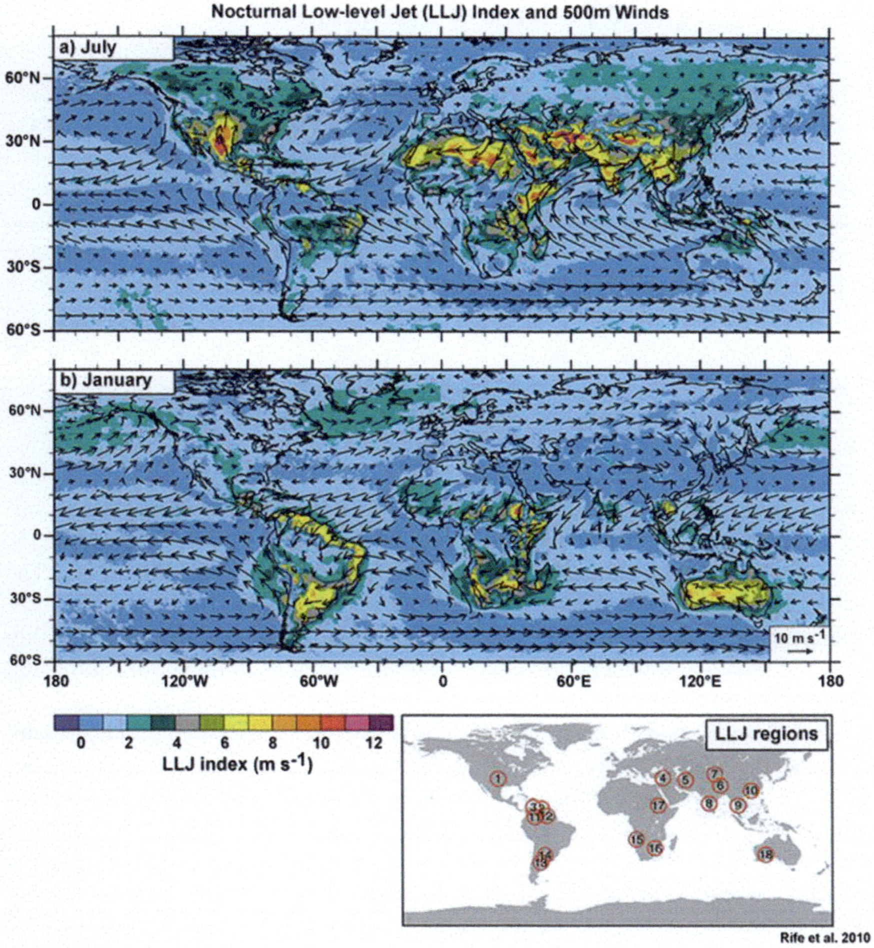

Fig. 12.12 500-mb winds and nocturnal low-level jets (*Note* the "1" label in the inset map refers to the Gulf of Mexico LLJ). *Source* UCAR CoMet cites Rife et al. (2010). Permission from Tara Torres

elsewhere) can influence the local circulation strongly. For example, just east of the Rockies, downslope winds that dry *adiabatically* as they sink can warm and desiccate an area intensively. These so-called *Chinook winds* can persist in some years and be intermittent or sporadic in others.

Influences on Spatial Patterns of Precipitation

The greatest precipitation totals in the continental U.S. and Canada are concentrated near the coasts. Rainfall is particularly abundant on the Pacific Coast on the western side of the coastal mountains. There, precipitation from old, occluded fronts is affected by orographic lifting, increasing totals. In western North America, average annual precipitation increases northward. As distance from the subtropical anticyclone—the Hawaiian-Pacific High—increases, the summer dry period is shortened and frontal activity occurs throughout the year. Precipitation is also abundant along the central and eastern Gulf of Mexico coast where winter frontal passages and autumn tropical cyclones supplement frequent convectional thunderstorms in the warmer seasons.

Throughout eastern North America, precipitation results from humidity carried inland from the Gulf of Mexico. The geography of the continent allows this moisture to extend much farther inland than in other continents. For example, St. Louis, Missouri, and Chicago, Illinois, are much more humid and wet than places similarly positioned in China. The greatest winter precipitation totals in the eastern U.S. and adjacent Canada follow the most common storm tracks described above. The tracks converge in the St. Lawrence River Valley of eastern Canada and the U.S. Northeast, leading to consistently high precipitation totals. In summer, the humidity supports convective thunderstorms, but not to the extent as in the more continental and southeastern coastal areas.

Moisture availability decreases westward, with the 100°W longitude usually considered the boundary between humid and dry areas. East of the Rockies and across the Great Plains, moderate amounts of precipitation occur when the Bermuda-Azores High expands and the flow around its western side can pump humid air from the Gulf of Mexico. This is most common in summer, when the Bermuda-Azores High is displaced farthest northward. The east-of-Rockies dry zone widens with increasing latitude northward, as Gulf of Mexico humidity cannot reach the northern parts of the continent as easily. A narrow dry zone in the south, which includes western Texas and western Oklahoma, gradually widens in the north across the Prairie Provinces of Canada.

The Intermontane Basin has the driest climates on the continent. There, a **double rain shadow** effect, which blocks humid air from both Pacific and Gulf of Mexico source regions, limits available moisture sources. The adiabatic descent of air on the rain shadow side creates warm and dry conditions throughout most of the year. The most common way to get precipitation is from a storm coming from the southwest, often associated with the miniature monsoon effect in the southwestern United States.

The atmospheric circulation systems, from general circulation features to semi-permanent anticyclones, to traveling mid-latitude wave cyclones and associated fronts, to regional circulations such as the LLJ, all influenced by the broad-scale teleconnections, act in concert to make the environment of the U.S. and Canada the way that it is. In the next chapter, the specific climatic types and mean temperatures and precipitation will be presented for representative locations in the U.S. and Canada.

Further Reading

Ashley, W. S. (2007). Spatial and temporal analysis of tornado fatalities in the United States: 1880–2005. *Weather and Forecasting, 22*(6), 1214–1228. https://doi.org/10.1175/2007WAF2007004.1

Burnett, A. (1994). Regional-scale troughing over the southwestern United States: Temporal climatology, teleconnections, and climatic impact. *Physical Geography, 15*(1), 80–98. https://doi.org/10.1080/02723646.1994.10642506

Dixon, P. G., Mercer, A. E., Choi, J., & Allen, J. S. (2011). Tornado risk analysis: is Dixie Alley an extension of Tornado Alley? *Bulletin of the American Meteorological Society, 92*(4), 433–441. https://doi.org/10.1175/2010BAMS3102.1

Gonçalves Neto, A., Langan, J. A., & Palter, J. B. (2021). Changes in the Gulf Stream preceded rapid warming of the Northwest Atlantic Shelf. *Earth Environment, 2*(1), Art. No. 74. https://doi.org/10.1038/s43247-021-00143-5

Leathers, D. J., Yarnal, B., & Palecki, M. A. (1991).The Pacific North-American teleconnection pattern and United-States climate .1. Regional temperature and precipitation associations. *Journal of Climate, 4*(5), 517–528. https://doi.org/10.1175/1520-0442(1991)004<0517:TPATPA>2.0.CO;2

Li, W. H., Li, L. F., Fu, R., Deng, Y., & Wang, H. (2011). Changes to the North Atlantic Subtropical High and its role in the intensification of summer rainfall variability in the southeastern United States. *Journal of Climate, 24*(5), 1499–1506. https://doi.org/10.1175/2010JCLI3829.1

Lydolph, P. E. (1985). *The climate of the earth.* Rowman and Littlefield. 978-0865981195.

Rife, D. L., Pinto, J. O., Monaghan, A. J., Davis, C. A., & Hannan, J. R. (2010). Global distribution and characteristics of diurnally varying low-level jets. *Journal of Climate, 23*(19), 5041–5064. https://doi.org/10.1175/2010JCLI3514.1

Rohli, R. V., & Li, C. (2021). *Meteorology for coastal scientists* (525 pp). Springer. 978–3–030–73093–2–2.

Smith, B. T., Thompson, R. L., Grams, J. S., Broyles, C., & Brooks, H. E. (2012). Convective modes for significant severe thunderstorms in the contiguous United States. Part 1: Storm classification and climatology. *Weather and Forecasting, 27*(5), 1114–1135. https://doi.org/10.1175/WAF-D-11-00115.1

Stensrud, D. J. (1996). Importance of low-level jets to climate: A review. *Journal of Climate, 9*(8), 1698–1711. https://doi.org/10.1175/1520-0442(1996)009%3c1698:IOLLJT%3e2.0.CO;2

Suriano, Z. J., & Leathers, D. J. (2017). Synoptic climatology of lake-effect snowfall conditions in the eastern Great Lakes region. *International Journal of Climatology, 37*(12), 4377–4389. https://doi.org/10.1002/joc.5093

Ting, M. F., & Wang, H. L. (2006). The role of the North American topography on the maintenance of the Great Plains summer low-level jet. *Journal of the Atmospheric Sciences, 63*(3), 1056–1068. https://doi.org/10.1175/JAS3664.1

Trewartha, G. T. (1981). *The earth's problem climates* (2nd ed.). University of Wisconsin Press.

Trewartha, G. T., & Horn, L. H. (1980). *Introduction to climate* (5th ed.). McGraw-Hill.

Uccellini, L. W., & Johnson, D. R. (1979). The coupling of upper and lower tropospheric jet streaks and implications for the development of severe convective storms. *Monthly Weather Review, 107*(6), 682–703. https://doi.org/10.1175/1520-0493(1979)107%3c0682:TCOUAL%3e2.0.CO;2

U.S. Global Change Research Program. (2017). In D. J. Wuebbles, D. W. Fahey, K. A. Hibbard, D. J. Dokken, B. C. Stewart, & T. K. Maycock (Eds.), *Climate science special report: Fourth national climate assessment* (Vol. I, 470 pp). U.S. Global Change Research Program. https://doi.org/10.7930/J0J964J6

Wallace, J. M., & Gutzler, D. S. (1981). Teleconnections in the geopotential height field during the Northern Hemisphere winter. *Monthly Weather Review, 109*(4), 784–812. https://doi.org/10.1175/1520-0493(1981)109%3c0784:TITGHF%3e2.0.CO;2

Chapter 13
United States, Canada, and Adjacent Oceans (Examples)

Abstract The U.S. and Canada have a wide range of climatological environments. The southeastern U.S. is dominated by the Bermuda-Azores High advecting moisture from the Gulf of Mexico, leaving the damp, sultry Humid Subtropical (Cfa) climate, which has no dry season and usually has a slight precipitation decrease in the autumn months. The U.S. Northeast and Continental Interior (generally corresponding to the Midwest) are influenced year-round by mid-latitude wave cyclone passages and tend to have Humid Continental (Dfa, Dfb) climates, also with little precipitation seasonality. The Great Plains and Prairies of Canada also experience humidity from the Gulf of Mexico, but to a lesser extent; they are a transition zone where the C climates give way to Semi-arid (BSh, BSk) climates. The Interior West is largely BSh, BSk, or True Desert (BWh, BWk), as it sits in a double rain shadow of mountain ranges. The Pacific coast is influenced by the stability associated with the cold California Current that is reinforced by the Hawaiian-Pacific High in summer, to create a Mediterranean (Csa, Csb) climate, with Marine West Coast (Cfb) farther north. Canadian examples are also shown, which continue the patterns of the northern United States.

Keywords Bermuda-Azores High · Chaparral · Climagraph · Double rain shadow · Hawaiian-Pacific High · Humid Continental (Dfa and Dfb) climates · Humid Subtropical (Cfa) climate · Marine West Coast (Cfb) climate · Mid-latitude wave cyclone · North American monsoon · Semi-arid (BSh and BSk) climates · True Desert (BWh and BWk) climates · Water balance diagram

General Regional Climatology

The vast area of the U.S. and Canada ensures that a wide variety of climatic conditions are found across the continent. All five of the major *Köppen climatic types* are present (Fig. 13.1). However, the only mainland representation of the A climate type is on the southern tip of Florida. The rest are all found more abundantly across the U.S. and Canada and adjacent waters.

R. V. Rohli et al., *Atmospheric and Oceanic Circulation*,
https://doi.org/10.1007/978-3-031-66142-6_13

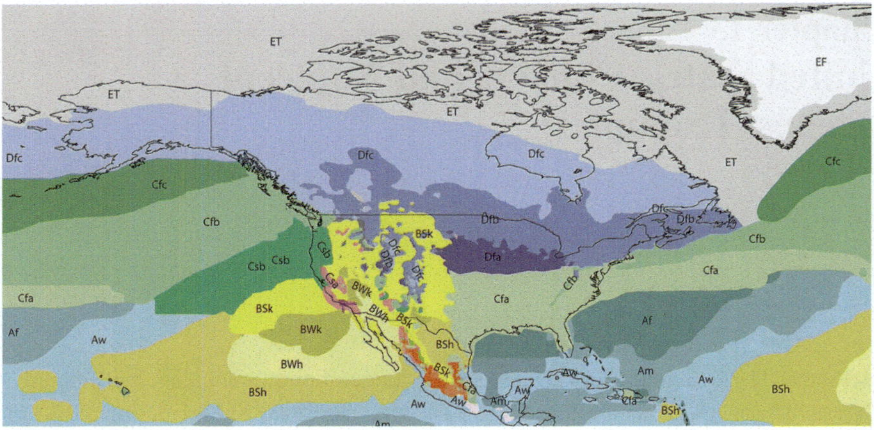

Fig. 13.1 Köppen climate types of the U.S., Canada, and surrounding marine environment. *Source* Modified from Reynolds, Rohli, Johnson, Waylen, and Francek, Exploring Physical Geography, McGraw-Hill

A line and bar chart that depicts average *temperature* and *precipitation*, respectively, for a given place, to provide a visual depiction of the climatic conditions at that place is known as a **climagraph** (also known as a climograph). Most of the climatological data used to construct the climagraphs and surface *water balance* diagrams in the chapters that follow are available on Wikipedia, for the 1991–2020 averaging period.

The Southeast

The southeastern U.S. includes the coastal region from near Philadelphia to just north of the *Tropic of Cancer*, nearly 3300 km (2050 mi). The Coastal Plain physiographic region comprises virtually all of Florida and much of the U.S. states of Louisiana, Mississippi, and Alabama. Along the Atlantic coast, the Coastal Plain extends from the shores of the North Atlantic westward to the piedmont—foothills—of the Appalachian Mountains from Georgia northward, and it narrows gradually with increasing distance northward to New York City, where it ceases to exist and the coastline becomes rocky. More than 118 million people reside in this area.

Nearly all of the southeastern U.S. is classified as a *Humid Subtropical* (Cfa) *climate*. Winters are generally mild, and summers are long, hot, humid, and uninterrupted for months. Precipitation is abundant through every month of the year, on average, but droughts do occur, with substantial impacts in this area occupied by water-intensive vegetation, cropland, and human activities.

The annual distribution of precipitation in the Southeast is generally even throughout the year, on average, but an autumn minimum is notable for most locations. Autumn is a time of transition between the primary precipitation-forcing mechanisms. At that time of year the precipitation enhancement by *convection* from summer and associated *air mass thunderstorms* and severe weather decrease with the cooling of the surface, yet the cold-season precipitation associated with *fronts* trailing from *mid-latitude wave cyclone* passages has not yet set in, as the *polar front* tends to remain farther north than the U.S. Southeast in autumn. Migratory mid-latitude wave cyclones that do occur in autumn are typically weak, because of the zonal flow conditions that dominate the autumn season. This situation leads to an autumn decrease in precipitation amounts in much of the region. Some coastal locations may have a high-precipitation autumn if the area is near the landfall area of *tropical cyclones* ranging from weak *easterly waves* to *tropical depressions*, to *tropical storms*, to *hurricanes*.

Climatic data for four representative cities across the U.S. Southeast (Fig. 13.2) are used to illustrate how the forcing mechanisms play out in terms of climate. The climagraph and water balance diagram for Miami Florida (Fig. 13.3), indicate the impressive summer precipitation and environmentally wet conditions that occur throughout the Florida Peninsula, and the influence of the winter *subsidence* from the Bermuda-Azores High on much drier conditions in winter. Nevertheless, despite the winter precipitation suppression, Florida has more annual average thunderstorm-days than any other part of the continental U.S. and Canada (see again Fig. 12.7). The climagraphs and water balance diagrams for Little Rock, Arkansas; Mobile, Alabama; and Raleigh, North Carolina (Figs. 13.4, 13.5 and 13.6, respectively), hint at the mild annual average temperatures and the complicated precipitation patterns. Generally wet conditions exist year-round, but with a slight suggestion of a late summer to autumn lull from the moisture-rich conditions at some sites in the region.

The Northeast

Most of the northeastern U.S. lies within the *Humid Continental—warm summer* (Dfb) *climate* region. This climatic type is similar to the Humid Subtropical type in the Southeast except that temperatures are reduced for the higher-latitude climate zone. Winters are harsher and summers are milder than the Southeast, but both regions have a fairly even distribution of precipitation throughout the year.

Temperatures are continental, despite the coastal location, because the *mid-latitude westerlies* push *continental polar* (cP) *air masses* and storm systems from inland into the area. During summer, the area is susceptible to heat waves if the Bermuda-Azores High expands or is displaced northwestward. The subsidence associated with the high pressure reduces clouds and precipitation while allowing abundant solar radiation to reach the surface. The resulting extreme heat has a detrimental effect on the elderly and people with preexisting health conditions.

For stations south of New York City, a slight precipitation maximum in summer is common. The region is most affected by moisture *advection* from the clockwise and outward circulation associated with the Bermuda-Azores High, which can even fuel the air mass thunderstorms that augment summer precipitation totals for these areas.

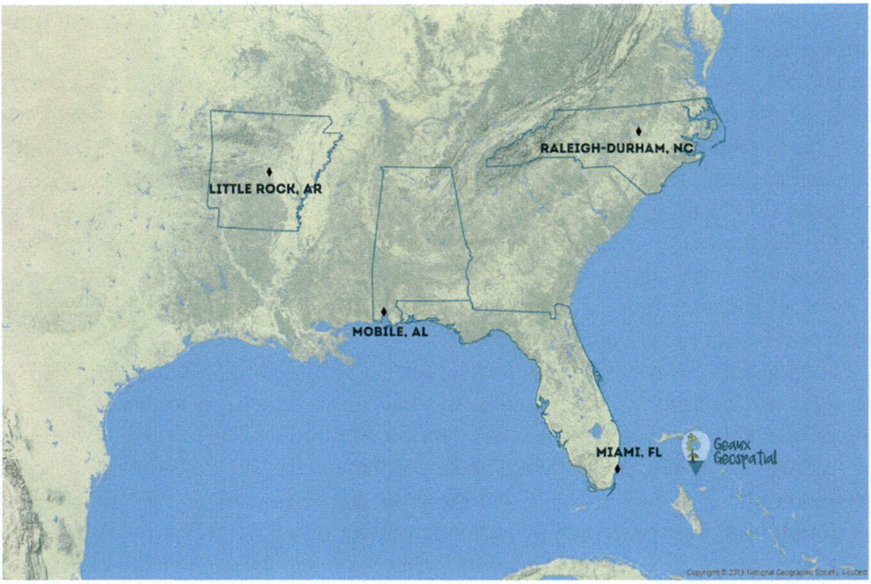

Fig. 13.2 Southeastern U.S. cities depicted in the corresponding climagraphs and water balance diagrams. *Source* Geaux Geospatial

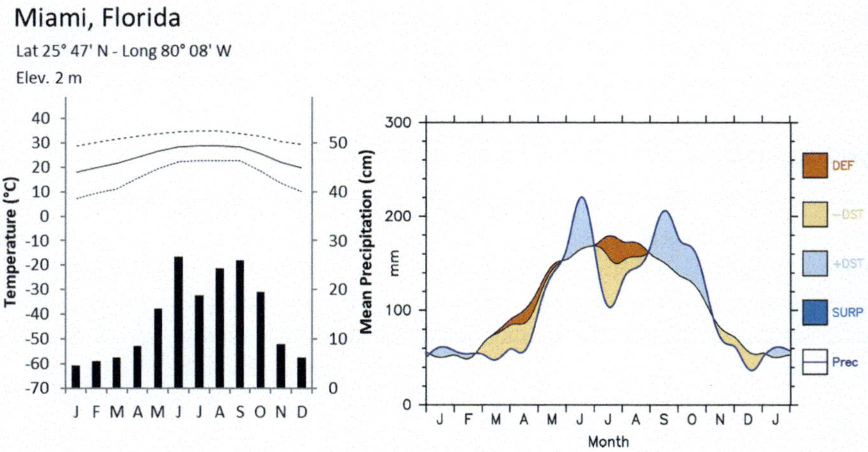

Fig. 13.3 Climagraph and water balance diagram for Miami, Florida. Data for 1991–2020 averaging period. *Source* Authors. https://weatherspark.com/, http://cyclops.deos.udel.edu/wimp/public_html/

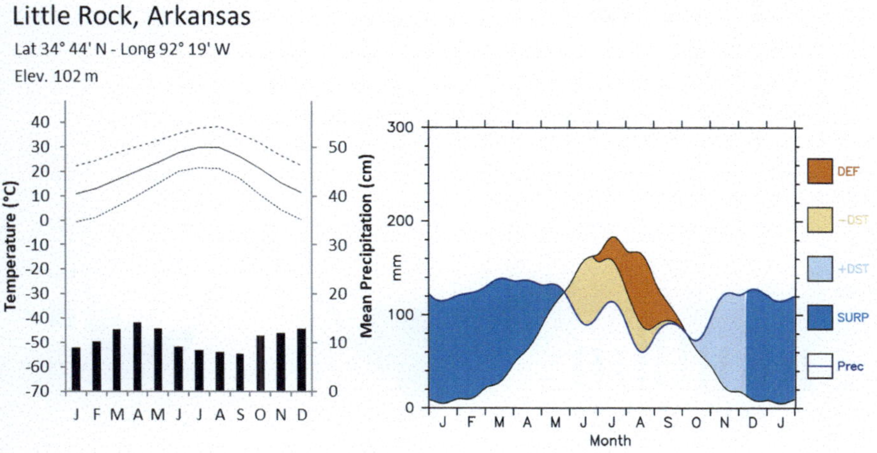

Fig. 13.4 Climagraph and water balance diagram for Little Rock, Arkansas. Data for 1991–2020 averaging period. *Source* Authors. https://weatherspark.com/, http://cyclops.deos.udel.edu/wimp/public_html/

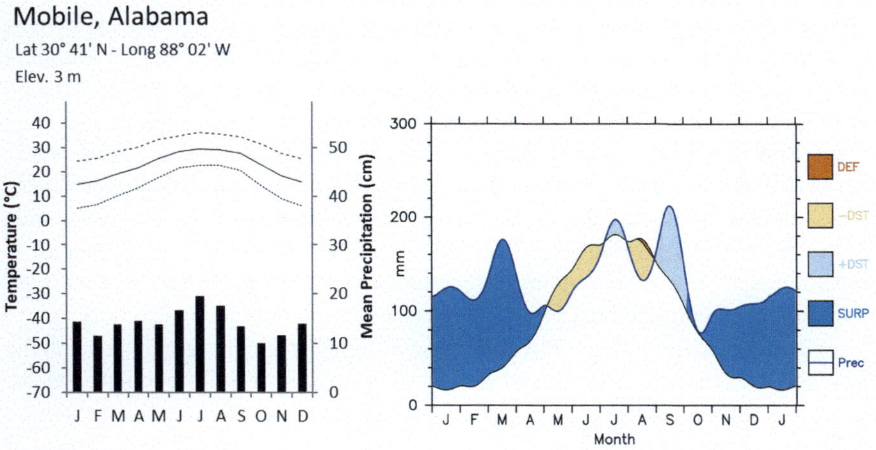

Fig. 13.5 Climagraph and water balance diagram for Mobile, Alabama. Data for 1991–2020 averaging period. *Source* Authors. https://weatherspark.com/, http://cyclops.deos.udel.edu/wimp/public_html/

North of New York City, precipitation peaks in winter because of the convergence of winter cyclone tracks. Regardless of the extent of the meridional or zonal configuration of the *polar front jet stream* west of the Northeast region, this jet stream tends to overlie the Northeast area often. Whereas other locations may receive more or less precipitation associated with migratory frontal cyclones and their corresponding tracks, this region maintains relatively consistent cyclone passages and precipitation from them.

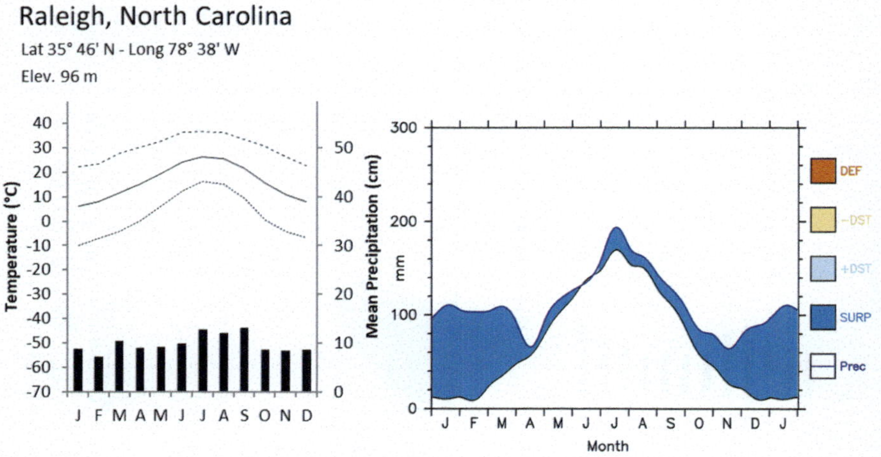

Fig. 13.6 Climagraph and water balance diagram for Raleigh, North Carolina. Data for 1991–2020 averaging period. *Source* Authors. https://weatherspark.com/, http://cyclops.deos.udel.edu/wimp/public_html/

The area is also prone to receiving strong mid-latitude wave cyclones known as *nor'easters*, which typically form in explosive *cyclogenesis* in winter. Nor'easters occur as decaying frontal cyclones move over the warm *Gulf Stream* off the Atlantic coast. The systems undergo rapid strengthening from the rapid influx of low-level energy. This results in a strong mid-latitude wave cyclone, termed a *Hatteras Low*, that tracks along the Atlantic seaboard. Under these conditions, the coast experiences winds from the northeast, hence the name. In addition to strong onshore winds and destructive surf, nor'easters may dump abundant snowfall along the coast if temperatures are low. The systems are particularly devastating because of the high population density of the area.

Nor'easters are difficult to predict because a slight shift in the track can mean the difference between rain and several feet of snow in very heavily populated areas. Because of their predictive uncertainty, municipalities must keep city workers on hand for snow removal and other city services. If the area receives rainfall instead of snow, the time and expense is wasted. But if the area experiences abundant snowfall unexpectedly, then transportation and other city services become hampered.

Climagraphs and water balance diagrams for representative cities of the U.S. Northeast (Fig. 13.7) include those for Washington, District of Columbia (Fig. 13.8), and Philadelphia, Pennsylvania (Fig. 13.9), which both show small summer maxima in precipitation. This pattern suggests moisture advection from the Bermuda-Azores High. The climagraphs for the higher-latitude Boston, Massachusetts (Fig. 13.10), and Portland, Maine (Fig. 13.11), display no such summer precipitation maxima, but instead a more steady pattern from month to month.

Temperatures for all representative cities generally decrease with increasing latitude. However, the climatic data and corresponding climagraphs and water balance

Fig. 13.7 Northeastern U.S. cities depicted in the corresponding climagraphs and water balance diagrams. *Source* Geaux Geospatial

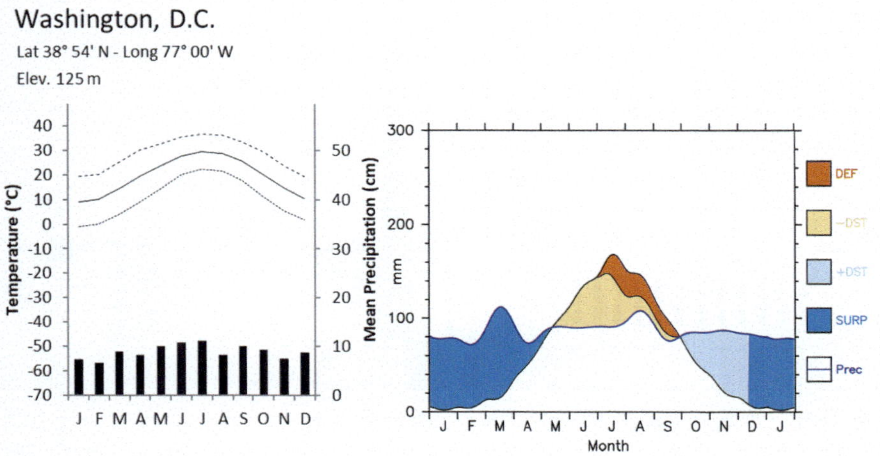

Fig. 13.8 Climagraph and water balance diagram for Washington, District of Columbia. Data for 1991–2020 averaging period. *Source* Authors. https://weatherspark.com/, http://cyclops.deos.udel.edu/wimp/public_html/

diagrams indicate summer water stress and *deficits* when *potential evapotranspiration* rates typically exceed precipitation during the warm season, as is also the case in much of the Southeast. Water *surpluses* are associated with the cooler months.

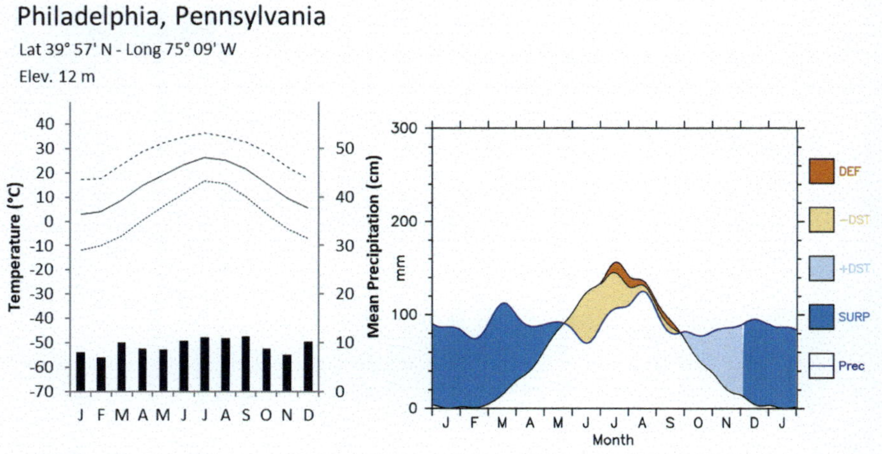

Fig. 13.9 Climagraph and water balance diagram for Philadelphia, Pennsylvania. Data for 1991–2020 averaging period. *Source* Authors. https://weatherspark.com/, http://cyclops.deos.udel.edu/wimp/public_html/

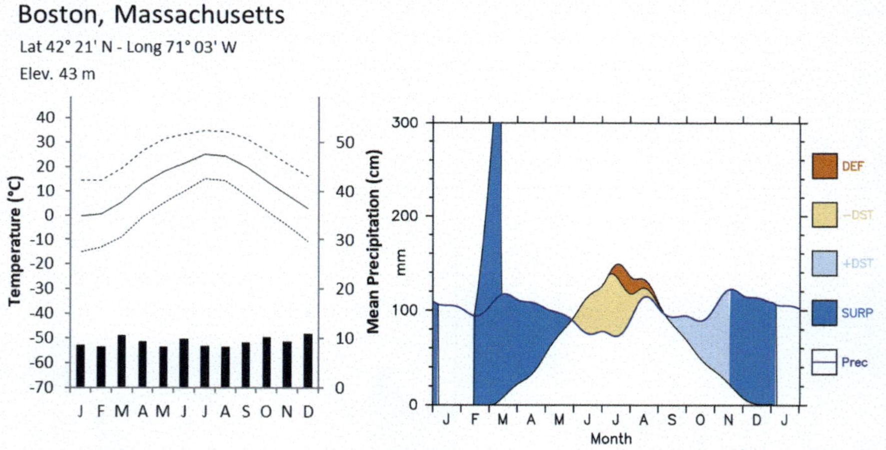

Fig. 13.10 Climagraph and water balance diagram for Boston, Massachusetts. Data for 1991–2020 averaging period. *Source* Authors. https://weatherspark.com/, http://cyclops.deos.udel.edu/wimp/public_html/

The Continental Interior

The Continental Interior of the U.S. includes the area between the spine of the Rocky Mountains to the west and the Appalachian Mountains to the east, near the Great Lakes, roughly corresponding to the American Midwest (Fig. 13.12). Climatic zones vary primarily by latitude, with the southern part of the region classified as Humid Subtropical (Cfa) climate, the central area as *Humid Continental—hot summer*

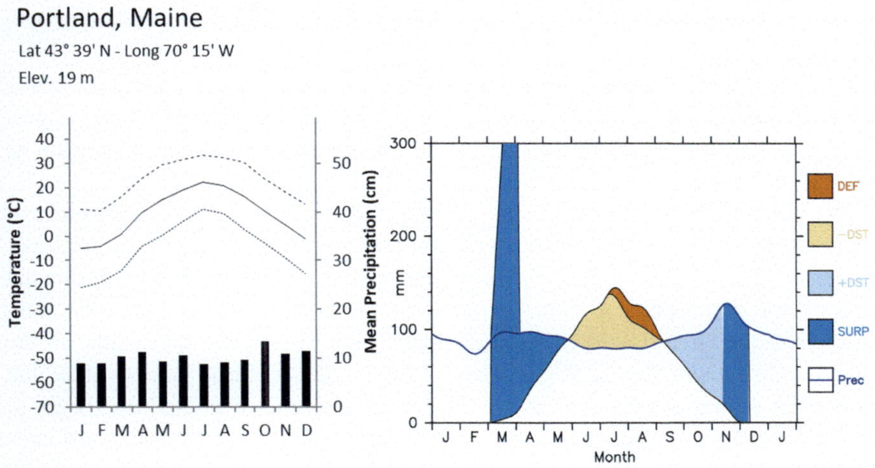

Fig. 13.11 Climagraph and water balance diagram for Portland, Maine. Data for 1991–2020 averaging period. *Source* Authors. https://weatherspark.com/, http://cyclops.deos.udel.edu/wimp/public_html/

Fig. 13.12 Midwestern U.S. cities depicted in the corresponding climagraphs and water balance diagrams. *Source* Geaux Geospatial

(Dfa) *climate* and the northern reaches as Humid Continental—warm summer (Dfb) climate.

The 22 °C (72 °F) *isotherm* for the warmest monthly mean temperature of the year separates the Dfa and Dfb climates. The warmer summer temperatures suggest more *continentality*, which leads to more extreme seasonal variations than locations closer to an ocean.

The polar front jet stream exerts a strong influence on both temperature and precipitation for this region. Cooler temperatures are north of the jet while warmer temperatures prevail to its south. The migratory frontal cyclones embedded within the trough-to-ridge side of the *Rossby wave* generate nearly all cool-season precipitation, except near the Great Lakes, where *lake effect snow* is extreme, especially on the eastern and southern shores, where some of the highest regional snow totals in the world exist. The lake effect snows are of reduced importance as the lakes begin to freeze. Whether from lake effects or *synoptic* weather systems, snowfall contributes significant additions to the surface *water balance* for this region. Water balance diagrams typically show a large spike in runoff in late spring associated with the "first flush" when abundant snowpack melts.

Summer precipitation is most commonly produced by convective air mass thunderstorms. Periodically in summer, the region also experiences frontal passage associated with a mid-latitude wave cyclone core moving north of the U.S./Canadian border. Typically, the trailing *cold front* of the system sweeps through the region, bringing drier and cooler air. The migratory systems provide short-term relief from the generally warm and often oppressively humid summer conditions that dominate the continental interior.

Warm-season precipitation is largely tied to moisture advection from Bermuda-Azores High circulation. The associated airflow advects humid air inland, primarily from the Gulf of Mexico. The moisture drives thunderstorm development. The convective activity may be exacerbated by the development of *mesoscale convective systems* (MCSs) or *mesoscale convective complexes* (MCCs) which bring copious precipitation and severe weather conditions over large areas.

Winter precipitation type in association with mid-latitude wave cyclone passages depends on the temperatures of the contrasting air masses. South of the frontal system core, rain, freezing rain, and sleet are common. North of the system core, and behind (i.e., west or northwest) the migrating system, snowfall often occurs.

Precipitation in this region is augmented by the presence of the Great Lakes. Areas l*eeward* of the lakes receive more precipitation than those on the windward side. This is especially true in winter, with the development of lake effect snows on the leeward side of the water bodies. For example, Minneapolis receives an average of approximately 129 cm (51 in) of snow annually and Detroit experiences 104 cm (41 in) on average, but Buffalo, which is more often on the leeward side of Lake Erie, collects an average exceeding 228 cm (90 in) per year. Similar snow belts exist on the other side of the lakes in Ontario, Canada.

Lake effect snow may occur with the passage of either low-pressure or high-pressure systems. Any subfreezing airflow over the unfrozen lake surfaces triggers lake effect snow processes on the leeward side as the airflow receives water vapor

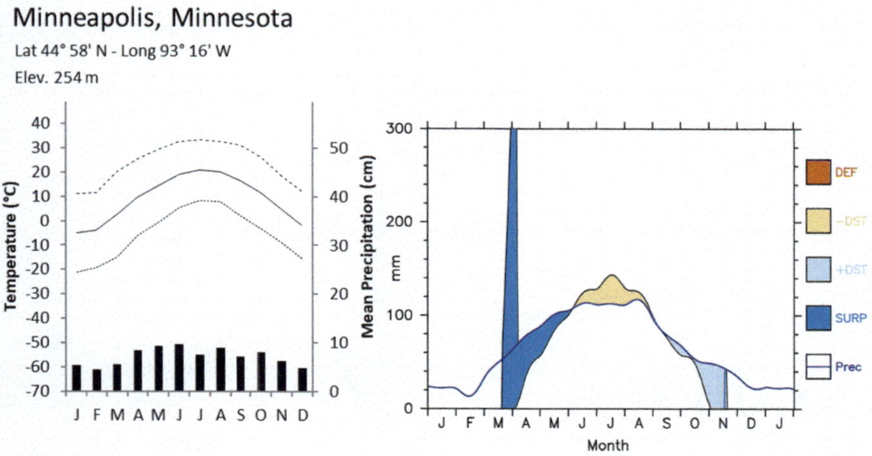

Fig. 13.13 Climagraph and water balance diagram for Minneapolis, Minnesota. Data for 1991–2020 averaging period. *Source* Authors. https://weatherspark.com/, http://cyclops.deos.udel.edu/wimp/public_html/

evaporated from the lake surface. Typical lake effect snow bands extend about 160 km (100 mi) from the lakes. The snows are most prominent in early winter, before ice begins to form on the lakes. Even though Lake Erie is the most southerly of the Great Lakes, it is usually the lake that accumulates the most ice because its shallow depth prohibits it from storing enough energy in summer to be released upward well into the winter. As a result, ice forms earlier in the winter season than in the other Great Lakes.

Climagraphs and water balance diagrams for four representative cities of the region (see again Fig. 13.12) include Minneapolis, Minnesota (Fig. 13.13), which has generally low temperatures from October through April. During that time, precipitation is primarily snowfall. The associated water balance indicates a surplus spike in March. This first flush occurs as snowmelt runs off the landscape and into streams when temperatures warm above the freezing mark. Lansing, Michigan (Fig. 13.14), and Akron, Ohio (Fig. 13.15), have similar first flush spikes. However, Louisville, Kentucky, which is farther south (Fig. 13.16), has an extensive winter water surplus because the higher winter temperatures produce liquid precipitation without significant snowfall. All representative cities have small late spring precipitation maxima, but precipitation is well distributed throughout the year.

The Great Plains

The part of the U.S. between the Rocky Mountains and the Mississippi River is termed the Great Plains because of the fairly uniform landscape and slight physiographic relief throughout the region. This sparsely-populated area is also notable for the

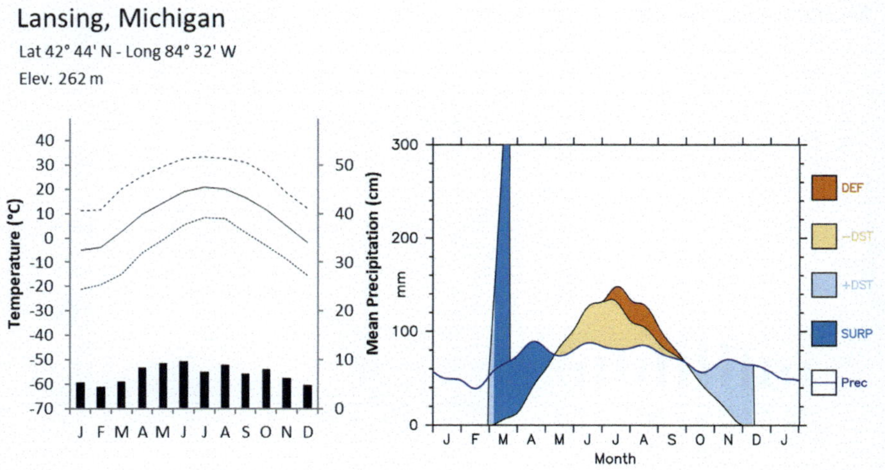

Fig. 13.14 Climagraph and water balance diagram for Lansing, Michigan. Data for 1991–2020 averaging period. *Source* Authors. https://weatherspark.com/, http://cyclops.deos.udel.edu/wimp/public_html/

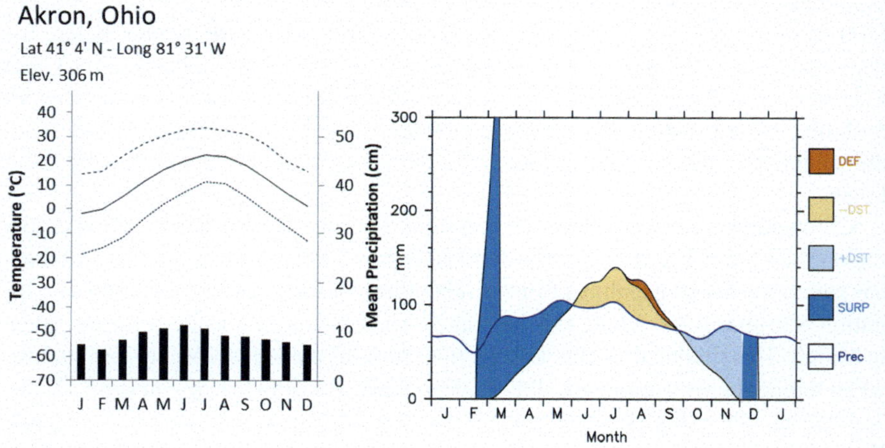

Fig. 13.15 Climagraph and water balance diagram for Akron, Ohio. Data for 1991–2020 averaging period. *Source* Authors. https://weatherspark.com/, http://cyclops.deos.udel.edu/wimp/public_html/

high-quality loam soils that allow the cultivation of grains such as wheat and corn. Climate is as important to the Great Plains economy as it is in any other region of the U.S. or Canada.

One general influence on climate within this area is that elevation gradually increases from east to west. The region terminates at the Rocky Mountains, where elevation increases abruptly. The mountain "Front Range" marks the western regional boundary.

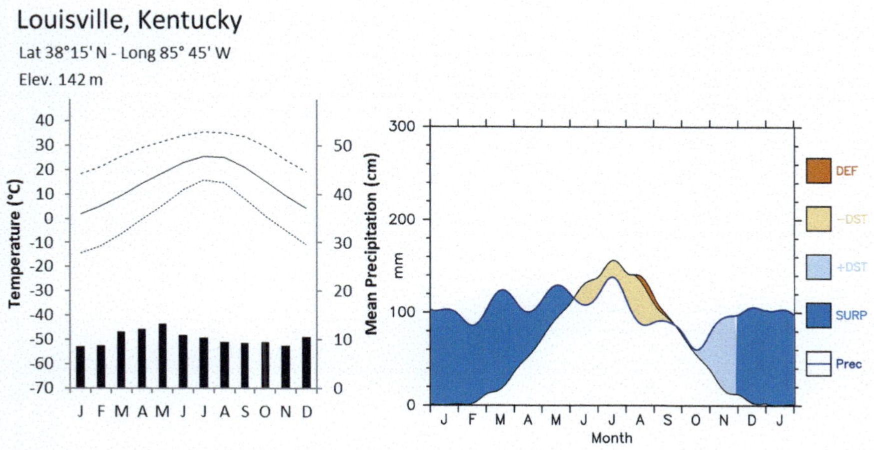

Fig. 13.16 Climagraph and water balance diagram for Louisville, Kentucky. Data for 1991–2020 averaging period. *Source* Authors. https://weatherspark.com/, http://cyclops.deos.udel.edu/wimp/public_html/

Precipitation totals decrease as elevation increases, with low totals occurring in the higher-elevation west and increasing toward the lower-elevation southeast. The Gulf of Mexico is the primary moisture source for the area.

Temperatures show great variation through the year. The region is significantly removed from water sources, which allows significant continentality to produce the large annual temperature range. Note the increase in the *annual temperature range* for the higher-latitude and more westerly representative cities in this region (Fig. 13.17) of Bismarck, North Dakota (Fig. 13.18), and Denver, Colorado (Fig. 13.19).

Precipitation also shows a geographic influence with the western cities of Amarillo, Texas (Fig. 13.20), and Denver being notably drier than the other examples. Kansas City, Missouri (Fig. 13.21)—the easternmost city—is markedly wetter than the others. Great Plains precipitation typically reaches a maximum in summer or late spring. Precipitation is primarily convective in nature with resulting *air mass thunderstorms* being particularly frequent and intense when water vapor is abundant. This scenario is most likely in the warmest months, when the Bermuda-Azores High is in its strongest and most poleward position. The water vapor feeds convective thunderstorms in addition to destabilizing the low-level environment during times of abundant surface heating. Large convective air mass thunderstorms and MCS or MCC development is common in such situations.

Low-pressure systems associated with fronts or mid-latitude wave cyclones originate and migrate through the region all year, but especially in winter and spring. These wave cyclones are associated with upper-level Rossby wave troughing which normally occurs leeward (eastward) of the Rocky Mountains. Despite the frequency of mid-latitude wave cyclones in winter and early spring, these months typically have the smallest precipitation totals, as moisture advection into the region is reduced at that time of year with the weakening and southward-displaced position of the

Fig. 13.17 Great Plains U.S. cities depicted in the corresponding climagraphs and water balance diagrams. *Source* Geaux Geospatial

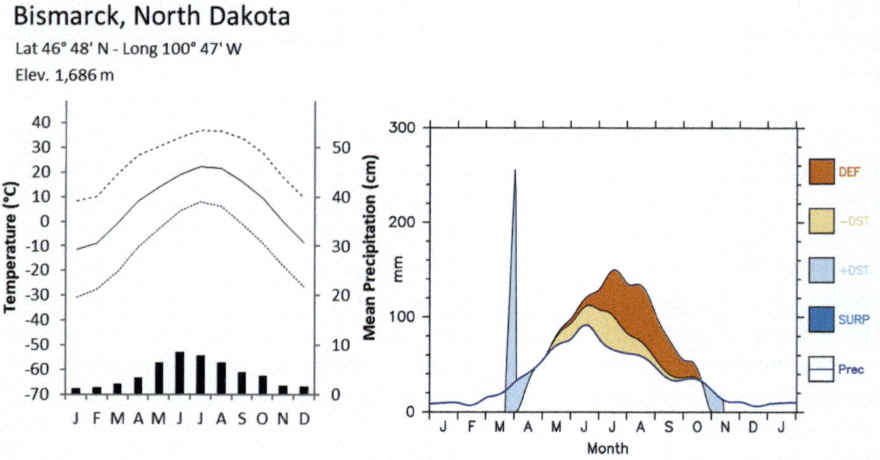

Fig. 13.18 Climagraph and water balance diagram for Bismarck, North Dakota. Data for 1991–2020 averaging period. *Source* Authors. https://weatherspark.com/, http://cyclops.deos.udel.edu/wimp/public_html/

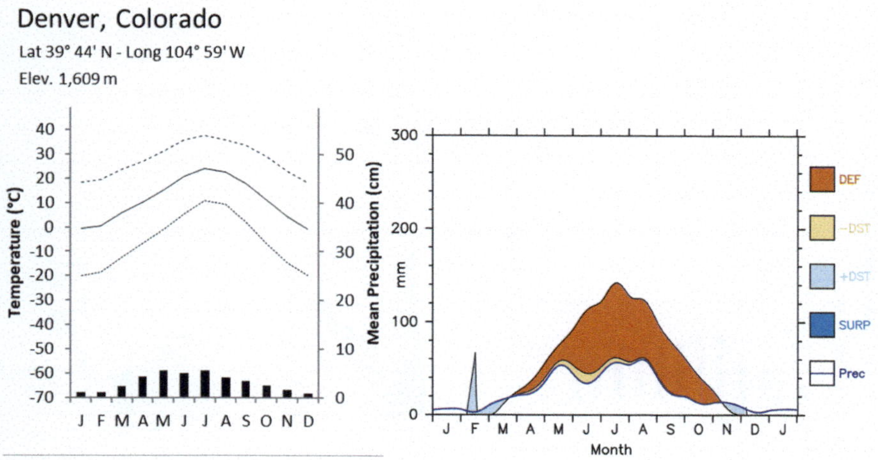

Fig. 13.19 Climagraph and water balance diagram for Denver, Colorado. Data for 1991–2020 averaging period. *Source* Authors. https://weatherspark.com/, http://cyclops.deos.udel.edu/wimp/public_html/

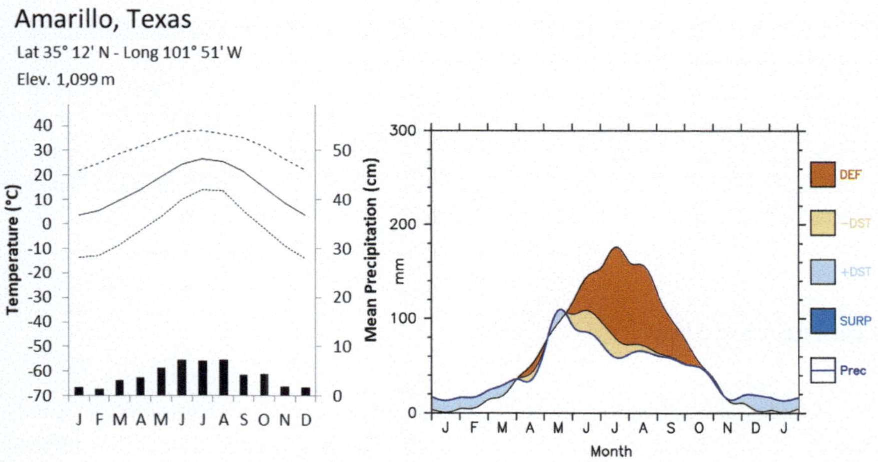

Fig. 13.20 Climagraph and water balance diagram for Amarillo, Texas. Data for 1991–2020 averaging period. *Source* Authors. https://weatherspark.com/, http://cyclops.deos.udel.edu/wimp/public_html/

Bermuda-Azores High. Nevertheless, powerful storms that drop intense precipitation are possible in winter and spring.

Precipitation may be enhanced by *orographic lifting* from the east in association with the rise of elevation. *Adiabatic* cooling occurs as the warm and humid air ascends the slopes, augmenting precipitation events in addition to causing widespread fogs. Other forms of precipitation, such as tropical cyclones, are of minor importance for the Great Plains. Remnants of a tropical cyclone may occasionally penetrate inland in

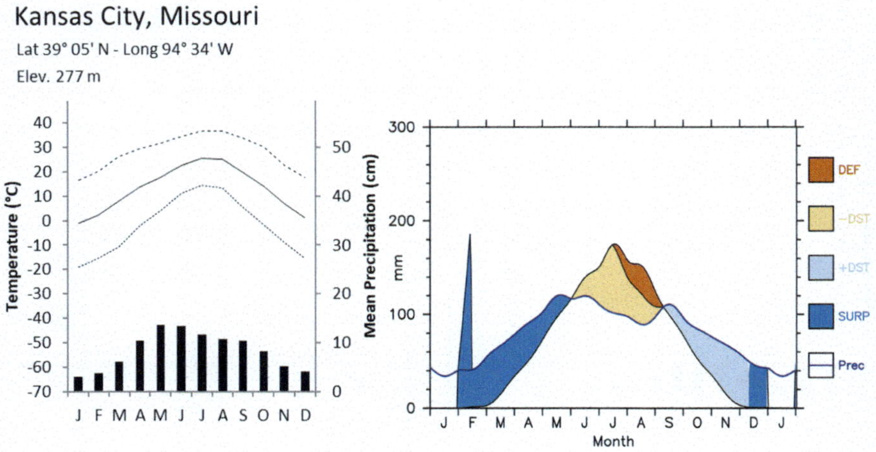

Fig. 13.21 Climagraph and water balance diagram for Kansas City, Missouri. Data for 1991–2020 averaging period. *Source* Authors. https://weatherspark.com/, http://cyclops.deos.udel.edu/wimp/public_html/

the region and contribute heavy precipitation totals, but usually such systems recurve to the northeast before reaching the Great Plains.

The continentality influence is also prominent in the water balance diagrams. Summer is a time of high temperatures and high evapotranspiration rates. Despite summer maxima in precipitation, summer moisture deficits are still common in all of the representative cities. In the higher-latitude locations, a spike in water balance surplus in late spring is coincident with first flush snowmelt.

The Interior West

Dry climates are defined as those in which potential evapotranspiration routinely exceeds precipitation. By this definition, dry climates cover much of the interior western U.S. and extend southwestward off the coast of California into the eastern North Pacific. Interior drainage in which evaporation leaves behind salts—as in Utah's Great Salt Lake—is characteristic of such regions. Four cities are selected as representative of this region (Fig. 13.22). A small section of adjacent interior western Canada also qualifies as dry.

The *True Desert* (BW) *climate* in North America is centered on the southwestern U.S. and adjacent northern Mexico. The *Semi-arid* (BS, or Steppe) *climate* is a transitional border between the very dry BW climate region and wetter surrounding locations. The BS climate surrounds the BW core except where disrupted by a coast. BS water balance deficits are persistent enough to preclude tree growth, except adjacent to lakes or streams, or in other local low spots, but are not as arid as the BW locations. The annual variability of precipitation in BS regions is also typically far

Fig. 13.22 Interior Western U.S. cities depicted in the corresponding climagraphs and water balance diagrams. *Source* Geaux Geospatial

greater than for BW areas. One year may see copious precipitation while the next may experience relatively little, but over long time scales the region is classified as arid. The Rio Grande, which forms much of the border between the U.S. and Mexico, carries a wide-ranging volume of water.

The North American BS climate is generally bordered abruptly by the Cascade and Sierra Nevada Mountains to the west and more gradually to the east by the eastern Great Plains. Latitudinally, the region extends from northwestern Mexico into southwestern Canada.

North America's dry climates result from specific circulation features. Subsidence extends in a belt globally, including the southwestern U.S. and adjacent northwestern Mexico, near the 30° parallel of latitude. This belt includes, but is not limited to, the subtropical anticyclones. Dry regions also exist in western continental locations and over adjacent eastern ocean basins because of the stable atmospheric conditions induced by cold ocean currents. Arid regions may also be induced by a *rain shadow* on the leeward sides of mountain chains as air subsides downwind of the mountain axis. And finally, regions may be dry because of their extreme interior continental locations, far removed from significant moisture sources. Any of the above, or any combination of the above factors, may create a dry region. Collectively, they produce the large area of B climates characteristic of much of the Interior West of the U.S. and Canada. Of the five major Köppen climate categories, the B climate is the most extensive terrestrial type, occupying 27 percent of the terrestrial Earth.

Despite the overall arid conditions, in summer the region may experience rainfall associated with the clockwise flow around the Bermuda-Azores High. These summer storms tend to be spotty but violent, and may produce heavy precipitation and frequent lightning. This phenomenon tends to occur during the so-called **North American monsoon**, caused by *convergence* of air from both the Gulf of California and the Gulf of Mexico to fill in air that ascended in a *thermal low* associated with intense daytime surface heating during the hottest months. The North American monsoon is noticeable in the Phoenix climagraph and water balance diagram (Fig. 13.23). Nevertheless, the monsoonal effect does not extend far inland and is small, weak, and short-lived compared to that in other parts of the world. Most notable are the extremely high temperatures and low precipitation totals that lead to large water deficits through most of the year in Phoenix.

The subdivisions of the True Desert (BW) climates—the *True Desert—hot* (BWh) *climate* and *True Desert—cold* (BWk) *climate* — are delineated primarily based on winter temperatures. Lower-latitude deserts rarely experience a month with an average temperature below 0 °C (32 °F), while higher-latitude areas can have monthly mean temperatures below freezing in winter. During winter, *Chinook winds* can be important sources of rapid local warming in the northern part of this region. Chinooks occur when air is forced over a mountain from a high-pressure system on either side, forcing air to descend adiabatically on the leeward side. In winter, the result is a substantial warm pulse of air capable of melting snowpack.

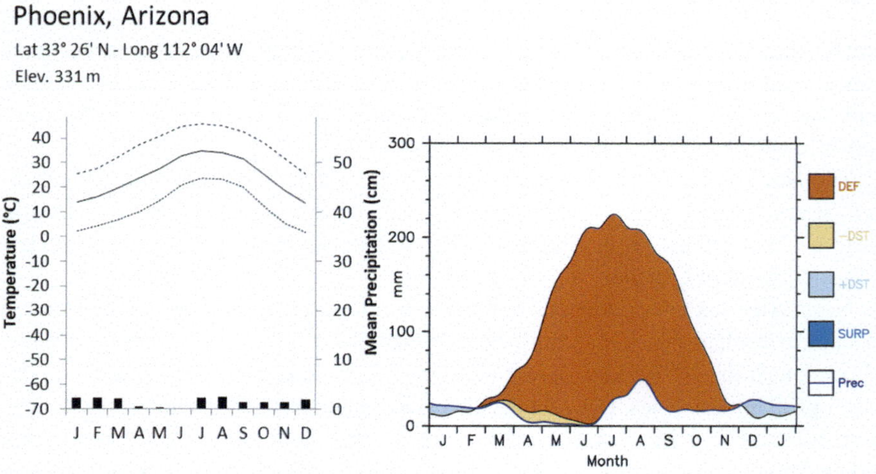

Fig. 13.23 Climagraph and water balance diagram for Phoenix, Arizona. Data for 1991–2020 averaging period. *Source* Authors. https://weatherspark.com/, http://cyclops.deos.udel.edu/wimp/public_html/

Places with arid climates typically have wide temperature ranges throughout the day. Daytime summer temperatures may be scorching, but just a few hours later after sunset, the temperature plummets. The diurnal temperature range typically exceeds the annual range in mean monthly temperatures throughout the year. The wide daily temperature range is caused by the lack of *humidity.* Without much available water, most of the daytime solar radiation (K↓) and Q* is used for sensible heating and little latent heating occurs, leading to a high *Bowen ratio.* Little energy is used to evapotranspire water and minimal cloud cover is present to reflect or absorb K↓. At night, however, the lack of humidity and cloud cover allows a high percentage of the longwave radiation to exit to space as *outgoing longwave radiation.* This cools the surface and leaves less water vapor (by far the most abundant *greenhouse gas*) to absorb that longwave radiation and re-radiate it back to the surface as *counter-radiation* (K↓).

Average summer temperatures frequently surpass 38 °C (100 °F) in Southwest desert areas, with daily maxima of 44 °C (110 °F) or higher occurring periodically. The highest recorded temperature in North America was 57 °C (134 °F) at Furnace Creek Ranch, Death Valley, California, on 13 July 1913. Seasonally, winter temperatures are relatively cool, with average monthly temperatures around 10 °C (50 °F) during January and February.

Situated north of the arid Southwest but still within the Interior West is the Intermontane Basin, which is sandwiched between the western ranges such as the Cascades and Sierra Nevada and the Rocky Mountains. Little precipitation falls in the Intermontane Basin because of this *double rain shadow* effect caused by the bordering mountains. Climagraphs and water balance diagrams appear in Figs. 13.24, 13.25 and 13.26 for Salt Lake City, Utah; Boise, Idaho; and Spokane, Washington; respectively. The climagraphs show extensive temperature continentality, typical of inland, moisture-scarce, mid-latitude locations.

Stations in the northern Intermontane Basin, including Boise and Spokane, often have a winter precipitation maximum because of the position of the polar front jet stream over the area at that time. Winter storms are from mid-latitude wave cyclones, but because of the milder *maritime polar* (mP) *air mass* to the west of the systems, the storms produce generally broad and gentle precipitation. Snow is uncommon in low-elevation areas, although it occurs. Higher elevations may experience significant snowfall. Extreme southern parts of the region are impacted somewhat by the North American monsoon. Salt Lake City has slightly greater precipitation totals than most nearby locations in the region, because of its high elevation. At all four sites in the Intermontane Basin, the water balance diagrams indicate substantial deficits during warm periods. The three northern sites have a spring first flush surplus spike.

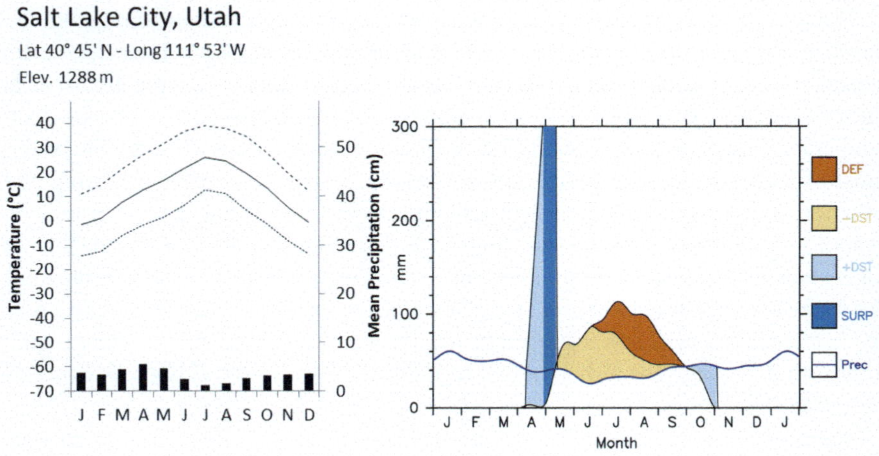

Fig. 13.24 Climagraph and water balance diagram for Salt Lake City, Utah. Data for 1991–2020 averaging period. *Source* Authors. https://weatherspark.com/, http://cyclops.deos.udel.edu/wimp/public_html/

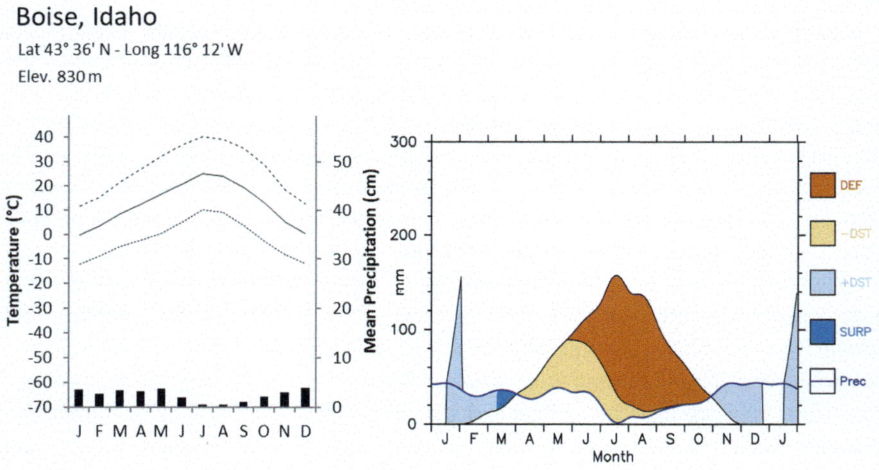

Fig. 13.25 Climagraph and water balance diagram for Boise, Idaho. Data for 1991–2020 averaging period. *Source* Authors. https://weatherspark.com/, http://cyclops.deos.udel.edu/wimp/public_html/

The West Coast

The U.S. and Canadian Pacific coast is one of the most complex climate regions on the continent, with a vast array of spatial variability and many micro-climates, because of the many mountain systems in the area. Temperatures are generally mild near the coast because of the Pacific Ocean's high *specific heat* and the consistently

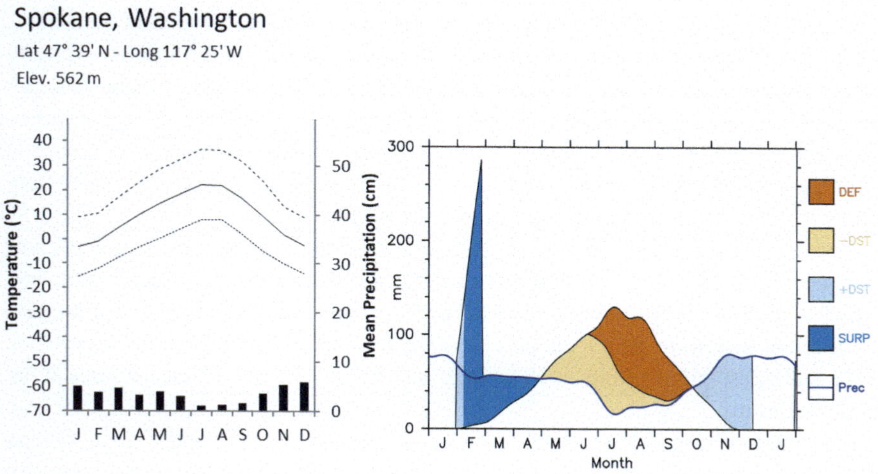

Fig. 13.26 Climagraph and water balance diagram for Spokane, Washington. Data for 1991–2020 averaging period. *Source* Authors. https://weatherspark.com/, http://cyclops.deos.udel.edu/wimp/public_html/

cold *Alaska Current* and *California Current*. Elevation keeps highland areas pleasant in summer and cold in winter, but the *slope aspect* and angle are also important determiners of local temperatures. Peak temperatures occur late in summer, especially west of the mountains, because water's ability to store solar radiant energy and release it slowly over time, both diurnally and seasonally.

Santa Ana winds are a southern California equivalent to the orographic effect in the Chinooks. Unlike the Chinooks, however, which occurs primarily during winter, the Santa Ana winds occur in summer, when the air is already dry from subsidence caused by the *Hawaiian-Pacific High* and an associated *temperature inversion* over the region. The Santa Ana is formed when an anticyclone over the Intermontane Basin forces dry air westward, first upslope and then downslope, causing hot, dry winds to reach very heavily populated areas of southern California.

Precipitation is sparse along the southwestern coast, which remains particularly sunny, in large part because the California Current stabilizes the atmosphere through the year, but especially in summer when the water is even colder compared to the air around it, and also because the sinking associated with the Hawaiian-Pacific High is nearby. Tropical cyclones are rare but possible. During summer, strengthening and expansion of this subtropical anticyclone induces subsidence, leading to very dry conditions. However, once the cold California Current finally warms, it does so quickly and the warmth of the surface water circulates rapidly throughout the ocean basin. The cold-warm boundary beneath the polar front jet stream bounces back northward suddenly in late summer, to occupy a position near Alaska. The return of at least some precipitation in the colder months leaves the southwestern coast in a classic *Mediterranean—warm summer* (Csb) *climate*, with the *Mediterranean—hot summer* (Csa) *climate* farther inland but still near the coast.

Vegetation is specifically adapted to the dry summers with abundant solar radiation. **Chaparral** vegetation, comprised of short, woody, hardy shrubs specially adapted to dry soils and high temperatures, dominates this area. The chaparral vegetation of the region is exposed and somewhat resistant to wildfires. Many wildfires today are caused by human interactions with the dry environment, especially during Santa Ana winds, which blow drier air into the region. Large areas are adversely affected as the fires spread rapidly in these conditions.

Precipitation generally increases from south to north, as the Csa climate gives way to the *Marine West Coast* (Cfb) *climate* in coastal Oregon, Washington, and British Columbia. Position on the *windward* versus leeward side of a mountain is important in determining local precipitation amounts. Windward sides experience much more precipitation and smaller water balance deficits than leeward sides at the same elevation.

Precipitation everywhere along the Pacific coast of the U.S. and Canada occurs primarily from mid-latitude wave cyclone passages. However, these passages are restricted to the cool season in the southwestern coastal area and become more abundant and more commonly year-round phenomena with distance northward. However, few thunderstorms occur anywhere in this region, as only weak air mass contrast, with maritime polar air dominating, is present.

El Niño alters the typical storm intensities and tracks, steering storm systems from the Hawai'i area to California and contributing snowpack in the mountainous parts of California. The associated mid-latitude wave cyclone tracks that begin near the Hawaiian Islands are termed the *Pineapple Express*. These systems may be quite strong and bring flash flooding and mudslides to the typically dry region.

Climagraphs and water balance diagrams for four cities (Fig. 13.27) are selected here to represent the Pacific coast of the U.S. and Canada. Moving from south to north, at San Diego and San Francisco, California, the climatic data (Figs. 13.28 and 13.29, respectively) reveal the small annual temperature range associated with the strong influence of the ocean in these Csa climates. The lack of summer precipitation suggests the influence of the stabilizing Hawaiian-Pacific High and particularly the cold California Current on its eastern side.

The climagraphs and water balance diagrams for two sites that are much farther north—Seattle, Washington (Fig. 13.30) and Yakutat, Alaska (Fig. 13.31), respectively—show lower temperatures, much larger annual temperature ranges, and much more abundant precipitation. Substantial water balance deficits occur throughout the warm season, especially on the southwestern coast, but not at Yakutat, where the lower temperatures and much greater precipitation totals create no water stress, even during the warmest months. The month of maximum precipitation occurs later in winter toward the south, but is most often in winter as the polar front jet stream gradually moves southward and Hawaiian-Pacific High retreats southward and weakens during the cool season.

The diversity of U.S. climates, as depicted by data and graphics, gives the U.S. and its adjoining ocean environs widely-ranging environmental conditions. But those environmental conditions at a given place vary through time. The same is true in adjacent Canada.

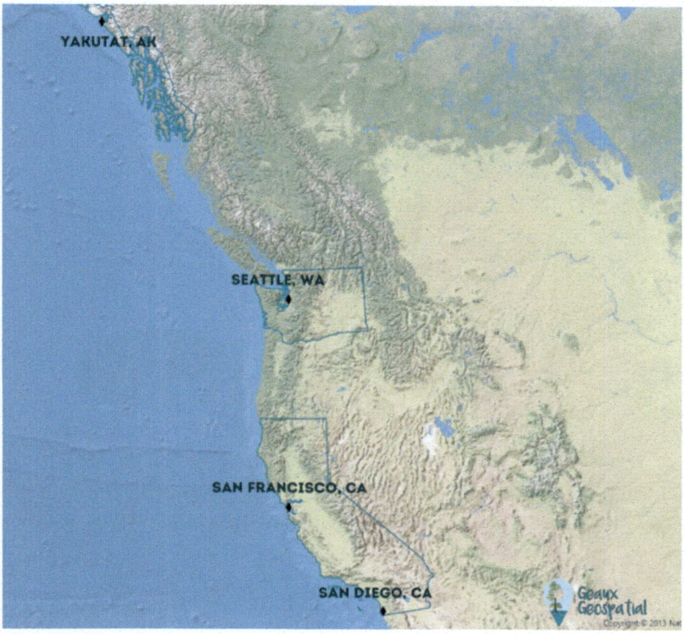

Fig. 13.27 Pacific coast U.S. cities depicted in the corresponding climagraphs and water balance diagrams. *Source* Geaux Geospatial

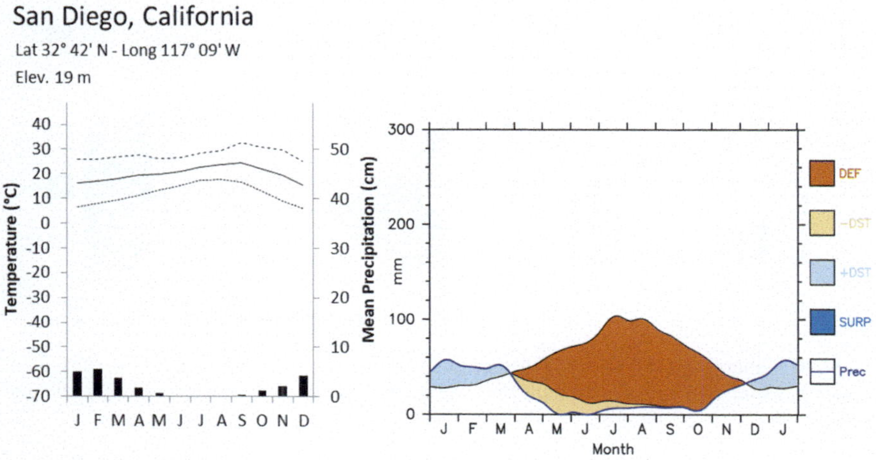

Fig. 13.28 Climagraph and water balance diagram for San Diego, California. Data for 1991–2020 averaging period. *Source* Authors. https://weatherspark.com/, http://cyclops.deos.udel.edu/wimp/public_html/

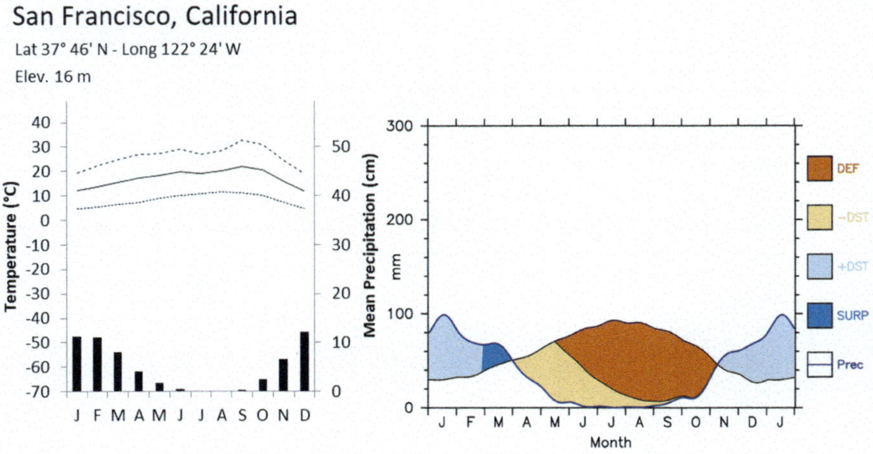

Fig. 13.29 Climagraph and water balance diagram for San Francisco, California. Data for 1991–2020 averaging period. *Source* Authors. https://weatherspark.com/, http://cyclops.deos.udel.edu/wimp/public_html/

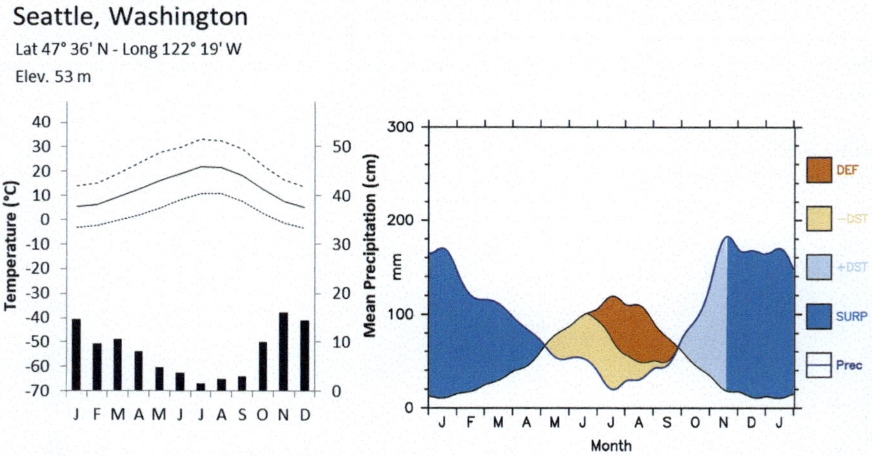

Fig. 13.30 Climagraph and water balance diagram for Seattle, Washington. Data for 1991–2020 averaging period. *Source* Authors. https://weatherspark.com/, http://cyclops.deos.udel.edu/wimp/public_html/

Canadian Examples

Climagraphs and water balance diagrams for three Canadian cities near the U.S.—Halifax, Nova Scotia; Quebec City, Quebec; and Winnipeg, Manitoba (Fig. 13.32)—are shown in Figs. 13.33, 13.34 and 13.35. These represent a continuation of the

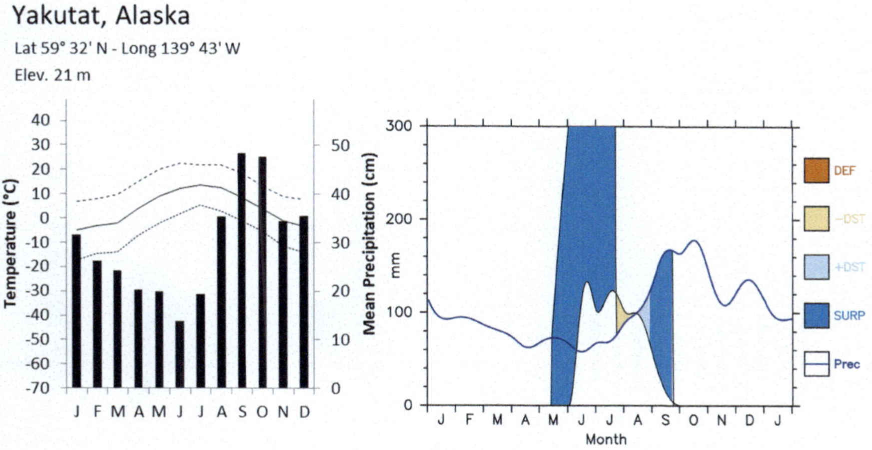

Fig. 13.31 Climagraph and water balance diagram for Yakutat, Alaska. Data for 1991–2020 averaging period. *Source* Authors. https://weatherspark.com/, http://cyclops.deos.udel.edu/wimp/public_html/

features of the U.S. Northeast and Continental Interior, but with slightly colder conditions that tend to reduce water balance deficits and increase the first flush compared to adjacent U.S. cities.

Conditions depicted on the Halifax water balance diagram may be slightly misleading—snow is fairly infrequent along the coast but it is much heavier a short distance inland. Fog is an important component of the climatology in the Maritime Provinces of Canada and the New England coast of the U.S., especially with easterly winds, such as after a mid-latitude wave cyclone passage to the south. Land cover also is important; a study to select the optimal site for a new airport in Halifax based

Fig. 13.32 Canadian cities for which climagraphs and water balance diagrams are shown. *Source* Geaux Geospatial

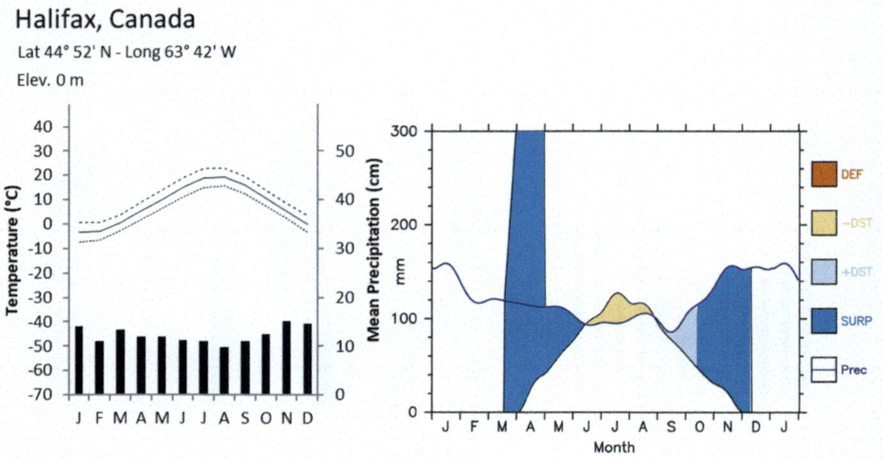

Fig. 13.33 Climagraph and water balance diagram for Halifax, Nova Scotia. Data for 1980–2016 averaging period. *Source* Authors. https://weatherspark.com/, http://cyclops.deos.udel.edu/wimp/public_html/

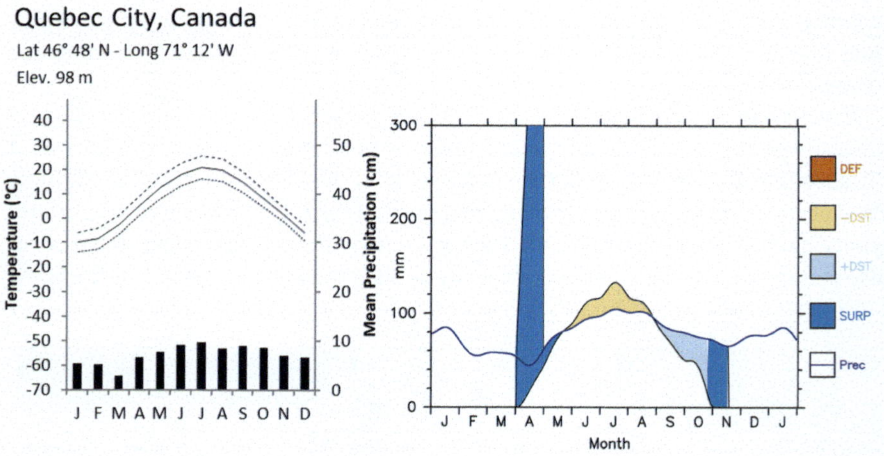

Fig. 13.34 Climagraph and water balance diagram for Quebec City, Quebec, Canada. Data for 1980–2016 averaging period. *Source* Authors. https://weatherspark.com/, http://cyclops.deos.udel.edu/wimp/public_html/

the selection in part on the fact that it was the least foggy location in the area, but after the area was deforested, it became one of the foggiest sites in the area. Climagraphs and water balance diagrams for other Canadian cities, especially those in *Subarctic* (Dfc, Dfd) *climates*, will be presented in Chap. 21—polar climates.

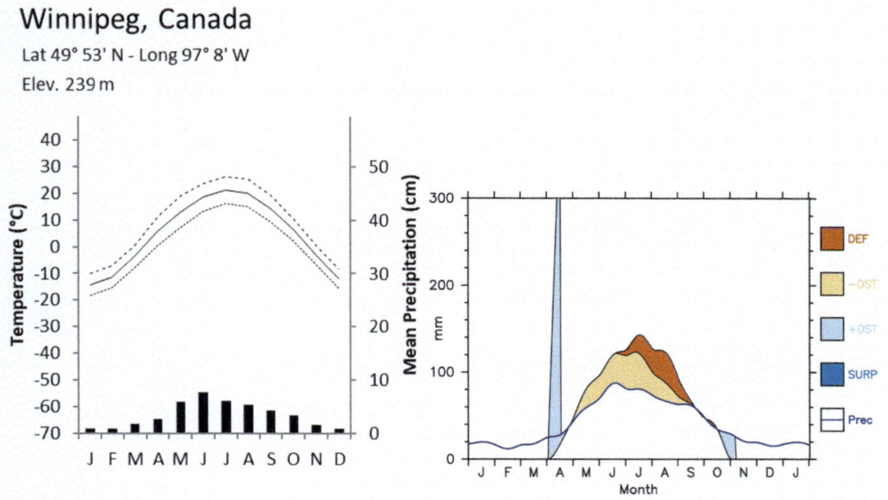

Fig. 13.35 Climagraph and water balance diagram for Winnipeg, Manitoba. Data for 1980–2016 averaging period. *Source* Authors. https://weatherspark.com/, http://cyclops.deos.udel.edu/wimp/public_html/

Further Reading

Adams, D. K., & Comrie, A. C. (1997). The North American monsoon. *Bulletin of the American Meteorological Society, 78*(10), 2197–2213. https://doi.org/10.1175/1520-0477(1997)078%3c2197:TNAM%3e2.0.CO;2

Collins, J. M., Rohli, R. V., & Paxton, C. (2017). *Weather and climate of Florida: More than just sunshine* (248 pp). University Press of Florida. 978–0–81–305444–5.

Guzman-Morales, J., Gershunov, A., Thiess, J., Li, H. Q., & Cayan, D. (2016). Santa Ana wind of Southern California: Their climatology, extremes, and behavior spanning six and a half decades. *Geophysical Research Letters, 43*(6), 2827–2834. https://doi.org/10.1002/2016GL067887

Ingram, K., Dow, K., Carter, L., & Anderson, J. (2013). *Climate of the Southeast United States: Variability, change, impacts, and vulnerability*. Island Press. https://doi.org/10.5822/978-1-61091-509-0.pdf

Schubert, S. D., Chang, Y., Suarez, M. J., & Pegion, P. J. (2008). ENSO and wintertime extreme precipitation events over the contiguous United States. *Journal of Climate, 21*(1), 22–39. https://doi.org/10.1175/2007JCLI1705.1

Vega, A. J., Grymes III, J. M., & Rohli, R. V. (2013). *Louisiana weather and climate* (340 pp). Jones and Bartlett. 0–7637–91628.

Yang, Q., Houze, R. A., Leung, L. R., & Feng, Z. (2017). Environments of long-lived mesoscale convective systems over the central United States in convection permitting climate simulations. *Journal of Geophysical Research-Atmospheres, 122*(24), 13288–13307. https://doi.org/10.1002/2017JD027033

Zielinski, G. A. (2002). A classification scheme for winter storms in the eastern and central United States with an emphasis on Nor'easters. *Monthly Weather Review, 83*(1), 37–52. https://doi.org/10.1175/1520-0477(2002)083%3c0037:ACSFWS%3e2.3.CO;2

Chapter 14
Europe and Adjacent Ocean Environments

Abstract Though considered a continent, Europe is largely a series of peninsulas attached to Asia. The nearby water bodies moderate the temperatures effectively, particularly considering the high-latitude location of the continent. The maritime influence of the North Atlantic Gyre, especially the Bermuda-Azores High and North Atlantic Drift, set up a Marine West Coast (Cfb) climate north of the Alps with frequent presence of maritime polar (mP) air masses and, except for cases of blocking highs, mid-latitude wave cyclone passages. The Bermuda-Azores High, along with the stabilizing influence of the Mediterranean Sea in summer and Canary Current, generate Mediterranean (Csa and Csb) climates south of the Alps. Eastward into central Europe, the maritime influence wanes, as the Siberian High and its continental Arctic (cA) and continental polar (cP) air masses become increasingly dominant. Eastern Europe, including European Russia, are dominated by Humid Continental (Cfa and Cfb) climates, as continentality becomes more pronounced. Northern Russia and Scandinavia have Humid Continental—mild summer (also known as Subarctic; Dfc) and Tundra (ET) climates. The entire Mediterranean arc has dry summers and humid winters, with localized topography that creates powerful and important local wind systems.

Keywords Alps Mountains · Bermuda-Azores High · Canary Current · Enthalpy · Humid Continental (Df) climate · Icelandic Low · Marine West Coast (Cfb) climate · Maritime polar (mP) air mass · Mediterranean (Cs) climate · Mid-latitude wave cyclone · North Atlantic Drift · Siberian High · Tundra (ET) climate

General

Europe is a logical next stop on the tour around the world because it has much in common climatologically with the U.S. and Canada. However, it is influenced more strongly than the U.S. and Canada by the many bays and inlets of the sea, giving it much less *continentality* than the U.S. and Canada. The second-smallest of the traditionally-defined continents, after Australia, Europe's climate is defined by the great influence of the ocean on the atmospheric circulation patterns and air masses.

R. V. Rohli et al., *Atmospheric and Oceanic Circulation*,
https://doi.org/10.1007/978-3-031-66142-6_14

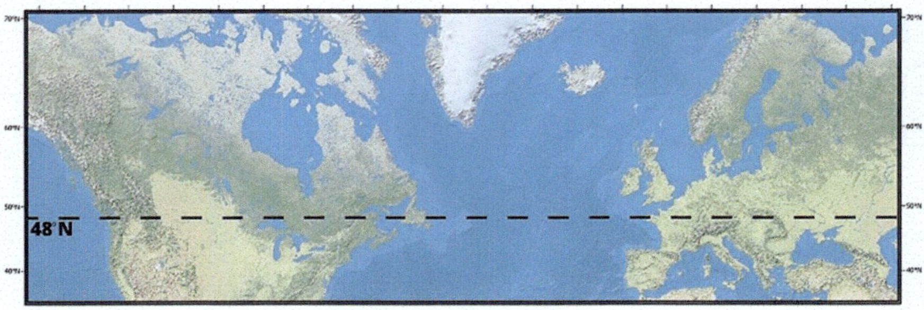

Fig. 14.1 Relative latitude of Europe compared to the continental United States. *Source* Geaux Geospatial

The maritime environment explains why countries like the United Kingdom and Norway have milder *temperatures* than many locations in the continental U.S. despite being located at higher latitudes than any place in the continental U.S. (Fig. 14.1).

The ocean circulation of surface currents in the Atlantic Ocean directs warmer water from the northern margins of the *Gulf Stream* toward Europe as the *North Atlantic Drift*. Sea surface temperatures in the eastern North Atlantic can be 8–10 °C (14–18 °F) warmer than on the western side near eastern Canada. These warm waters moderate regional air masses; the *maritime polar* (mP) *air masses* are more important here than anywhere else on Earth, as the *mid-latitude westerlies* push them deep into the continent and provide the moisture that feeds European storm systems.

In addition to the Atlantic Ocean, the water bodies that have the strongest influence on the atmospheric circulation in Europe—the Mediterranean, Aegean, Black, Baltic, Adriatic, Ionian, and Caspian Seas (although the latter is really a lake)—all produce similar maritime effects (Fig. 14.2). Most of these seas are relatively shallow, so they warm and cool faster than the open ocean and adjacent locations, displaying slightly more seasonality than adjacent locations near the ocean. Still, the Mediterranean Sea is 4–6 °C (7–11 °F) warmer than the adjacent air in winter. This difference has the effect of promoting an *unstable atmosphere*, but because Europe is on the side of the *Bermuda-Azores High* that is stabilized by the cold **Canary Current** – named for the nearby Canary Islands offshore of northwestern Africa, a *stable atmosphere* generally prevails, especially in the western part of the Mediterranean basin.

The maritime influence on Europe is displayed by the *Marine West Coast* (Cfb) *climate type* in the system of *Köppen climate types*. The physical geography of the continent allows the ocean influence to extend well inland from the North Atlantic (Fig. 14.3). Moving eastward, continentality only begins to characterize the climate in central European countries such as Switzerland, Austria, and southern Germany. The eastern European countries of Poland, Ukraine, and Russia have more seasonal climate types, most notably the *Humid Continental—warm summer* (Dfb) *climate*. The *Humid Continental—hot summer* (Dfa) *climate* is sparsely represented, with only localized sections of southeastern Europe and adjacent western Asia.

To the south, the *Mediterranean—warm summer* (Csb) *climate* and the *Mediterranean—hot summer* (Csa) *climate* suggest the temperate environment, but with

Fig. 14.2 Selected topographic features of Europe. *Source* Geaux Geospatial

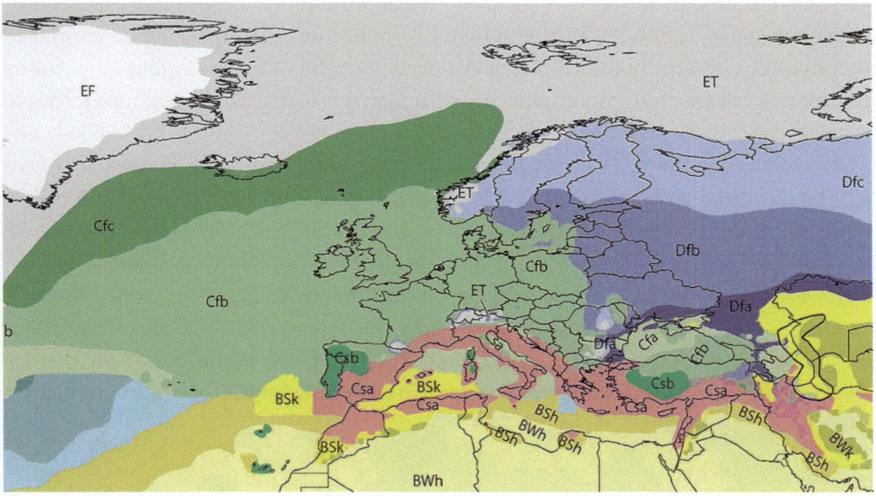

Fig. 14.3 Köppen climate types of Europe and surrounding marine environment. *Source* Modified from Reynolds, Rohli, Johnson, Waylen, and Francek, Exploring Physical Geography, McGraw-Hill

strongly winter-dominant *precipitation* patterns that occur near the Mediterranean Sea. The area covered by the European and adjacent North African and Southwest Asian Cs climates surrounding the Mediterranean Sea justifies the Mediterranean climate name.

Topographic Influences (Site)

The most striking of Europe's local *site* characteristics are the east–west oriented mountain chains. The Pyrenees between Spain and Portugal; the Alps between France, Switzerland, Austria, and Italy; and the Carpathians and Balkans in southeastern Europe (see again Fig. 14.2) effectively separate the climates of the north from those of the Mediterranean region (see again Fig. 14.3). The mountain ranges hinder the north–south exchange of air masses, unlike the situation in the U.S. described in earlier chapters. As a result, frontal zones in Europe are generally weaker because bitter cold air rarely comes into contact with much warmer air.

Just as important as the mountain ranges that do occur is the absence of high north–south-running mountain ranges over western Europe. The only north–south mountain chains are smaller and mostly in northern Europe such as the Pennines and Grampian Highlands in the United Kingdom and the modest mountains in Scandinavia (see again Fig. 14.2). These highlands produce an *orographic effect* similar to the mountains of northwestern North America, but the effects are generally not as pronounced because the mountains are sufficiently low to allow moisture to move eastward over them.

The relatively low elevations allow maritime polar (mP) air masses to move far inland, unlike in North America's western coast, from the Atlantic (Fig. 14.4) along with humid conditions. These air masses are warmer than would otherwise be expected for their latitude because of the prevalence of the North Atlantic Drift.

Because of the high latitude of Europe and the blocking Alps, the effects of *maritime tropical* (mT) *air masses* are largely restricted to the Mediterranean basin region. Moreover, the relatively cold Canary Current often modifies mT air masses substantially, mitigating their sweltering, uncomfortable conditions before reaching Europe.

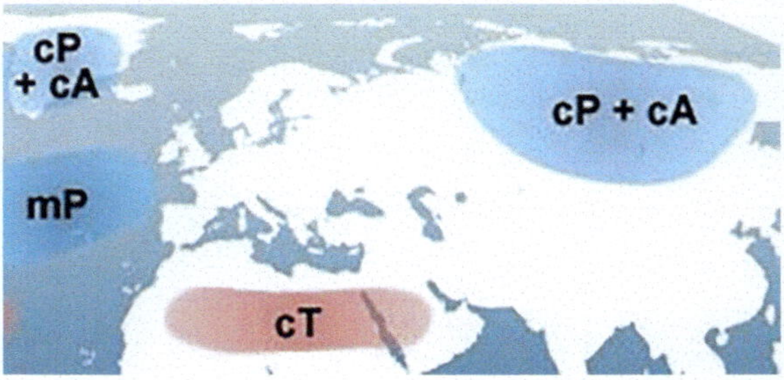

Fig. 14.4 Air masses affecting Eurasia. *Source* UCAR CoMet

Other air mass types are less dominant but also important for European climate. *Continental polar* (cP) *air masses* and *Arctic* (cA or A) *air masses* associated with the *Siberian High* create very cold wintertime conditions across northern Europe. Also, *continental tropical* (cT) *air masses* originating in Africa occasionally affect southern Europe, particularly Italy, Spain, and France.

Circulation Influences (Situation)

As described in earlier chapters, *situation* refers to the broader-scale forces that act upon, and exert implications on, the atmospheric and oceanic features of a place. Most of Europe is in the zone of the mid-latitude westerlies. Moderate air masses with mild temperatures contribute to stable conditions in most areas. Little free *convection*-generated precipitation occurs, as the lack of continentality over western and southern Europe hinders intense surface heating and spontaneous rising motion. Precipitation-forming processes are much more influenced by regional circulation features.

The Bermuda-Azores High, *Icelandic Low*, and Siberian High are the three important semi-permanent pressure cells that dominate Europe's atmospheric circulation. The Bermuda-Azores High strengthens and drifts northward over the subtropical North Atlantic Ocean nearer to Europe in summer. The clockwise and outward flow around this surface high-pressure area sends maritime air from the North Atlantic from west to east, far into Europe. The counter-clockwise and inward flow associated with the Icelandic Low forces onshore flow to its south, which is north of the Bermuda-Azores High, reinforcing the dominant westerly flow of maritime air. Between the Icelandic Low and the Bermuda-Azores High, although winds flow across Europe generally from west to east, their exact orientation is determined by the *amplitude* and position of the *Rossby waves* at any particular time.

Eastern Europe is influenced by the Siberian High, which forms in winter in northern Eurasia because of the bitter cold, and therefore dense, continental air there, which induces sinking. Even though the general circulation suggests that air should have low pressure, not high pressure, in these subpolar areas—as in the case of the Icelandic Low and *Aleutian Low*—the cold air is so intense that it induces *subsidence* and high pressure over this huge land mass. The clockwise and outward flow around the Siberian High pumps cold, dry air into eastern Europe, against the westerlies, and prevents the maritime influence from extending any farther inland. As summer approaches, the long hours of daylight and continentality over western Asia cause the Siberian High to dissipate.

In winter, northeasterly flow from the Siberian High meets southwesterly flow from the Bermuda-Azores High, producing a frontal zone at the cP-mT or A-mT air mass boundary. When the atmospheric properties differ sharply along this boundary, vigorous *cyclogenesis* can occur. But more common, especially in western Europe, is air mass interaction between mT and mP air masses, or between two mP air masses that are only slightly different in atmospheric properties. In such circumstances, only weak fronts and storms form, because of the relatively subtle differences in temperature and density. Thus, Europe is characterized by relatively weak and rare thunderstorms and severe weather. However, regardless of the strength of the air mass contrast, if a trough-to-ridge side of the Rossby wave overlies this air mass boundary and abundant humidity is present in the air, intensification of a frontal zone is likely.

An important source of winter precipitation for western Europe comes from old, *occluded fronts* associated with decaying *mid-latitude wave cyclones* moving from North America across the North Atlantic. These systems can be reinvigorated when they cross the Icelandic Low area, destabilized by warm oceanic waters and characterized by extremely intense ocean–atmosphere heat exchange and subsequent *latent heat* release in the *condensation* and *deposition* processes. The sum of latent and sensible heat—the total energy content of the atmosphere—is known as **enthalpy**. This phenomenon of vigorous enthalpy exchange as dynamic mid-latitude systems become reinvigorated over warm water is similar to the rebirth of dying systems that can occasionally become *nor'easters* off the North American coast, especially when underlain by the trough-to-ridge side of the Rossby wave, but a smaller percentage of the European systems become violent. Nevertheless, strong reinvigoration can occur, creating wind disasters, particularly in France. The peak European storm season is from October to March, but such storms can occur at any time of year.

In 2015, Europeans officially began the practice of naming these violent mid-latitude windstorms, similar to the way that *tropical cyclones* are named. However, different countries may use different names for the same storm, and some countries, such as Norway, name storms that affect their country but the storms may not be named at all by other European countries. Unlike tropical cyclone naming conventions, no universally accepted threshold wind speed or other meteorological criteria exist for the decision of whether to name the storm.

Despite these inconsistencies, some mid-latitude wave cyclone-induced windstorms have been so notorious that the same name for them is recognized almost universally. These include Storm Dennis in February 2020, which generated central pressures of 920 millibars—almost 100 mb below standard sea-level pressure, along with wind gusts of 230 km hr^{-1} (140 mph), as it struck Ireland, the United Kingdom, Scandinavia, the Netherlands, and Germany. Even in the years before such storms were named, some were so violent that the unofficial name has been accepted universally. Examples include Storm Xynthia (February–March 2010) in southwestern Europe; Storm Lothar followed closely by Martin (December 1999) in France, Switzerland, and Germany, which caused over 100 deaths; and "Lothar's Big Brother" (March 1876) that victimized France, the Low Countries of Belgium, Luxembourg, and the Netherlands, and Germany.

Winter cyclogenesis regions also occur elsewhere in Europe. The two other main regions are the western Mediterranean (near Sardinia) and the eastern Mediterranean (near Cyprus). Cyclogenesis in these areas occurs by *advection* of cold air from northern Europe into the Mediterranean region through the lowland areas of southern France and/or the eastern Greece/western Turkey area. The warm water surfaces provide enthalpy and water vapor necessary for cyclogenesis. However, cyclogenesis is rather sporadic compared to the consistent preferred mid-latitude cyclogenesis zones and tracks of the U.S. and Canada.

By the time the Siberian High dissipates with the onset of summer, the Mediterranean Sea begins to be relatively cool, compared to the air above it, stabilizing the atmosphere. This temperature profile creates a *mesoscale* high-pressure belt over the Mediterranean Sea, reinforced by the subsidence side of the Bermuda-Azores High. The generally sunny, clear, dry summer conditions in this Mediterranean climate throughout the basin makes for ideal recreation and tourism.

General Temperature Patterns

Northern Europe is largely a product of the moderating influence of the Atlantic Ocean, while southern Europe is a much warmer product of the Mediterranean Sea. During the warm season, the strong and northward-displaced Bermuda-Azores High advects maritime air inland to produce *zonal isotherms* that correspond to latitude. The farther poleward the location, the cooler the air.

In the cool season, the easterly outflow from the strong Siberian High acting in opposition to the mid-latitude westerlies and preventing maritime air from penetrating into eastern Europe disrupts the simple zonal isotherm pattern. Instead, these conditions tend to produce a *meridional* isotherm pattern that increases wave amplitude and allows deeper penetration of warm air poleward in ridges and cold air equatorward in troughs. The effect in Europe is to advect frigid cP or A air from eastern Europe southward into the deepened Rossby wave troughs. In some cases, a place farther north may be warmer than another place farther south just a few degrees of longitude away.

The occasional anomalous atmospheric circulations notwithstanding, Europe is generally warmer than other locations at the same latitude. For example, Norway is 30–40 °F warmer than locations at similar latitude in Siberia or Canada/Alaska. The strong oceanic influence driven by the warm North Atlantic Drift keeps temperatures moderate for the latitudes, despite Siberian High influence.

As in nearly all other places, temperature decreases more per unit distance vertically than it does latitudinally. Thus, experiencing a different *Köppen climate type* in Europe may be as simple as going up the nearest hill or mountain. This concept of changing Köppen climate types upslope is known as vertical zonation, with the effects often visible on the landscape. The highland regions also support widely complicated growing seasons, which shorten with both elevation and latitude. In some locations, vegetation transitions from deciduous trees (i.e., those that lose their leaves) to tundra over a vertical change of only a few thousand feet.

The farthest poleward or elevational extent of tree survival roughly corresponds to the 10 °C (50 °F) isotherm of the warmest month and is known as the *treeline*. Because of the mild summers in most of Europe, the treeline tends to be relatively low in Europe's mountainous areas. *Tundra* (ET) *climates*, which cannot support trees, occur poleward or upslope of the treeline. Even higher elevations have *Ice Cap*

(EF) *climates*, where not even the low-to-the-ground tundra types of vegetation can survive.

General Precipitation Patterns

As with temperature, the Alps provide a legitimate separation of precipitation regimes over Europe. Areas north of the Alps are generally characterized by year-round precipitation, owing to the continual proximity to the *polar front* and its embedded mid-latitude wave cyclones. The prevalence of relatively cool mP air limits convective activity in the northern zone. However, in some inland regions such as in Poland, Belarus, the Ukraine, and Russia, surface heating and convective activity are sufficient to produce a summer precipitation maximum.

The region south of the Alps is dominated by stability associated with the extension of the Bermuda-Azores High over the often-stable Mediterranean basin. In these Mediterranean climates, summers are dry with a distinct precipitation maximum in winter. Hotel rooms in Italy, Spain, Greece, and similar destinations are much more expensive in summer for a reason!

North of the Alps

For the part of Europe north of the Alps, temperatures decrease slowly but steadily with distance eastward and inland, as the terrain allows deep penetration of maritime air from the North Atlantic. As described previously, this air is much warmer than most other locations at similarly high latitudes because of the influence of the North Atlantic Drift, which is an extension of the Gulf Stream. The moderate temperatures allow dense and varied vegetation, despite very high latitudes (Fig. 14.5).

Fig. 14.5 Vegetated landscapes of Scandinavia: **a** near Vik, Iceland; **b** near Bergen, Norway. *Source* Authors

Northwestern Europe's moderated conditions are evidenced by climatological data from several representative sites shown in Fig. 14.6. Annual average temperatures are around 10 °C (50 °F) at Dublin, Ireland, and Brussels, Belgium (Figs. 14.7 and 14.8). The annual temperatures of London, United Kingdom and Paris, France, are only slightly warmer, at 12 °C (54 °F) (Figs. 14.9 and 14.10). The winds in this area are persistent and strong, offering possibilities for wind energy generation in the North Sea and adjacent waters and coastal areas. The climagraphs and water balance diagrams for all four of these sites reveal several similar characteristics climatologically: relatively small *annual temperature ranges*, consistent precipitation throughout the year, and abundant water for *surplus*. Water balance *deficits* are virtually non-existent in the very green landscape of Dublin, with small summertime deficits in Brussels and London, and larger but still small deficits in Paris. Stream runoff allows for several important rivers, such as the Rhine which flows from the Alps northward to the North Sea via the Netherlands, the Loire of France, and the Elbe of Czechia and Germany.

However, Scandinavia is much colder, as shown in the climagraphs and water balance diagrams for the locations shown in Fig. 14.11 and as shown by the climagraphs of Reykjavik, Iceland (Fig. 14.12); Oslo, Norway (Fig. 14.13); and Stockholm, Sweden (Fig. 14.14). In the high-latitude marine areas and in parts of Iceland, the rare-over-land *Humid Subtropical—mild summer* (Cfc) *climate* results from strongly moderated summer and winter temperatures.

The North Atlantic Drift elevates the average annual temperature of Reykjavik, Iceland, the world's northernmost national capital, to a reasonably comfortable 6.6 °C (43.9 °F) for a location near the *Arctic Circle*. Mean annual temperatures at Oslo and Stockholm are approximately 7 °C (44 °F) and 8 °C (46 °F), respectively. The "first flush" of snowmelt shows a sharp spike in runoff in all of these northern locations.

For both the western European and Scandinavian sites, precipitation is relatively well-distributed throughout the year. However, precipitation decreases noticeably in summer in Reykjavik and in spring at Oslo and Stockholm, and to a lesser extent, in Brussels and Paris. The delayed onset of the seasons leaves these cities, especially the Scandinavian ones, well north of the polar front. Thus, most of northern Europe is also north of the preferred tracks of mid-latitude wave cyclones that the polar front steers and the associated trailing cold fronts, while little convectional heating is occurring so far northward. Moreover, in spring, the still-cold ocean water surface creates a *temperature inversion*, or at least reduces the *environmental lapse rate*, which stabilizes the lower atmosphere. Nevertheless, precipitation remains adequate year-round to keep water balance deficits non-existent at Reykjavik and small to moderate in mainland Scandinavia. Autumn tends to have slightly more precipitation than other seasons, particularly at coastal stations, because of the destabilization of the atmosphere from an ocean surface that warmed all through the summer. Coastal locations also tend to have a slightly extended wet season into winter due to the prevalence of the still-warm ocean surface.

In north-central and northeastern Europe, as represented by the three cities shown in Fig. 14.15, temperatures decrease with distance northeastward and inland. For example, the mean annual temperature of Budapest, Hungary, is 11 °C (51 °F), while

Fig. 14.6 Northwestern European cities for which climagraphs and water balance diagrams are shown. *Source* Geaux Geospatial

that of Warsaw, Poland, is near 9 °C (48 °F), and that of Moscow, Russia, is 6 °C (42 °F) (Figs. 14.16, 14.17 and 14.18, respectively). At these more continental locations, summer and autumn usually have the most precipitation. Warm-season precipitation is enhanced by thermal convection that promotes low-level instability and abundant water vapor capacity, in addition to mid-latitude wave cyclone passages. The water balance diagrams of all three of these sites show evidence for a sudden spring snowmelt.

Deficits remain modest but increase with sites that are more inland, eastward, and nearer to the Mediterranean climate type, as in the case of Budapest. Nevertheless, the temperate, generally humid conditions allow for several important rivers, such as the Dnieper of Ukraine, Belarus, and Russia; the Dniester of Ukraine; the Danube, which winds its way through southeastern and central Europe; the Vistula of Poland;

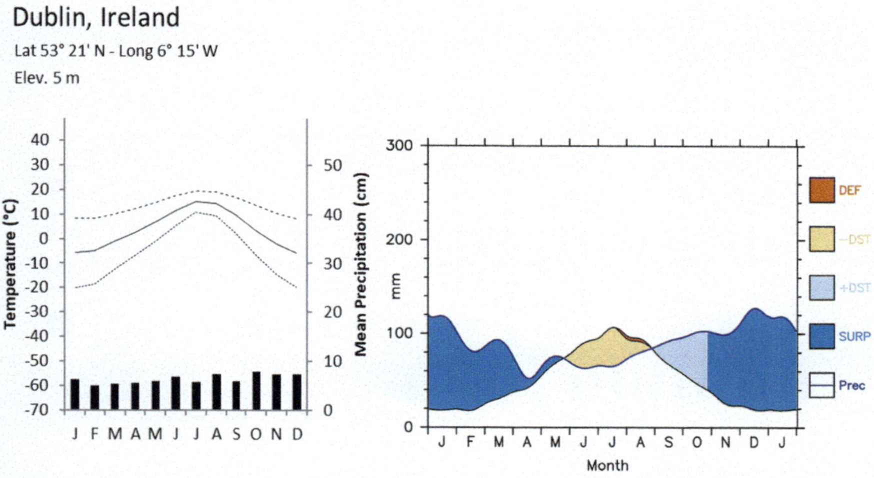

Fig. 14.7 Climagraph and water balance diagram for Dublin, Ireland. Data for 1991–2020 averaging period. *Source* Authors. https://weatherspark.com/, http://cyclops.deos.udel.edu/wimp/public_html/

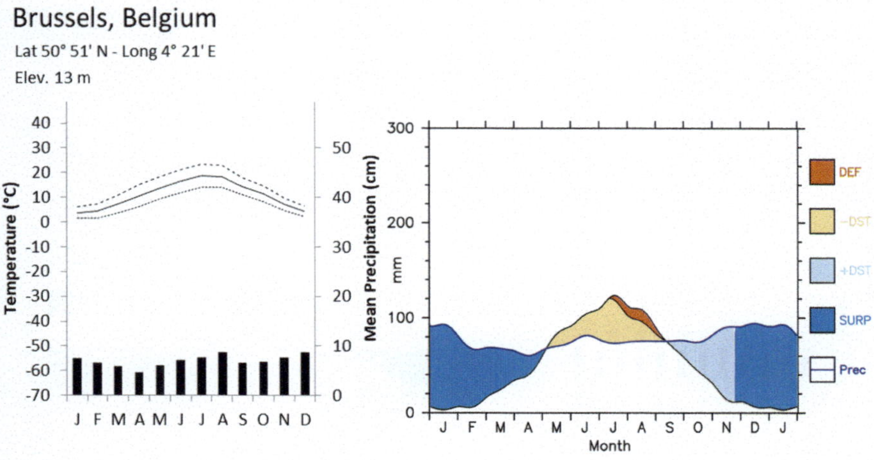

Fig. 14.8 Climagraph and water balance diagram for Brussels, Belgium. Data for 1991–2020 averaging period. *Source* Authors. https://weatherspark.com/, http://cyclops.deos.udel.edu/wimp/public_html/

and Europe's longest river, the Volga in Russia, that are important economically in this generally densely populated region.

In far northeastern Europe, the brutal winters and cool summers create an extensive *Humid Continental—mild summer* (also known as *Subarctic*, Dfc) *climate*. Even farther north and/or upslope, the summers are too cool to allow tree growth. These

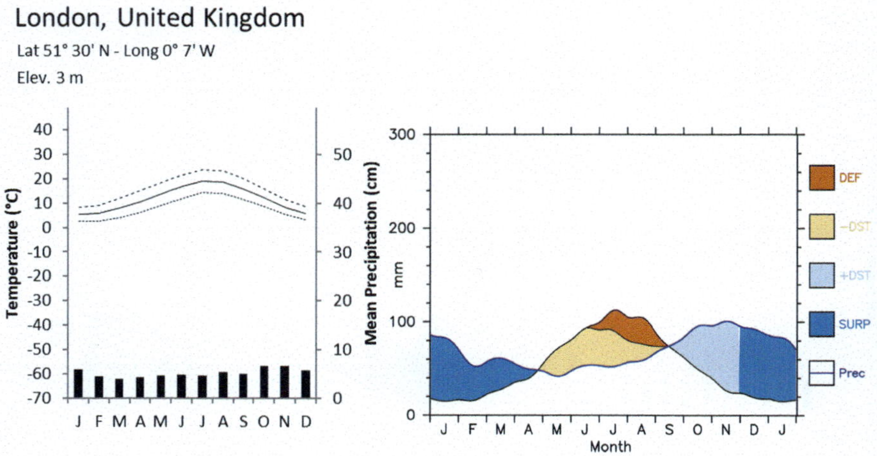

Fig. 14.9 Climagraph and water balance diagram for London, United Kingdom. Data for 1991–2020 averaging period. *Source* Authors. https://weatherspark.com/, http://cyclops.deos.udel.edu/wimp/public_html/

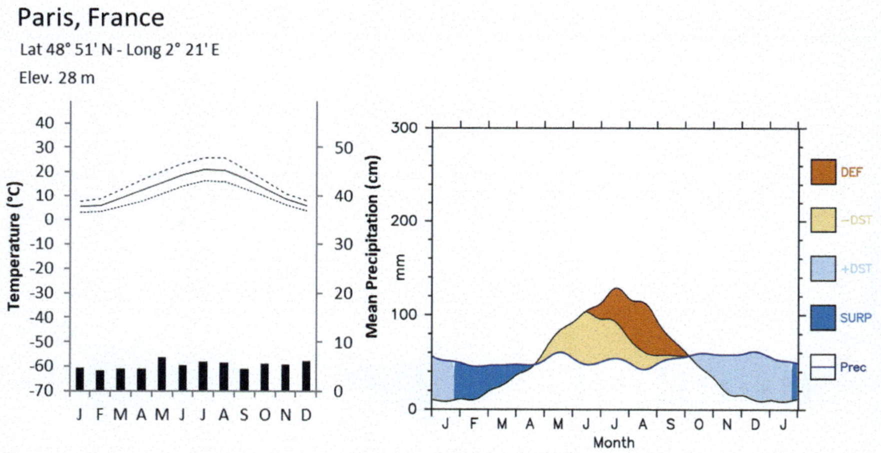

Fig. 14.10 Climagraph and water balance diagram for Paris, France. Data for 1991–2020 averaging period. *Source* Authors. https://weatherspark.com/, http://cyclops.deos.udel.edu/wimp/public_html/

locations poleward and/or upslope of the treeline have ET climate. However, except for the summits of the highest mountains, the Dfc and ET climates are restricted to northeastern Europe; the moderating influence of the North Atlantic Drift prevents ports from freezing even as far north as Murmansk, Russia.

In general across Europe north of the Alps, the growing season is fairly long, especially considering the high-latitude location. At Falmouth, on the south coast of the United Kingdom, freezes occur on a few days per year from November to March

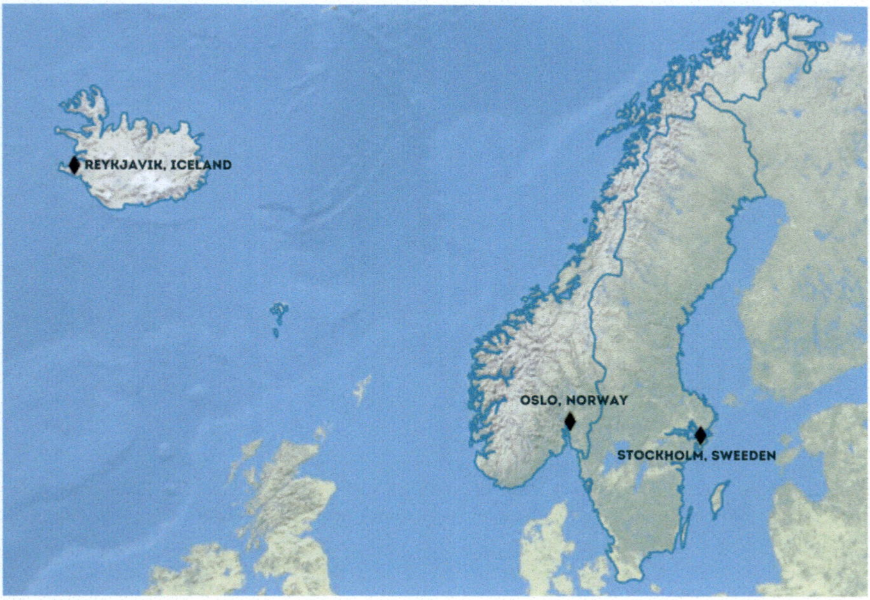

Fig. 14.11 Scandinavian cities for which climagraphs and water balance diagrams are shown. *Source* Geaux Geospatial

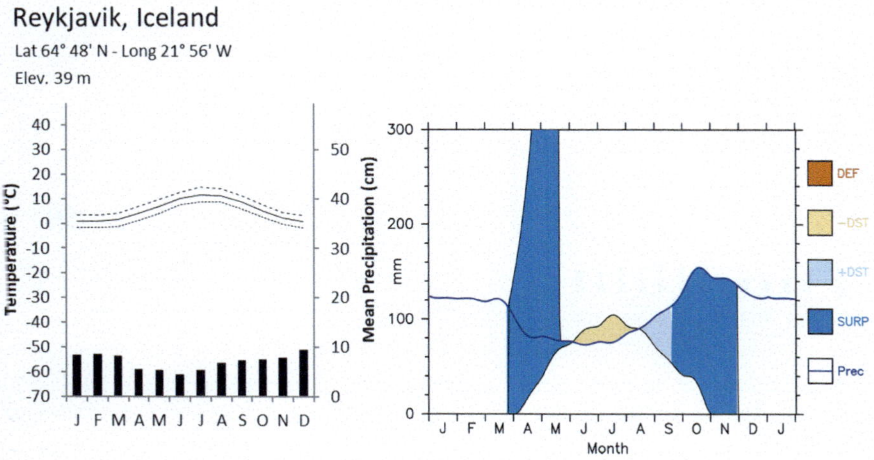

Fig. 14.12 Climagraph and water balance diagram for Reykjavik, Iceland. Data for 1991–2020 averaging period. *Source* Authors. https://weatherspark.com/, http://cyclops.deos.udel.edu/wimp/public_html/

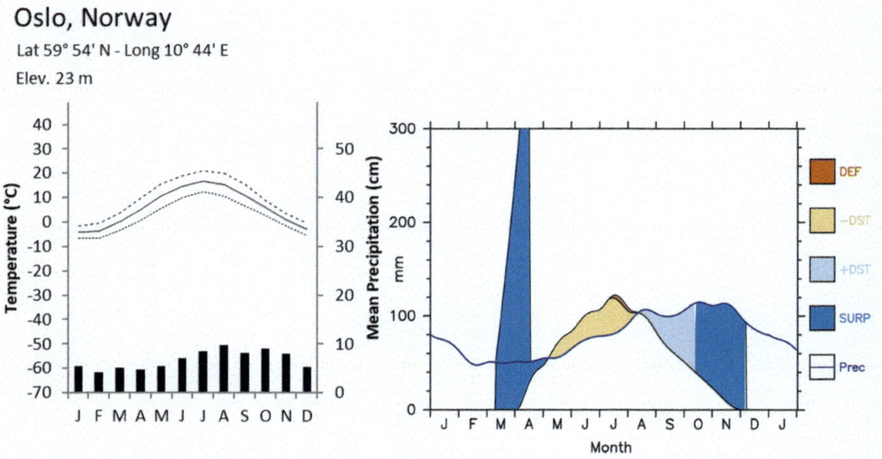

Fig. 14.13 Climagraph and water balance diagram for Oslo, Norway. Data for 1980–2016 averaging period. *Source* Authors. https://weatherspark.com/, http://cyclops.deos.udel.edu/wimp/public_html/

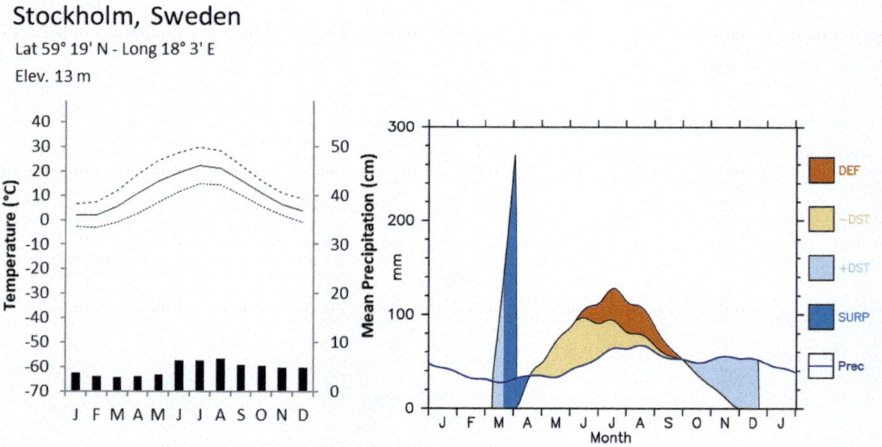

Fig. 14.14 Climagraph and water balance diagram for Stockholm, Sweden. Data for 1991–2020 averaging period. *Source* Authors. https://weatherspark.com/, http://cyclops.deos.udel.edu/wimp/public_html/

as the island experiences a substantial maritime influence. Palm trees can grow there and on the southern coast of Ireland, even at latitudes of 50 to 52° N. In continental Europe, growing seasons are shorter but nevertheless are impressively long. For example, Berlin (Germany), at 52° N latitude, experiences freezing temperatures on an average of only about 20 days per year, compared to 26 days per year when the temperature never climbs above freezing at any point in the day in Kansas City,

Fig. 14.15 North-central and northeastern European cities for which climagraphs and water balance diagrams are shown. *Source* Geaux Geospatial

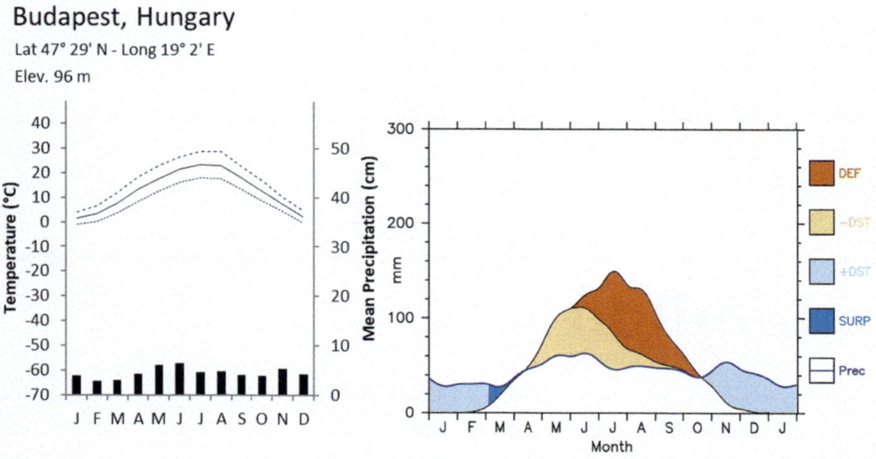

Fig. 14.16 Climagraph and water balance diagram for Budapest, Hungary. Data for 1991–2020 averaging period. *Source* Authors. https://weatherspark.com/, http://cyclops.deos.udel.edu/wimp/public_html/

Missouri (39° N). Farther inland in eastern Europe, in Bucharest, Romania (44.4° N), January is the only month with a mean temperature below freezing.

With such a strong maritime influence, it may be surprising that precipitation totals are moderate. Brussels, London, and Paris experience only 84 cm (33 in), 62 cm (24 in), and 64 cm (25 in) per year, respectively, and Reykjavik, Oslo, and

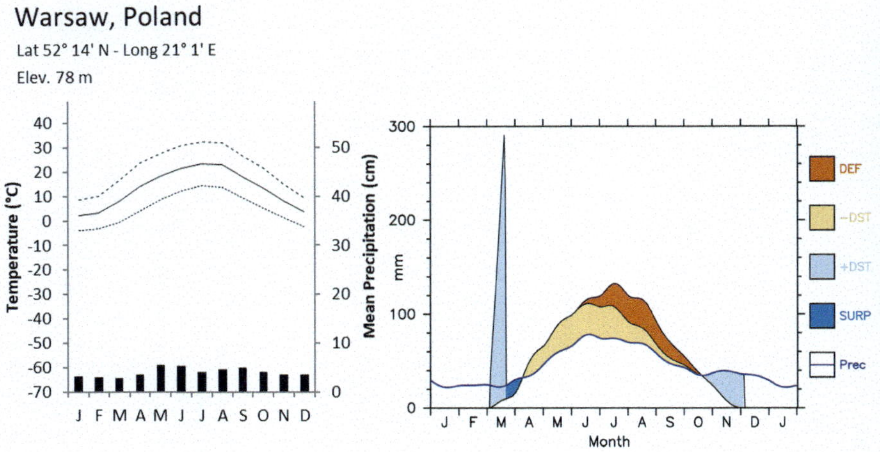

Fig. 14.17 Climagraph and water balance diagram for Warsaw, Poland. Data for 1991–2020 averaging period. *Source* Authors. https://weatherspark.com/, http://cyclops.deos.udel.edu/wimp/public_html/

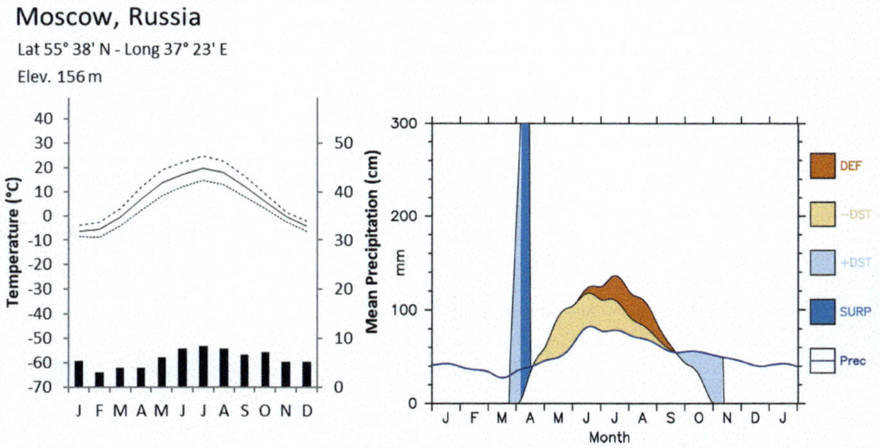

Fig. 14.18 Climagraph and water balance diagram for Moscow, Russia. Data for 1980–2016 averaging period. *Source* Authors. https://weatherspark.com/, http://cyclops.deos.udel.edu/wimp/public_html/

Stockholm get an average of only 84 cm (33 in), 80 cm (32 in), and 55 (22 in). Budapest, Warsaw, and Moscow experience an average of 53 cm (21 in), 48 cm (19 in), and 71 cm (28 in), respectively. The number of precipitation days, though not necessarily precipitation totals, gradually decreases inland. By comparison, nearly every site in the southeastern U.S. receives substantially more precipitation than any of these sites.

Less surprising is the fact that the number of precipitation days per year is high in Europe north of the Alps. Like the Pacific Northwest of the U.S., frequent passages of old, occluded storms bring generally small totals because of the weak air mass contrast, with mP air typically found behind the system. The small thermal gradient of the air masses across the frontal boundary forces little uplift of warmer air, and therefore, wide, blankety clouds such as nimbostratus are a common sight. Light, widespread precipitation occurs, similar to that characteristic of U.S. and Canadian warm fronts but occurring commonly in Europe north of the Alps in association with all types of fronts.

The driest areas of the region typically occur on the *rain shadow* side of highlands and mountains and usually in northern areas where cold mP air has less water vapor in the cooler temperatures. *Orographic lifting* causes wetter areas to appear on the windward sides of mountains. This pattern is especially evident in Ireland, Scotland, Wales, and western Norway. For example, the western coast of Ireland experiences about 117 cm (46 in) of precipitation per year, while only 76 cm (30 in) per year falls on Dublin, on the Irish east coast. Coastal zones (e.g. Netherlands, Denmark, Belgium) are notably wet as horizontal *convergence* occurs with the deceleration of maritime air upon encountering extra *friction* at the coastline.

Unlike the Pacific coast of North America, northern Europe does not experience a southward progression of maximum precipitation months from autumn to late winter. The less consistent and more complicated polar front jet stream migrations in Europe make the seasonality of precipitation more complex. The complexity of the polar front behavior is affected by frequent, abundant, and rapidly-changing departures from the typical flow, as described in the next section.

Blocking Highs Over Europe

Ridges and troughs change in amplitude over time. If the ridge amplitude increases to a great extent and remains stationary or quasi-stationary for days or even weeks at a time, a *blocking pattern* occurs in which the anticyclone beneath the amplified ridge persists. Figures 4.12 and 4.13 showed hypothetical examples of an *omega block* and a *Rex block* from North America. However, unlike North America, where the position of the ridge-trough configuration associated with *Pacific-North American* (PNA) *pattern* tends to anchor ridges over the western mountain cordillera and robust circulations frequently bring different air mass contrasts which dissipate blocking patterns relatively quickly, blocking patterns in Europe persist even in the absence of a major mountain range. They are more frequent and intense than in North America primarily because of a weaker prevailing circulation.

European blocking patterns can persist for several weeks, most commonly occurring between 50°–70° N and 0°–10° E (Fig. 14.19), near or over the United Kingdom, and affecting short and medium-term weather upstream and downstream. Seasonally, blocking patterns are more frequent in winter and spring, but they may occur over Europe in any season.

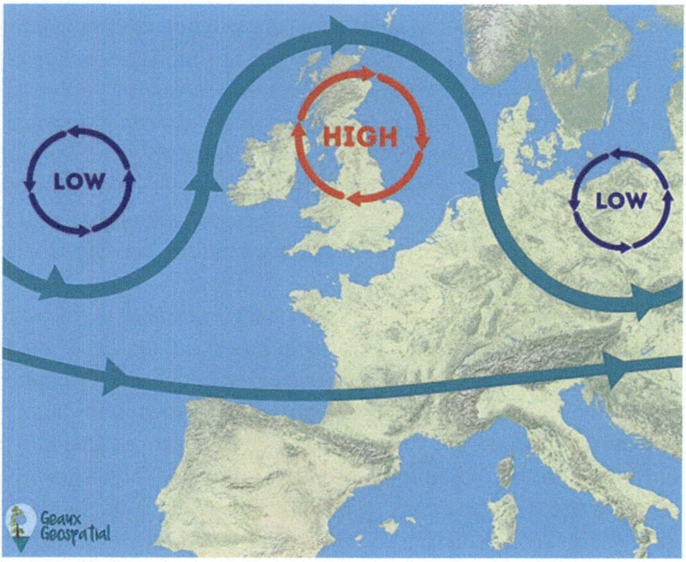

Fig. 14.19 Typical blocking pattern over Europe. *Source* Geaux Geospatial. https://www.metoffice.gov.uk/weather/learn-about/weather/how-weather-works/high-and-low-pressure/blocks

Blocking patterns typically split the *polar front jet stream* into two sectors, as shown by the blue arrows in Fig. 14.19. The polar front jet pulls migrating mid-latitude wave cyclones either to the north or to the south of the block. Thus, blocking is associated with wetter-than-normal conditions in northern Europe and southern Europe as the migrating cyclones pick up Atlantic moisture and initiate frontal precipitation. However, the rest of Europe tends to be particularly dry during blocking patterns, especially directly under the influence of the anticyclone.

The bifurcation of the polar front jet stream also means that little temperature contrast exists across Europe during blocking episodes. Depending on the exact position of the block, maritime tropical (mT) air masses from the subtropical North Atlantic can cause anomalously high temperatures in northernwestern Europe while mP and cP air masses give downstream areas in southern Europe below-normal temperatures. Beneath the block, droughts can cause water shortages in this densely-populated continent.

Blocking patterns in summer may lead to excessive heat and heat-related stresses, including fatalities. For example, during the blocking episodes of 1975, dry conditions led to increased insect damage, agricultural losses, increased pollution, low stream levels and other water shortages. The situation prompted legislation for increased conservation measures across Europe as the negative impacts of extreme climatic variability were recognized in earnest for perhaps the first time. The blocking high and associated heat wave during August 2018 caused heat-related illness and death, extensive crop damage, and widespread wildfires, especially through the

Iberian Peninsula (i.e., Portugal, Spain, and the landlocked principality of Andorra in the Pyrenees Mountains).

Mediterranean Europe

The region south of the Alps, including the Iberian, Italian, and Balkan peninsulas, and many smaller ones, borders the Mediterranean Sea. The associated climate is termed the *Mediterranean* (Cs) *climate type*, regardless of where in the world similar climates occur. The Cs climate is, arguably, the most desirable climate type of all. It should come as no surprise that the clear, dry summers with pleasant temperatures year-round attract many residents and visitors. Unlike in some other regions, the Cs climate type in southern Europe can extend far inland.

The complex and diverse topography and land-sea-air interactions/feedbacks play havoc with localized atmospheric circulation in Mediterranean Europe. Many localized winds occur, with regional names (Fig. 14.20). The best-known include the Mistral winds of the French Riviera, the Etesian winds over the Aegean Sea, the Bora of the Adriatic Sea region, and the western Zephyr which extends through the Mediterranean region.

The close connection to the vast Mediterranean Sea, including through local winds, contributes to the mild temperatures in Mediterranean Europe, particularly at these latitudes. For example, Rome and New York City are located along the same line of latitude and are both on a large body of water, yet Rome has much more moderated temperatures than New York, the latter of which is downwind in the mid-latitude westerlies from a large continent.

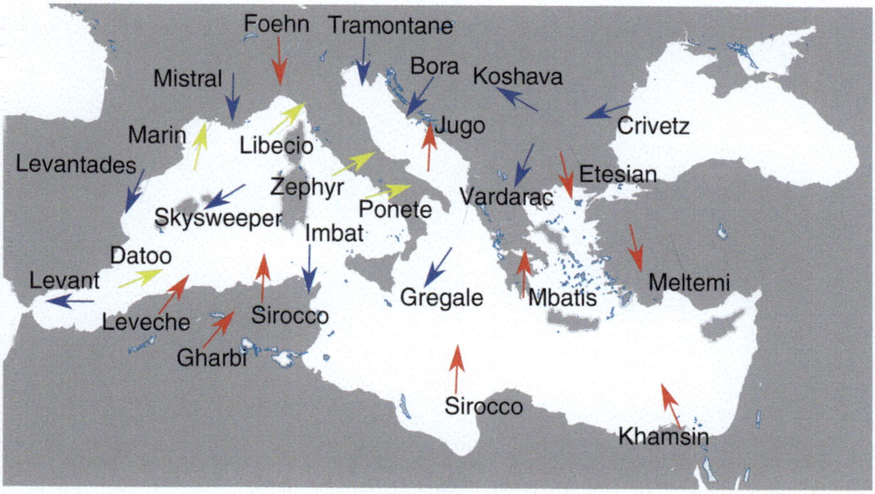

Fig. 14.20 Local winds of the Mediterranean region. *Source* Romanic (2018)

In general, temperatures decrease along a west-to-east gradient in winter, with distance from the moderating influence of the Atlantic Ocean. mP air masses are more dominant and more moderated in the western Mediterranean. The eastern part of the basin is influenced more by cP air from the Asian landmass in winter.

Summers may be considerably hot. The Mediterranean Sea has less influence in moderating temperatures in summer compared to winter because its shallowness allows for extensive heating of the sea during the high-Sun season. *Sea breezes* have less cooling influence on daytime high temperatures than in most other places affected by sea breezes.

Exceptions to the summer heat abound in the higher-elevation areas, such as the foothills of the Alps. Portugal is also cooler than the rest of Mediterranean Europe, on average, as the larger and deeper Atlantic Ocean absorbs energy. Therefore, Portuguese summers are more similar to those along the central and northern California coast, while the remainder of the region is more similar to the Baja Peninsula of Mexico. Cooler temperatures also occur, in both summer and winter, in Mediterranean Europe downwind of steep mountain gaps, especially through southern France and the Aegean Sea areas. The climagraphs of several example sites in Mediterranean Europe shown in Fig. 14.21 are ordered from west to east in the climagraphs for Ponta Delgada, Azores Islands; Lisbon, Portugal; Madrid, Spain; Barcelona, Spain; Naples, Italy; Venice, Italy; and Athens, Greece, in Figs. 14.22, 14.23, 14.24, 14.25, 14.26, 14.27 and 14.28. Water balance diagrams are available for all except for Ponta Delgada.

The precipitation and water balance characteristics of Mediterranean Europe are more complicated than on initial impression. Although the region is considered as "summer dry," some locations experience drier conditions than others, and a few, such as Venice (see again Fig. 14.27) and locations with orographic enhancement,

Fig. 14.21 Southern European cities for which climagraphs and (except in the case of Ponta Delgada) water balance diagrams are shown. *Source* Geaux Geospatial

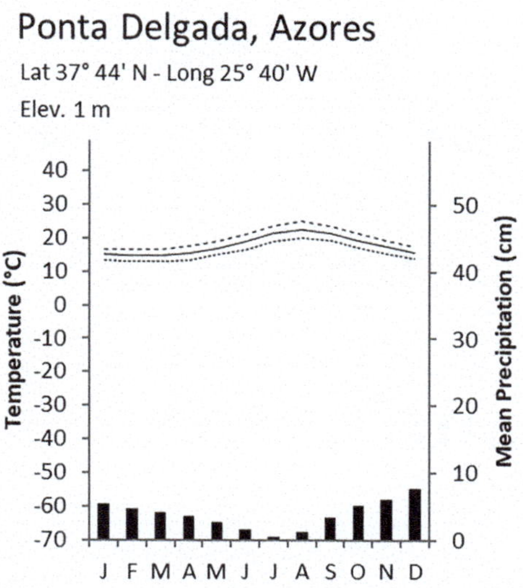

Fig. 14.22 Climagraph for Ponta Delgada, Azores Islands. Water balance data are unavailable. Data for 1980–2016 averaging period. *Source* Authors. https://weatherspark.com/, http://cyclops.deos.udel.edu/wimp/public_html/

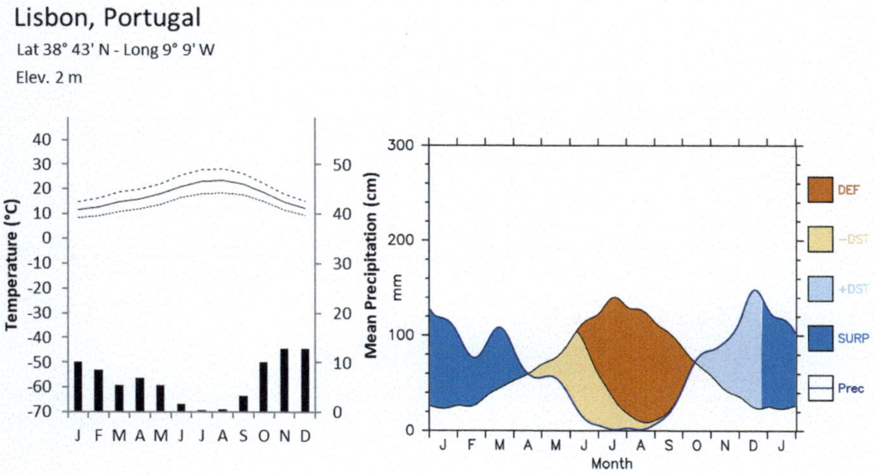

Fig. 14.23 Climagraph and water balance diagram for Lisbon, Portugal. Data for 1980–2016 averaging period

experience almost as much summer precipitation as in other times of the year. Nevertheless, the water balance diagrams for all locations having Mediterranean climates have the common characteristic of a substantial water balance deficit in the warm months, during the most intense *shortwave radiation* (K ↓) from the Sun (see again Figs. 14.23, 14.24, 14.25, 14.26, 14.27 and 14.28).

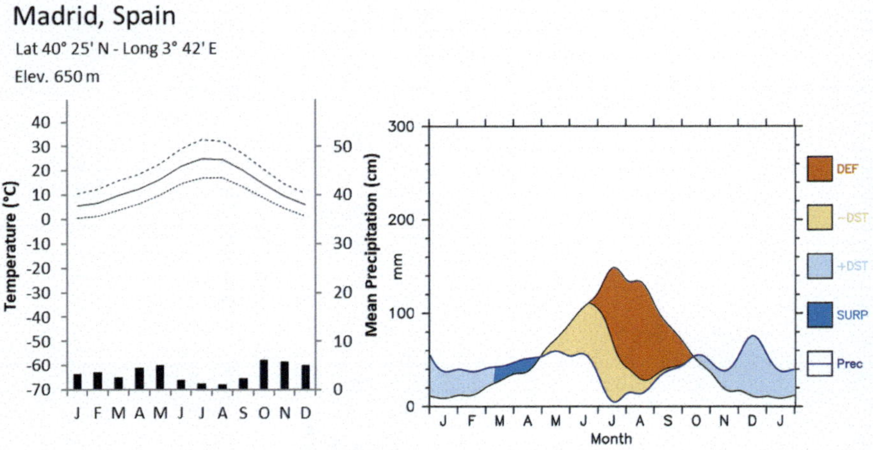

Fig. 14.24 Climagraph and water balance diagram for Madrid, Spain. Data for 1980–2016 averaging period. *Source* Authors. https://weatherspark.com/, http://cyclops.deos.udel.edu/wimp/public_html/

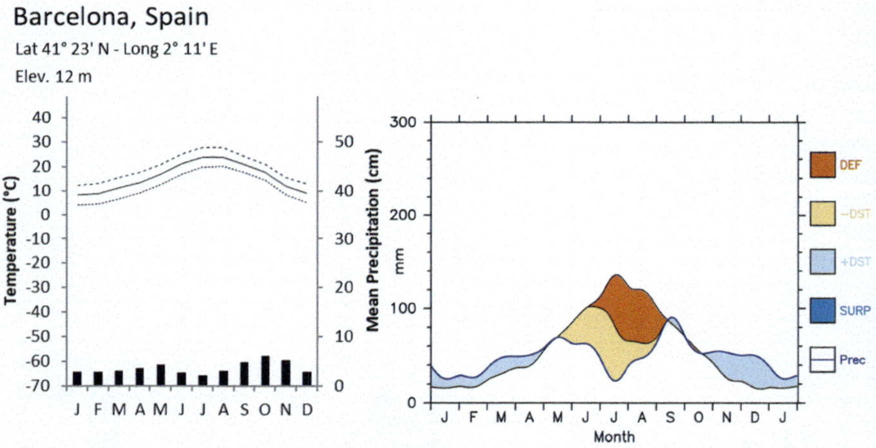

Fig. 14.25 Climagraph and water balance diagram for Barcelona, Spain. Temperature data are for the 1991–2021 averaging period and precipitation data are for 1999–2021 period. *Source* Authors. https://weatherspark.com/, http://cyclops.deos.udel.edu/wimp/public_html/

Precipitation in Mediterranean Europe is governed by a few primary factors. During winter, cP air advected from the east by the clockwise and outward circulation around the Siberian High may collide with much warmer, humid mP air moved from the west by the Bermuda-Azores High. These air mass interactions, or more commonly, weaker interactions between mP air and somewhat warmer mP air, initiate mid-latitude wave cyclones and their associated fronts. Precipitation is common in

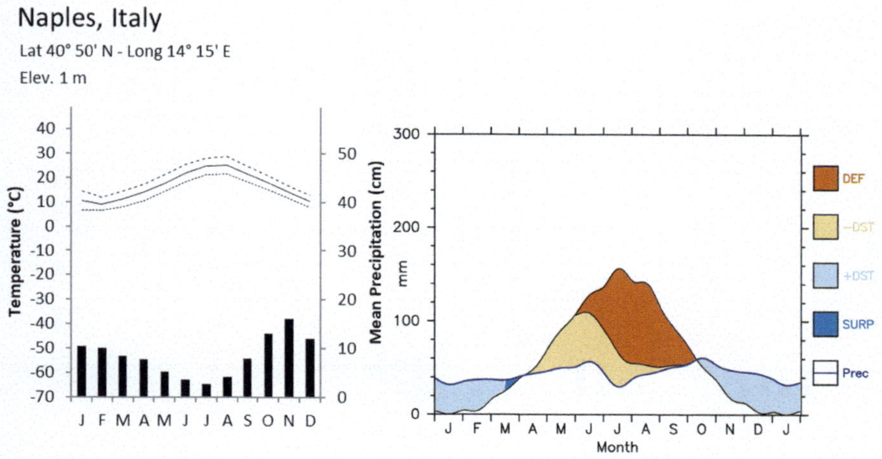

Fig. 14.26 Climagraph and water balance diagram for Naples, Italy. Temperature data are for the 1991–2021 averaging period and precipitation data are for 1999–2021 period. *Source* Authors. https://weatherspark.com/, http://cyclops.deos.udel.edu/wimp/public_html/

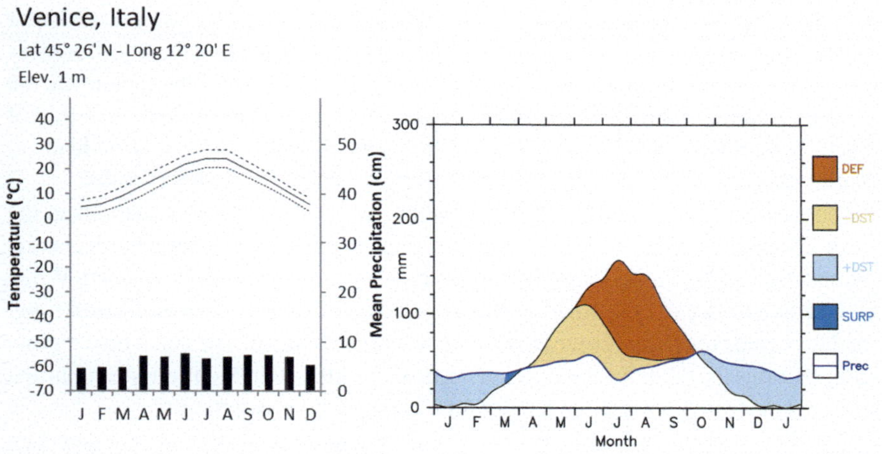

Fig. 14.27 Climagraph and water balance diagram for Venice, Italy. Temperature data are for the 1991–2021 averaging period and precipitation data are for 1999–2021 period. *Source* Authors. https://weatherspark.com/, http://cyclops.deos.udel.edu/wimp/public_html/

such situations, particularly as the relatively warm Mediterranean Sea provides low-level instability and water vapor. Troughs embedded within the longwave flow are common upper-level atmospheric features.

The combination of moisture advection from the mid-latitude westerlies and frontal forcing enhanced by orographic uplift makes the southern and western sides of mountain ranges wetter than the northern and eastern slopes in winter. By contrast, the clockwise and outward circulation from the Siberian High causes the northern

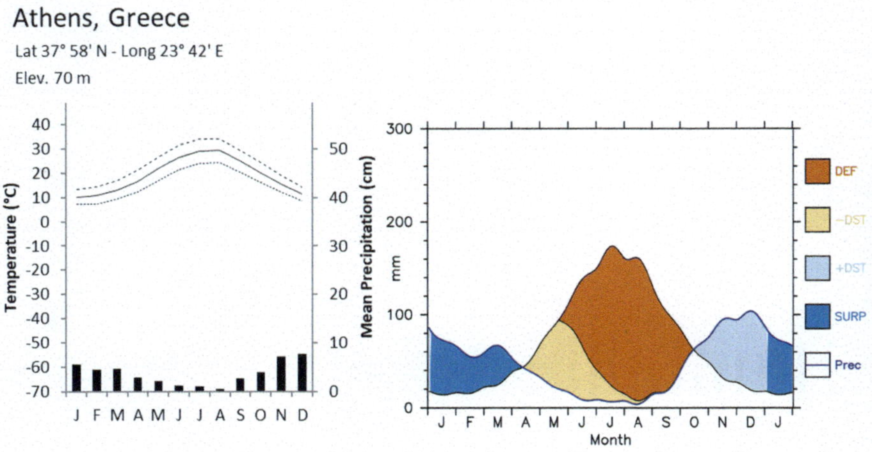

Fig. 14.28 Climagraph and water balance diagram for Athens, Greece. Data for 1991–2020 averaging period. *Source* Authors. https://weatherspark.com/, http://cyclops.deos.udel.edu/wimp/public_html/

and eastern sides of mountain ranges to be drier and in the rain shadow of water vapor originating from the Mediterranean Sea and Atlantic Ocean. Localized differences from the pattern are common, however. For example, in the area of the Ionian sea, the persistent counter-clockwise mesoscale circulation channels air through the Strait of Otranto which lies between the "heel" of Italy and Albania. This situation leads to volatile sea conditions and enhanced precipitation during the winter months.

The frequency and intensity of these frontal systems is tied to fluctuations in the *North Atlantic Oscillation* (NAO), which is a teleconnection between the Icelandic Low and Bermuda-Azores High. The opposing pressure systems act in concert. When one is strong (or weak), the other will also be strong (or weak). When both are strong, zonal flow (indexed as a positive NAO index value) is initiated. The zonal flow regime often positions the polar front jet stream north of the Alps, with Mediterranean Europe only rarely affected in terms of temperature change and precipitation.

However, if both the Icelandic Low and Bermuda-Azores High are weaker than normal, meridional flow (indexed as a negative NAO index value) occurs. Meridional flow involves amplified ridges and troughs, which, in addition to the large temperature differences across small longitudinal areas, initiates differences in upper-air spin. The areas that gain *positive vorticity*, or cyclonic spin, are where mid-latitude wave cyclones strengthen. As the longwave flow is amplified, troughs extend over the Mediterranean basin. The counter-clockwise (in the Northern Hemisphere) flow around a trough is coincident with cyclonic spin, so the region under the trough—Mediterranean Europe, in this case—experiences wetter conditions under such flow regimes.

If a ridge is positioned over the eastern Atlantic Ocean or British Isles, northwesterly flow and associated cold air advection that may occur in the eastern Mediterranean basin over the warm surface waters creates an unstable atmosphere conducive to cyclogenesis and precipitation. The situation is the opposite of what would normally be expected, as the probability of cyclogenesis is actually enhanced when Mediterranean Europe underlies the ridge-to-trough, rather than the trough-to-ridge, side of the Rossby wave. However, the position beneath the ridge-to-trough side of the Rossby wave during cyclogenesis has the effect of restricting intensity and duration of precipitation.

If a ridge is centered over the eastern Mediterranean Sea, with the associated trough positioned over western Europe, the southwesterly flow over the western Mediterranean Sea advects warm air instead of cold air over much of the basin. In this situation, southern Europe experiences enhanced frontal cyclone frequency because of the humidity, enthalpy, and position on the trough-to-ridge side, but atmospheric stability from the relatively warm air aloft restricts the vertical cloud development and storm severity.

Similar to the *Alberta Clippers* and *Colorado Wave Cyclones* that form the *leeward* side of the Rocky Mountains in North America, a preferred cyclogenesis zone exists on the leeward side of the Atlas Mountains in northwestern Africa. The cyclogenesis in this case is also caused by the *conservation of potential vorticity* and by the acceleration of steering flow as it is "squeezed" between the higher mountain elevations and the *tropopause*. The acceleration strengthens the *Coriolis effect*, which then deflects the flow equatorward. The resulting trough creates a cyclogenesis zone for storms to cross the Mediterranean basin. Other preferred cyclogenesis areas associated with mountain systems include the western Mediterranean Sea, the region of Sardinia, Corsica, western Italy, and the eastern Mediterranean Sea near Cyprus.

Resulting storm tracks are associated with the upper-air flow pattern in concert with the cyclogenesis area. Frontal cyclone trajectories are generally from west to east in the mid-latitude westerlies. The tracks bifurcate either northward or southward around the Anatolian Plateau and associated mountains of Turkey. The northern track forces cyclones into the Black Sea basin, and the southerly track moves cyclones into the Levant—the area immediately east of the Mediterranean Sea (Fig. 14.29). The northern track is more likely to have upper-level support as the region tends to lie beneath the trough-to-ridge side of the Rossby wave. The Black Sea can reinvigorate these systems with water vapor. The southern track is less likely to generate robust precipitation totals because the zonal Rossby wave pattern necessary to steer cyclones along this trajectory fails to supply positive vorticity and upper-level divergence on the trough-to-ridge side of the Rossby wave. As a result, frontal systems are weakened and the area is largely arid.

The annual number of precipitation days in Mediterranean Europe varies by latitude, with the southern Mediterranean region experiencing the fewest precipitation days and the frequency increasing northward. In the northern Mediterranean zone, the precipitation day isolines parallel the coastlines. Approximately 80 to more than 100 precipitation days per year occur in the northern Mediterranean, with half or less of that total in southern Mediterranean Europe.

Seasonally, the number of precipitation days in northern Mediterranean Europe is less than 10 in summer, about 20 to 25 during the transition seasons of autumn and spring, and around 40 in winter. Mountainous regions experience a higher total

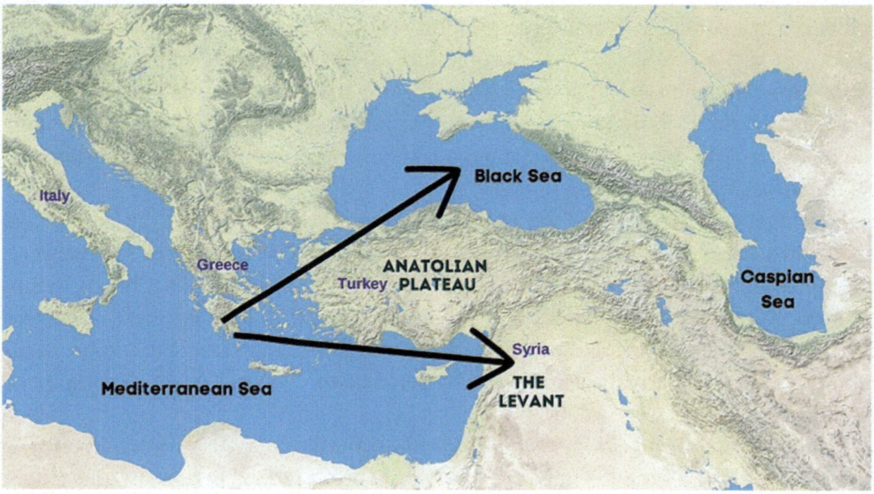

Fig. 14.29 Mid-latitude wave cyclone tracks over the eastern Mediterranean region. *Source* Geaux Geospatial

of precipitation days than reduced elevation areas. Very few precipitation days occur during summer for the area southeast of the Mediterranean basin.

The cool season shows a more variable pattern of precipitation day frequencies in Mediterranean Europe because of the nuances of land/sea thermal contrasts. Most precipitation days occur in the cyclogenesis regions of the central and eastern Mediterranean region. The fewest precipitation day totals occur in large land masses, including Turkey, central Spain, and northern Italy, where land is cooler than water, stabilizing the atmosphere.

In general, the frequency of precipitation days aligns with precipitation totals in Mediterranean Europe, as each storm tends to contribute similar precipitation amounts. The most precipitation and severe weather associated with mid-latitude wave cyclones generally occur after passage of the cold front, unlike in the U.S. and Canada, where the warm sector—the area between the warm front and cold front—often has the most active weather features. Especially in the colder half of the year, Mediterranean European precipitation tends to happen in these conditions, when the warm air in the warm sector is pushed up over the cold air, with its precipitation falling behind the surface cold frontal boundary. The preference for such precipitation originates from the weak air mass contrast, in which the warmer, more humid air mass from the Mediterranean slides over the colder air mass rather than being lifted vigorously at the surface frontal boundary. Cold air surges through the mountains out to sea can also destabilize the atmosphere over the relatively warm Mediterranean and can augment precipitation intensities.

Maximum precipitation events in Mediterranean Europe typically occur during the spring and autumn seasons, when convection is possible in the warmth, but beyond the time when the influence of the Bermuda-Azores High is strongest. Moving from northwest to southeast, the percentage of precipitation occurring in winter increases. Local anomalies can occur, especially where western slopes are wetter and eastern slopes are drier.

The combination of precipitation amount and intensity, vegetation cover, and slope steepness yields risks for landslides and related environmental hazards in this region. The dry summer leads to sparse vegetation, which destabilizes slopes. Mudslides or landslides can occur after heavy precipitation events that add weight and associated gravitational pull in addition to breaking the cohesion between soil particles. Mudslides are often most common in the drier regions, where less vegetation is in place, exposing bare soil. The next chapter describes adjacent areas that are even more arid than Mediterranean Europe.

Further Reading

Burlando, M. (2009). The synoptic-scale surface wind climate regimes of the Mediterranean Sea according to the cluster analysis of ERA-40 wind fields. *Theoretical and Applied Climatology, 96*(1–2), 69–83. https://doi.org/10.1007/s00704-008-0033-5

Hurrell, J. W., & van Loon, H. (1997). Decadal variations in climate associated with the North Atlantic oscillation. *Climatic Change, 36*(3–4), 301–326. https://doi.org/10.1023/A:1005314315270

Panagos, P., Ballabio, C., Borrelli, P., Meusburger, K., Klik, A., Rousseva, S., Tadić, M. P., Michaelides, S., Hrabalíková, M., Olsen, P., Aalto, J., Lakatos, M., Rymszewicz, A., Dumitrescu, A., Beguería, S., & Alewell, C. (2015). Rainfall erosivity in Europe. *Science of the Total Environment, 511*, 801–814. https://doi.org/10.1016/j.scitotenv.2015.01.008

Pfahl, S. (2014). Characterising the relationship between weather extremes in Europe and synoptic circulation features. *Natural Hazards and Earth System Sciences, 14*(6), 1461–1475. https://doi.org/10.5194/nhess-14-1461-2014

Rex, D. F. (1950). Blocking action in the middle troposphere and its effect upon regional climate. *Tellus, 2*(4), 275–301. https://doi.org/10.1111/j.2153-3490.1950.tb00339.x

Romanic, D. (2018). Local winds of Balkan Peninsula. *International Journal of Climatology, 39*(1), 1–17. https://doi.org/10.1002/joc.5743

Trewartha, G. T. (1981). *The Earth's Problem Climates*. University of Wisconsin Press. 978-0299082307

Trigo, I. F., Biggs, G. R., & Davies, T. D. (2002). Climatology of cyclogenesis mechanisms in the Mediterranean. *Monthly Weather Review, 130*(3), 549–569. https://doi.org/10.1175/1520-0493(2002)130%3c0549:COCMIT%3e2.0.CO;2

Xoplaki, E., González-Rouco, J. F., Luterbacher, J., & Wanner, H. (2003). Mediterranean summer air temperature variability and its connection to the large-scale atmospheric circulation and SSTs. *Climate Dynamics, 20*(7–8), 723–739. https://doi.org/10.1007/s00382-003-0304-x

Chapter 15
North Africa and Southwest/Central Asia and Adjacent Oceans

Abstract The western subtropical Atlantic Ocean, North Africa, Southwest Asia (commonly called the Middle East) and adjacent waters, and Central Asia predominantly have arid (B) climates. In North Africa, Canary Current-induced stability and persistent, intense subsidence on the eastern side of the Bermuda-Azores High and poleward side of the Hadley Cell creates clear skies with moderately long daylight hours in summer. This continental tropical (cT) air mass source region supports oppressively high temperatures and suppresses precipitation, creating moisture stress and water balance deficits year-round, even though humidity may be abundant. Most of Southwest Asia experiences northwesterly airflow from the Bermuda-Azores High and its Mediterranean extension in summer, with subsidence promoting a stable atmosphere, with hot, dry landscapes. However, in winters across the entire region, mid-latitude wave cyclones bring cold fronts that lower temperatures and perhaps produce a modicum of precipitation. Central Asia is arid because of the extreme continentality, rain shadow effects, and distance from moisture sources, but its cold winters with Siberian High-advected continental polar (cP) air masses reduce deficits at that time of year. Yet this vast region is far from uniform climatologically.

Keywords Canary Current · Continental polar (cP) air mass · Continental tropical (cT) air mass · Continentality · Desert pavement · Mediterranean (Cs) climate · Red Sea · Rub-al-Khali · Sahara Desert · Semi-arid (BS) climate · Siberian High · Subsidence · True Desert (BW) climate · Water balance deficit

General

The region of North Africa and Southwest/Central Asia involves pieces of two continents, but together they form a distinctive environmental region. The region spans across the wide, extensive swath from the eastern North Atlantic Ocean eastward between the Mediterranean Sea and the tropical climates of Africa, including the Arabian Peninsula and into inland Asia (Fig. 15.1). North Africa is distinguished from the rest of Africa by its association with the world's largest *True Desert* (BW)

R. V. Rohli et al., *Atmospheric and Oceanic Circulation*,
https://doi.org/10.1007/978-3-031-66142-6_15

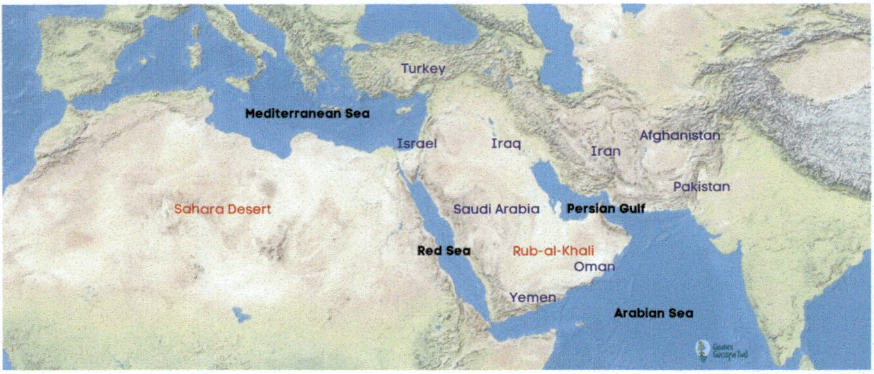

Fig. 15.1 Physiographic regions of North Africa and Southwest/Central Asia. *Source* Geaux Geospatial

climate—the Sahara. Southwest Asia is popularly known as "Middle East"—a term derived from Westerners' perspectives—and is the cradle of many ancient civilizations and three of the world's great religions. Central Asia largely consists of the landlocked nations of Asia, mostly surrounded by the "big three" of Russia, China, and India. The feature that unites this vast territory of North Africa and Southwest/Central Asia is dryness.

Most, but not all, of the region falls into the arid (B) climate types (Fig. 15.2). The area bordering the Mediterranean Sea experiences a *Mediterranean—hot summer* (Csa) *climate* similar to southern Europe. Unlike in the U.S., the *Mediterranean—warm summer* (Csb) *climate* extends inland, as in Turkey, where elevations are high. The Csa and Csb climates shift abruptly into the *Semi-arid* (BS) *climate* over a narrow zone before transitioning into truly arid regions, including the Sahara. Mountains such as the Caucasus between the Black Sea and Caspian Sea, and other mountains and high plateaus in the Levant, Iran, Afghanistan, and elsewhere, provide environments that have too much precipitation and/or too little *potential evapotranspiration* to fall into any of the B climatic types.

Topographic Influences (Site)

The desert landscapes of North Africa and Southwest/Central Asia have various local names, but the two major ones are the Sahara, which dominates North Africa, and the Rub-al-Khali, in the southern one-third of the Arabian Peninsula. The latter is one of the world's largest sandy areas. Most deserts contain little or no sand. Over time, persistent winds, uninterrupted by *friction* associated with topography, forests, or urban areas, transport the sand elsewhere, leaving behind a hard, rocky surface called **desert pavement**. The Sahara is mostly desert pavement interspersed with smaller dune fields where sand is deposited. On the Arabian Peninsula though, the

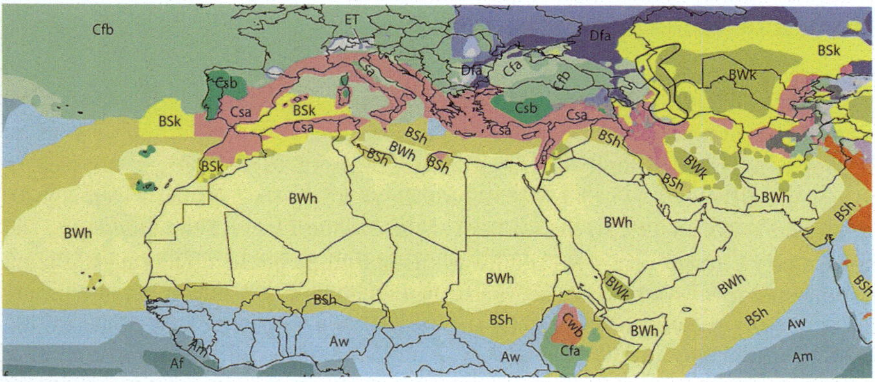

Fig. 15.2 Köppen climatic types of North Africa, Southwest/Central Asia, and the surrounding marine environment. *Source* Reynolds, Rohli, Johnson, Waylen, and Francek, Exploring Physical Geography, McGraw-Hill, modified

combination of topography and local winds limits transport and forms the sandy desert of the Rub-al-Khali.

The relatively flat desert floors are interrupted by large mountain ranges in several areas, especially in the eastern part of the region. The Atlas Mountains border the northern coast of Morocco and Algeria in northwestern Africa with elevations rising over 4000 m (13,000 ft). Along the western coasts of Saudi Arabia and Yemen, the Hijaz and Sarawat Mountains rise 2000 m (6500 ft) above the Red Sea. Farther north are the rugged Caucasus Mountains of Armenia, Azerbaijan, Georgia, and southern Russia, where Europe's highest mountain—Mount Elbrus—is located. South of the Caucasus, the Zagros Mountains rise to peaks of 4000 m (13,000 ft) in Iran, marking a transition zone to the higher plateaus of eastern Iran, Afghanistan, and Central Asia. Farther east are some of the highest ranges in the world—the Hindu Kush of Afghanistan, Pakistan, and Tajikistan; the Pamir Mountains in a swath from Afghanistan to China; the Karakoram of Pakistan, India, and China, which contains K2—the world's second highest mountain; and the Tian Shan of Kazakhstan, Uzbekistan, and Kyrgyzstan.

The extreme aridity of North Africa and Southwest/Central Asia is exemplified by the fact that there are no rivers in Saudi Arabia, and few elsewhere in the region. However, historically important rivers originating elsewhere—known as exotic rivers—do pass through the region. The Nile—the longest river on Earth—flows from the lower-latitude humid regions of Africa through Egypt to the Mediterranean Sea. Runoff from the higher elevations of Turkey form the Tigris and Euphrates Rivers in the "Fertile Crescent" of Iraq. These exotic rivers are critical water resources and have played an important role in the development of civilizations.

Circulation Influences (Situation)

The aridity of North Africa and Southwest/Central Asia can be explained by the atmospheric circulation of the region. Clouds and *precipitation*-forming processes are suppressed by persistently high atmospheric pressure. In North Africa, the eastern side of the *Bermuda-Azores High* extends eastward in the Northern Hemisphere winter and shifts slightly to the north over the Mediterranean Sea in summer. The cold *Canary Current* continues southward from Europe but its effect on stabilizing the atmosphere of North Africa is even stronger than its stabilizing effect on Europe. The *temperature inversion* created by such warm air over much colder water off the North African coast allows the sinking air at the top of the air column to compress and warm *adiabatically* much more than the air at the bottom of the air column can compress and warm. This process reinforces the inversion and is known as a **subsidence inversion**. Sinking air thus persists all the way to the surface on the eastern side of the North Atlantic and also in other basins where subsidence inversions occur in association with cold water currents beneath much warmer air (Fig. 15.3).

In Southwest and Central Asia, the strong Siberian High of winter brings higher pressure as far as the Arabian Peninsula. Consistently strong *subsidence* also caps atmospheric *convection* on the Arabian Peninsula, making clouds and precipitation scarce. Subsidence is enhanced even more than in most other parts of the Earth near the sinking side of the *Hadley cell* around 30°N latitude over the Arabian Peninsula, because of the additional implications of the area's position at the sinking side of a latitudinal circulation adjacent to the Walker circulation (see again Fig. 6.2) of the Pacific Ocean.

The lack of precipitation does not imply that the air is always dry. Desert air does contain *humidity*. The Mediterranean Sea, Red Sea, and the Persian Gulf are important moisture sources. While their water *temperatures* are not low enough to moderate air temperature significantly, they are a source of water vapor through *evaporation* and *latent heating*. Proximity to water bodies keeps *specific humidity* relatively high,

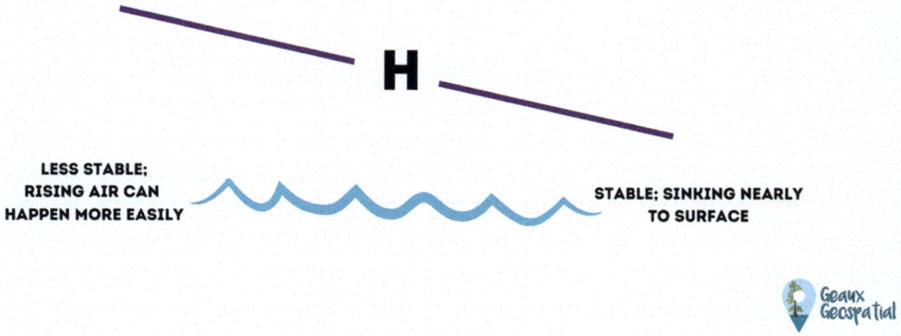

Fig. 15.3 Schematic representing the tilting of the subtropical anticyclones, reinforced by stability and subsidence inversion on their eastern sides (western sides of continents). *Source* Geaux Geospatial

especially near the Red Sea, which is relatively shallow in a hot environment. Coastal areas can experience excessive humidity levels, contributing to extreme heat stress.

North Africa

North Africa is overwhelmingly covered by the True Desert (BW) climates of the Sahara. Precipitation and cloud cover are limited by high-pressure conditions. The clear skies, high *Sun angles*, and high *Bowen ratio* result in extremely high summer temperatures. Figure 15.4 depicts the relative locations of North African locations for which climagraphs and water balance diagrams are shown in subsequent pages.

Mean summer temperatures reach 29 °C (85 °F) in Cairo, Egypt, and remain high to moderate throughout the year (Fig. 15.5). Across the Sahara, daily maximum temperatures often exceed 38 °C (100 °F). High temperatures, in combination with very little precipitation, result in moisture stress and very large water balance *deficits* throughout the year.

The only parts of North Africa that receive some relief from the hot, dry conditions are the immediate coastline, especially in the Northwest. Winter *mid-latitude wave cyclones* passing over the Mediterranean Sea in winter provide a modicum of seasonal precipitation. Isolated locations can be wetter. For example, on the windward slopes of the Atlas Mountains, precipitation totals can exceed 80 cm (31 in). These passing winter storms create a small area of Csa climate along the coast. Although a lack of long-term reliable data precludes the inclusion of a good example of these

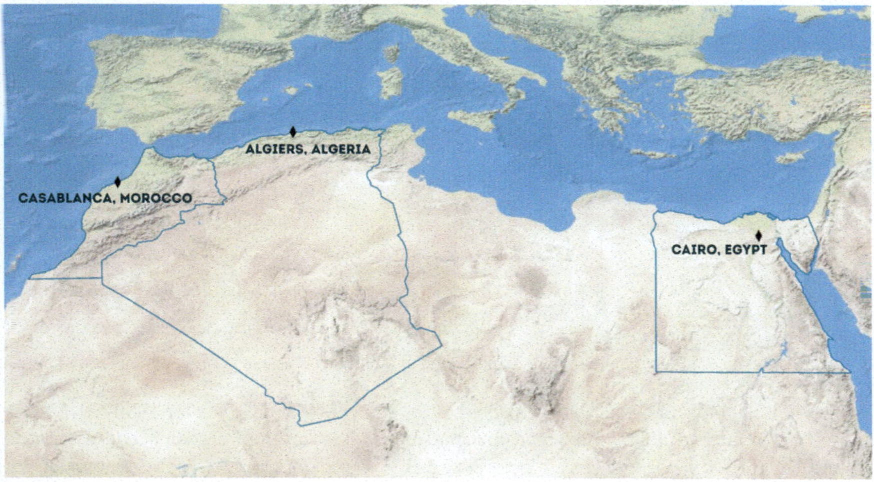

Fig. 15.4 North African cities for which climagraphs and water balance diagrams are shown. *Source* Geaux Geospatial

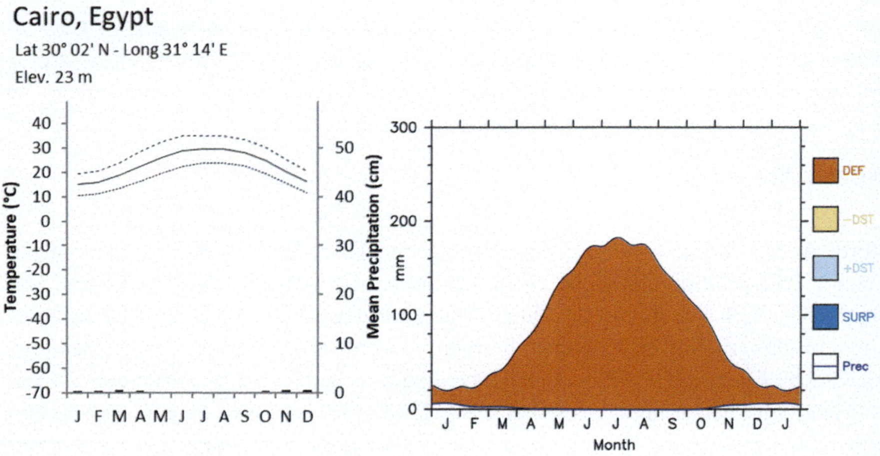

Fig. 15.5 Climagraph and water balance diagram for Cairo, Egypt. Data for 1980–2016 averaging period. *Source* Authors. https://weatherspark.com/, http://cyclops.deos.udel.edu/wimp/public_html/

Csa climates of North Africa using climagraphs and water balance diagrams, the coastal area is represented by Algiers, Algeria (Fig. 15.6), and Casablanca, Morocco (Fig. 15.7).

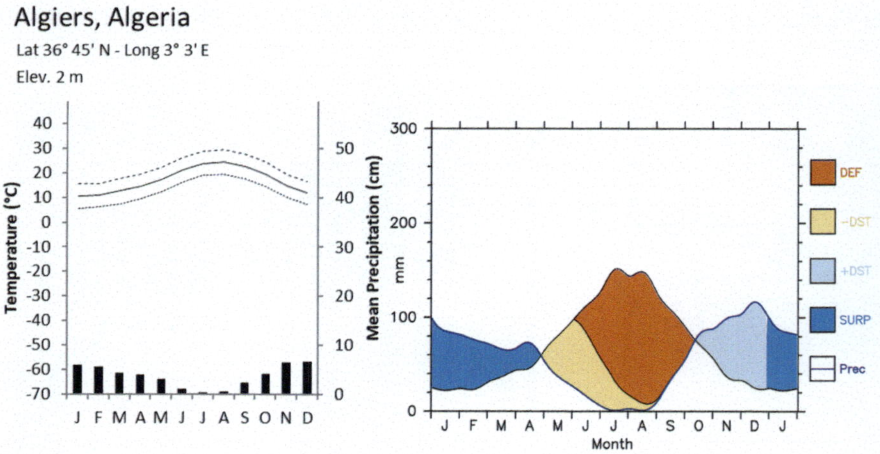

Fig. 15.6 Climagraph and water balance diagram for Algiers, Algeria. Data for 1980–2016 averaging period. *Source* Authors. https://weatherspark.com/, http://cyclops.deos.udel.edu/wimp/public_html/

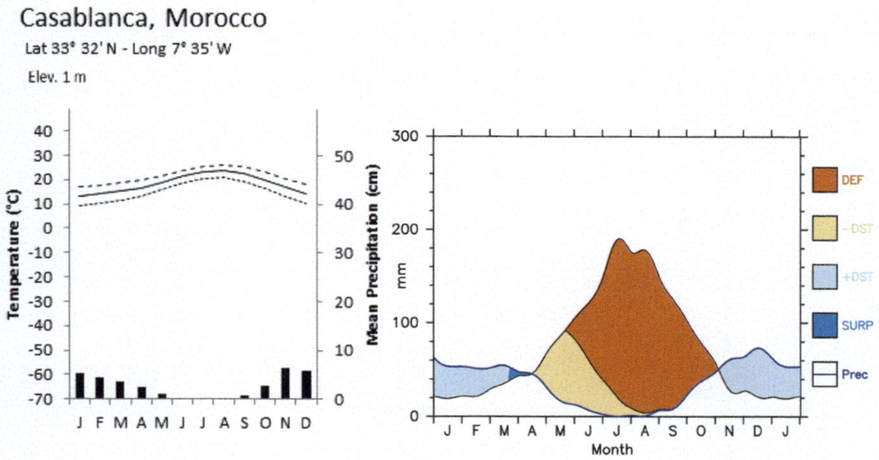

Fig. 15.7 Climagraph and water balance diagram for Casablanca, Morocco. Data for 1980–2016 averaging period. *Source* Authors. https://weatherspark.com/, http://cyclops.deos.udel.edu/wimp/public_html/

Southwest and Central Asia

Excessively hot summers virtually devoid of precipitation continue eastward, producing massive water balance deficits in Southwest and Central Asia. Most of Southwest Asia is under northwesterly flow from the Bermuda-Azores High and its extension over the Mediterranean Sea in summer. The general subsidence promotes a *stable atmosphere*, which allows for very hot and very dry surface environments across the region. Southwestern and Central Asian cities represented by climagraphs and water balance diagrams are labeled in Fig. 15.8. In Baghdad, Iraq, the average temperature is 34 °C (94 °F) in summer. The hottest climate is over the Arabian Peninsula. Cities like Riyadh and Mecca, Saudi Arabia, have an extremely harsh summer, with mean monthly temperatures at or above 36 °C (96 °F) and virtually no precipitation from June through September (Figs. 15.9 and 15.10). The combination of oppressive heat and humidity at coastal Mecca poses a significant human health threat during the Islamic pilgrimages to the city.

Elevation can provide relief, such as over the Yemen Highlands, in the Levant, and in other highland pockets of Csa climate of northern Iran, Afghanistan, and Central Asia. For instance, the 750 m (2460 ft) of elevation at Jerusalem, Israel (Fig. 15.11), allows a summer average temperature of 25 °C (77 °F). Damascus, Syria (Fig. 15.12), has a similar temperature moderation, and Ankara, Turkey (Fig. 15.13), is even cooler, at 23 °C (74 °F).

Winters in Southwest and Central Asia are controlled by the Siberian High and its clockwise and outward surface flow, which is strong enough to bring the area welcome relief from the heat of the high-Sun season. Winter temperatures are similar to those of the southwestern United States. While not exceedingly cold, as in Central Asia,

Fig. 15.8 Southwestern and Central Asian cities for which climagraphs and water balance diagrams are shown. *Source* Geaux Geospatial

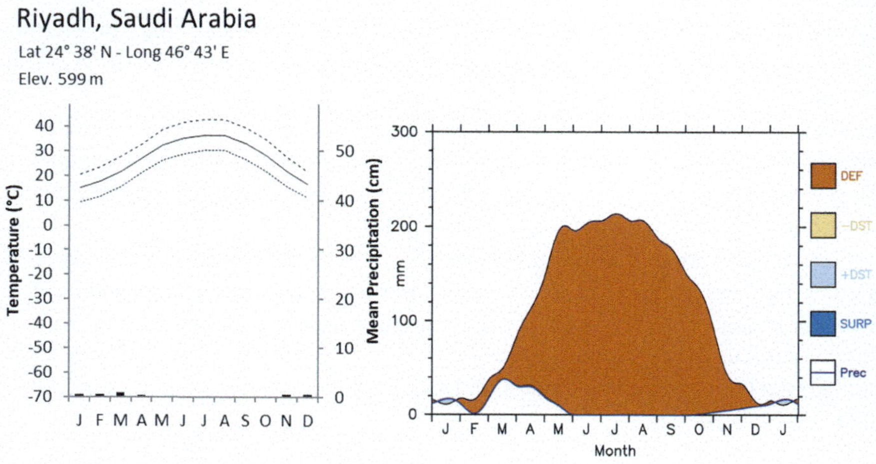

Fig. 15.9 Climagraph and water balance diagram for Riyadh, Saudi Arabia. Data for 1980–2016 averaging period. *Source* Authors. https://weatherspark.com/, http://cyclops.deos.udel.edu/wimp/public_html/

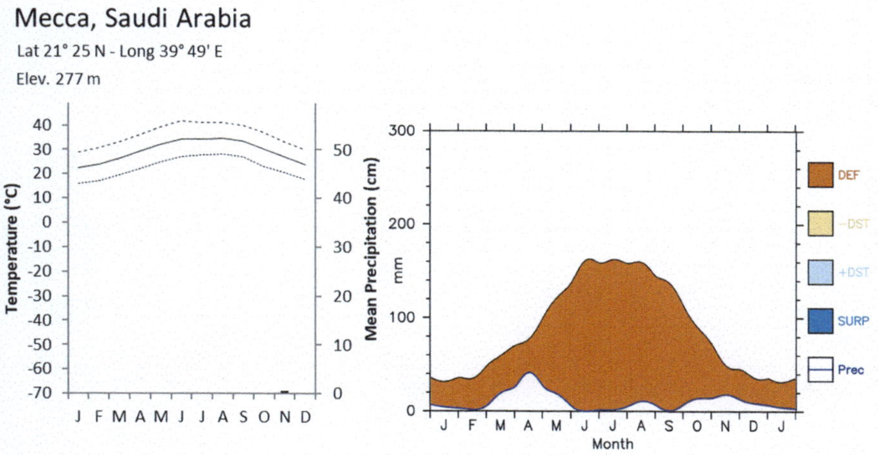

Fig. 15.10 Climagraph and water balance diagram for Mecca, Saudi Arabia. Data for 1980–2016 averaging period. *Source* Authors. https://weatherspark.com/, http://cyclops.deos.udel.edu/wimp/public_html/

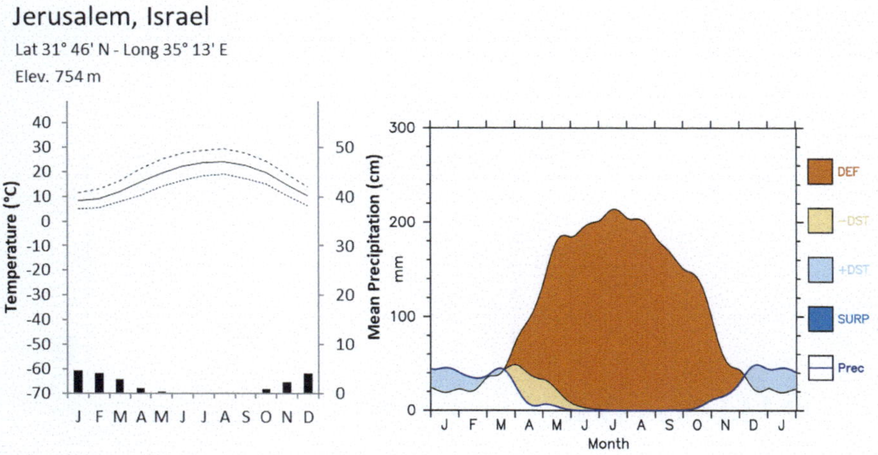

Fig. 15.11 Climagraph and water balance diagram for Jerusalem, Israel. Data for 1980–2016 averaging period. *Source* Authors. https://weatherspark.com/, http://cyclops.deos.udel.edu/wimp/public_html/

Southwest Asia is cooler than expected given its latitude. Representative climagraphs for parts of the region most influenced by the Siberian High are shown for a west-to-east transect from Baghdad, Iraq; to Mashhad, Iran; to Kabul, Afghanistan in Figs. 15.14, 15.15, and 15.16, respectively. Note the influence of the Siberian High on winter precipitation at the sites and the effect of elevation on temperature at the latter two locations.

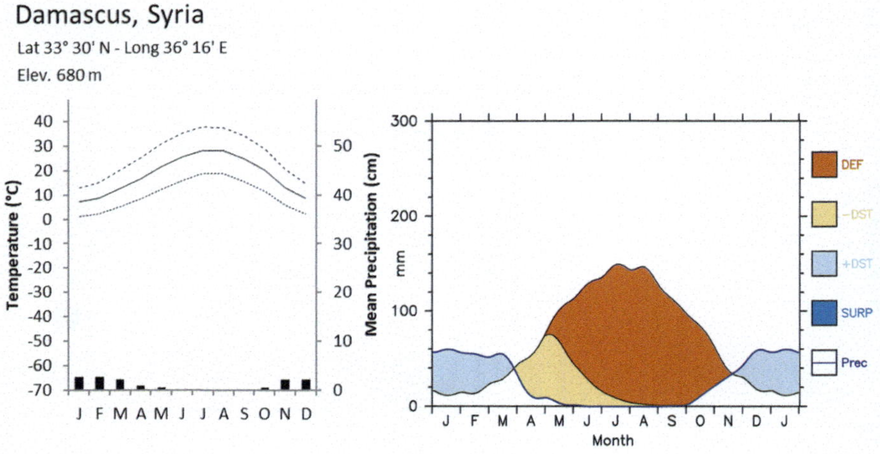

Fig. 15.12 Climagraph and water balance diagram for Damascus, Syria. Data for 1991–2020 averaging period. *Source* Authors. https://weatherspark.com/, http://cyclops.deos.udel.edu/wimp/public_html/

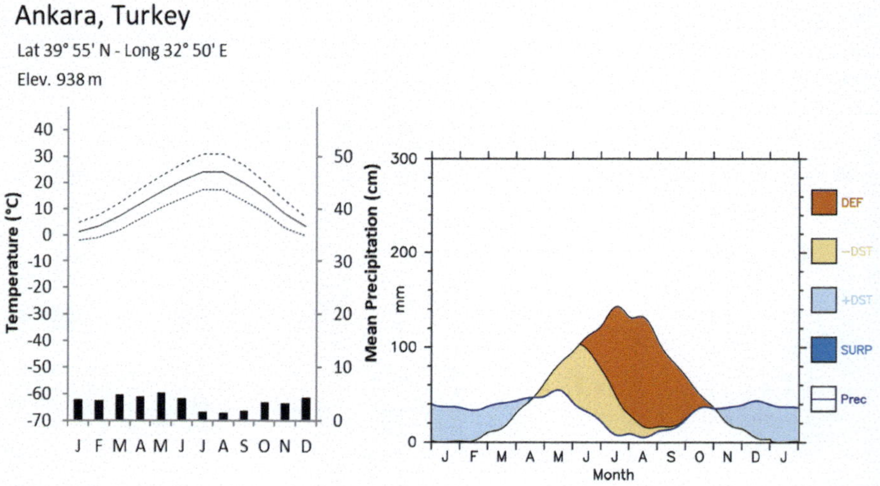

Fig. 15.13 Climagraph and water balance diagram for Ankara, Turkey. Data for 1991–2020 averaging period. *Source* Authors. https://weatherspark.com/, http://cyclops.deos.udel.edu/wimp/public_html/

Although winter precipitation may reach Turkey, Syria, Iraq, and Iran from occasional *mid-latitude wave cyclones* originating from the Mediterranean basin, the Siberian High also suppresses precipitation. Many of the systems that eventually bear rain move southward around the mountains of Turkey, thus avoiding the stability of the Siberian High, even though the track seldom underlies the trough-to-ridge side of the Rossby wave which would enhance uplift. Approximately 30 cyclones per

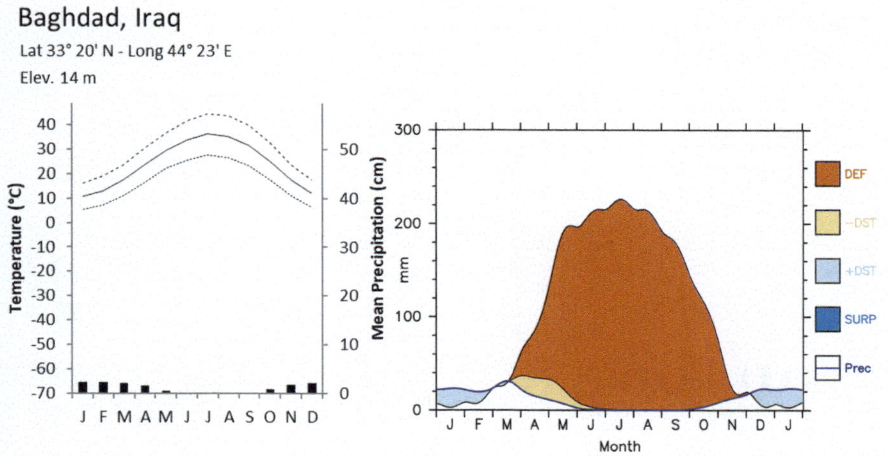

Fig. 15.14 Climagraph and water balance diagram for Baghdad, Iraq. Data for 1980–2016 averaging period. *Source* Authors. https://weatherspark.com/, http://cyclops.deos.udel.edu/wimp/public_html/

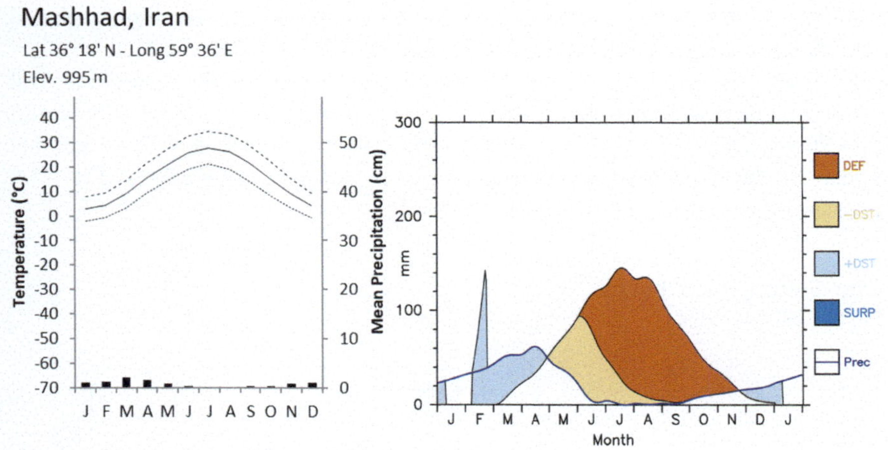

Fig. 15.15 Climagraph and water balance diagram for Mashhad, Iran. Data for 1980–2016 averaging period. *Source* Authors. https://weatherspark.com/, http://cyclops.deos.udel.edu/wimp/public_html/

year move along this southward route and about 20 take the northward route. "Lake effect" rains and occasional snows at higher elevations are produced on the *leeward* side of the water bodies from the frontal lifting. Precipitation amounts decrease from west to east with distance from the Mediterranean Sea.

Even including these winter events, paltry precipitation amounts continue eastward and precipitation-generating mechanisms are virtually absent in summer. In addition, the continentality and simple distance from major moisture sources become

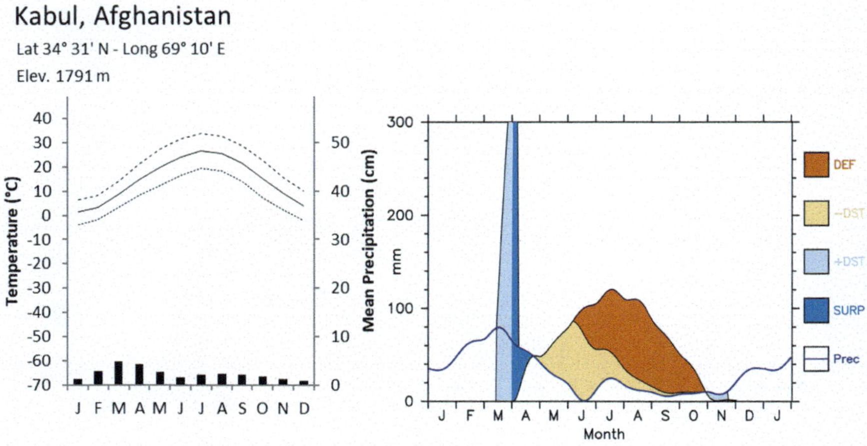

Fig. 15.16 Climagraph and water balance diagram for Kabul, Afghanistan. Data for 1980–2016 averaging period. *Source* Authors. https://weatherspark.com/, http://cyclops.deos.udel.edu/wimp/public_html/

major factors in the climates of Central Asia, as is suggested by the climagraph and water balance diagram for Nur-Sultan, Kazakhstan (Fig. 15.17).

Land use practices, most famously the diversion of water in the case of the Aral Sea, can exacerbate water shortages. The Aral Sea was among the world's largest lakes 70 years ago, but irrigation projects have virtually eliminated it. The water that appears on the Kasakhstan-Uzbekistan border east of the Caspian Sea in Fig. 15.8 represents the remnants of the mostly dry Aral Sea lake bed.

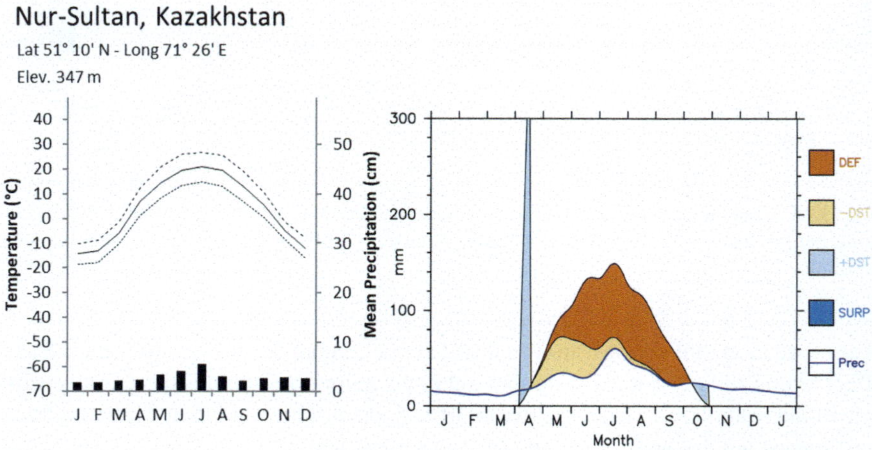

Fig. 15.17 Climagraph and water balance diagram for Nur-Sultan, Kazakhstan. Data for 1991–2020 averaging period. *Source* Authors. https://weatherspark.com/, http://cyclops.deos.udel.edu/wimp/public_html/

However, a few sources of relief exist in parts of Central Asia. The spring melting of winter snows at higher elevations contributes to brief first flushes of runoff. Furthermore, while continentality ensures that summers remain hot, convection entices small amounts of summer rain associated with the thermally-driven low atmospheric pressure over interior Asia, mitigating the overall levels of moisture stress compared to other parts of Southwest Asia. Nevertheless, climatological summer water balance deficits remain universally high in the entire region from Morocco to Kazakhstan and into Mongolia and western and central China. Eventually, proximity to the Pacific Ocean becomes a factor that brings more precipitation, but only in the eastern part of the next region and the topic of the next chapter—East Asia.

Further Reading

Ammar, K., El-Metwally, M., & Wahab, M. M. A. (2014). A climatological analysis of Saharan cyclones. *Climate Dynamics, 43*(1–2), 483–501. https://doi.org/10.1007/s00382-013-2025-0

Fontaine, B., Janicot, S., & Monerie, P. A. (2013). Recent changes in air temperature, heat waves occurrences, and atmospheric circulation in northern Africa. *Journal of Geophysical Research—Atmospheres, 118*(15), 8536–8552. https://doi.org/10.1002/jgrd.50667

Labban, A. H., Mashat, A. W. S., & Awad, A. M. (2020). The variability of the Siberian High ridge over the Middle East. *International Journal of Climatology, 41*(1), 104–130. https://doi.org/10.1002/joc.6611

Patlakas, P., Stathopoulos, C., Flocas, H., Kalogeri, C., & Kallos, G. (2019). Regional climatic features of the Arabian peninsula. *Atmosphere* 10(4). https://doi.org/10.3390/atmos10040220

Rohli, R. V., & Vega, A. J. (2018). *Climatology* (Fourth edition, 418 pp). Jones and Bartlett. 978–1–284–11998–5.

Chapter 16
East Asia and Adjacent Ocean Environments

Abstract The intensely-populated East Asia region is largely influenced by mid-latitude atmospheric circulation features, but also with polar and tropical elements. Unlike the U.S., Canada, and Europe, East Asia has a strong monsoon circulation influence and lacks an upstream ocean to provide moisture. Three subsections of East Asia can be identified. Interior East Asia is an extension of arid central Asia, with the Gobi as its most prominent desert. The Himalaya Mountains and Tibetan Plateau block moisture flowing onshore from reaching central Asia, and the westerlies have already precipitated most of their moisture before reaching this area. From north to south, northeastern Asia is largely covered by the cool-, mild-, warm-, and hot-summer varieties of the microthermal Winter Dry (Dwd, Dwc, Dwb, and Dwa, respectively) climate and the hot summer variety of the Mesothermal Winter Dry (Cwa) climate. The Siberian High and winter monsoon generate the winter dryness. Japan and adjacent waters are dominated by Marine West Coast (Cfb) and Humid Subtropical (Cfa) climates. Southeastern China has Cfa, or Cwa climates where the monsoon produces winter aridity, but both are influenced strongly by typhoons and a quasi-persistent, nearly stationary, east–west-oriented weak baroclinic zone known as the Mei-yu or Baiu phenomenon.

Keywords Arid (B) climates · Humid Subtropical (Cfa) climate · Mei-yu (Baiu) phenomenon · Mesothermal and microthermal Winter Dry (Cw, Dw) climates · Jet streak · Kuroshio Current · Mindanao Current · Monsoon · North Pacific Gyre · Oyashio Current · Siberian High · Subarctic (Df) climate · Tibetan Low · Typhoon

General

Asia is by far the largest continent, spanning from south of the Equator low-latitude tropics to north of the *Arctic Circle*, and stretching over 160 degrees of longitude. Therefore, it is not surprising that its varied environments support every major *Köppen climate type*. Continentality ensures that arid (B) climate types are common in the interior while the high latitudes also contain extensive *microthermal* (D)

R. V. Rohli et al., *Atmospheric and Oceanic Circulation*,
https://doi.org/10.1007/978-3-031-66142-6_16

climates. South and East Asia are affected by tropical processes including the *intertropical convergence zone* (ITCZ) as a source of climatic variation. *Mesothermal* (C) *climates* occur along the coast.

Topographic Influences (Site)

The most important climate influence for Asia is *continentality*. Interior locations are as far removed as anyplace else on Earth from significant water sources, resulting in extreme seasonal *temperature* variations. The continentality effect creates hot summers and bitterly cold winters for the interior. Adding to the continentality are the extreme environments of Asia's significant topographic features, including the highest mountains in the world in the Himalaya and the Tibetan Plateau, known as the world's "rooftop". The effect of the Earth's largest continent even extends offshore to the mountainous islands of the Japanese Archipelago.

Yet the continent is also adjacent to the world's largest ocean—the Pacific, and also the Indian Ocean. Extensions of the oceans include the Sea of Okhotsk, Sea of Japan, Yellow Sea, South China Sea, East China Sea, Gulf of Tonkin, Gulf of Thailand, Andaman Sea, Bay of Bengal, and Arabian Sea (Fig. 16.1). The marine influence complicates climates in coastal areas and islands. Although all of these oceanic extensions are water vapor sources, the main source of *humidity* and *enthalpy* to fuel *precipitation* is the Indian Ocean. Some of the world's most important and impressive rivers are in this region, from the Ob, Yenisey, and Lena Rivers that flow northward from Siberia into the Arctic Ocean, to the Yellow (or Huang He), Yangtze, and Pearl Rivers that flow eastward through intensely populated eastern China into the Pacific Ocean.

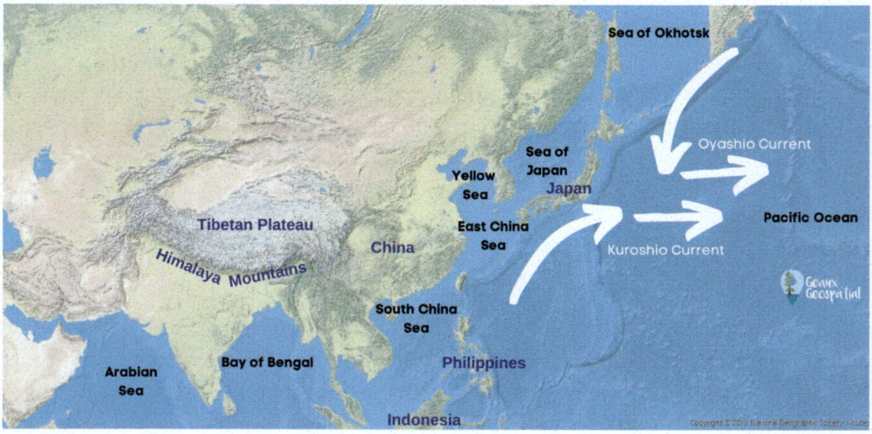

Fig. 16.1 Physiographic regions of East Asia and the adjacent oceans. *Source* Geaux Geospatial

Circulation Influences (Situation)

The extreme continentality of Asia includes bitterly low winter temperatures, with the density of this frigid air producing some of the highest surface atmospheric pressures on Earth. The highest recorded atmospheric pressure of 1083.8 mb was measured at Agata, Siberia, on 31 December 1968. The *Siberian High* develops from this extreme chilling of the interior Asian landmass. This strong anticyclone produces clockwise surface winds that swirl outward from the high-pressure center and carry continental air toward the southern and eastern coasts, promoting dry conditions across much of Asia in winter.

Not only does the Siberian High produce exceedingly high pressures and dry conditions, but it also influences the temperatures across the continent. Interestingly, the coldest regions are not the highest in latitude. Although ice-covered through most of the year, polar Asia remains near the open ocean and the warmth of the deeper ocean. Thus, the lowest temperatures occur in the more southerly interior locations where extreme continentality combines with still-short periods of daylight and oblique *Sun angles* to generate the coldest air in the Northern Hemisphere. This cold air spirals out from the Siberian High, bringing low winter temperatures across most of central and eastern Asia. Southern Asia is only spared by the presence of the Himalaya Mountain range that effectively blocks the air from penetrating farther south.

The opposite conditions occur in summer when the **Tibetan Low** replaces the Siberian High. Despite the increased frequency of cloud cover and precipitation, high average temperatures in the continental interior are common in summer as solar radiant energy is converted to *sensible heat* when water to evaporate is in short supply. The sensible heating with little latent heating is a recipe for the atmosphere over interior China and the rest of central Asia to warm quickly. The *annual temperature range* in these areas is a remarkable 60 °C (~108 °F). The thermally-driven Tibetan Low forms in the area of the warmest, most unstable air, drawing air inland from the Indian and Pacific Oceans. The strength of the Tibetan Low causes the ITCZ to shift farther poleward in summer than anywhere else on Earth (Fig. 16.2), bringing extensive storminess to southern Asia.

Viewed over the course of a year, the warm season is dominated by damp, onshore winds and wet conditions, and the winter brings cold offshore flow and dry conditions. This seasonal cycle creates the famous Asian *monsoon*. A monsoon refers to a seasonal reversal of wind, but not a storm event or a wet season *per se*. The summer season winds bring warm, humid, marine air into the continent, often stimulating heavy rainfall events. In winter, the pattern reverses and cold, stable air reaches the coasts from the Asian interior. Each phase exists for approximately five months, separated by a transition period. The winter winds from the Siberian High cross East Asia from the northwest, while southerly summer winds from the tropics are turned to the right by the *Coriolis effect* and become southwesterly. The seasonal winds of the Asian monsoon dominate the climates of East and South Asia (Fig. 16.3). The

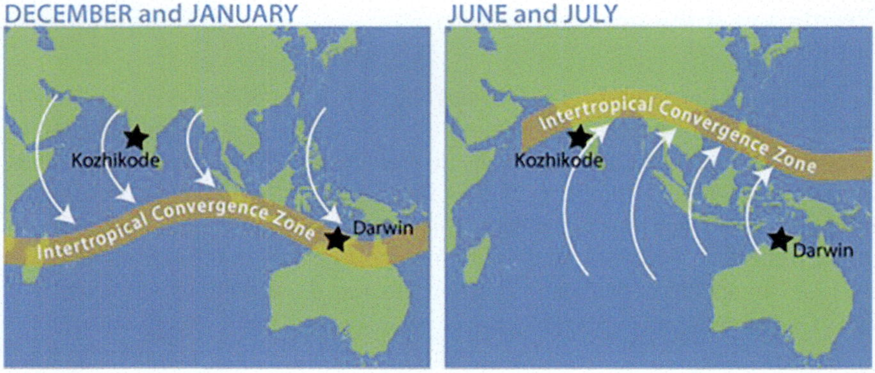

Fig. 16.2 Typical position of the inter-tropical convergence zone in winter and summer over Asia. *Source* UCAR. Permission from Tara Torres

Asian monsoon system affects a great number of Earth's inhabitants and is covered in more detail in Chap. 17.

Two major ocean currents lie off the East Coast of Asia. These currents, namely, the warm *Kuroshio Current* and the cold **Oyashio Current** (see again Fig. 16.1), have key implications for the climate in the archipelagos. The Oyashio Current is analogous to the *Labrador Current* of North America that extends from high latitudes along the coast. It slips southward on the landward side of the warm Kuroshio Current as the latter is deflected eastward into the deeper North Pacific, much as the *Gulf Stream* is deflected northeastward into the deep North Atlantic Ocean. The shape of the ocean basins restricts the northern extent of the *North Pacific Gyre*, forcing the Kuroshio Current eastward at lower latitudes than the Gulf Stream in the Atlantic. The warmer waters of the Kuroshio Current affect the coast from southeastern China to

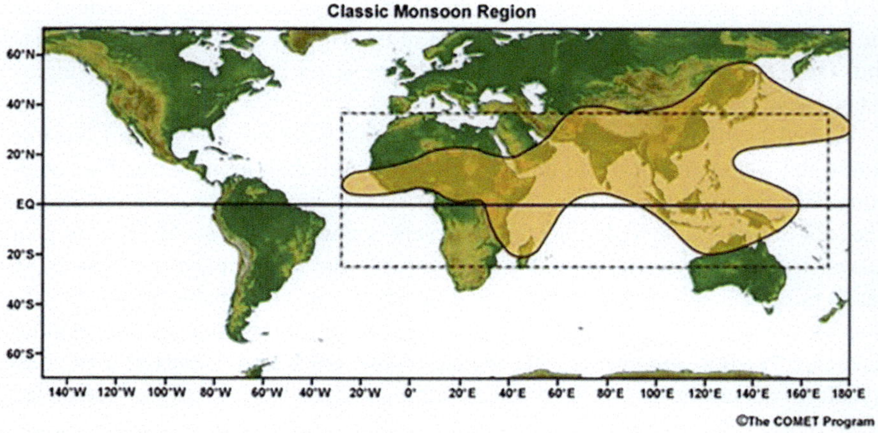

Fig. 16.3 The region of the Asian monsoon. *Source* UCAR CoMet. Permission from Tara Torres

southern Japan, while northern Japan and areas northward are affected more directly by the cold Oyashio waters. To the south, a branch of tropical water, known as the **Mindanao Current**, splits from the Kuroshio Current and carries warm water past the Philippines toward Indonesia.

Interior East Asia

As was described in Chap. 15, the temperature climatology of Interior East Asia is controlled by its extreme continentality. Wide annual temperature ranges are common. With little cloud cover or humidity in the air to absorb *outgoing longwave radiation*, the nocturnal winter temperatures under the influence of the Siberian High are frigid. Yet the summer daylight hours and much higher Sun angle allow summer afternoons to be pleasant to warm in most of Interior East Asia, most of the time, if not outright hot in some places. For instance, the all-time high temperature record is 43 °C (109 °F) in Kabul, Afghanistan, and 42 °C (108 °F) in Bishkek, Kyrgyzstan, despite elevations of 1790 m (5870 ft) and 800 m (2600 ft), respectively, and a latitude of nearly 43° N for Bishkek.

The precipitation climatology of Interior Asia is also controlled by continentality and the Siberian High, but in addition, topography and monsoon circulation are key drivers. The Himalaya Mountains and Tibetan Plateau induce a strong *rain shadow* effect on Interior Asia, resulting in extensive dryness on the *leeward*, northern sides of the mountains, even in the summer monsoon season. As humid air from the Indian Ocean encounters the mountain axis and descends along the northern side, the air warms adiabatically. As air temperature and *relative humidity* are inversely related, cloud formation and precipitation processes are stymied.

The combination of the temperature and precipitation, and the seasonality of each, along with formidable mountains and Tibetan Plateau along southern interior Asia, causes an extensive *True Desert* (BW) *climate* in Interior East Asia (Fig. 16.4). The Gobi Desert of western China and Mongolia is one of the driest large areas on Earth. The rain shadow effect exacerbates the aridity caused by its great distance from surface water sources and the winter influence of the Siberian High. In addition, the dry swath of Interior East Asia is also the product of many thousands of kilometers of the wringing out of humid air from the Atlantic Ocean and Mediterranean Sea as the *mid-latitude westerlies* continue their trek across the vastness of Asia. In addition to the Gobi, interior East Asia also includes the Kyzylkum Desert in Kazakhstan and Uzbekistan, and the Karakum Desert from the China border westward through central Asia. The interior also includes Afghanistan and Iran and much of Turkmenistan to the Caspian Sea and extending northward into Russia. Figure 16.5 shows the locations of selected Interior East Asia stations for which climagraphs and water balance diagrams appear. Figure 16.6 depicts the climagraph and water balance diagram for the world's coldest capital city—Ulaanbaatar, Mongolia—which has huge water balance *deficits* throughout the year despite the low temperatures that mitigate the need for water through evapotranspiration.

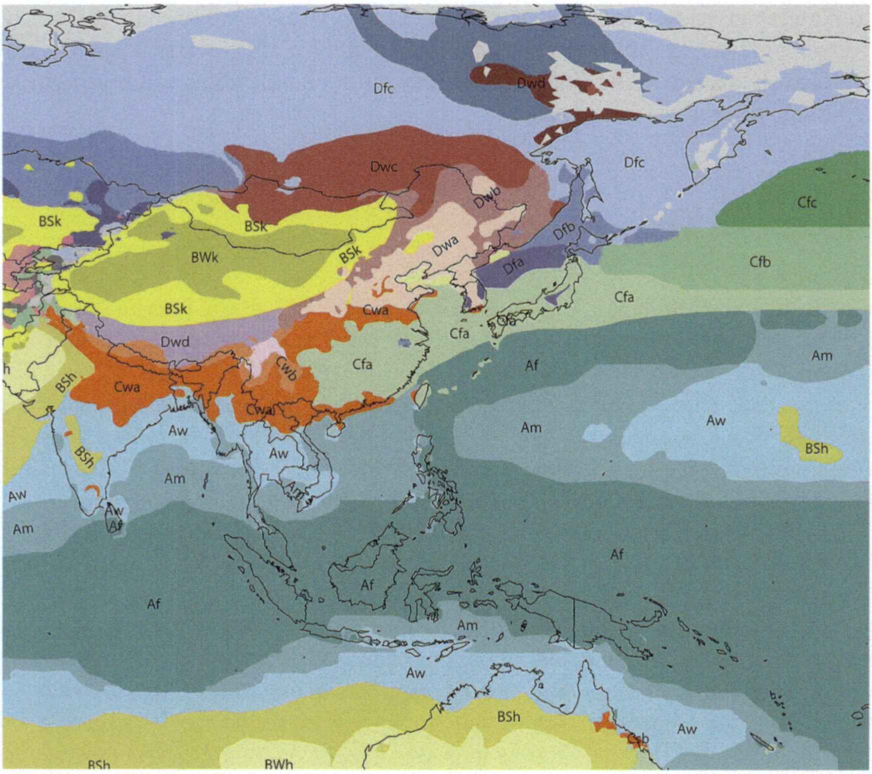

Fig. 16.4 Köppen climatic types of East Asia and surrounding marine environment. *Source* Modified from Reynolds, Rohli, Johnson, Waylen, and Francek, *Exploring Physical Geography*, McGraw-Hill

Yet Interior East Asia is not entirely covered by Arid climates. Outside of the most arid regions, persistent winter subsidence, when the Siberian High is in place, creates the *Microthermal Winter Dry* (Dw) *climate* (denoted by pink, plum, and purplish colors in Fig. 16.4). An example is Xining, China (Fig. 16.7), where the high elevation of 2275 m (7465 ft) on the Tibetan Plateau creates a "warm summer" variety of Dw (Dwb). The "mild summer" (Dwc) and "cool summer" (Dwd) varieties are at even higher elevations and latitudes. In the Köppen system of climate types, a "winter dry" climate requires ten times more precipitation during the summer than in winter. This pattern occurs here because of the strong effect of the East Asian monsoon.

Farther from the core of the dryness are places that are far enough poleward to have small enough *potential evapotranspiration* so as to avoid large water balance deficits in summer, as the snowmelt into late spring recharges the soil with water. Instead, the winters are dry from the strong influence of the Siberian High. Most of the southern part of the Russian Far East and adjacent northern Mongolia fall into this Dwb category. Farther west at these high latitudes, enough precipitation falls to

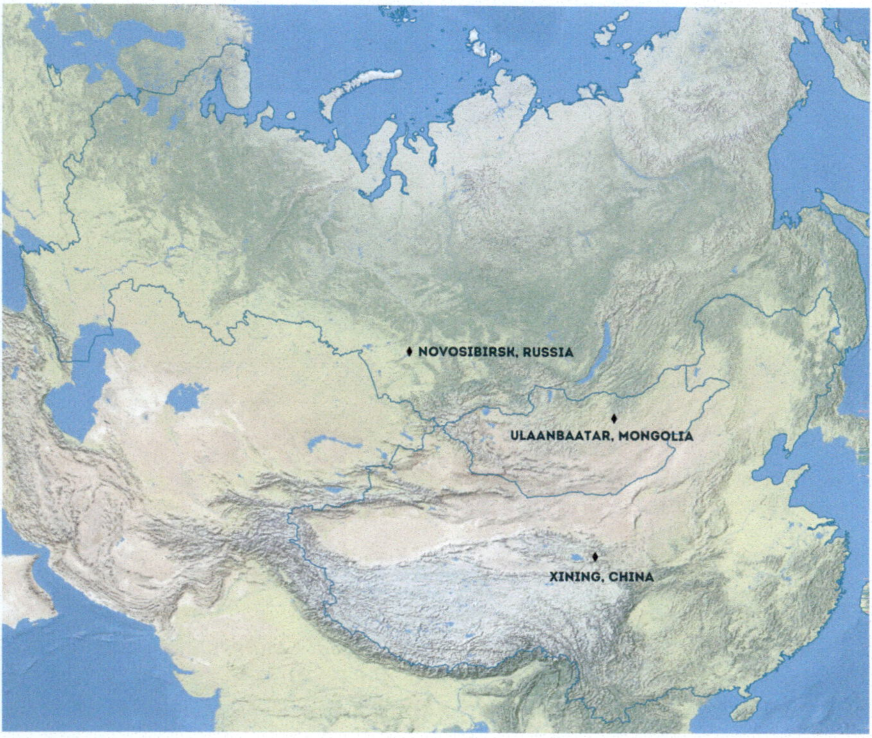

Fig. 16.5 Interior East Asian cities for which climagraphs and water balance diagrams are shown. *Source* Geaux Geospatial

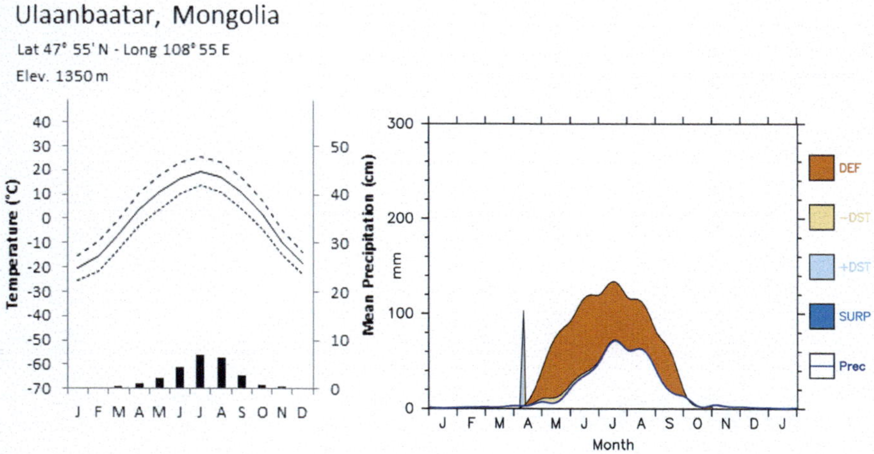

Fig. 16.6 Climagraph and water balance diagram for Ulaanbaatar, Mongolia. Data for 1991–2020 averaging period. *Source* Authors. https://weatherspark.com/, http://cyclops.deos.udel.edu/wimp/public_html/

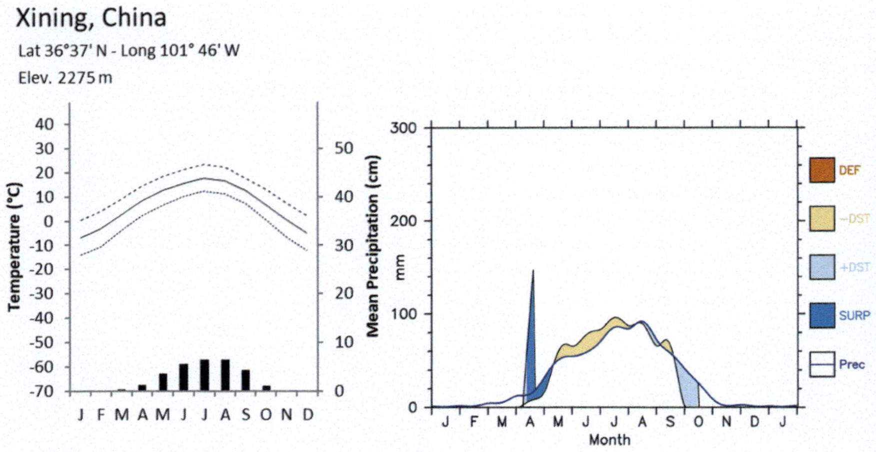

Fig. 16.7 Climagraph and water balance diagram for Xining, China. Data for 1980–2016 averaging period. *Source* Authors. https://weatherspark.com/, http://cyclops.deos.udel.edu/wimp/public_html/

save the climate from being arid. One example is Novosibirsk, Russia, which lies at 55° N, well north of northwestern China, where *Humid Continental—warm summer* (Dfb) *climate* (denoted as pinkish/purplish areas in Fig. 16.4) prevails (Fig. 16.8). In extreme northern Asia, the monsoon effect withers. The *Humid Continental—warm summer* and *Humid Continental—cool summer*, both of which are also known as *Subarctic* (Dfc, Dfd), *climates* at these latitudes is considered in Chap. 21 on polar climates.

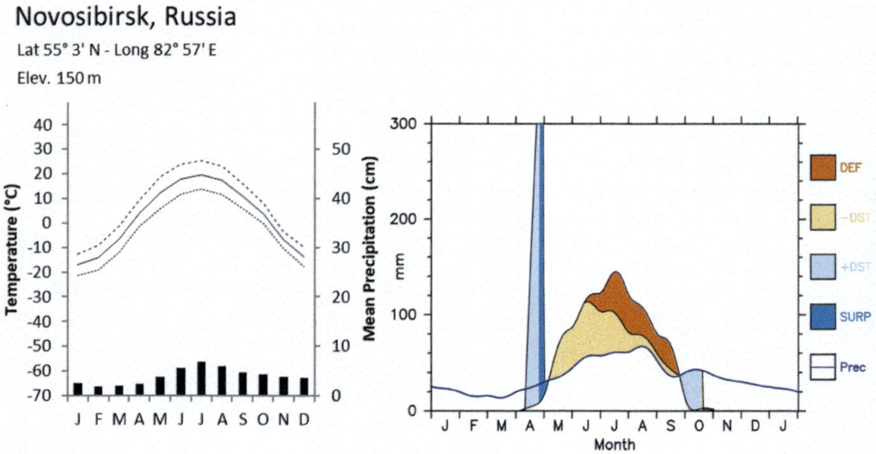

Fig. 16.8 Climagraph and water balance diagram for Novosibirsk, Russia. Data for 1991–2020 averaging period. *Source* Authors. https://weatherspark.com/, http://cyclops.deos.udel.edu/wimp/public_html/

Northeastern Asia

Continentality, exacerbated by the influence of the Siberian High in winter and the wet monsoon in summer, remains a dominant theme of the temperature regime in northeastern Asia. In winter, Arctic air outbreaks are common. Often, one outbreak will follow another within only a few days, reinforcing cold conditions, particularly if the preceding system left a snow cover that would increase the surface *albedo*. Along the northeastern Asian coast including the Manchuria region of northeastern China, the warm Kuroshio Current contributes to a sharp temperature gradient between the frigid interior and the open ocean, as it flows northward along the Pacific coast. Because of the dominant offshore airflow from the Siberian High, the warmth is confined to a narrow coastal strip.

Coastal Northeast Asian winters are less maritime-influenced than in most other coastal zones because outflow from the very strong northerly winds circulating around the Siberian High supplies *continental polar* (cP) and *continental Arctic* (cA) *air masses*. In addition, the Sea of Okhotsk is frozen in winter and reinforces the frigid properties of the overlying air more like a land surface than a water body. The high frequency of clouds from *orographic lifting* of the mid-latitude westerlies impacts the mountainous terrain of western Japan. There, the cloud cover insulates the surface such that nighttime temperatures remain higher than would occur during clear conditions that promote radiational cooling of the surface. For this reason, in Japan, a greater temperature range exists in the east than in the west.

In most of Japan, including the largest and most populous island of Honshu, August tends to be the hottest month. Seasons are delayed in maritime areas, including much of Japan, because the ocean stores heat through summer and releases heat during the cool season. The cold Oyashio Current gives Hokkaido—the northernmost of Japan's four main islands—cool summers. Hokkaido is too cool and not wet enough for rice cultivation, so this agricultural region thrives on livestock and dairy farming. However, the thermal influence of the Oyashio Current is less important for the other Japanese islands, where the Kuroshio Current exerts more influence.

Winter precipitation in northeastern Asia is primarily generated in frontal storms, particularly near the coast. However, the dominant winter offshore airflow from the clockwise and outward flow around the Siberian High produces only weak frontal boundaries between successive cold surges. Rather than having warm and cold air masses interacting, cold and colder air masses interact. The resulting fronts typically produce little rain or snow, even as the leading air mass warms somewhat as it advances over a slightly warmer surface nearer to coastlines. Furthermore, little evaporation occurs from the frozen and/or dry inland surfaces to enhance the moisture in the storm systems. Moisture *advection* from the oceans is sparse, because of the overwhelming offshore flow from the robust Siberian High. As a result, winter precipitation associated with fronts trailing from *mid-latitude wave cyclones* is light, even though precipitation days may be frequent.

Location and magnitude of warm-season precipitation is influenced by the poleward retreat of the *polar front* as the *circumpolar vortex* contracts. Although air mass contrast is much greater in these high-latitude summers than in winters, summer fronts trailing from the mid-latitude wave cyclones are generally weak and comparable to those occurring across the northern tier of U.S. states and in southern Canada in summer. However, despite the generally weak updrafts, these systems are fed by significant moisture advection from the south as the counter-clockwise rotation around the cyclone pulls Pacific moisture into the storm system, and heavy precipitation can occur. These warm-season systems provide approximately two-thirds of the precipitation in Manchuria, with 3 to 6 cyclone passages per summer. As elsewhere, the combined influence of temperature and precipitation, and the seasonality of each, gives the characteristic *Köppen climate type.* Northern China and North Korea are among the few large, contiguous areas of the hot summer variety of *Mesothermal Winter Dry* (Cwa) *climate* and the hot or warm summer varieties of *Microthermal Winter Dry* (Dwa, Dwb) *climates* (orange, pink, and plum-colored areas, respectively, in Fig. 16.4). During winter, the northwesterly wind system combines with the offshore flow regime initiated by the Siberian High to create very dry continental airflow over the region. In summer, the system reverses to encourage the advection of warm, humid, maritime air. This reversal produces a pronounced wet season in summer that offsets the extreme winter aridity.

The extensive and very densely-populated Dwa climate region in northeast Asia is represented here by the very large cities shown in Fig. 16.9, and includes climagraphs and water balance diagrams of Harbin, China (Fig. 16.10); Beijing, China (Fig. 16.11); and Seoul, South Korea (Fig. 16.12). A distinctive feature of the water balance diagram in much of northeastern China, as shown for Harbin and Beijing, is the deficit in spring, just before the onset of the summer monsoon. Nearby water bodies give Seoul abundant moisture year-round to avoid the deficits.

To the east, precipitation is more likely over Japan. The polar front jet stream steers the mid-latitude wave cyclones eastward, with some cyclones tracking north of the Himalaya Mountains and others to the south through southern China. The upper-level support above the trough-to-ridge side of the *Rossby waves* means that regardless of whether the cyclones follow the northern or southern trajectory, they tend to track over Japan where they occlude. Thus, climates of Japan have much in common with those of the U.S. Pacific Northwest, the U.S. New England states, and the Maritime Provinces of Canada. However, the Japanese occluded cyclones more typically have less upper-level support than their North American counterparts and therefore tend to produce less precipitation. They are more likely to be positioned too close to the ridge downstream from the trough-to-ridge side that originally advected them toward Japan. The mountain spine of Japan not only contributes to orographically-enhanced precipitation but also reinforces the locking in place of these ridges, similar to those over the Rocky Mountains of the U.S. and Canada. Storms often track across the entire length of Japan, from south to north, with *sea ice* extending the continental influence of the land surface seaward (Fig. 16.13).

Mid-latitude wave cyclones can "explode" over the warm offshore waters before or after tracking over Japan, if they are finally sufficiently far eastward to be on the trough-to- ridge side of the preferred position of the Rossby waves. Moreover, near Japan the cyclones are energized and destabilized by the warm Kuroshio Current

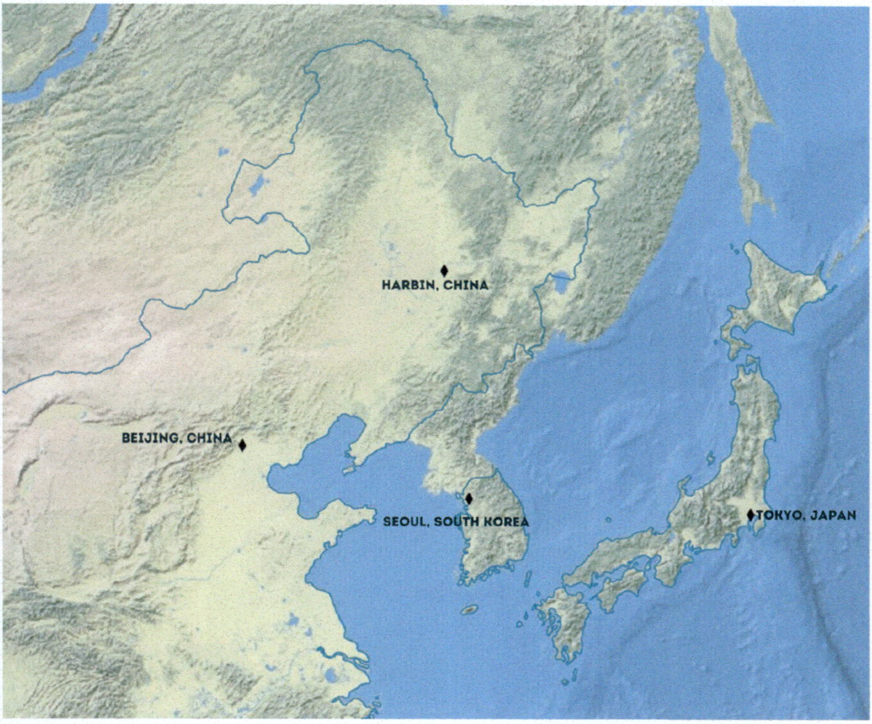

Fig. 16.9 Northeastern Asian cities for which climagraphs and water balance diagrams are shown. *Source* Geaux Geospatial

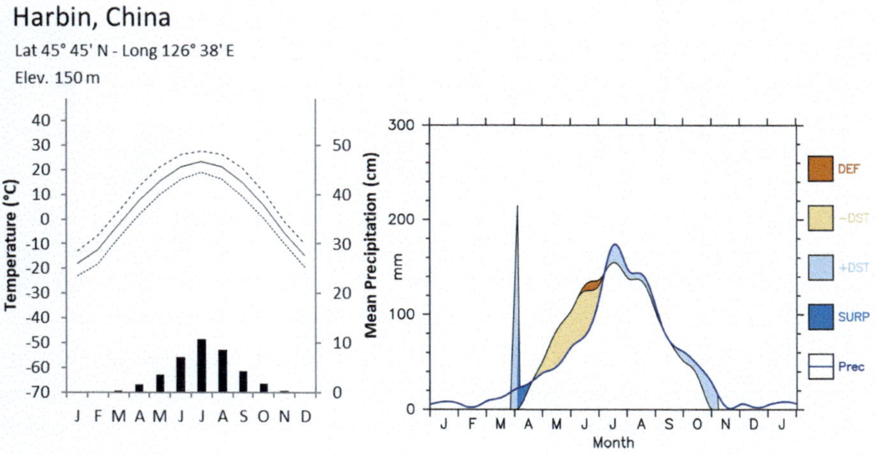

Fig. 16.10 Climagraph and water balance diagram for Harbin, China. Data for 1980–2016 averaging period. *Source* Authors. https://weatherspark.com/, http://cyclops.deos.udel.edu/wimp/public_html/

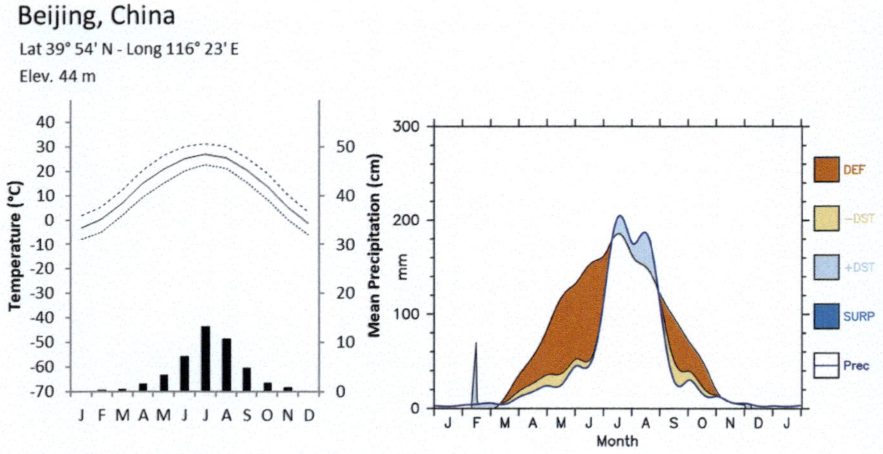

Fig. 16.11 Climagraph and water balance diagram for Beijing, China. Data for 1980–2016 averaging period. *Source* Authors. https://weatherspark.com/, http://cyclops.deos.udel.edu/wimp/public_html/

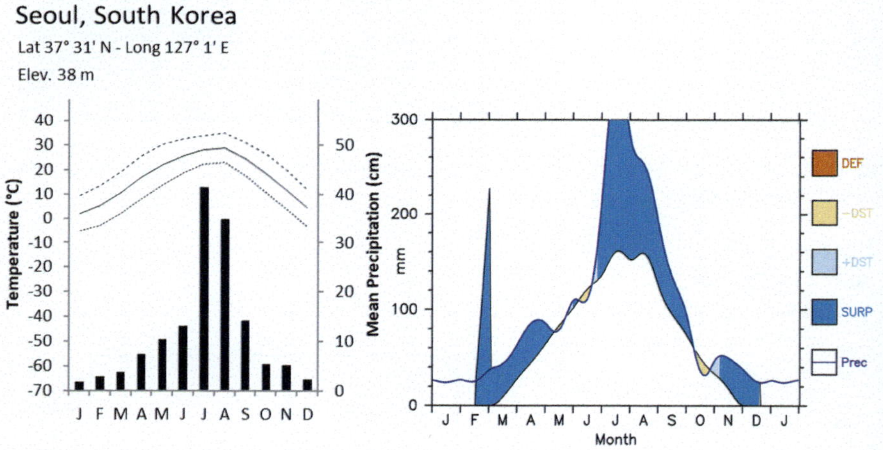

Fig. 16.12 Climagraph and water balance diagram for Seoul, South Korea. Data for 1991–2020 averaging period. *Source* Authors. https://weatherspark.com/, http://cyclops.deos.udel.edu/wimp/public_html/

and enthalpy is added by abundant humidity over the coast. The sharp heat gradient at the boundary between the cold Asian landmass and the warm current provides a dynamic setting that also supports rapid intensification. Such mid-latitude wave cyclones are analogous to the nor'easters of North America's Atlantic coast and not surprisingly, they are also called "nor'easters" in Asia. Japan is more affected by these mid-latitude wave cyclones and nor'easters than China, where all of these ingredients are less likely to be present simultaneously.

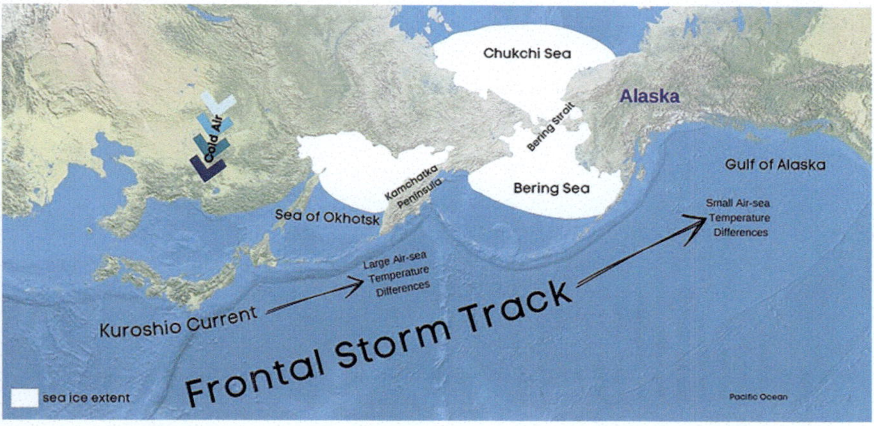

Fig. 16.13 Relationship between polar air masses, sea ice, the Kuroshio Current, and mid-latitude wave cyclone tracks. Ice extent varies seasonally and is only shown for the nearby Arctic Ocean. *Source* Geaux Geospatial

The influence of the polar front jet stream throughout the year and orographic enhancement over the rugged terrain can create substantial precipitation totals. Many Japanese stations report double to triple the precipitation of adjacent eastern China in all seasons. For example, Tokyo receives 150 cm (59 in) of precipitation per year, which is even more than in southern China and plunges the area into a *Humid Subtropical* (Cfa) *climate* (Fig. 16.14). Yet the spatial distribution of precipitation is far from even. The topographically-induced microclimates generally contain wet windward sides and dry leeward sides. Orographic influences may be extreme, with some locations receiving more than 250 cm (100 in) per year. Interestingly, such locations are on the eastern and southern side of the southern islands, where the precipitation-generating air flow is primarily southeasterly, from the *Hawaiian-Pacific High*. Yet the northern islands have significant orographic effects solely on the western side of the highlands.

Southeastern China

Several unique and prominent topographic and atmospheric circulation features impact the climate of southeastern China substantially. As described earlier, the *polar front jet stream* bifurcates because of the east–west orientation of the Himalaya Mountains and adjacent Tibetan Plateau causes (Fig. 16.15). One jet branch tends to remain north of the mountains while the other is usually positioned south of the mountains. The northern jet's position fluctuates wildly, but the southern jet flows over northern India with persistence, ultimately curving northward over southern China and Japan.

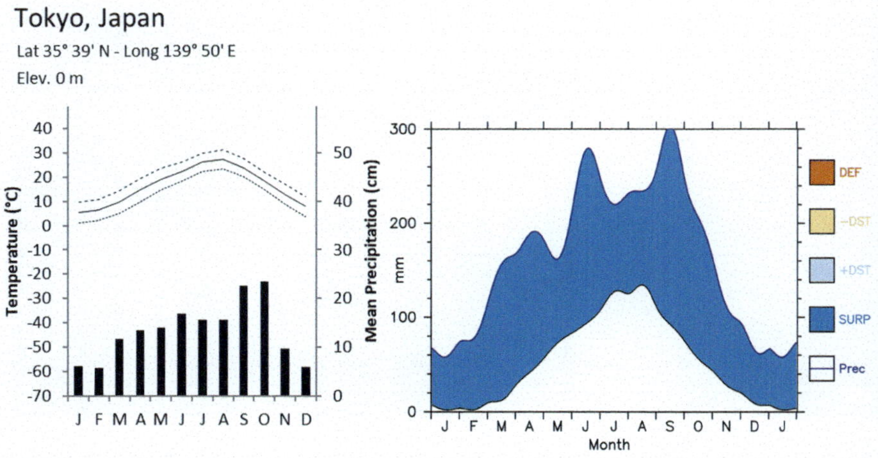

Fig. 16.14 Climagraph and water balance diagram for Tokyo, Japan. Data for 1991–2020 averaging period. *Source* Authors. https://weatherspark.com/, http://cyclops.deos.udel.edu/wimp/public_html/

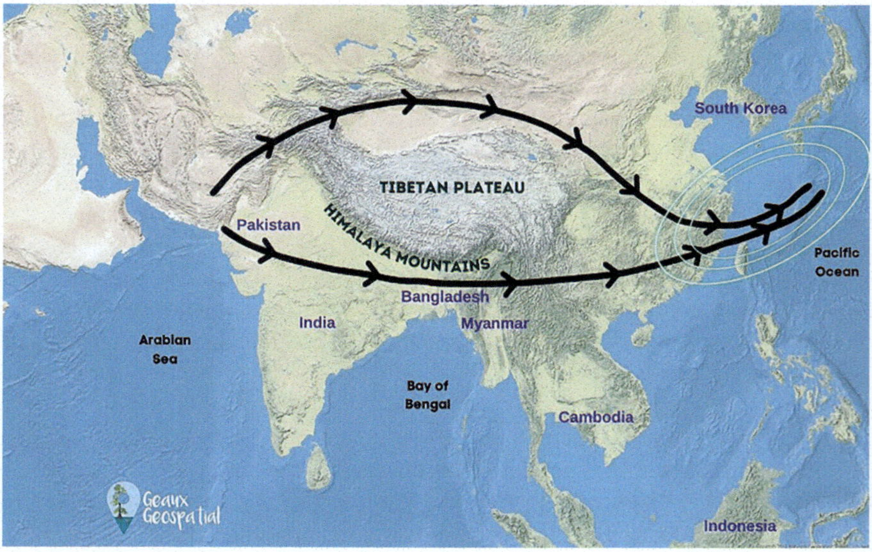

Fig. 16.15 Average Asian polar front jet stream bifurcation around the Tibetan Plateau. *Source* Geaux Geospatial

Strong cool season mid-latitude wave cyclones occur in association with the interaction between the coast, the ocean waters, and the overlying jet stream. As the northern branch of the polar front jet stream exits the continent, it encounters a strong *baroclinic zone* between the cold continent and the warm adjacent waters of the East China Sea and the Kuroshio Current, particularly as the air accelerates overhead in winter and spring (see the red areas, denoting fast winds aloft, in Fig. 16.16a and, to some extent, Fig. 16.16b), providing upper-level *divergence* that supports surface uplift. Recall from Chap. 3 that the front right side of the *jet streak* exit region is where severe storms are most common, and these are offshore in Fig. 16.16. The strong thermal contrast and upper-level support combines with *positive vorticity advection* to create strong mid-latitude wave cyclones that form offshore, occasionally pass through southern Japan, and then move across the northern Pacific Ocean toward North America. To the west, China remains in the ridge-to-trough axis of the northern jet and exposed to drier cP air masses from the interior. This upper-level divergence zone disappears in summer (Fig. 16.16c) and re-appears in autumn, but farther north than in winter or spring (Fig. 16.16d).

In spring, a quasi-persistent, nearly stationary, east–west-oriented, weak baroclinic zone forms in the lower troposphere from the coast of China eastward into the Pacific, south of Japan, known as the **Mei-yu** or **Baiu** phenomenon and shown by the blue, wavy lines in Fig. 16.17. This zone forms in association with the relatively fast polar front jet stream and *subtropical jet stream* over central Asia and its associated ridges and troughs.

The East Asian Mei-yu baroclinic zone (Fig. 16.17) is enhanced by low-level, warm air advection on its equatorward flank. This low-level flow occurs beneath the *jet streak* exit region and forms a small trough-to-ridge bend over the region. The abundant water vapor and low-level instability can fuel *mesoscale convective systems* (MCSs) or *mesoscale convective complexes* (MCSs) that migrate eastward along the baroclinic zone.

The resulting Mei-yu (or Baiu) fronts are particularly impressive considering that the thermal contrast is small between the adjacent air masses. These Mei-yu (or Baiu) fronts, called "plum rains" in Chinese, are a major rain-producer in southeastern Asia from mid-spring to mid-summer. They may be considered as simply low-pressure systems embedded within the southwesterly monsoon winds of the early summer. The Mei-yu (or Baiu) fronts can also be associated with local low-level jets, such as near Taiwan. They can also provide a weak baroclinic environment that can promote tropical cyclogenesis associated with *easterly waves* and advect the disturbances northeastward in the southwesterly winds along the Mei-yu front.

During summer, the southern branch of the jet disappears as the Asian monsoon flow dominates. The disappearance of the southern jet allows southwesterly winds to dominate. The southwesterly summer monsoon winds advect moisture from the *Pacific Warm Pool*, boosting rainfall totals. During late summer, southeasterly winds occur in association with better-developed clockwise and outward flow around the Hawaiian-Pacific High, as it reaches its maximum size and intensity. The strengthening occurs as the ITCZ begins to retreat equatorward with the vertical solar rays.

Autumn can bring extreme wind, *storm surge*, and heavy rainfall events caused by *tropical cyclones*, the strongest of which are called *typhoons* in East Asia and simply *cyclones* in Southeast Asia, where all cyclones are tropical in origin. The immense area of warm, tropical waters in the Pacific Ocean, especially near the Pacific Warm Pool, supports about 30 percent of the world's tropical cyclones in the western North

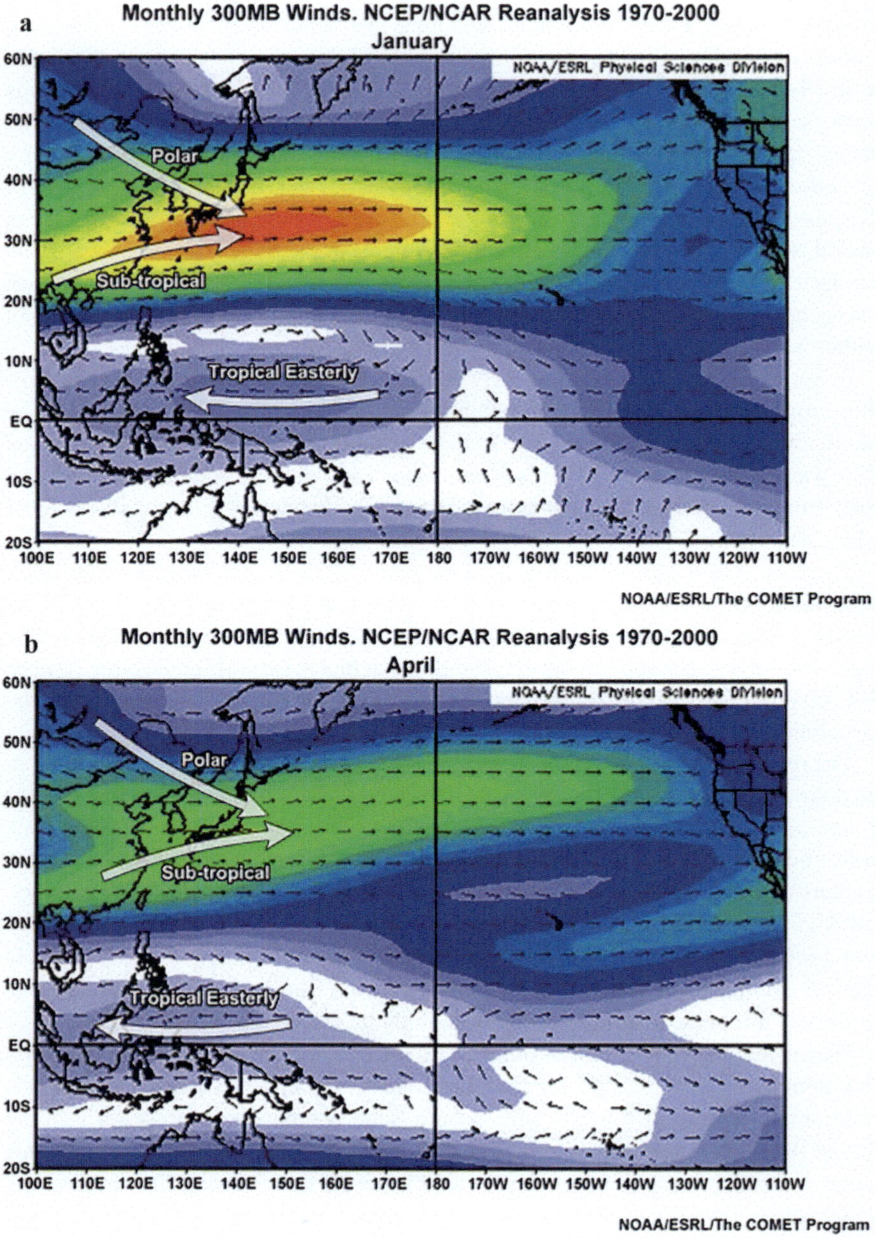

Fig. 16.16 Monthly 300-mb flow regimes in **a** January, **b** April, **c** July, and **d** October, showing the fastest winds in red and yellow, with wind vectors, and the strong upper-level divergence and baroclinic zone in cool months mostly offshore from East Asia. *Source* UCAR CoMet/NOAA. Permission from Tara Torres

c

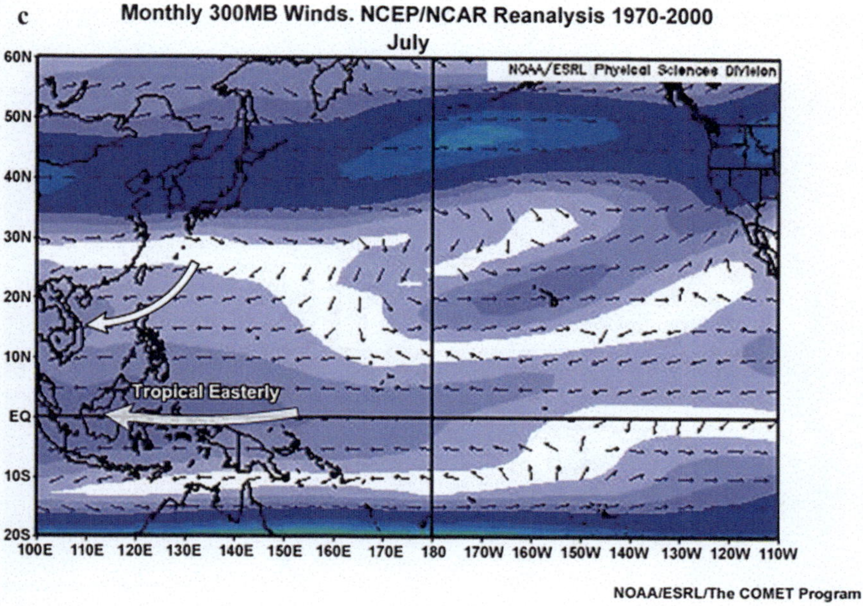

d

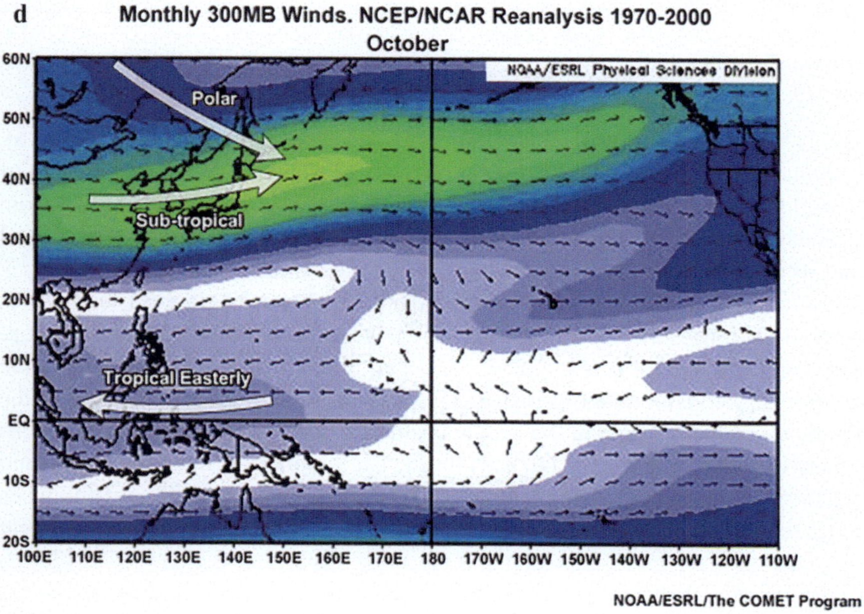

Fig. 16.16 (continued)

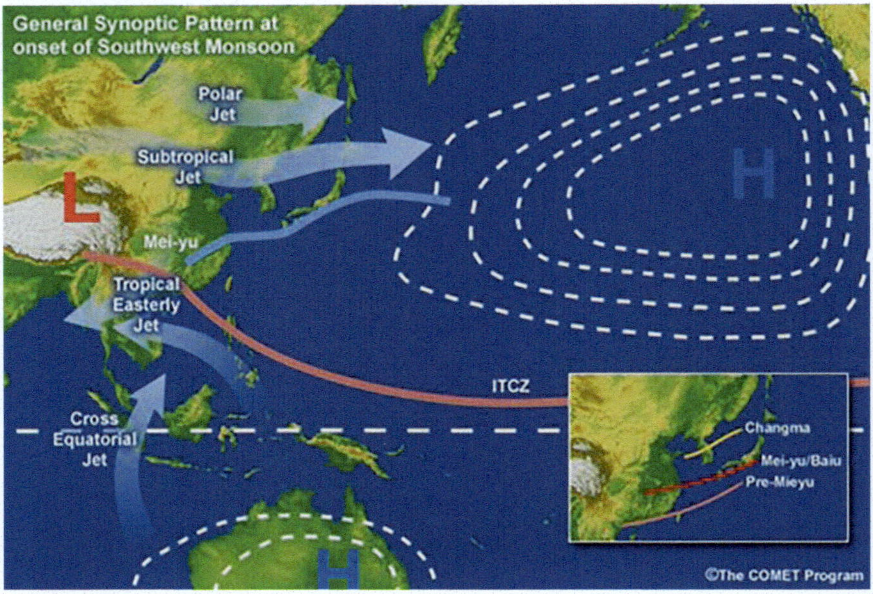

Fig. 16.17 Relationship between Mei-yu and other *synoptic-scale* atmospheric circulation features just before southwesterly monsoon circulation sets in over the region. *Source* UCAR CoMet. Permission from Tara Torres

Pacific Ocean basin, the most active tropical cyclogenesis area on Earth (Fig. 16.18). The high frequency and overall stronger intensity of tropical cyclones in this part of the world creates a formidable hazard across all of coastal southeastern Asia from the Malay Peninsula to Japan. Many of the Earth's most densely populated coastal areas are at substantial risk (Fig. 16.19).

Similar to the *hurricanes* of the North Atlantic, typhoons begin as easterly waves that cross the low-latitudes from east to west in the *trade wind* flow. But the environment is so favorable for development in this part of the world that in addition to the average of 19.3 typhoons per year, there is less seasonality of these events than in the Atlantic or even elsewhere in the Pacific basin (Fig. 16.20). Nevertheless, typhoons increase in both frequency and intensity during late summer and autumn (August and September in the Northern Hemisphere). This timing coincides with maximum sea surface temperatures, which provide the required evaporation and subsequent latent heat to fuel the systems. Ocean temperatures must reach a threshold of 26 °C (79 °F) to generate enough latent heat to maintain tropical cyclone development.

These atmospheric circulation features combine to give southeastern China winter temperatures that can be lower than expected for the latitude, because of the persistence of the offshore flow from the continental interior. For instance, the mean January temperature in Shanghai, China, is 5 °C (41 °F). Winter accounts for a small percent of precipitation totals (Table 16.1), particularly in the north and south, where the

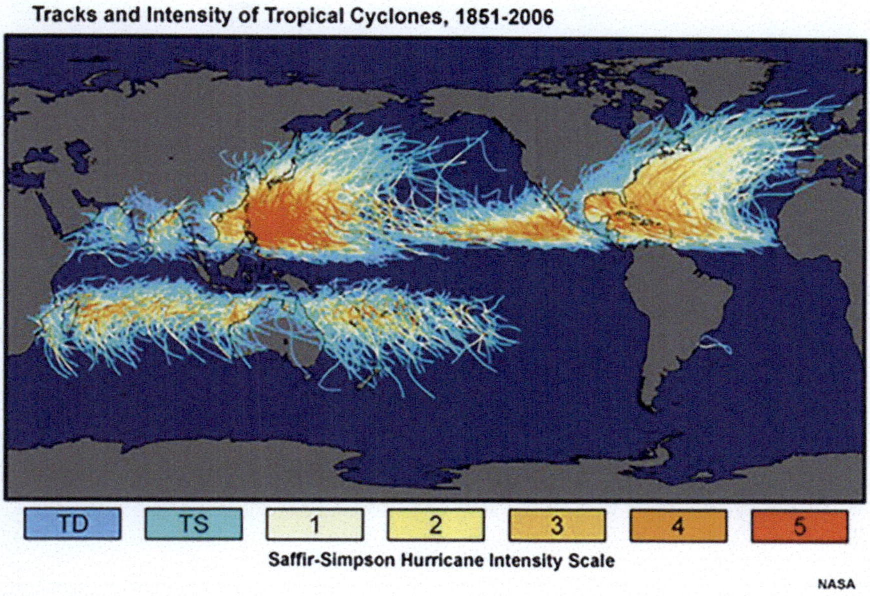

Fig. 16.18 Global tropical cyclone tracks and intensities by cyclogenesis region, 1851–2006. *Source* NASA

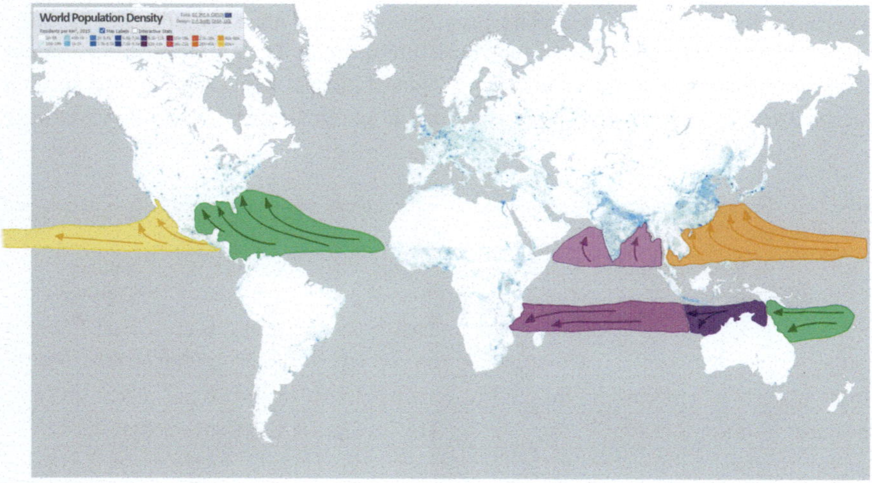

Fig. 16.19 Global tropical cyclone regions relative to population. *Source* Modified from UCAR CoMet; Lumino City

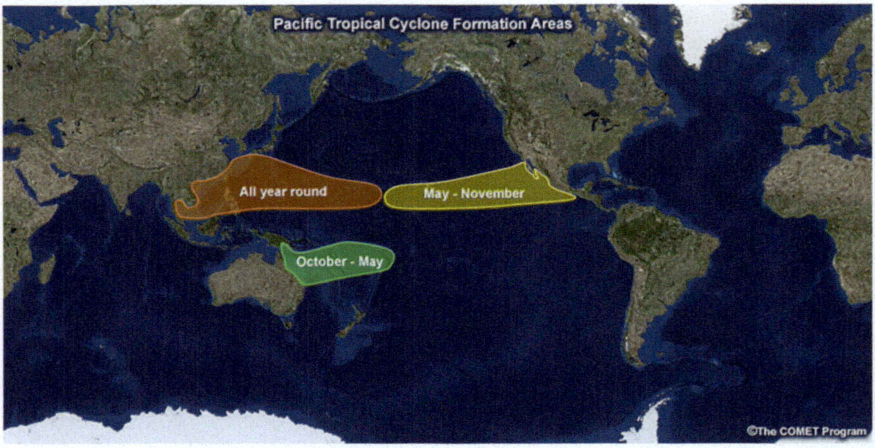

Fig. 16.20 Pacific Ocean tropical cyclogenesis regions by time of year. *Source* UCAR CoMet. Permission from Tata Torres

Siberian High and subtropical highs play important roles in suppressing precipitation. Winter cyclogenesis tends to occur to the east of southeastern China.

The region warms quickly in the summer months, with Shanghai reaching 29 °C (84 °F) in July. The circulation also brings a much wetter winter climate than in most other parts of East Asia. The high temperatures are accompanied by large *water vapor capacity* and proximity to moisture sources. The rapid warming induces rising air and a *thermal low* that draws moisture inland, with *convection*-enhanced precipitation, as suggested by Table 16.1. But there remains a wide spatial variability

Table 16.1 Seasonal precipitation (mm) and as a percentage of annual precipitation in selected cities of eastern China

City	Climate type	Latitude	Winter (Dec–Feb)		Spring (Mar–May)		Summer (Jun–Aug)		Autumn (Sep–Nov)	
			Total precip	% of annual	Total precip	% of annual	Total precip	% of annual	Total precip	% of annual
Harbin	Dwa	45.75	16	3	82	15	350	65	91	17
Shenyang	Dwa	41.80	28	4	116	17	427	62	118	17
Beijing	Dwa	39.94	10	2	66	13	356	67	96	18
Qingdao	Cwa	26.07	41	6	116	17	389	57	140	20
Shanghai	Cfa	31.22	186	15	271	22	542	44	246	20
Wuhan	Cfa	30.58	247	17	379	25	603	40	263	18
Hangzhou	Cfa	30.27	150	11	389	30	575	44	197	15
Xiamen	Cfa	24.48	155	12	390	30	549	42	207	16
Guangzhou	Cwa	23.13	141	7	624	32	878	45	308	16

Source Wiki

of precipitation totals, and abrupt decreases with distance inland. Sharp seasonal precipitation contrasts—a characteristic feature of monsoon circulation—play an even larger role in the tropical areas of Asia south of the Himalaya Mountains, the focus of the next chapter.

The temperature and moisture regimes cause most of the heavily-populated southeastern part of China to be classified as a *Humid Subtropical* (Cfa) *climate* or Cw climate. Three representative examples of the Cfa climate are the cities labeled in Fig. 16.21. From west to east at similar latitudes, these are Wuhan, Hangzhou, and Shanghai (Figs. 16.22, 16.23 and 16.24, respectively). Consistent features in all three climagraphs are the moderate winter temperatures and prominent summer precipitation maxima. Also common to the three sites is the spring and early summer water balance *surpluses* and minimal deficits, as abundant precipitation arrives at the time of year when potential evapotranspiration peaks. It is no surprise that this region is bustling with both urban and agricultural activity.

Farther south, the offshore continental flow from the Siberian High is still evident, making winters dry. In these southeastern China locations with a Cw climate, winter temperatures are too cool to be classified as tropical (A) climates as are found at similar latitudes on the equatorward margins of continents, such as at Miami, Florida (see again Fig. 13.3). And the monsoon effect makes the wet season even wetter than at most other similarly-positioned locations, including Miami. Another Chinese megacity—Guangzhou—serves as a representative example of the Cw climate in southeastern China (Fig. 16.25).

Overall, the sheer size and diversity of the Asian continent and adjacent oceans support diverse climates in East Asia. Similar features cause the Cw climate of southeastern China to transition across space, with similar climatic realms in South Asia, the focus of the next chapter.

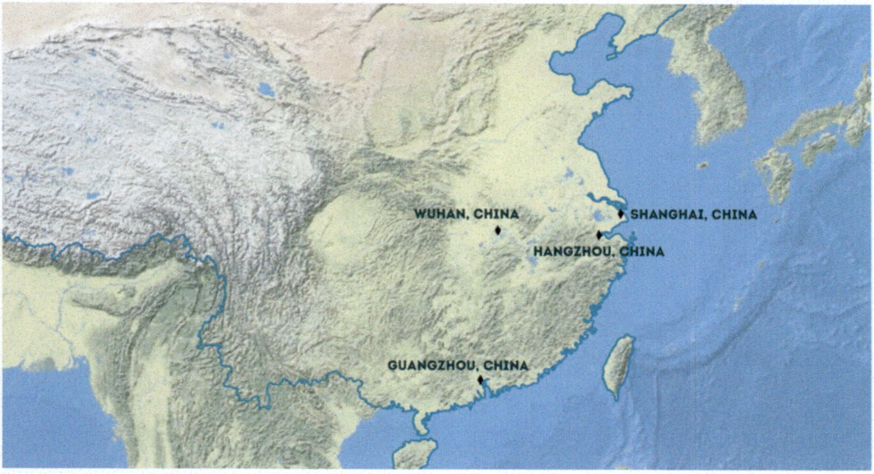

Fig. 16.21 Southeastern Chinese cities for which climagraphs and water balance diagrams are shown. *Source* Geaux Geospatial

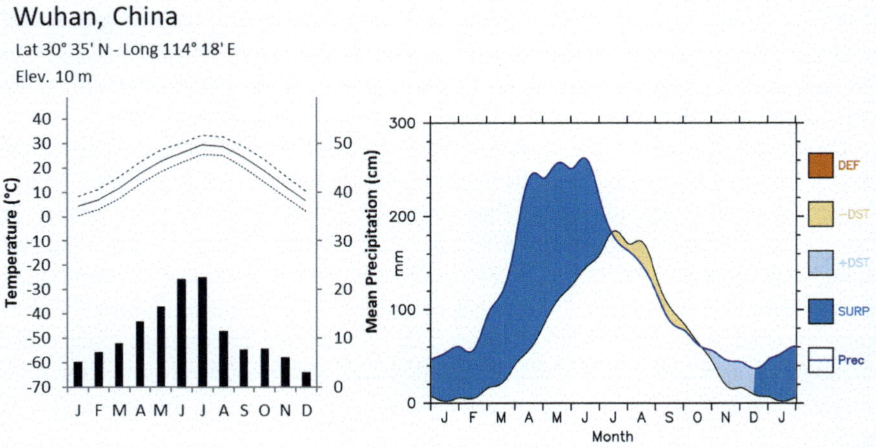

Fig. 16.22 Climagraph and water balance diagram for Wuhan, China. Temperature data are for the 1991–2021 averaging period and precipitation data are for 1999–2021 period. *Source* Authors. https://weatherspark.com/, http://cyclops.deos.udel.edu/wimp/public_html/

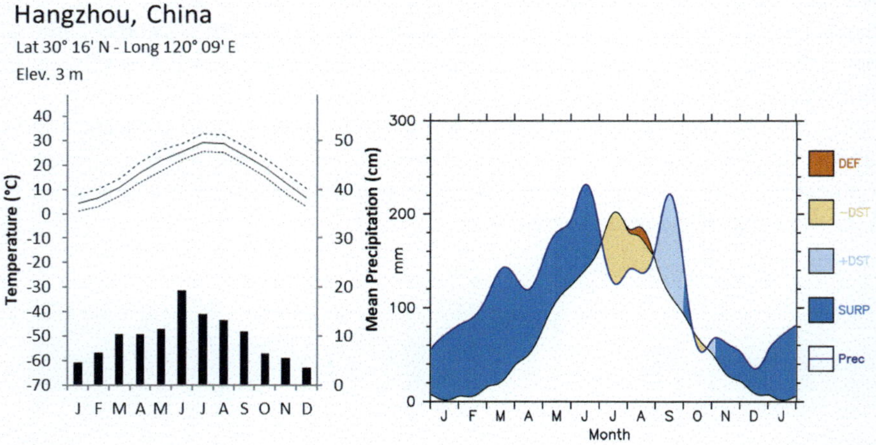

Fig. 16.23 Climagraph and water balance diagram for Hangzhou, China. Data for 1980–2016 averaging period. *Source* Authors. https://weatherspark.com/, http://cyclops.deos.udel.edu/wimp/public_html/

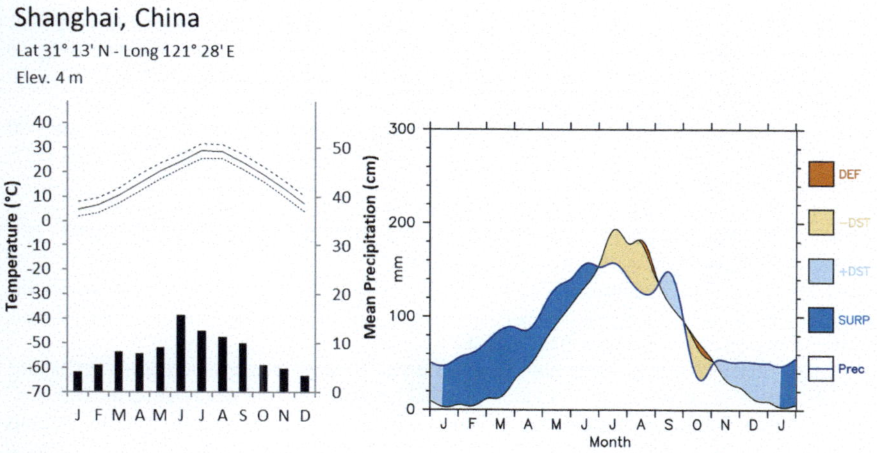

Fig. 16.24 Climagraph and water balance diagram for Shanghai, China. Data for 1980–2016 averaging period. *Source* Authors. https://weatherspark.com/, http://cyclops.deos.udel.edu/wimp/public_html/

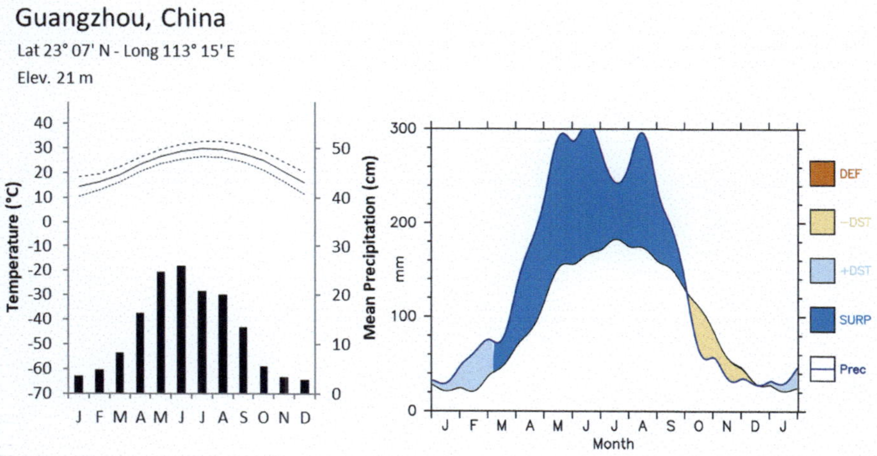

Fig. 16.25 Climagraph and water balance diagram for Guangzhou, China. Data for 1980–2016 averaging period. *Source* Authors. https://weatherspark.com/, http://cyclops.deos.udel.edu/wimp/public_html/

Further Reading

Chen, G. T. J., Wang, C. C., & Lin, D. T. W. (2005). Characteristics of low-level jets over northern Taiwan in mei-yu season and their relationship to heavy rain events. *Monthly Weather Review, 133*(1), 20–43. https://doi.org/10.1175/MWR-2813.1

Chen, G.T.-J., & Yu, C. C. (1988). Study of low-level jet and extremely heavy rainfall over northern Taiwan in the mei-yu season. *Monthly Weather Review, 116*(4), 884–891. https://doi.org/10.1175/1520-0493(1988)116%3c0884:SOLLJA%3e2.0.CO;2

Chen, H., Teng, F., Zhang, W., & Liao, H. (2017). Impacts of anomalous mid-latitude cyclone activity over East Asia during summer on the decadal mode of East Asian summer monsoon and its possible mechanism. *Journal of Climate, 30*, 739–753. https://doi.org/10.1175/JCLI-D-16-0155.1

Chen, S. J., Kuo, Y. H., Zhang, P. Z., & Bai, Q. F. (1992). Climatology of explosive cyclones off the East-Asian coast. *Monthly Weather Review, 120*(12), 3029–3035. https://doi.org/10.1175/1520-0493(1992)120%3c3029:COECOT%3e2.0.CO;2

Ho, C. H., Baik, J. J., Kim, J. H., Gong, D. Y., & Sui, C. H. (2004). Interdecadal changes in summertime typhoon tracks. *Journal of Climate, 17*(9), 1767–1776. https://doi.org/10.1175/1520-0442(2004)017%3c1767:ICISTT%3e2.0.CO;2

Hsu, H. H., Zhou, T. J., & Matsumoto, J. (2014). East Asian, Indochina, and Western North Pacific summer monsoon – An update. *Asia-Pacific Journal of Atmospheric Sciences, 50*(1), 45–68. https://doi.org/10.1007/s13143-014-0027-4

Wang, B., & Linho. (2002). Rainy season of the Asian-Pacific summer monsoon. *Journal of Climate, 15*(4), 386–398. https://doi.org/10.1175/1520-0442(2002)015<0386:RSOTAP>2.0.CO;2

Yihui, D. (1990). Buildup, air-mass transformation and propagation of Siberian High and its relations to cold surge in East-Asia. *Meteorology and Atmospheric Physics, 44*(1–4), 281–292. https://doi.org/10.1007/BF01026822

Part VI
Regional Responses in the Tropics and Southern Hemisphere

Chapter 17
South Asia and Adjacent Ocean Environments

Abstract Shielding from frigid air by the Himalaya Mountains and the Tibetan Plateau causes continental tropical (cT), maritime equatorial (mE), and maritime tropical (mT) air masses to dominate South and Southeast Asia. Monsoonal circulation creates a strong summer precipitation maximum in most of the region, with temperatures commonly peaking immediately before the wet monsoon onset. Much of Arid (B climate) northwest and south-central parts of South Asia receive a somewhat higher percentage of precipitation in winter than elsewhere in the region. Northern Indian precipitation totals increase from west to east as the northwestward bending of the southwesterly counter-trades advects Bay of Bengal moisture onshore. Southern Indian precipitation totals decrease from the west coast to the south-central peninsula, because of orographic and rain shadow effects of the Western Ghat Mountains. From north to south, Mesothermal Winter Dry–hot summer (Cwa), Tropical Wet-dry (Aw), Tropical Monsoon (Am), and Tropical Rain Forest (Af) climates dominate South Asia and adjacent waters. Southeast Asian temperature and precipitation patterns resemble those of adjacent South Asian latitudes, but some eastern coastal areas are partially blocked from summer monsoonal flow. The Indian Ocean Dipole (IOD) and Madden–Julian Oscillation (MJO) are important sources of inter- and intra-annual climate variability.

Keywords Himalaya Mountains · Humid Subtropical (Cfa) climate · Indian Ocean Dipole (IOD) · Inter-tropical convergence zone (ITCZ) · Madden–Julian Oscillation (MJO) · Mesothermal Winter Dry–hot summer (Cwa) climate · Northeast and southeast trade winds · Semi-arid (BS) climate · Southwesterly counter-trades · Tibetan Plateau · Tropical Monsoon (Am) climate · Tropical Rain Forest (Af) climate · Tropical Wet-dry (Aw) climate · True Desert (BW) climate

R. V. Rohli et al., *Atmospheric and Oceanic Circulation*,
https://doi.org/10.1007/978-3-031-66142-6_17

General

As with the discussion of East Asia in the previous chapter, the interaction of the monsoon circulation with regional and local topography controls the climatic patterns of South and Southeast Asia and adjacent water bodies. The monsoon technically refers to a seasonal reversal of the circulation system with winter winds blowing dry air offshore from a continental interior and summer winds blowing onshore from the adjacent oceans. The onshore flow coinciding with periods of heavy rain is only one of many important consequences for the sixty percent of humanity who live in Asia. These two modes of seasonal circulation interact with the extensive mountain ranges of South Asia in several ways. Most notably, when humid air ascends the *windward* slopes of high mountain ranges, the resulting *orographic lifting* supplements other *precipitation*-generating mechanisms to produce some of the largest rainfall totals on Earth. The magnitude and seasonality of *temperature* and especially precipitation, and therefore climatic types, vary widely across South Asia and adjacent water bodies. This spatial variability can be explained by both the site (topography) and situation (atmospheric and oceanic circulation), and the interplay between them, in South Asia.

Topographic Influences (Site)

The South Asian region is bordered on the north by the Himalaya Mountains and the Tibetan Plateau (Fig. 17.1). These physiographic features formed approximately 50 million years ago—just a short time ago in geologic time—when the Australian-Indian Plate collided with the Eurasian Plate, and these features are still rising today, at a similar rate as the rate that your toenails are growing. The uplifted land that resulted from that collision contains the highest peaks on Earth (~ 8000–9000 m; 26,000–29,500 ft). Mountains and complex terrain extend southeastward through the Southeast Asian nations of Myanmar, Laos, and Vietnam. South of the Himalaya Mountains, a lowland basin extends across the northern Indian subcontinent, including much of India and Bangladesh, where the Indus and Ganges River basins capture the runoff from the highlands. The Ganges River has the second-greatest discharge of any river on Earth.

The western coast of southern India is bordered by a geologically-old and eroded mountain range (1000–2000 m; 3200–6500 ft) known as the Western Ghats, while the Eastern Ghats are even lower and more irregular on the eastern coast of the country. To the east, the Ganges and Brahmaputra Rivers form a large lowland delta comprising most of Bangladesh, on their way to the Bay of Bengal. Smaller lowland deltas, most notably from the Mekong River, comprise parts of Thailand, Cambodia, and southern Vietnam.

Off the southern coast of the region lie the warm waters of the Indian Ocean and associated Arabian Sea and Bay of Bengal (Fig. 17.1). Uniformly high *latent heat*

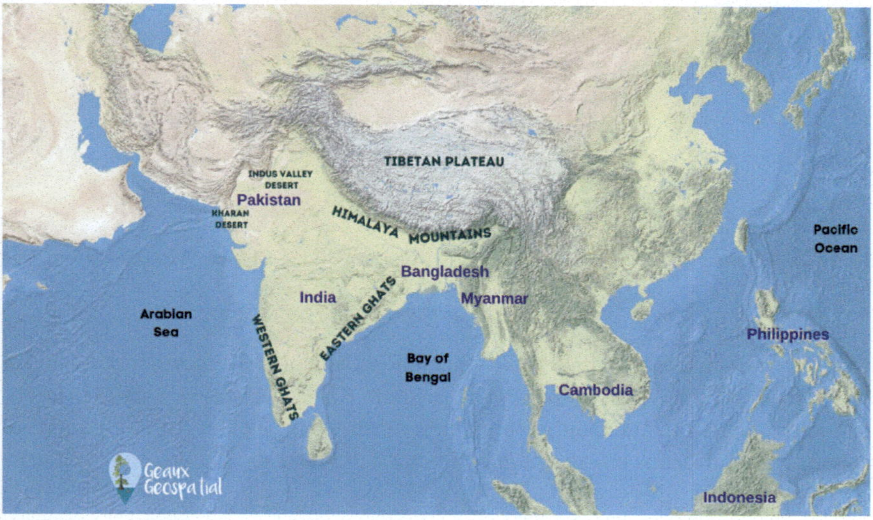

Fig. 17.1 Physiographic regions of South Asia. *Source* Geaux Geospatial

fluxes produce excessive *humidity* and a constant supply of water vapor to the mainland. However, the highest sea surface temperatures are farther east in the vicinity of the *Maritime Continent* consisting mostly of Indonesia, where the *convergence* of the tropical trade winds from the western Pacific accumulates energy-rich waters of the *Pacific Warm Pool*. Sea level is 10–20 cm (4–8 in) higher adjacent to the Maritime Continent than in the eastern Pacific. The bathwater-warm water is linked to strong uplift in the *inter-tropical convergence zone* (ITCZ) in most years.

Circulation Influences (Situation)

Situated between warm tropical oceans and the interior of the world's largest continent, South Asia is influenced by air masses from drastically different sources. While the high mountains that line the northern edge of the region effectively block true *continental polar* (cP) *air masses* from reaching South Asia, drier air in the form of *continental tropical* (cT) *air masses* originating over the interior areas of Pakistan and the Tibetan Plateau exert an important influence (Fig. 17.2). To the south, very warm and humid *maritime equatorial* (mE) *air masses* and *maritime tropical* (mT) *air masses* are drawn inland from the Indian Ocean and western Pacific Ocean.

The movement of these air masses and the resulting patterns of climate are determined by the Asian monsoon circulation. The reversal of the primary steering circulation between winter and summer resembles a much-larger and seasonal-scale version of a *land breeze* and *sea breeze* system, with flow toward lower-pressure zones offshore in winter and flow from the ocean onto the land mass in summer (Fig. 17.3).

Fig. 17.2 Air mass source regions for South Asia. *Source* UCAR CoMet. Premission from Tara Torres

While other continents, and also East Asia, have a monsoonal circulation, it is most pronounced and influential in the culture in South Asia. The seasonal shifts in atmospheric circulation in association with the uniformly warm waters of the Indian and western Pacific Oceans create unique climatic patterns across South and Southeast Asia.

Farther equatorward, South and (especially) Southeast Asia are influenced increasingly by the ITCZ. As in all parts of the world affected by the ITCZ, the solar radiant energy associated with the more overhead *Sun angles* provides buoyancy and an unstable atmosphere, which encourages *convection*. If water vapor is abundant, the result is a line of convective thunderstorms initiated in the afternoon—the ITCZ. Because the more-or-less continuous belt of thunderstorms that form the ITCZ is driven by *shortwave radiation*, its position migrates latitudinally with the direct rays of the Sun, approximating the migration of the vertical solar rays as they oscillate between the *Tropic of Cancer* and *Tropic of Capricorn* through the year—northward as Northern Hemisphere summer approaches and southward as Southern Hemisphere summer approaches. However, the migration of the ITCZ lags slightly behind the latitude of the overhead Sun angle and it never quite reaches the latitude of the Tropics of Cancer and Capricorn.

The ITCZ not only causes summer rainfall—with "summer" defined more by the Sun angle than by the local temperature—but also induces thermal variations throughout the region. The high-Sun period and the *net radiation* (Q*) increases associated with more direct solar radiation *absorption* induce thermal convection that peaks daily during the afternoon. The convection results in cumulus cloud formation, ultimately leading to the prevalence of large *cumulonimbus* thunderstorms. The combination of thick cumulus clouds and associated precipitation cools the surface. Evaporation of the precipitated water at the surface further cools the landscape as surface energy absorption occurs in the form of the *latent heat flux* instead of the *sensible heat flux*. Paradoxically, the net effect in some places is a slightly cooler "summer" than the rest of the year.

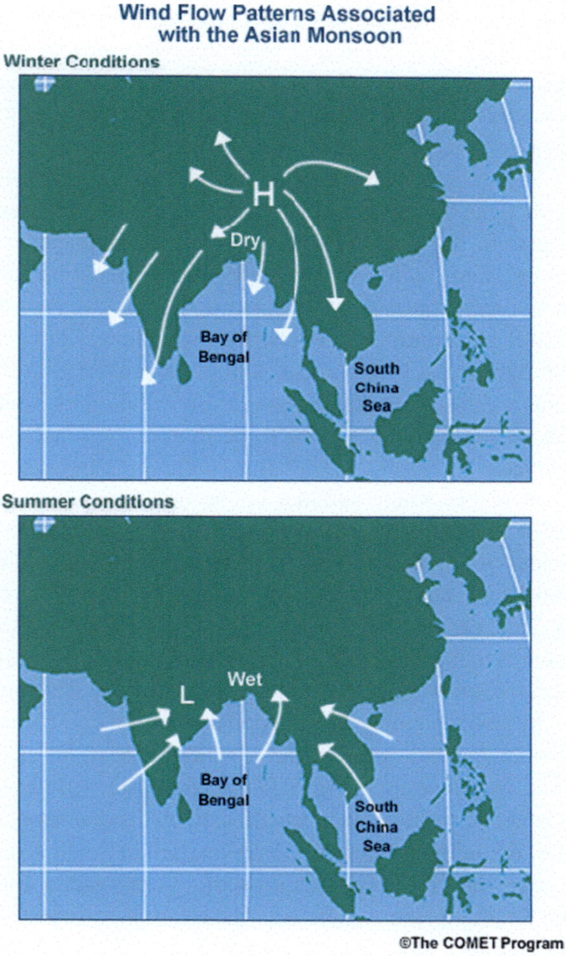

Fig. 17.3 Asian monsoon circulation highlighting the seasonal reversal of wind direction over South Asia. *Source* UCAR CoMet. Premission from Tara Torres

The South Asian Monsoon

The seasonal reversal of wind associated with the South Asian monsoon occurs as strong thermal gradients exist between the large continental interior of Asia and the surrounding Pacific and Indian Oceans. In winter, the interior of Asia is chilled significantly, resulting in the formation of the very strong *Siberian High*, producing dry offshore flow throughout the Asian landmass (Fig. 17.3). At such a broad spatial scale, the *Coriolis effect* deflects the winds to the right. This period marks the onset of the dry winter monsoon when parts of South Asia may go three to five months with no significant precipitation.

During summer, the *Tibetan Low* forms over the heated Asian interior and produces strong onshore flow along the Pacific and Indian Ocean coasts as air rushes inland to "fill" the low, after again being deflected by the Coriolis effect (see again

Fig. 17.3). *Advection* of warm, humid air from the ocean toward the interior provides water vapor for cloud and precipitation processes. The orographic effect, sloping surfaces, and reduced *potential evapotranspiration* in high-elevation areas leaves not only abundant precipitation but also relatively little *evapotranspiration*, giving the region large water *surpluses* in summer.

This monsoonal surface atmospheric pressure oscillation, or seesaw, between the Indian and southwestern Pacific Oceans and the Asian continental interior over the course of the year is the key driver of the atmospheric circulation in South Asia. As pressures build over the cooler surface (continental interior in winter and oceans in summer), pressures will correspondingly fall over the warmer surface (oceans in winter and continental interiors in summer). The ocean temperatures change little throughout the year compared to the continental interiors. The summer monsoonal rains are the most notable part of the oscillation for South and Southeast Asia because the pressure gradient tends to be more intense and uninterrupted by the mountain barrier.

In places where summer convectional precipitation is enhanced by orographic uplift, extreme precipitation results. For example, Cherrapunji, in northeastern India, holds the world record for the greatest monthly precipitation recorded (929.99 cm, or 366.14 in), and for the greatest annual precipitation recorded, at 2646 cm (1042 in) or 26.5 m (87 ft) of water. This record has stood since 1860–1861. It is even more astounding to consider that this "annual" precipitation record overwhelmingly occurred over the four months comprising the wet summer monsoon phase.

Seasonal changes in monsoon circulation coincide with the seasonal migration of the ITCZ, which moves over a wider range of latitudes here than anywhere else in the tropics (see again Fig. 3.12). As the easterly trade winds follow the ITCZ into the warmer hemisphere they are affected by that hemisphere's Coriolis effect, which can alter their direction as they make landfall in South Asia. In the Northern Hemisphere winter, the ITCZ, suggested by the green-shaded area in Fig. 17.4, is shifted into the Southern Hemisphere near the coast of Australia, and South Asia remains in the northeasterly flow of the *northeast trades*. But in the Northern Hemisphere summer, when the ITCZ has shifted far to the north, as in Fig. 17.5, the *southeast trades* cross the Equator from the Southern Hemisphere and begin to be turned toward their right after they move into the Northern Hemisphere. The resulting areas of southwesterly winds that affect South and Southeast Asia (see again Fig. 17.5) are called the **counter-trades** or **southwesterly counter-trades**. As a result, most of South and Southeast Asia experiences northeasterly air flow in winter and southwesterly flow in summer.

While a simple wind reversal pattern explains the South Asian monsoon on a seasonal time scale, variations in its strength over time are a result of more complicated global circulation features. At interannual time scales, the summer monsoon circulation that produces most of the region's precipitation is influenced by the global *Walker circulation*. In its normal phase, when neither El Niño nor La Niña are occurring, the Walker circulation helps to maintain the southwesterly counter-trade surface flow across South Asia in summer. In the ocean, the airflow forces the surface water across the tropical Pacific and it warms in the process. The result is the Pacific Warm

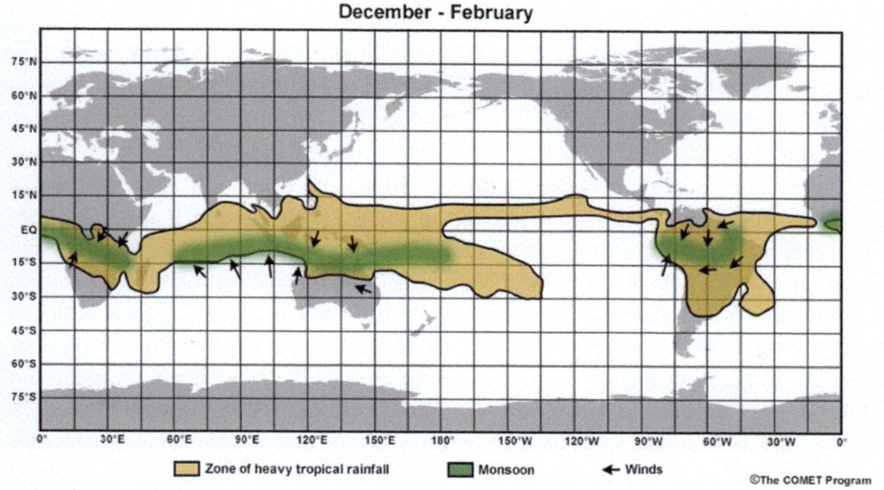

Fig. 17.4 December–February monsoon-influenced areas and counter-trade wind flow. *Source* UCAR CoMet. Premission from Tara Torres

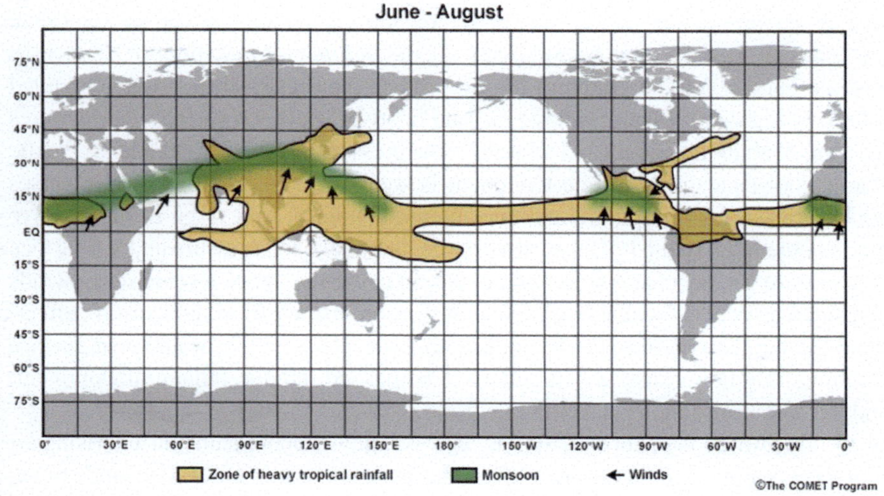

Fig. 17.5 June–August monsoon-influenced areas and counter-trade wind flow. *Source* UCAR CoMet. Premission from Tara Torres

Pool near the Maritime Continent. These warm waters enhance the convection of low-density air, strengthening the ITCZ in this region.

At the same time, on the other side of the Pacific off the South American coast, cold bottom waters rise up to the surface to take the place of the "missing" water. This cold water, centered near Peru, encourages a *stable atmosphere* and even a *temperature inversion*, resulting in sinking air and relatively high air pressure. The east–west-oriented Walker circulation cell of air is thus created between the relative pressure differences on either side of the tropical Pacific (see again Fig. 6.2). Air rising over the Maritime Continent crosses the Pacific as tropical westerlies high in the *troposphere* above the surface tropical easterlies and sinks near South America before returning westward near the surface as the trade winds.

The Walker circulation is significant for South and Southeast Asia because in addition to affecting tropical Pacific Ocean air and water circulation, it also influences circulation cells adjacent to the tropical Pacific Ocean, including the South Asian monsoon. Some of the air from the rising branch over the Maritime Continent goes to the west, becoming upper-level easterly winds over the tropical Indian Ocean to East Africa, where it sinks near the Arabian Peninsula. Then, the surface air flows eastward, returning to the low-pressure area of the Maritime Continent and reinforcing the westerly summer monsoon flow over South Asia (see again Fig. 6.2). The Walker Circulation thus influences South Asia indirectly by enhancing surface air movement from west to east in the Indian Ocean between the Arabian Peninsula and Indonesia. Surface winds from the Arabian Peninsula toward South and Southeast Asia are the southwesterly counter-trades that characterize the summer monsoon season.

This Indian Ocean circulation cell undergoes variability in its strength across time. The extreme phases—weaker than the "normal" circulation and stronger than the "normal" circulation—are referred to as the *Indian Ocean Dipole* (*IOD*), which is sometimes called the "Indian El Niño." When the seesaw forms by the coupled relatively high air pressure, sinking air, and relatively cool sea surface temperatures over the Arabian Peninsula and relatively low pressure, rising air, and a large Pacific Warm Pool near the Maritime Continent is weaker than normal or even reverses temporarily, this atmospheric and oceanic circulation is said to be in the positive mode of the IOD. In the IOD's positive mode, the weakened pressure and sea surface temperature gradients across the Indian Ocean cause the upper-level easterly winds and surface westerly winds (with slight deflection of each from the Coriolis effect) to be weaker than normal too. Weakened upper-level easterly flow and near-surface southwesterly flow cause less humid air to be pushed into the Indian Subcontinent, and the wet monsoon is weaker than normal.

At other times, the "normal" sinking air over the Arabian Peninsula and rising air near the Maritime Continent strengthen. This change marks the negative mode of the IOD. During the negative IOD, the upper-level easterly and near-surface southwesterly winds strengthen, and the onshore monsoon flow of humid air into South Asia is intensified. The characteristically heavy rains of summer are stronger than normal.

As described in Chap. 6, positive IOD events often but not always, occur during an El Niño event. When the two coincide, the impacts of each tend to be magnified, as the altered circulation in the air and ocean from one of these circulation cells reinforces the anomalous circulation in the other, like gears in a machine. Thus, El Niño events are likely to correlate with a weakened South Asian monsoon circulation, particularly when reinforced by a strongly positive IOD. Such impacts can be catastrophic, as the weakened wet summer monsoon delivers less of the much-needed precipitation for crops that feed large populations. When either an El Niño or the positive IOD event is occurring, but not both, the impacts, such as to the South Asian monsoon, tend to be less drastic. Similarly, La Niña and negative IOD events often, but not always, overlap. When they coincide, the impacts of each in South and Southeast Asia tend to be stronger, and when they occur separately, their impacts are often weaker.

While it is tempting to oversimplify and overgeneralize, the ENSO-IOD relationship is quite complex, and each event is unique. Research continues about which phenomenon is more likely to initiate changes in the other, and how the lag process varies by season and location. It is likely that other modes of variability, such as ENSO-like features in the Atlantic Ocean, as described in Chap. 18, are also connected to the changes in atmospheric and oceanic circulation associated with ENSO and the IOD. It is difficult to generalize about the impacts in this complicated Earth-ocean–atmosphere feedback system. Nevertheless, discoveries are being made that will continue to be useful in seasonal weather and short-term climate variability forecasts.

To make matters even more complicated, at intra-annual time scales, such as over the course of a few weeks, the strength of the South Asian monsoon is also affected by the **Madden–Julian Oscillation (MJO)**, also known as the 30–60 Day Oscillation or 40–50 Day Oscillation. The characteristic weather feature of the MJO is a set of linear bands of convective thunderstorms that crosses the tropical Pacific Ocean from west to east. The bands occur similarly to the thunderstorm lines in *squall line* development of *mid-latitude wave cyclones* and in rainbands in *tropical cyclones* with the downdrafts of one line feeding updrafts of the adjacent line MJO bands begin as surface heating causes widespread evaporation over the tropical ocean surface, with a low-level *unstable atmosphere* and enhanced convection. Cloud cover and convective thunderstorms then form. These become oriented linearly as the westerly winds aloft in the tropics push them along, similarly to the way a rake tends to accumulate leaves in a linear orientation. Clearer skies prevail in advance of the line, but the heating promotes atmospheric instability. As the upper-air westerlies steer the line eastward over the clear area, the *unstable atmosphere* causes the thunderstorm line to maintain itself for weeks at a time, unlike *air mass thunderstorms* in mid-latitude wave cyclones.

The MJO consists of eight phases that together last 30–60 days. Each phase is associated with the primary zone of convection in a different part of the equatorial ocean around the world, with Phases 2 through 5 having the primary convective zone, shown in blue in Fig. 17.6, in the Indian Ocean nearest to South and Southeast Asia. During these phases, the South Asian summer monsoon is "excited" by the convection. However, in other MJO phases, especially Phase 6 through 8, convection near South and Southeast Asia is suppressed and the South Asian monsoon weakens.

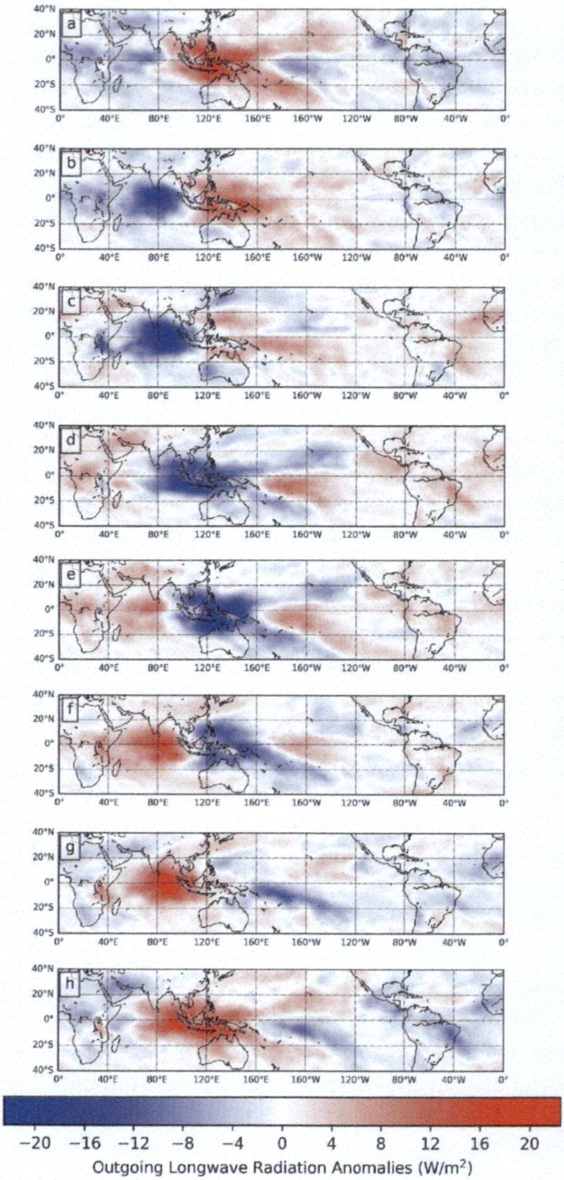

Fig. 17.6 Phases of the Madden–Julian Oscillation from Phase 1 (a) through 8 (h), as represented by *outgoing longwave radiation* anomalies. Red indicates cloud-free conditions because *longwave radiation* escapes to space relatively easily, while blue indicates heavy cloud cover because longwave radiation does not escape to space as readily. *Source* Caparotta (2018)

Seasonal Weather

A review of the temperature and precipitation patterns of South Asia by season suggests that, not surprisingly, this monsoon-dominated part of the world supports large seasonal precipitation differences. As a broad generalization, the wettest regions of South Asia generally receive a high percentage of the annual precipitation total in summer, whereas the limited precipitation occurring in drier areas tends to be distributed more evenly throughout the year. Although *annual temperature ranges* across South Asia are small to moderate, the season of warmest temperature is tied more closely to the monsoon than to the Sun angle in parts of South Asia.

In winter, high pressure over Asia directs northeasterly surface prevailing winds across South and Southeast Asia. The Himalaya Mountains and other highland regions block most of the coldest air from extending into the lower latitudes, as cT air masses bring dry conditions to the majority of the region. At upper levels, the southern branch of the *polar front jet stream*—which often merges with the *subtropical jet stream* in that part of the world—pulls cold air into far northern Pakistan and northwestern India, setting up a weak front in the vicinity. The weak front and humidity from the Arabian Sea allow for some precipitation from eastward-moving mid-latitude wave cyclones that can survive the eastward trip across dry lands. A few can even reach Myanmar and northern sections of Southeast Asia. While rare, these systems are critically important as they produce the snowpack at higher elevations that ultimately provides water during the dry winter season. Across southern India, the prevailing winter northeast surface winds most often follow a continental track and remain dry and stable. For most of India and Southeast Asia, winter is the dry season. The only exceptions are mountainous areas such as the Eastern Ghats of India, along with Vietnam and the Philippines, where the northeast trade winds may pass over the ocean surface, gather moisture, and then generate an orographic effect when they return over land (Fig. 17.7).

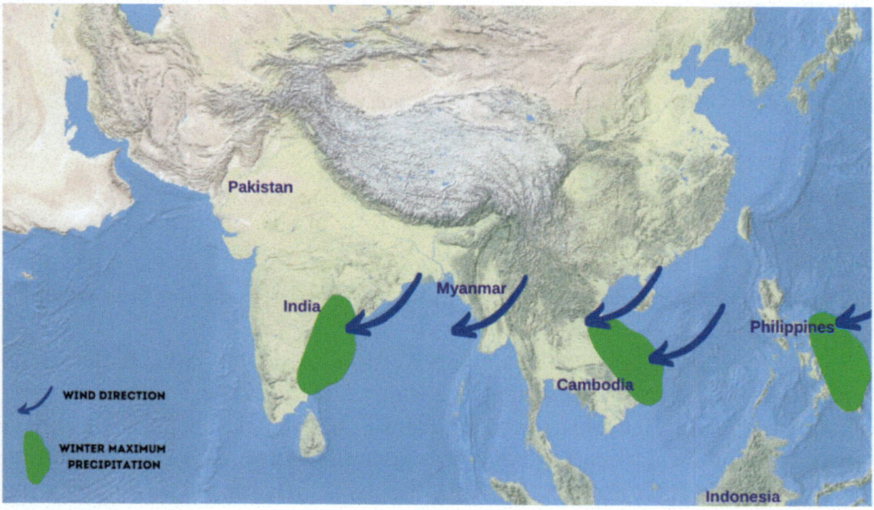

Fig. 17.7 Winter monsoon circulation emphasizing areas of onshore flow of humid air on the eastern coasts of South Asia. *Source* Geaux Geospatial

Spring is generally the hottest time of year in South Asia. Increasing Sun angles and available radiant energy enhance regional circulation effects. Subsidence in association with the high-pressure regions near 30° N compresses and warms air *adiabatically* as it sinks. Northeasterly and outward flow from the moderately intense *Siberian High* bring continued dry conditions, as in winter. Lack of cloud cover allows even more *shortwave radiation* (K↓) to reach the surface. The resulting *Bowen ratio* is high and temperatures rise rapidly. Mean daily temperatures may exceed 30 °C (86 °F). Toward the end of spring, the southern tip of India begins to develop *thermal lows* and some convection, signaling the start of the oncoming summer monsoon season. These thermal lows are relatively shallow in the atmosphere and restricted to the southernmost coastal margins of India rather than over the interior subcontinent. They tend to be more likely to produce dust storms than rain in terrestrial South Asia.

Surface winds shift suddenly from spring to summer as the Siberian High weakens and the ITCZ migrates northward. As the Siberian High dissipates, the southern branch of the polar front jet stream, which is often merged with the subtropical jet, disappears gradually and intermittently. When this jet weakens, the ITCZ surges northward relatively suddenly and extensively. As air flows in toward the ITCZ over South and Southeast Asia, it is deflected to the right by the Coriolis effect and is manifested as the southwesterly counter-trades. This southwesterly flow dominates the Northern Hemisphere summer months over most of the region and advects humid air from the Indian Ocean, which becomes unstable over the warming land, leading to heavy rainfall. The southern and western sides of the peninsulas, such as Thailand and the Ganges Delta region, receive maximum precipitation in summer, especially where mountains such as the Western Ghats contribute orographic lifting (Fig. 17.8). In these locations, as much as 80–90% of precipitation occurs in summer. The rainy season begins in early May in Southeast Asia, in early June for southern India, and in early July for northern India. In summer, the northward movement of the ITCZ and its associated rainfall tends to surge in sudden shifts separated by several days during which the ITCZ remains in a steady position. These surges may be linked to the MJO, which affects the strength of tropical convection as the associated convective storms circle the globe eastward every 30–60 days. In various parts of South and Southeast Asia, the MJO phases modulate moisture conditions, the monsoonal circulation, and/or tropical cyclone activity in the Northern Hemisphere summer but not winter (Fig. 17.9).

Summer precipitation can take a number of forms in South Asia. As the ITCZ surges northward, wind speeds increase, causing a horizontal *convergence* of air that enhances convection and produces squall lines of heavy rain. Spatially, this effect forms a series of distinct rainfall lines that surge northward during summer monsoon onset (Fig. 17.10). Rainfall is enhanced by orographic lifting on the westward-facing (i.e., *windward*) slopes of mountains in the region, and especially as the surface winds turn northward into the Himalaya range in northeastern India. Finally, small cells of low pressure, related to *easterly waves*, move inland periodically through the summer, especially near the Bay of Bengal as the tropical cyclone season begins.

In autumn, the monsoon transition begins. The southwesterly surface winds weaken as the ITCZ retreats southward and northeasterly winds return in their place.

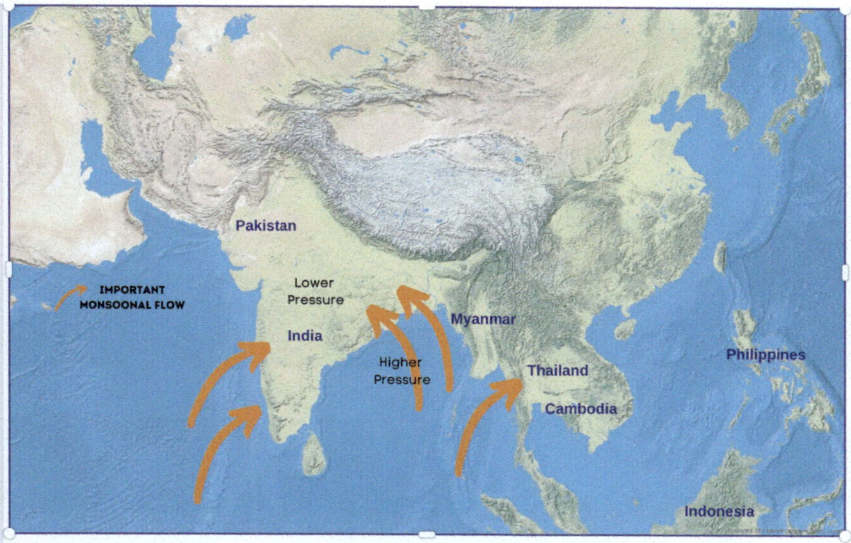

Fig. 17.8 Summer monsoon circulation emphasizing areas of onshore flow of humid air on the western coasts of South Asia. *Source* Geaux Geospatial

Even as the period of heaviest rainfall recedes, the possibility of strong storms still exists in the form of tropical cyclones. These intense storms are most likely to impact the eastern sides of India and the islands of Southeast Asia during the Northern Hemisphere autumn months, and they are responsible for most of the rainfall from May to December in some islands, such as the Philippines. In South and Southeast Asia, tropical cyclones are called simply "cyclones," because the vast majority of their cyclones are of tropical origin. The Indian Ocean is a high-frequency region for tropical cyclogenesis, with cyclones affecting the entire coast of the Bay of Bengal and into the Arabian Sea. The extremely high population density of the Bay of Bengal region combines with the low -elevation coastal geography with deadly consequences. For example, the Bohla Cyclone of 1970 caused over 500,000 deaths in Bangladesh (which was known as "East Pakistan" at the time).

South Asia

Taken holistically, the circulation features give characteristic climatological features to South Asia and its surrounding marine environment. Climagraphs and water balance diagrams are shown for the sites labeled in Fig. 17.11. Temperatures are uniformly high, but with some seasonal variability related to both Sun angles and cloud cover. Over most of the region, the warmest time of year is spring (March–May) when Sun angle increases combine with the dry monsoon to allow little absorption,

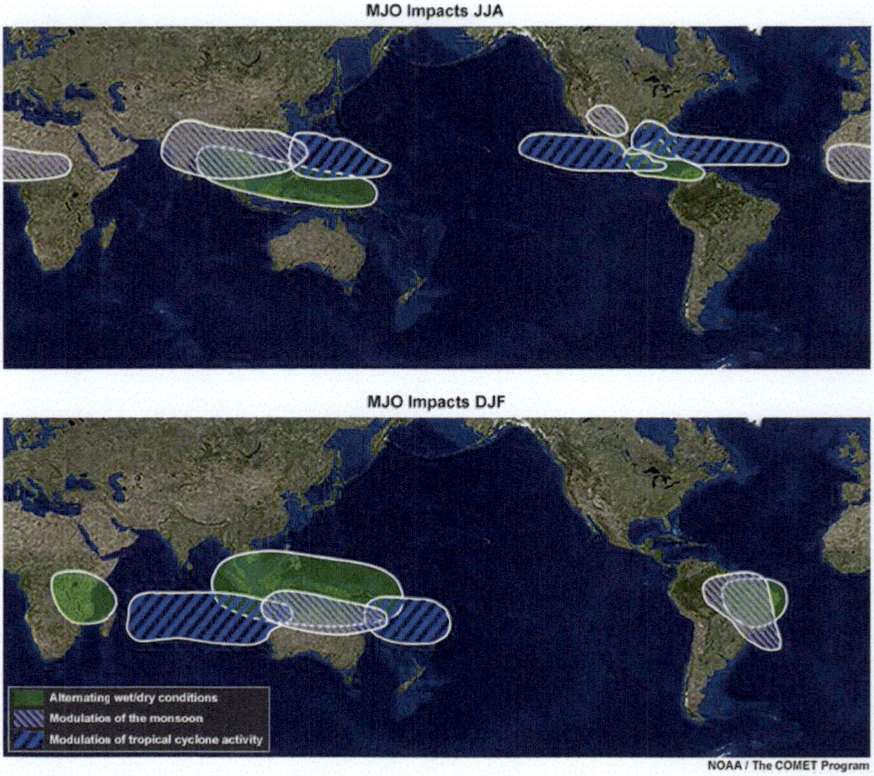

Fig. 17.9 Regions of MJO influence regarding monsoon strength and tropical cyclone activity in **a** June–August; and **b** December–February. *Source* UCAR CoMet

scattering, and *reflection*—collectively known as *attenuation*—of shortwave radiation. Mean daily temperatures typically fall around 30 °C (86 °F) on average, but can be as high as 33 °C (91 °F) in May in places such as Chennai, India, in the southernmost Indian state of Tamil Nadu, on the southeastern coast (Fig. 17.12), where *deficits* are high through the dry season, which lasts until September.

Ironically, the time of year with the highest Sun angle and more hours of daylight—near the *June solstice*—have lower peak temperatures. Daytime high temperatures dip slightly to 30 °C (86 °F) but remain above 25 °C (77 °F) at night. Cloud cover and precipitation of the wet monsoon suppress daytime heating and absorb *outgoing longwave radiation* at night. For many locations across South Asia, mean daily temperatures in summer and winter are similar. Autumn and winter remain warm as Sun angles decrease but skies clear. Even in winter, mean daily temperatures remain around 18 °C (64 °F) in Kolkata, in the Indian state of West Bengal, in extreme eastern coastal India (Fig. 17.13). The Himalaya block the coldest outbreaks of winter air from reaching the continent.

A summer precipitation maximum is prominent in almost all of South Asia, but especially in the wettest parts of India (Fig. 17.14). For example, in New Delhi in the northern state of Uttar Pradesh, an average of 21 cm (8 in) of precipitation falls in

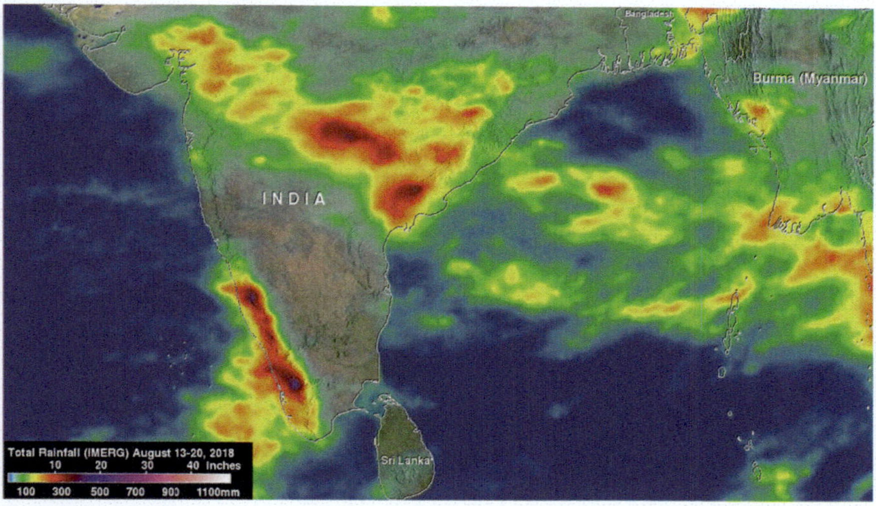

Fig. 17.10 Rainfall accumulation for 13–20 August 2018 over India showing monsoon surge/break periods. *Source* NASA/JAXA/SSAI/Hal Pierce. https://www.nasa.gov/feature/goddard/2018/nasa-captures-monsoon-rains-bringing-flooding-to-india

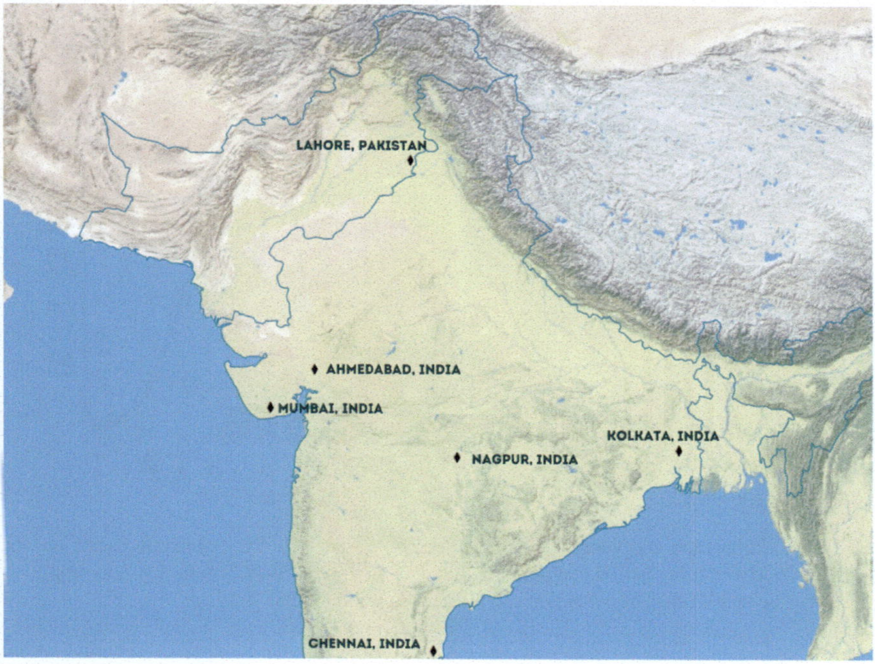

Fig. 17.11 South Asian cities for which climagraphs and water balance diagrams are shown. *Source* Geaux Geospatial

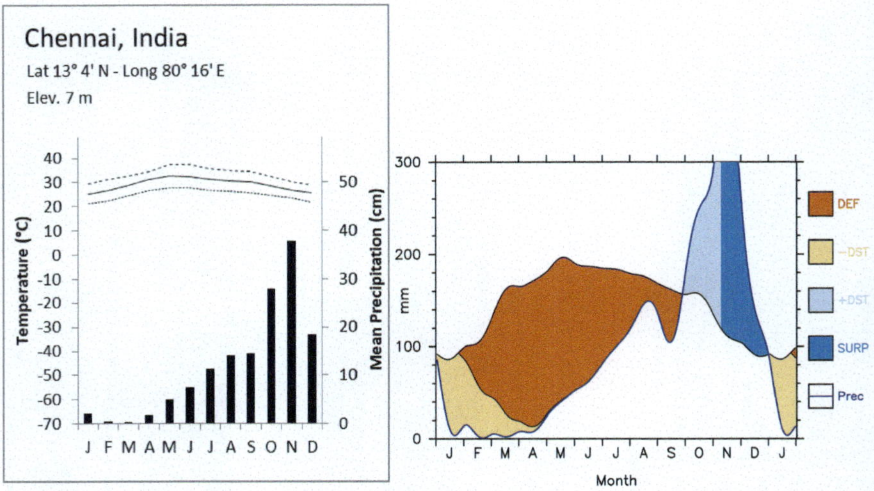

Fig. 17.12 Climagraph and water balance diagram for Chennai, India. Data are for the 1991–2020 averaging period. *Source* Authors. https://weatherspark.com/, http://cyclops.deos.udel.edu/wimp/public_html/

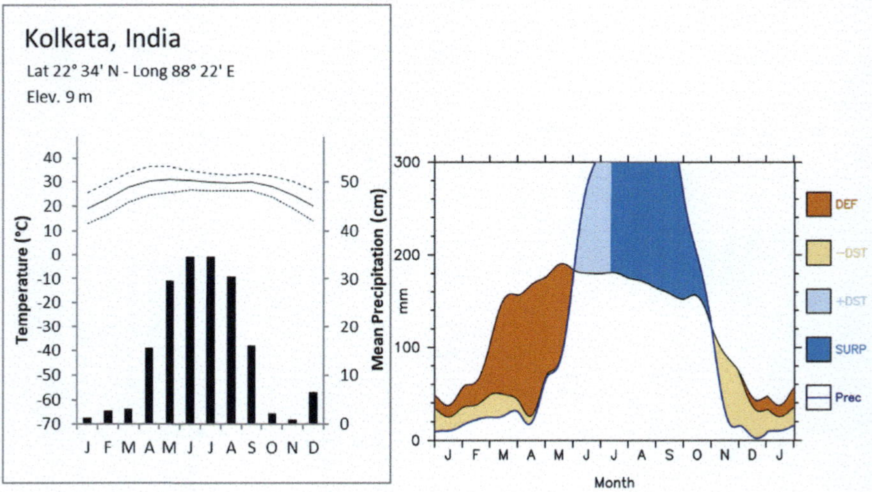

Fig. 17.13 Climagraph and water balance diagram for Kolkata, India. Temperature data are for the 1991–2020 averaging period and precipitation data are for 1999–2020 period. *Source* Authors. https://weatherspark.com/, http://cyclops.deos.udel.edu/wimp/public_html/

July—roughly 30% of its annual total. An even more extreme example is Cherrapunji in the northeast, where an average of 290 cm (114 in)—roughly 26% of its astounding annual total—falls in June.

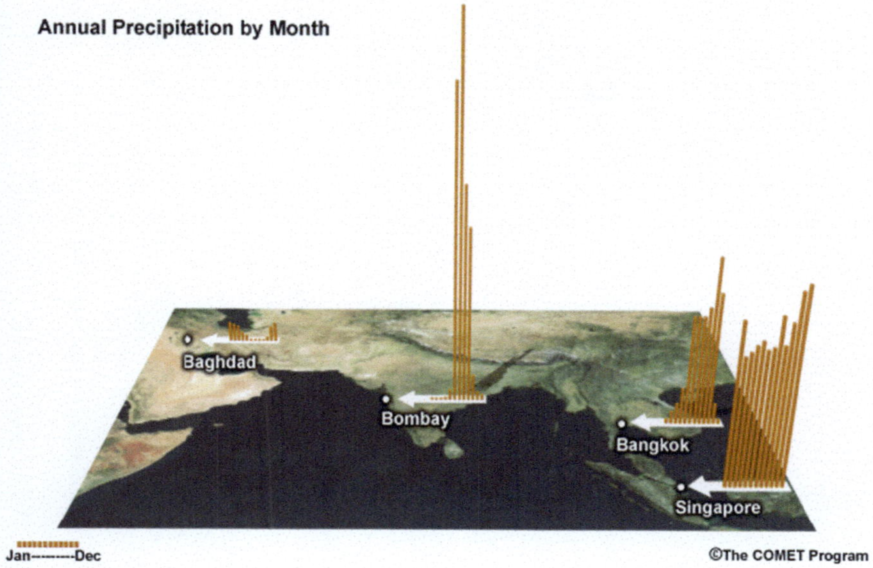

Fig. 17.14 Magnitude and seasonality of precipitation maximum across South and Southeast Asia. *Source* UCAR CoMet. Premission from Tara Torres

These patterns of temperature and precipitation create a *Tropical Wet-dry* (Aw) *climate* in most of peninsular India, coastal Southeast Asia, and adjacent ocean environments, indicating that they are much drier in the winter half of the year than in the half centered on summer (Fig. 17.15). Areas on the immediate southern coast extending out into the Bay of Bengal and Arabian Sea tend to have a *Tropical Monsoon* (Am) *climate*, which experience their precipitation maxima in summer, but with only one to three dry months in the low-Sun season. Over undisturbed land, these areas would be covered by tropical rain forest vegetation, as the dry season is too short to desiccate soils for a sufficiently long period to prevent trees from acquiring enough water to survive.

Some higher-elevation sites and places on the poleward edge of the Aw zones are too cool in winter to technically have tropical (A) climates, but winter remains dry. Much of the northeastern part of India falls into this category and has a *Mesothermal Winter Dry–hot summer* (Cwa) *climate*. Small sections of northern Nepal and adjacent localized areas have elevations and poleward latitudes that keep summers warm rather than hot; these places have a *Mesothermal Winter Dry–warm summer* (Cwb) *climate*. The western parts of South Asia dry out at the northern limits of summer ITCZ extent, as a continuation of the arid (B) climates of Southwest and central Asia (Chap. 15). These desert climates of Pakistan and northwestern India extend southwestward over the Arabian Sea (see again Fig. 17.15).

In northern India, precipitation totals increase from west to east. The southwesterly surface winds of the summer monsoon reach the northern end of the Bay of Bengal and are deflected north and westward by topography, bringing high precipitation

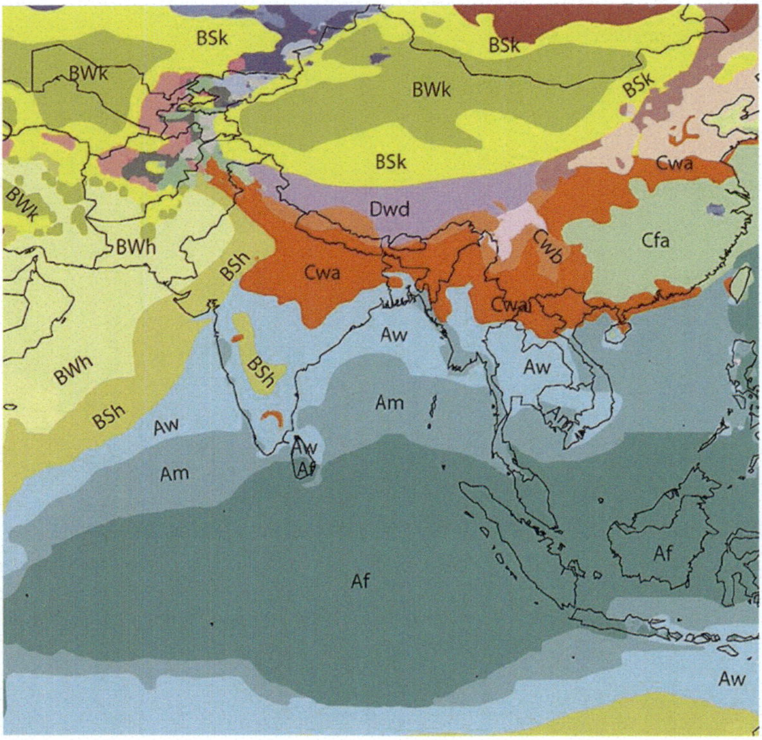

Fig. 17.15 Köppen climatic types around South Asia and surrounding marine environment. *Source* Modified from Reynolds, Rohli, Johnson, Waylen, and Francek, Exploring Physical Geography, McGraw-Hill.

totals to northeastern India and Bangladesh. Orographic lifting enhances convection and leads to some of the highest monthly precipitation totals on Earth. The extra warmth of the water in the western Indian Ocean during a negative IOD phase, particularly when accompanied by El Niño conditions in the Pacific, occurs because the pile-up of warm water limits *upwelling* in the coastal western Indian Ocean. These conditions also tend to bring more humid air into South Asia when the southwest monsoon is in force. Similarly, the colder than normal waters in the western Indian Ocean during a negative IOD tends to bring less moisture into India with the southwest monsoon, or delays the onset of the monsoon circulation, contributing to interannual variability. Precipitation totals decrease to the west, gradually transitioning to desert climates in the northwestern parts of India and northern Pakistan.

In southern India, the opposite pattern prevails, with precipitation totals decreasing from the west coast to the south-central peninsula. At these latitudes the southwest monsoon circulation brings the humid air onshore in the vicinity of the Western Ghats, leading to higher precipitation totals on the western coast. The south-central peninsula is shielded from this precipitation by the mountains. In this area and in the southeastern peninsula, the northeast trades make winter the dominant precipitation season (see again Fig. 17.7), even though totals are sparse, particularly in the south-central peninsula. Central India is dry because it sits in a *double rain shadow* of the Western and Eastern Ghats; humid air from the Arabian Sea and Bay of Bengal is largely blocked from entering central India.

In summary, India is wettest in the southwest and northeast, and drier in the northwest, south-central, mid-continent, and southeast (Fig. 17.16). The wettest areas, such as Mumbai in the southwest peninsula and Kolkata on the northeast peninsula (Fig. 17.17 and see again Fig. 17.13, respectively) receive the majority of their rainfall during the summer monsoon season when the combination of onshore southwesterly flow, humid air, lower pressure over land, and orographic lifting produce extremely large monthly totals. Mumbai exceeds 40 cm (16 in) of rain per month in June, July, and August, while Kolkata receives two-thirds of its rain from May to August. These extremely high rainfall totals lead to large water balance surpluses that produce considerable runoff to the rivers of the region. Moisture stress is common in the hot spring season before the summer monsoon rains arrive, especially in interior areas such as Nagpur, India (Fig. 17.18). Substantial deficits also occur for most of the year even during the monsoon season in the semi-arid northwestern peninsula, as at Ahmedabad, India (Fig. 17.19), and all year toward the northwest such as at Lahore, Pakistan (Fig. 17.20).

Many of the more remote, rural, Arid (B climate) parts of South Asia—in the northwest and south-central—tend to receive a higher percentage of their precipitation in winter than the wetter and more coastal sites. Extreme northwestern India and northern Pakistan and Afghanistan are beyond the reach of the most intense, onshore, humid summer monsoon flow. Much of the paltry precipitation in such areas tends to

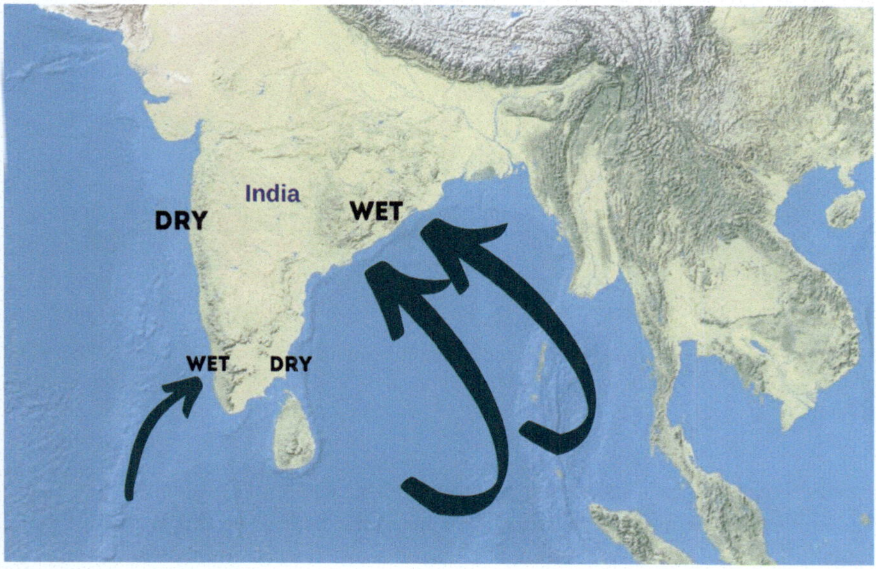

Fig. 17.16 Summary of moisture regimes across India. *Source* Geaux Geospatial

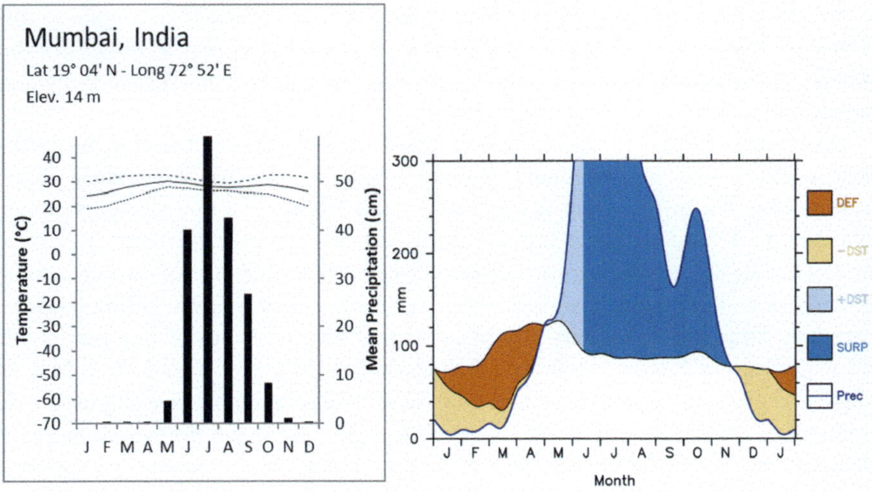

Fig. 17.17 Climagraph and water balance diagram for Mumbai, India. Data for 1981–2016 averaging period. *Source* Authors. https://weatherspark.com/, http://cyclops.deos.udel.edu/wimp/public_html/

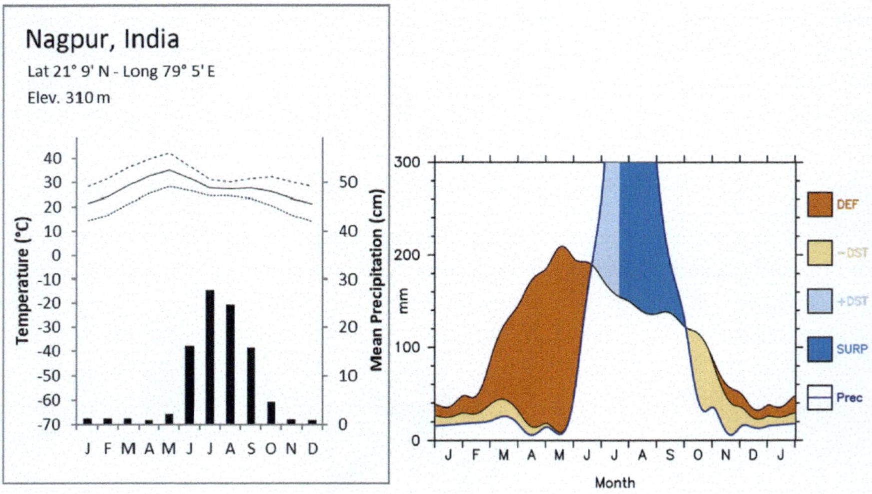

Fig. 17.18 Climagraph and water balance diagram for Nagpur, India. Data for 1981–2016 averaging period. *Source* Authors. https://weatherspark.com/, http://cyclops.deos.udel.edu/wimp/public_html/

be caused by the occasional winter precipitation-bearing mid-latitude wave cyclone. The south-central Indian Peninsula is in the rain shadow of the Western Ghats, reducing precipitation from summer monsoon flow. Instead, moisture increases in these areas and in the more humid peninsular areas to the east of the B climates

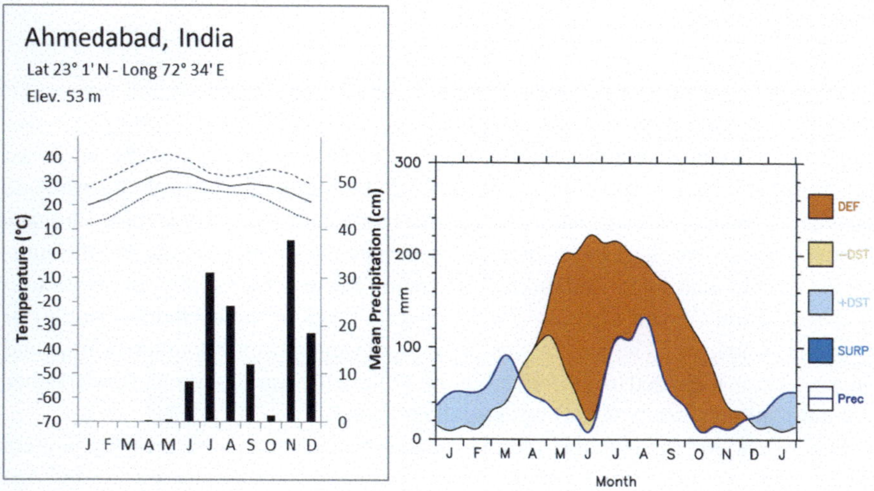

Fig. 17.19 Climagraph and water balance diagram for Ahmedabad, India. Data are for the 1991–2020 period. *Source* Authors. https://weatherspark.com/, http://cyclops.deos.udel.edu/wimp/public_html/

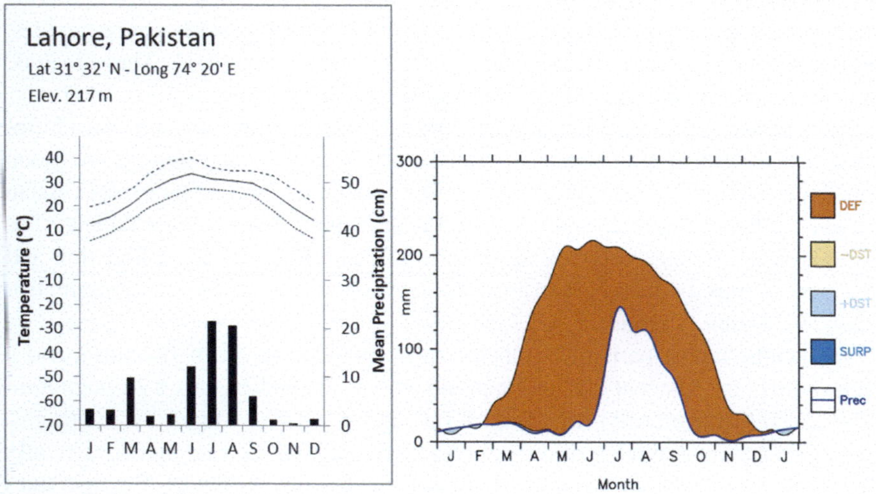

Fig. 17.20 Climagraph and water balance diagram for Lahore, Pakistan. Temperature data are for the 1961–1990 averaging period and precipitation data are for 1991–2020 period. *Source* Authors. https://weatherspark.com/, http://cyclops.deos.udel.edu/wimp/public_html/

during the periods of northeasterly flow passing over the Bay of Bengal in the low-Sun season. For example, Chennai receives roughly half of its annual 138 cm (54 in) of precipitation in October and November (see again Fig. 17.12).

Southeast Asia

Temperature patterns of the Southeast Asia peninsula and nearby islands are similar to those of South Asia to the west. In addition to the reasons for the spring heat over South Asia, the heat in Southeast Asia is also related to the trek of the northeast trades over the Pacific Warm Pool before landfall. Temperatures in Southeast Asia and adjacent water bodies remain warm throughout the year and tropical (A) climates dominate. Rare cool surges in winter may penetrate the smaller mountains of Southeast Asia and reach into Thailand and Vietnam. These cooler spells, in addition to higher elevations, result in areas of Cwa climate types in the northernmost part of Myanmar, Laos, and Vietnam. A tiny area of adjacent extreme southern China and some peaks too localized to appear on Fig. 17.15 have the warm-summer (Cwb) rather than hot-summer variety (Cwa) because of elevation.

Also like southern Asia, the majority of Southeast Asia experiences dry winters and wet summers associated with the South Asian monsoon, producing Aw and Am climates. For the huge populations of monsoon-affected East, South, and Southeast Asia, along with some other parts of the tropical world—well over half of the world population, the word "season" is more likely to evoke thoughts of "wet" or "dry" than "hot" versus "cold." The proportion of summer contributions to precipitation totals are especially high in Myanmar, Thailand, and the western slopes of Indonesia where mountain ranges intercept the onshore flow of southwesterly counter-trades. Representative examples are in the locations shown in Fig. 17.21. These include Naypyidaw, Myanmar (Fig. 17.22) and Manila, Philippines (Fig. 17.23).

By contrast, the eastern coasts of Southeast Asia such as Vietnam, the Philippines, and the Indonesian islands are blocked from some of the summer monsoon flow. In autumn and winter, they receive more precipitation than most other Asian monsoon-influenced places and only small if any winter water balance deficits. The "wrap around" northeasterly flow around the Siberian High creates onshore flow and wetter conditions for these coastal areas during the winter monsoon. A representative example is Hanoi, Vietnam (Fig. 17.24).

The southernmost parts of Southeast Asia, including most of Indonesia and a huge swath of the adjacent Indian Ocean and other water bodies, have a *Tropical Rain Forest* (Af) *climate* more typical of equatorial regions outside of monsoon influences. The nearly constant number of daylight hours throughout the year and consistently high noon Sun angles in these low latitudes cause a near-absence of temperature seasonality. Temperature ranges are small. In fact, the temperature difference between the mean of the warmest and coldest months is often less than the difference between the daily high and low temperatures.

These Af climates also have relatively even, well-distributed monthly mean rainfall totals throughout the year, with no climatologically dry months. Twice-yearly precipitation maxima occur as the ITCZ passes overhead. The lower the latitude, the more likely the precipitation maxima are for the spring and autumn months, because the direct Sun angles occur on the Equator at the *vernal equinox* and *autumnal equinox*. Rains fall overwhelmingly in the afternoon hours, on perhaps four to six

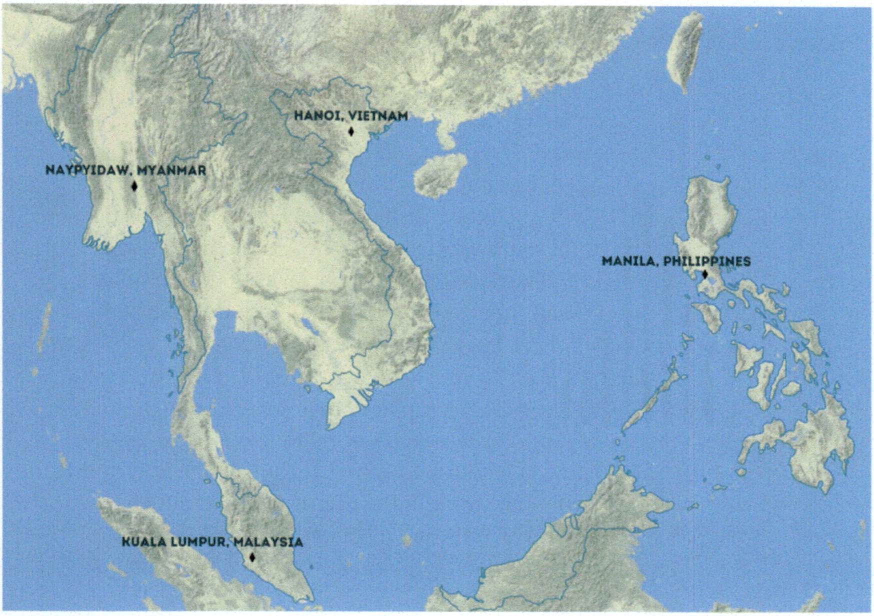

Fig. 17.21 Southeast Asian cities for which climagraphs and water balance diagrams are shown. *Source* Authors. https://weatherspark.com/, http://cyclops.deos.udel.edu/wimp/public_html/

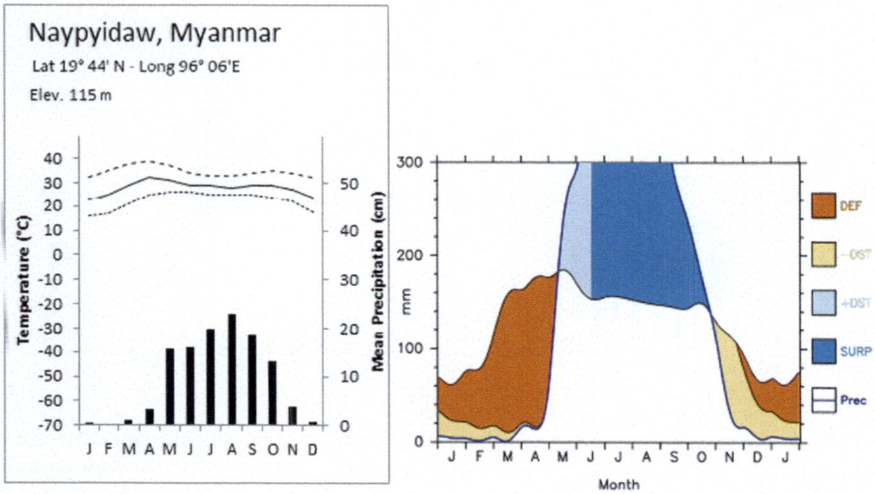

Fig. 17.22 Climagraph and water balance diagram for Naypyidaw, Myanmar. Data for 1981–2016 averaging period. *Source* Authors. https://weatherspark.com/, http://cyclops.deos.udel.edu/wimp/public_html/

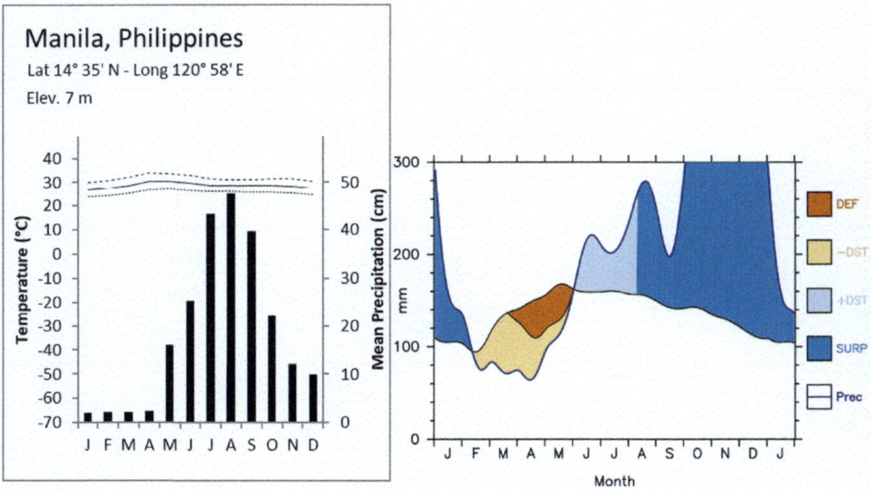

Fig. 17.23 Climagraph and water balance diagram for Manila, Philippines. Data for 1991–2020 averaging period. *Source* Authors. https://weatherspark.com/, http://cyclops.deos.udel.edu/wimp/public_html/

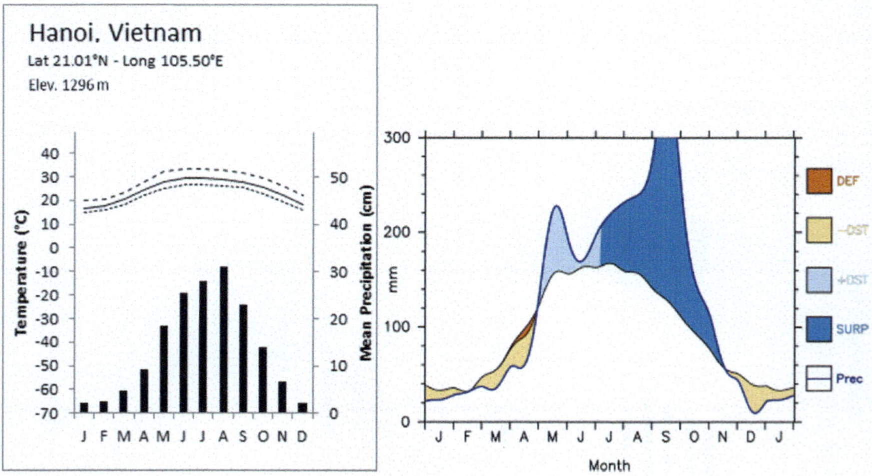

Fig. 17.24 Climagraph and water balance diagram for Hanoi, Vietnam. Data for 1981–2016 averaging period. *Source* Authors. https://weatherspark.com/, http://cyclops.deos.udel.edu/wimp/public_html/

days per week, in most Southeast Asian Af climates. A representative example of the Af climate type is Kuala Lumpur, Malaysia (Fig. 17.25).

Collectively, the *mesothermal* (C) and tropical (A) *climates* that pervade South and Southeast Asia and adjacent water bodies also appear elsewhere in the low latitudes

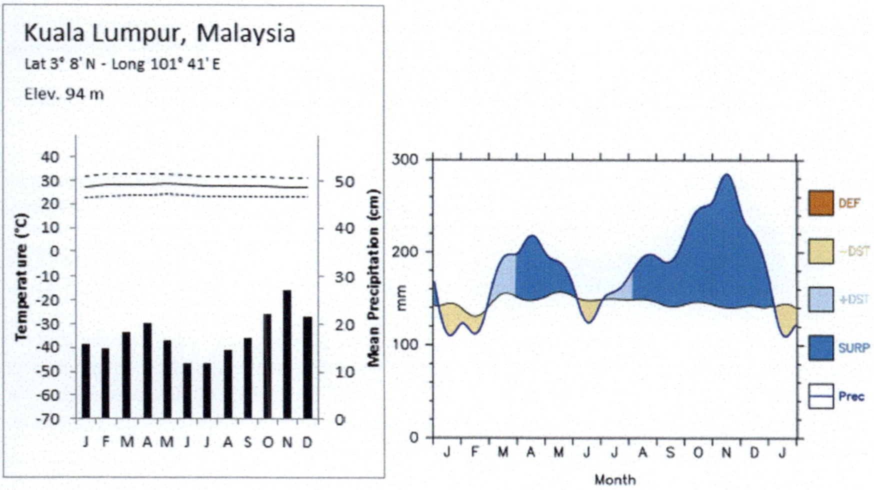

Fig. 17.25 Climagraph and water balance diagram for Kuala Lumpur, Malaysia. Data for 1981–2016 averaging period. *Source* Authors. https://weatherspark.com/, http://cyclops.deos.udel.edu/wimp/public_html/

around the world, including the topics of the next three chapters. Relatively large expanses of arid (B) climates also characterize the next world region—Africa.

Further Reading

Anil, N., Kumar, M. R. R., Sajeev, R., & Saji, P. K. (2016). Role of distinct flavours of IOD events on Indian summer monsoon. *Natural Hazards, 82*(2), 1317–1326. https://doi.org/10.1007/s11069-016-2245-9

Caparotta, S.P. (2018). *Madden-Julian oscillation relationships with cool season cyclogenesis, daily precipitation, and cool season severe weather frequencies in the Gulf of Mexico Region.* Louisiana State University. Ph.D. Dissertation.

Dimri, A. P., Yasunari, T., Kotlia, B. S., Mohanty, U. C., & Sikka, D. R. (2016). Indian winter monsoon: Present and past. *Earth-Science Reviews, 163*, 297–322. https://doi.org/10.1016/j.earscirev.2016.10.008

Findlater, J. (1981). An experiment in monitoring cross-equatorial airflow at low level over Kenya and rainfall of western India during the northern summers. In J. Lighthill & R. P. Pearce (Eds.), *Monsoon dynamics* (pp. 309–319). Cambridge University Press. ISBN: 0-521-22497-7.

Gadgil, S. (2003). The Indian monsoon and its variability. *Annual Review of Earth and Planetary Sciences, 31*, 429–467. https://doi.org/10.1146/annurev.earth.31.100901.141251

Kishore, P., Jyothi, S., Basha, G., Rao, S. V. B., Rajeevan, M., Velicogna, I., & Sutterley, T. C. (2016). Precipitation climatology over India: validation with observations and reanalysis datasets and spatial trends. *Climate Dynamics, 46*, 541–556.https://doi.org/10.1007/s00382-015-2597-y

Misra, V., & DiNapoli, S. (2014). The variability of the Southeast Asian summer monsoon. *International Journal of Climatology, 34*(3), 893–901. https://doi.org/10.1002/joc.3735

Nur'utami, M. N., & R. Hidayat,. (2016). Influences of IOD and ENSO to Indonesian rainfall variability: Role of atmosphere-ocean interaction in the Indo-Pacific sector. *Procedia Environmental Sciences, 33*, 196–203. https://doi.org/10.1016/j.proenv.2016.03.070

Rana, S., McGregor, J., & Renwick, J. (2015). Precipitation seasonality over the Indian subcontinent: An evaluation of gauge, reanalyses, and satellite retrievals. *Journal of Hydrometeorology, 16*(2), 631–651. https://doi.org/10.1175/JHM-D-14-0106.1

Rohli, R. V., Snedden, G., Martin, E., & DeLong, K. L. (2022). Impacts of atmosphere-ocean teleconnections on the south-central United States. *Frontiers in Earth Science, 10*, Art. No. 934654. https://doi.org/10.3389/feart.2022.934654

Taraphdar, S., Zhang, F. Q., Leung, L. R., Chen, X. C., & Pauluis, O. M. (2018). MJO affects the monsoon onset timing over the Indian region. *Geophysical Research Letters, 45*(18), 10011–10018. https://doi.org/10.1029/2018GL078804

Trewartha, G. T. (1981). *The Earth's problem climates* (2nd ed.). University of Wisconsin Press.

Ul Hasson, S., Pascale, S., Lucarini, V., & Böhner, J. (2016). Seasonal cycle of precipitation over major river basins in South and Southeast Asia: A review of the CMIP5 climate models data for present climate and future climate projections. *Atmospheric Research, 180*, 42–63. https://doi.org/10.1016/j.atmosres.2016.05.008

Chapter 18
Africa and Adjacent Ocean Environments

Abstract Africa is a land of tropical (A), arid (B), and mesothermal (C) climates affected by the Bermuda-Azores High, South Atlantic High, and South Indian Ocean High (also known as the Mascarene High), and the major ocean currents that they drive. From north to south are the world's largest desert—the Sahara, the semi-arid Sahel region, the Tropical Wet-dry, also known as the Tropical Savanna (Aw) climate, and the Tropical Monsoon (Am) and Tropical Rain Forest (Af) climates, with Aw and higher-elevation Mesothermal—Winter dry (Cw) climates, B climates, and finally Humid Subtropical (Cfa) and Mediterranean (Cs) climates in the southern tip of the continent. For a variety of reasons, the African Af climates are not as rainy in other places, and the massive Sahara Desert can exert influences elsewhere in Africa and beyond, such as through the Harmattan winds in west Africa, deserts of the Horn of Africa, and the Saharan air layer in the Americas.

Keywords African easterly jet · Agulhas Current · Benguela Current · Canary Current · Desertification · East African low-level jet · Equatorial currents and counter-currents · Harmattan wind · Inter-tropical convergence zone (ITCZ) · Northwesterly and southwesterly counter-trades · Somali Current · Somali jet · South Indian Ocean High and Gyre · Turkana jet

General

Africa is unique among the continents in that it straddles the Equator, with extensions into the mid-latitudes of both hemispheres. Its wide expanse of latitudes, from about 37° N to about 34° S, gives Africa, along with its surrounding ocean environs, influence from both mid-latitude and tropical air masses and weather systems. The northern section includes the largest desert on the planet—the Sahara, which was covered in Chap. 15. The annual migration of the *inter-tropical convergence zone* (ITCZ) has an important influence on much of Africa. As in Asia, the sheer size of the landmass creates a monsoon flow that enhances ITCZ effects in West Africa. All factors combine to create a continent with adjacent waters that is largely arid in

R. V. Rohli et al., *Atmospheric and Oceanic Circulation*,
https://doi.org/10.1007/978-3-031-66142-6_18

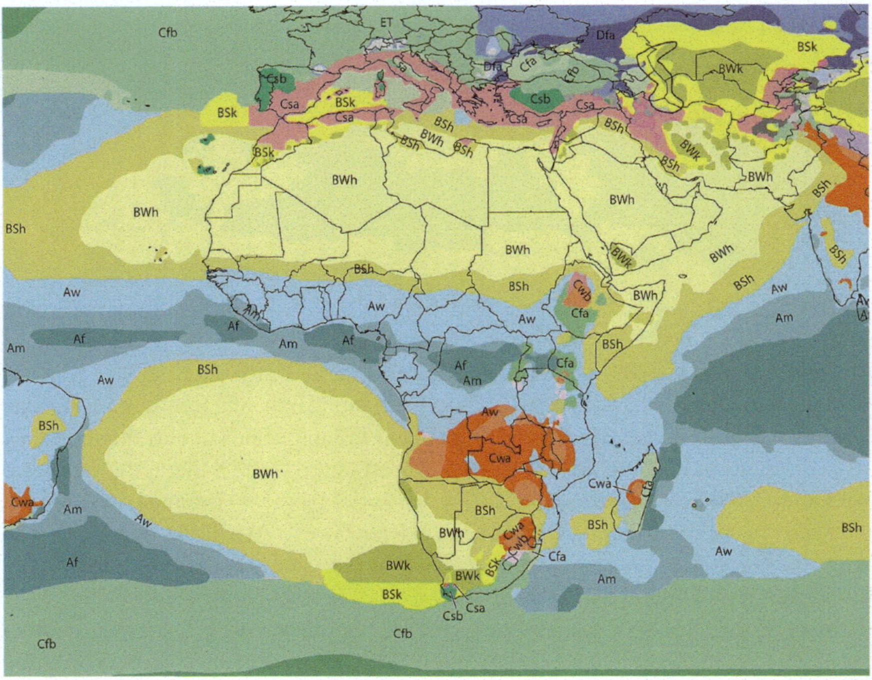

Fig. 18.1 Köppen climatic types around Africa and surrounding marine environment. *Source* Modified from Reynolds, Rohli, Johnson, Waylen, and Francek, Exploring Physical Geography, McGraw-Hill

the Northern Hemisphere mid-latitudes, hot and wet in the middle tropical region, and temperate with regions of aridity, particularly in the subtropical ocean, in the Southern Hemisphere middle and low latitudes (Fig. 18.1).

Topographic Influences (Site)

Africa is nearly bisected by the Equator and extends beyond 30° of latitude in both the Northern and Southern Hemispheres. The northern reaches of Africa approximate the latitude of the U.S. mid-Atlantic states while the southern extent is as far south as central Argentina in South America. However, because of its shape, Africa is mostly a Northern Hemisphere continent.

Most of Africa is a plateau; the coastal plain is generally narrow. The plateaued interior and the numerous waterfalls hinder inland river navigation. The rather young, emerging geology of the continental margins makes natural harbors, bays, peninsulas, estuaries, and other irregular coastal features also rare.

Mountain ranges are few and tend to be modest in height. Major terrain features include the Atlas Mountains of northwestern Africa (Fig. 18.2). The chain, geologically equivalent to the Appalachian Mountains of the U.S., has peaks that are less than 3050 m (10,000 ft) high. Other highlands are found in East Africa, centered on Ethiopia and southward along the Great Rift Zone (see again Fig. 18.2), where tectonic activity is slowly splitting the continental crust. Volcanoes such as Mt. Kilimanjaro in Tanzania are present where magma from deep within the Earth reaches the surface. The Ethiopian Highlands are rugged, but elsewhere, much of eastern and southern Africa forms a high plateau. Over a wide area south of the Great Rift Zone, from Uganda and Kenya to Zimbabwe and northern South Africa, average elevations exceed 1000 m (3300 ft). The topography of the East African and Southern African plateaus primarily influences regional climates by decreasing *temperature* at higher elevations, but few of Africa's topographic features are formidable and continuous enough to block broad-scale atmospheric circulation.

Lake Victoria lies just to the west of the Great Rift Zone. As one of the Earth's largest (but fairly shallow) lakes, Lake Victoria contributes humidity to the local

Fig. 18.2 Selected physical features and countries of Africa. *Source* Geaux Geospatial

atmosphere and increases precipitation in the nearby central and East African countries of Uganda, northern Tanzania, and southwestern Kenya. Major river basins include the Nile (see again Fig. 18.2)—the longest river on Earth. The Nile flows northward from the high continental inland plateau into the Mediterranean Sea. Much of the river flows through extremely arid regions, with its headwaters, or source, in a wet, tropical region. The Nile and its tributaries provide water for the vast majority of northeastern Africa's people.

The Congo River (see again Fig. 18.2)—the second-longest river in Africa—flows westward into the Atlantic Ocean from the equatorial interior of the continent, where the *Tropical Rain Forest* (Af) *climate* type provides abundant rainfall, giving the Congo the third-greatest discharge of any river on Earth. Africa's third-longest river—the Niger (see again Fig. 18.2)—lies northwest of the Congo River. The Niger drains the wetter regions of West Africa and reaches the Gulf of Guinea, an extension of the Atlantic Ocean, forming a large delta in coastal Nigeria.

Circulation Influences (Situation)

Because Africa straddles the Equator, the ITCZ is an important contributor to the continent's climates. The migration of the ITCZ through the year generates the wet/dry seasons that comprise African climates north and south of the tropical rain forests near and along the Equator. When the ITCZ is nearby, the cloud cover and afternoon *precipitation* suppresses daytime maximum temperatures. When and where the ITCZ is farthest away, the clear skies and lack of surface water allow solar radiation to be converted directly to *sensible heat* rather than *latent heat*.

In Africa, the latitudinal range and onset of ITCZ migrations are particularly inconsistent across space and time. This variation creates situations of environmental stress when the summer rains are delayed or diverted to other locations. Interannual variations in the position, movement, and strength of the ITCZ over Africa are related to broad-scale teleconnections such as *El Niño-Southern Oscillation* (ENSO) and the *North Atlantic Oscillation* (NAO). During periods when the ITCZ fails to move over a region or is weaker than normal, water-induced stress ensues as water balance *deficits*, and therefore droughts, accumulate.

If these droughts occur on an extensive scale, the long-term regional water balance may become altered. Healthy vegetation adds water vapor to the local atmosphere through transpiration. However, when the plants wilt under moisture stress during a strong drought, local *humidity* and precipitable water are reduced. This *positive feedback* loop depletes surface water and near-surface groundwater sources. In areas like the Sahel, immediately south of the Sahara Desert (see again Fig. 18.2), these natural variations have been exacerbated by overgrazing, which compacts the soil and strips it of vegetation. The result is the loss of water-holding capacity of the soil and an inability to regulate the soil water loss by the presence of vegetation. Afterwards, a higher percentage of precipitation becomes runoff immediately following precipitation events, at the expense of soil–water recharge. This increases the deficits even more, producing **desertification**, or the expansion of deserts. The physical landscape, altered to a drier state, increases the risk of resource conflict, water shortages, animal losses, famine, and disease.

Northern and Southern Africa lie within the zones of the *northeast trades* and *southeast trades*, respectively. Trade winds are driven primarily by the presence of *subtropical anticyclones* centered near 30° latitude in both hemispheres, but oscillating latitudinally with the seasons. The clockwise flow in the Northern Hemisphere and counter-clockwise flow around them in the Southern Hemisphere are only slightly outward, as the *Coriolis effect* is weak at these low latitudes. Nevertheless, the flow around the equatorward sides of the subtropical anticyclones are manifested as the northeast and southeast trades. The *southwesterly counter-trades* occur when the ITCZ is over the Northern Hemisphere, in Northern Hemisphere summer. These winds bend back northeastward in the Northern Hemisphere (Fig. 18.3). Likewise, the **northwesterly counter-trades** occur when the ITCZ is over the Southern Hemisphere, in Southern Hemisphere summer, as the northeast trades bend back southeastward on crossing the Equator into the Southern Hemisphere.

The subtropical anticyclones also have a direct effect on associated surface climates through *subsidence*. Much of the Sahara Desert and a substantial part of the North Atlantic Ocean lie within one such zone of strong subsidence that produces particularly strong and persistently high surface air pressure. Airflow around the *Bermuda-Azores High*, its counterpart in the Southern Hemisphere—the *South Atlantic High*, and the **South Indian Ocean High**, also known as the **Mascarene High**, cause *convergence* of the trade winds into the ITCZ.

The persistent and strong trade winds resulting from the Bermuda-Azores High are called the **Harmattan winds** over northwestern Africa. These winds are hot and dry. When the consistent flow of the Harmattans is particularly strong, dust from the

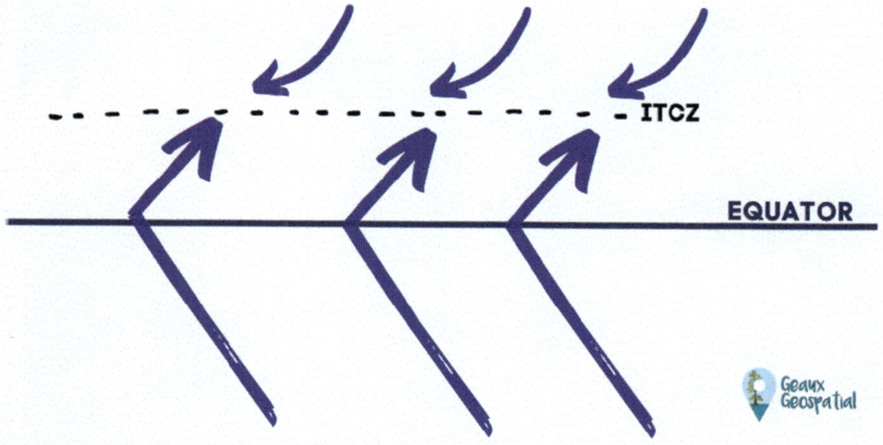

Fig. 18.3 Schematic of the idealized southeast trades (south of the Equator), southwesterly counter-trades (between the Equator and ITCZ), and northeast trades (north of the ITCZ), characteristic of Northern Hemisphere summer. *Source* Geaux Geospatial

Sahara Desert can be transported across the North Atlantic and into northern South America and the Caribbean Sea. This flow of dry air high in the northeast trades, from northwestern Africa to the Caribbean Sea and northern South America, known as the **Saharan air layer** (SAL), is guided along by the **African easterly jet** between the northeast and southeast trades over the ITCZ, during a period of monsoonal flow over West Africa, peaking in September (Fig. 18.4). The SAL has numerous impacts, including the alteration of *atmospheric stability* and cooling the Atlantic Ocean and Caribbean Sea sufficiently to prevent or weaken Atlantic *tropical cyclones*. Mineralogical analysis confirms that the SAL involves significant sand transport, but research about the temporal variability of the strength and impacts of the SAL is ongoing.

Other atmospheric circulations are also important elsewhere in Africa. For instance, *advection* of hot, dry air over much of East Africa by the northeast trades is reinforced by outflow around the southern side of the subsidence-induced anticyclone over the Arabian Peninsula related to the *Indian Ocean Dipole* (IOD). Westerly flow and forced *convection* from *mid-latitude wave cyclones* are present at the northern and southern margins of Africa and generally only influence the climate of these

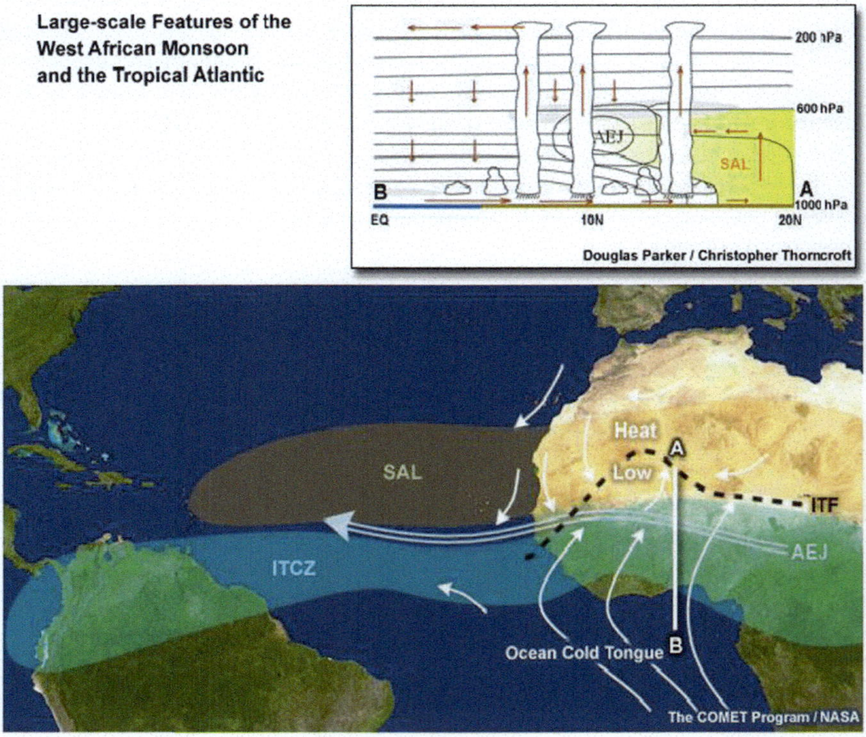

Fig. 18.4 The Saharan air layer (SAL) and its relationship to the African easterly jet and ITCZ, with the peak transport in September. *Source* UCAR CoMet/NASA. Permission from Tara Torres

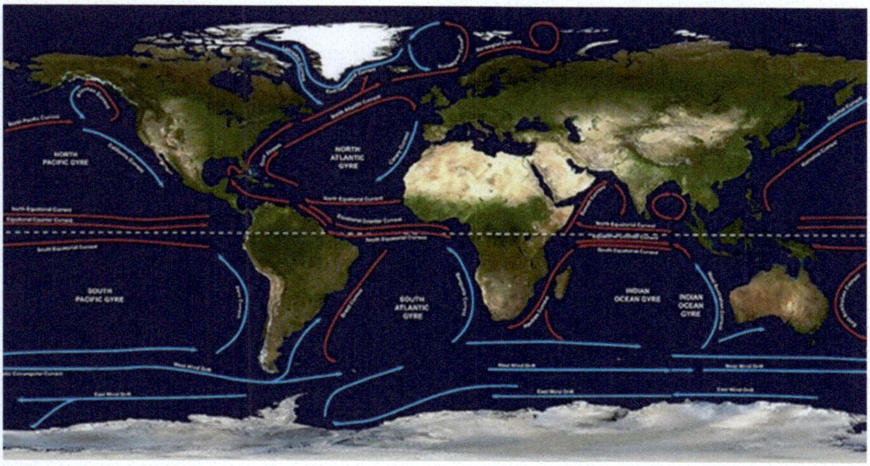

Fig. 18.5 Gyres and currents of the Atlantic and Indian Ocean basins. *Source* NOAA

areas in winter. And low-level counter-trades in both hemispheres converge into the ITCZ.

Atmospheric and oceanic near-surface circulations are strongly connected in this part of the world, with important consequences for Africa's climate patterns. The subtropical anticyclonic-rotating ocean surface *gyres* are driven by the atmospheric circulation around the subtropical anticyclones of the Northern and Southern Hemispheres, steering the major ocean currents (Fig. 18.5). In turn, the currents of the North and South Atlantic Oceans, as well as those in the Indian Ocean, have significant effects on atmospheric stability, humidity, and the movement of the African ITCZ.

The clockwise flow around the Bermuda-Azores High, creates the *North Atlantic Gyre*. The cold *Canary Current* flows along the northwestern quadrant of Africa and continues westward away from Africa in the form of the *North Equatorial Current*, enhancing cold-water *upwelling* near the northwest African coast. The cold water also frequently produces low-level *temperature inversions*, whereby temperatures increase upward from the surface, along the western coast of northern Africa. Inversions limit cloud formation and precipitation processes as the cooler, denser air near the surface will not rise. Instead, the subsidence supported by the cold Canary Current suppresses convective activity and related cloud and precipitation formation processes. Arid conditions from the Sahara Desert extend far beyond the Atlantic coast, well into the subtropical North Atlantic Ocean. Also in the northwestern quadrant is the *North Equatorial Counter-current*, not to be confused with the North Equatorial Current. The *North Equatorial Counter-current* accumulates warm, equatorial water from the South American coast toward the Gulf of Guinea in west Africa.

Circulation in the southwestern quadrant of Africa is influenced by the cold **Benguela Current**, which flows counter-clockwise around the South Atlantic High. The **South Atlantic Gyre** spins in the opposite direction of the Northern Hemisphere gyres, as the Coriolis effect deflects flow to the left in the Southern Hemisphere. Like its Northern Hemisphere counterpart, the South Atlantic Gyre drives cold water from high latitudes toward the Equator in the eastern ocean basins. The Benguela Current evokes a similar response in suppressing coastal cloud cover and precipitation.

Because of the shape of Africa, the Benguela Current extends farther northward than the Canary Current does southward. This difference pulls cold water northward to the Equator and beyond, keeping the overlying atmosphere stable. Consequently, the South Atlantic High is the only of the three subtropical anticyclones influencing Africa that remains important for Africa's climate year-round. The persistent high restricts seasonal migration of the ITCZ in West Africa, keeping it in the Northern Hemisphere even in January. ITCZ rainfall remains over the West African coast year-round.

The southeastern quadrant of the continent is influenced by the **Agulhas Current**, which flows on the western side of the counter-clockwise Southern Hemisphere **Indian Ocean Gyre**. Like all western ocean basin gyre-driven currents, the Agulhas is a warm current, as low-latitude waters are advected toward higher latitudes on the western side of the gyres. These waters produce an important source of atmospheric moisture for precipitation in Southern Africa.

Finally, off the northeastern quadrant of Africa is the **Somali Current**, which flows northward along the coast of the Horn of Africa and the country of Somalia, analogous to the Gulf Stream near North America. The North Equatorial Current flows westward along the low-latitude Indian Ocean to feed into the Somali Current. Although the Somali is a warm current, it fails to support precipitation-generating processes because the *Ekman spiral* initiates cold-water upwelling along the coast that chills and stabilizes the atmosphere, reinforcing high atmospheric pressure and stymying cloud cover and precipitation. The Somali Current supports the only major upwelling zone in the world on a western ocean basin.

North Africa

The bulk of Africa lies in the Northern Hemisphere where *continental tropical* (cT) *air masses* dominate the northern part of the continent (Fig. 18.6). As Chap. 15 showed, the region is extremely dry, with some humidity in the northwest where moisture is advected from the Atlantic Ocean around the Bermuda-Azores High. Although water vapor may be abundant, the rising motion needed to produce precipitation is suppressed by the region's position on the eastern side of the subtropical anticyclone, where the atmosphere is stabilized by the cold Canary Current. Warmer air overlies the cool surface layer, forming a temperature inversion.

The Bermuda-Azores High stretches across northern Africa, causing subsidence associated with the *Hadley cell* circulation. This subsidence is compounded by sinking air associated with the IOD connected to the *Walker circulation*. Thus, despite very hot surfaces, convection is limited across northern Africa and the Arabian peninsula. Even so, coastal air is humid from the adjacent water bodies, which continually

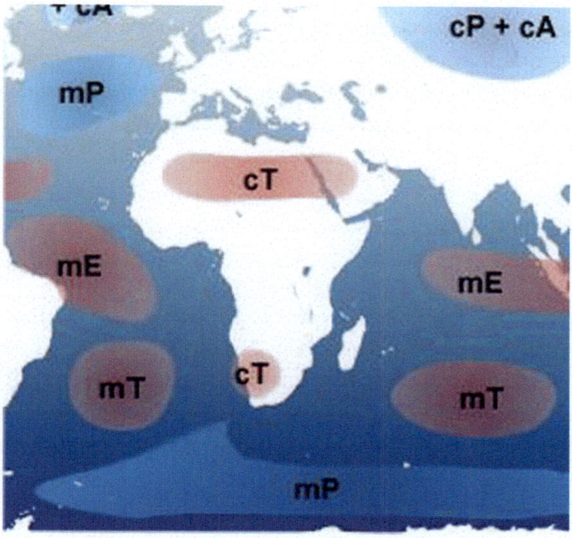

Fig. 18.6 Air masses source regions of Africa. *Source* UCAR CoMet. Permission from Tara Torres

evaporate surface water. Intense high-pressure subsidence restricts the ascent of this humid air. The region around the Red Sea is an example of a location with excessive atmospheric humidity that gets little precipitation.

The overall result of strong subsidence is the Sahara Desert, the largest desert in the world. Chapter 15 detailed the patterns of the *True Desert–hot* (BWh) *climate* across the Sahara and the narrow area of *Mediterranean–hot summer* (Csa) *climate* in coastal Morocco, Algeria, and Tunisia. In this chapter, attention is focused on the region from the Sahara southward.

The southern Sahara remains extremely arid, as it is too far north for the ITCZ to provide rain and too far south for frontal precipitation. For example, Nouadhibou (formerly known as Port Étienne), on the coast of Mauritania (20° 56′ N, 17° 03′ W) receives an average of less than 2.5 cm (1 in) of precipitation per year. Even more remarkable is the fact that most extremely arid locations like Nouadhibou have less than the average precipitation in most years, because a few extreme events tend to skew the average upward to an unrealistically high total in most years.

The southern Sahara has frequent wind and dust storms that are related to *thermal lows* developing from extreme daytime heat. However, clouds cannot form easily because of the lack of atmospheric water vapor in these inland areas or position in the *rain shadow* of the Ethiopian Highlands. On the southern margin of the desert, a small amount of rain falls in summer, as the ITCZ extends nearby as it migrates with the vertical solar rays. However, rain along the ITCZ itself is limited in this region. The cooler and more humid air of the southwesterly counter-trades meets with the hotter and drier air advected by northeast trades (locally known as Harmattan winds) at the ITCZ. Therefore, the African ITCZ often takes the form of a front, as the hotter air from the northeast trades is forced upward over the cooler southwesterly counter-trades. Most of the rain falls in a zone approximately 150–300 km (90–180 mi) south of this frontal boundary, as the hot air is moving southward. Precipitation is far south of the surface boundary because the extremely dry air must be lifted high in the air before it cools to its *dew point temperature* and begins the condensation and precipitation-producing processes.

The Sahel

South of the Sahara Desert is the Sahel, a zone that includes, from west to east, parts of Mali, Burkina Faso, Niger, Nigeria, Chad, southern Sudan, and Ethiopia. The Sahel is a climatic transition zone between the BWh and the tropical climates and generally corresponds to the *Semi-arid* (BS) *climate* (see again Fig. 18.1). Rainfall in the region comes in the Northern Hemisphere summer when the ITCZ is overhead. Temperatures are slightly reduced despite the nearly overhead noon Sun angles as cloudiness reflects incoming radiation. Precipitation also provides water to the surface, which cools the area through evaporative cooling.

Precipitation is variable from year to year. Most of the rain comes in July when the ITCZ is over the area, so a late arrival or early departure of the ITCZ could mean that the region will face nearly a year until rainfall returns. Factors that lead to rainfall failure include atmospheric and oceanic circulation variability in the form of *teleconnections*. For example, the weakened Walker circulation during warm-ENSO (i.e., El Niño) years affects the Sahel by weakening the *subtropical jet stream* that brings moisture. Figure 18.7 depicts the idealized positions of normal Walker circulation-related high- and low-pressure cells across the Equator from West Africa to South America. During an El Niño event, the pattern weakens or reverses, with the west-to-east arrow aloft between West Africa and East Africa representing the subtropical jet stream. At the same time that a weakened subtropical jet stream reduces summer rainfall in the Sahel, East Africa experiences more convection and precipitation, as the normal sinking overhead is weakened too (see again Fig. 18.7).

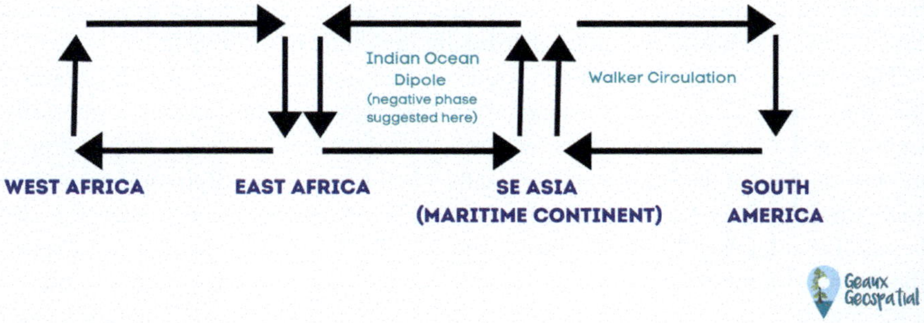

Fig. 18.7 Normal Walker circulation pressure distribution from West Africa to South America. *Source* Geaux Geospatial

The delay or absence of northward ITCZ migration may be nature-induced, but impacts can be exacerbated by human activities. For instance, water withdrawal and exposure of the bare soil by devegetation heats and dries the surface faster. Satellite evidence of re-greening in some areas offers hope.

West Africa

The climate of West Africa is characterized by wet summers and dry winters. A small monsoon effect contributes to winter dryness and summer precipitation. Furthermore, the ITCZ and the summer rain it produces remains in place over an extended period.

In winter, the cold water pulled northward by the strong South Atlantic High and South Atlantic Gyre keeps the ITCZ farther north than in most other tropical areas. Thus, the Gulf of Guinea and immediate coast continue to receive some winter precipitation, creating the lush, vegetated Af climates and *Tropical Monsoon* (Am) *climates.* However, winter dryness from the monsoon circulation contributes to the *Tropical Wet-dry* (Aw) *climate*, also known as *Tropical Savanna climates*, across most of West Africa south of the Sahel, extending offshore and into Central and East Africa, and the Indian Ocean.

Climate impacts of the ITCZ across West Africa can be divided into five idealized sectors from north to south as depicted in Fig. 18.8. Sector 1 is dominated by the northeast trades, dry cT air masses from the Harmattans, hot days because of abundant *shortwave radiation* (K↓), and cool nights as L↑ escapes out to space as *outgoing longwave radiation* under the mostly clear skies. The region is affected by the ITCZ "front" with the cT air carried by the northeast trades converging with the milder southwesterly counter-trades. The result is low-level cloud cover as depicted by the "1" in Figs. 18.8 and 18.9.

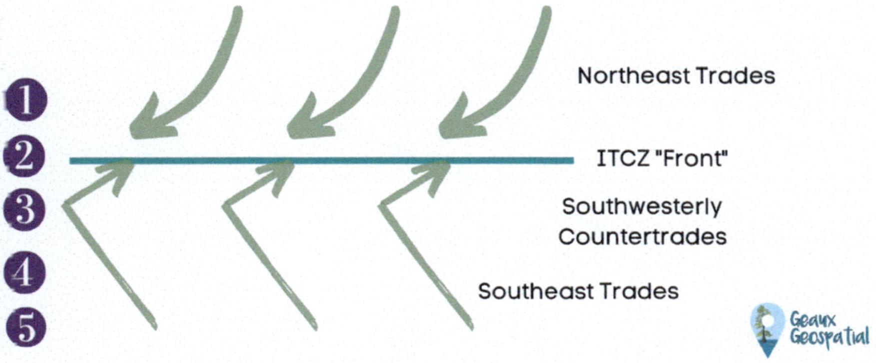

Fig. 18.8 Generalized latitudinal inter-tropical convergence zone sectors of Africa. *Source* Geaux Geospatial

Sector 2 is marked by southwesterly winds of the counter-trades, a shallow humid layer, and sporadic thunderstorm development, as shown in Figs. 18.8 and 18.9. Sector 3, located south of the ITCZ "front," has southeasterly winds, a deeper humid layer, and the development of heavy thunderstorms which are especially associated with *easterly waves*. Sector 4 contains southeasterly winds, a very deep humid layer, numerous precipitation days, and longer and more intense precipitation events. Sector 5 involves largely ocean areas where strong southeasterly flow produces clouds but little precipitation because of the cold water surface. Precipitation is limited over the ocean but more common over the heated land mass. The southernmost ITCZ sectors generate precipitation and slightly lower temperatures from cloud cover when the ITCZ is nearby.

The influence of the ITCZ sectors on temperature and precipitation as they shift with the seasons is evident in the climagraphs and water balance diagrams for representative locations shown in Fig. 18.10, at Bamako, Mali (Fig. 18.11); Monrovia, Liberia (Fig. 18.12); and Abuja, Nigeria (Fig. 18.13). As expected in tropical (A) climates, all three cities have narrow daily and *annual temperature ranges*, abundant rainfall and water *surplus* all summer (especially Monrovia and Abuja), and substantially less water in winter when the city is closer to the ITCZ "front." The winter dry period is shorter in the Am climate of Monrovia and longer with more substantial moisture stress in the Aw climate of Bamako and Abuja, where the ITCZ shifts farther south in the low-Sun season, leaving these cities positioned in Sector 1. In West African places with similar Aw climates, the high and low temperatures for the day often fall within a range of about 24–30 °C (75–86 °F), except in elevated areas.

West Africa is also an important source region for tropical cyclone formation. Easterly waves containing lines of convective thunderstorms that form in ITCZ Sector 3 migrate from east to west in the trade wind flow. As they move over the Atlantic Ocean, evaporation fuels thunderstorm development. The easterly waves are

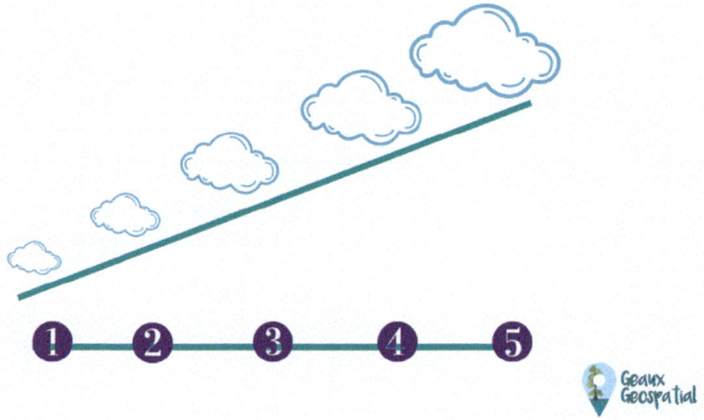

Fig. 18.9 Generalized vertical profile of the inter-tropical convergence zone in Africa with respect to surface sectors. *Source* Geaux Geospatial

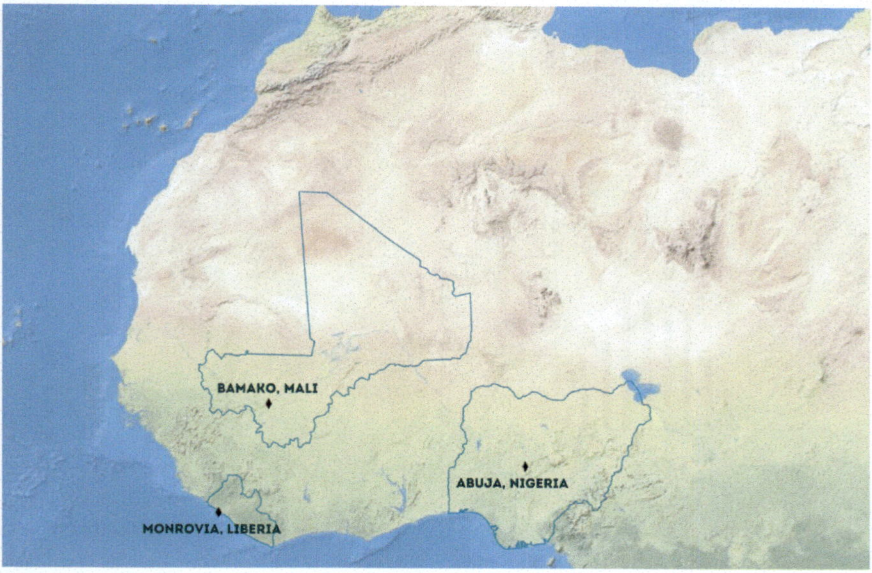

Fig. 18.10 West African cities for which climagraphs and water balance diagrams are shown. *Source* Geaux Geospatial

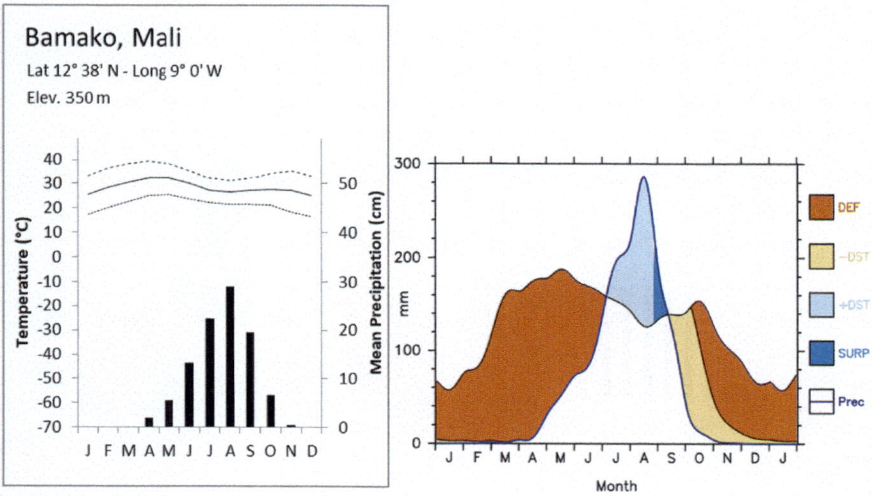

Fig. 18.11 Climagraph and water balance diagram for Bamako, Mali. Data for 1951–2000 averaging period. *Source* Authors. https://weatherspark.com/, http://cyclops.deos.udel.edu/wimp/public_html/

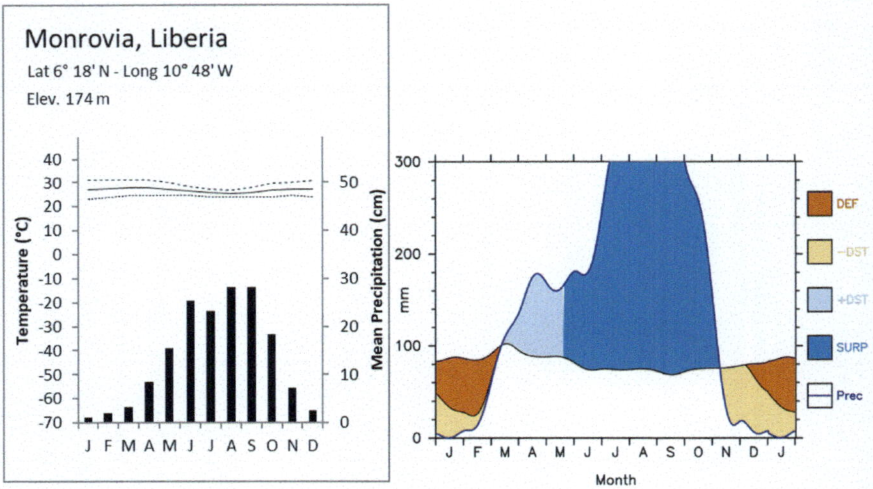

Fig. 18.12 Climagraph and water balance diagram for Monrovia, Liberia. Data for 1981–2016 averaging period. *Source* Authors. https://weatherspark.com/, http://cyclops.deos.udel.edu/wimp/public_html/

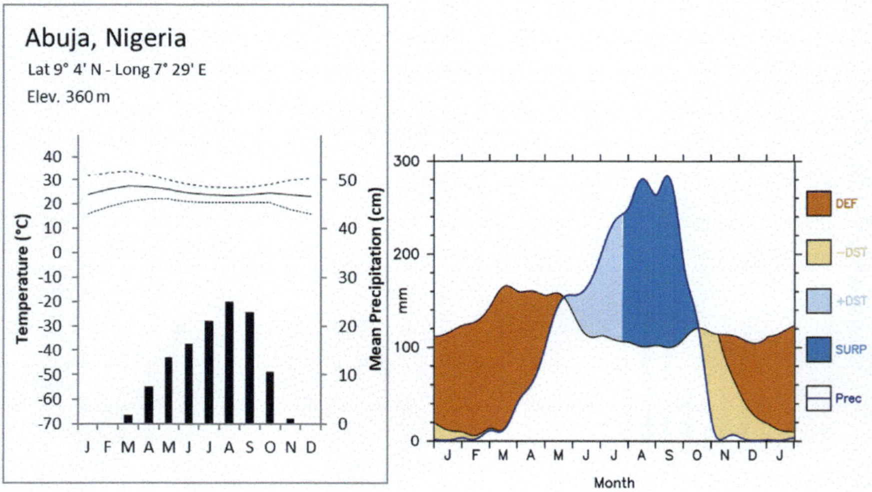

Fig. 18.13 Climagraph and water balance diagram for Abuja, Nigeria. Data for 1981–2016 averaging period. *Source* Authors. https://weatherspark.com/, http://cyclops.deos.udel.edu/wimp/public_html/

seedlings for North Atlantic tropical cyclones that typically affect the Caribbean and North America. An average of about 64 easterly waves move off the African coast through the 1 June to 30 November Atlantic *hurricane* season. Of these, about ten develop into named tropical storms and six into hurricanes each year, but with wide interannual variability. Tropical cyclones may strengthen quickly over the eastern North Atlantic, if they avoid the Canary Current, after originating as thunderstorms over West Africa. These tropical cyclones are known as **Cabo (Cape) Verde storms** because of their African origin and are often among the most intense Atlantic tropical cyclones. A research group from Colorado State University uses rainfall in the Sahel/West Africa as a predictor of the number of North Atlantic tropical cyclones in the upcoming season.

Central Africa

The core of Central Africa is classified as an Af climate type (Fig. 18.14). Cities with climagraphs and water balance diagrams in this region are labeled in Fig. 18.15. Abundant rainfall in the region results from water vapor advected inland by the trade winds. For example, in Entebbe, Uganda (0° 3′ N, 32° 27′ E) the average temperature remains around 21–22 °C (70–72 °F) in each month, and precipitation occurs year-round but is greatest in April and May (25–28 cm; 10–11 in) and November (18 cm; 7 in) when the ITCZ is overhead.

However, the Af climate type does not extend all the way eastward to the Indian Ocean. The northeast trades advect air from very dry regions in Southwest Asia, limiting the extent of the Af climates in Central Africa, and mountains associated with the East African rift zone hinder Indian Ocean moisture from being transported inland by the southeast trades. Therefore, the zone of Af climates in Africa is narrow. West of the Rift Valley, moisture advection occurs around the southern side of the South Atlantic High, which produces enough westerly flow to provide some humidity to Central Africa. Humidity waxes and wanes with the strength and position of the high-pressure cell. The Af climate does, however, resume over the equatorial

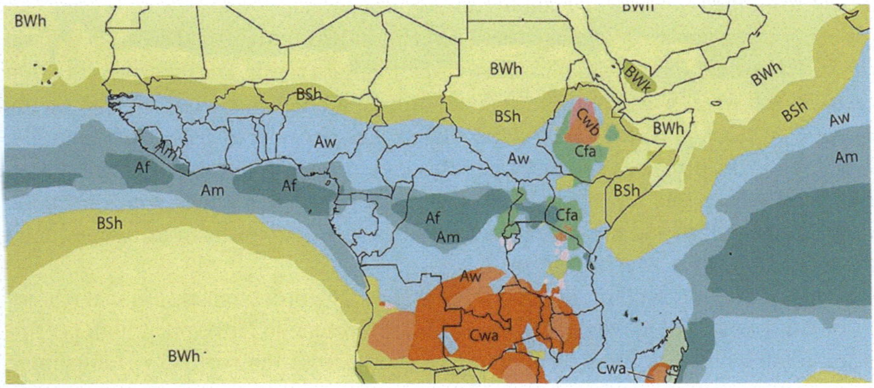

Fig. 18.14 Köppen climatic types around Central Africa and surrounding marine environment. *Source* Modified from Reynolds, Rohli, Johnson, Waylen, and Francek, Exploring Physical Geography, McGraw-Hill

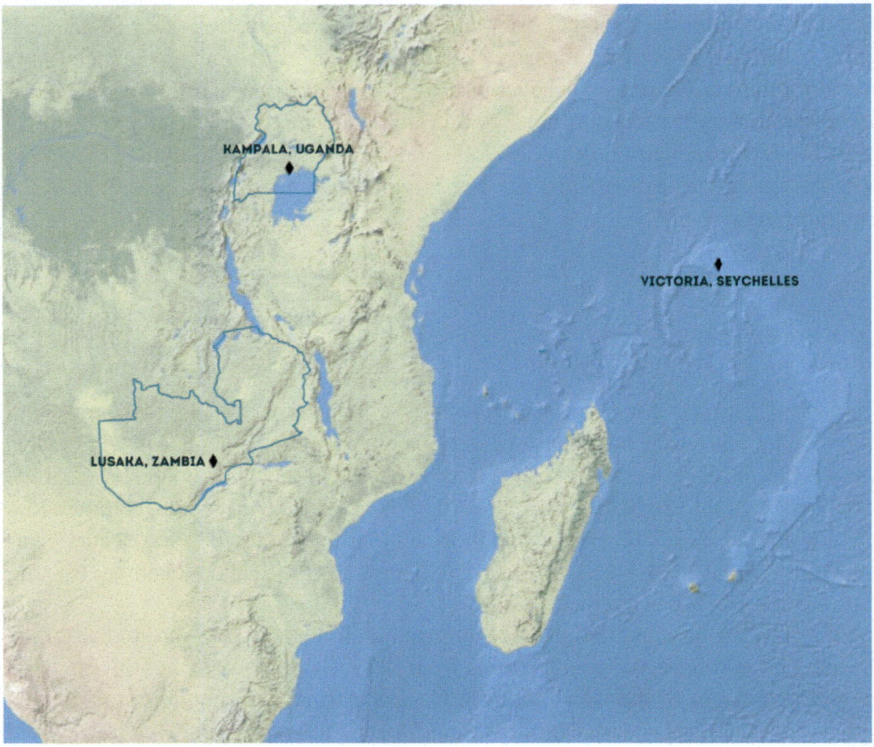

Fig. 18.15 Central African cities for which climagraphs and water balance diagrams are shown. *Source* Geaux Geospatial

Indian Ocean, represented by Victoria, Seychelles (Fig. 18.16). While water balance information is unavailable for Victoria, the drier Southern Hemisphere winter is evident from the climagraph, even in this Af climate.

Precipitation generally averages between 150 and 200 cm (60 and 80 in) in Central Africa. These totals are the smallest of any Af climate on Earth. The most precipitation occurs in the west or where there is some orographic uplift, such as in Cameroon. In east-Central Africa, the precipitation decreases because of a lack of humidity. Atmospheric disturbances and precipitation occur primarily through free convection of heated surfaces and also from the convergence between westerly winds from the South Atlantic High and the easterly trade winds. *Squall lines* form from horizontal convergence in the westerly air flow.

To the north and south of the Af climate region, the pronounced seasonal pattern of rainfall follows the movement of the ITCZ. From June to November, high precipitation totals and associated surplus can be seen across the Northern Hemisphere region as far east as Uganda (Fig. 18.17) and western Ethiopia.

The Southern Hemisphere's summer season draws the ITCZ south of the Equator from November through March. The climagraphs of *Mesothermal Winter Dry–hot*

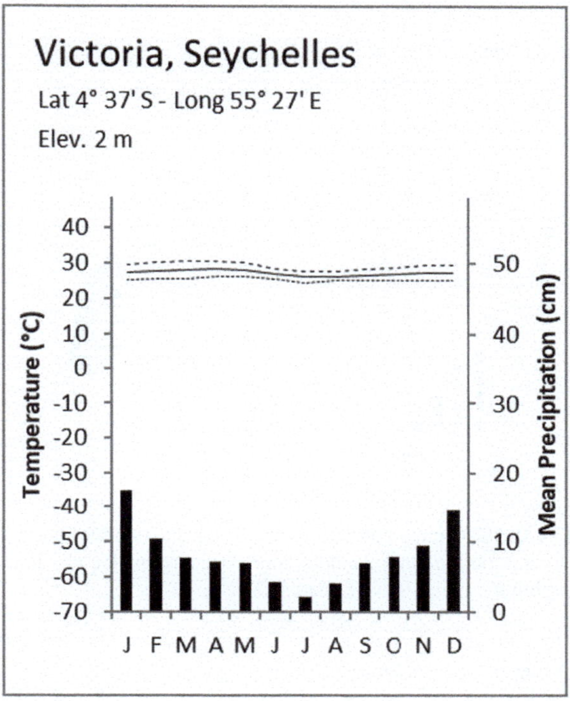

Fig. 18.16 Climagraph for Victoria, Seychelles. Data for 1981–2016 averaging period. *Source* Authors. https://weatherspark.com/

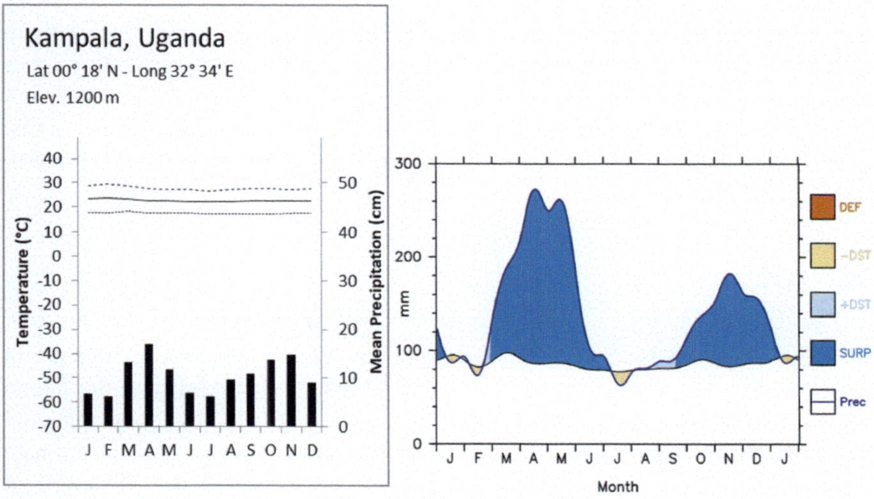

Fig. 18.17 Climagraph and water balance diagram for Kampala, Uganda. Data for 1991–2020 averaging period. *Source* Authors. https://weatherspark.com/, http://cyclops.deos.udel.edu/wimp/public_html/

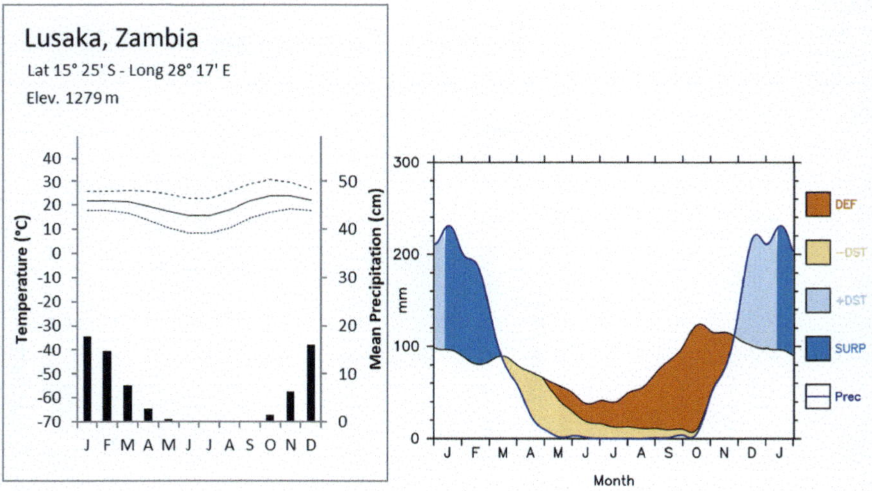

Fig. 18.18 Climagraph and water balance diagram for Lusaka, Zambia. Data for 1981–2016 averaging period. *Source* Authors. https://weatherspark.com/, http://cyclops.deos.udel.edu/wimp/public_html/

summer (Cwa) *climate* and *Mesothermal Winter Dry–warm summer* (Cwb) *climate* cities on the upland plateaus such as Lusaka, Zambia (Fig. 18.18), are less oppressive than the Aw climates in adjacent lowlands and depict this reversal of the wet and dry months from those Aw and Cw climates in the Northern Hemisphere. The Southern Hemisphere winter is extremely dry in these places, as the subtropical anticyclones sit overhead, with moisture stress evident in the water balance diagram. This moisture seasonality causes the natural landscape to be dominated by savanna vegetation that can go dormant during the dry season and spring back to life when the rains return.

East Africa

The East African landscape is arid. The region known as the Horn of Africa might be expected to be wet because of its tropical latitude and proximity to the Indian Ocean. However, this part of the continent is largely influenced by topography and local-scale effects of the atmosphere and ocean circulation to create a unique climate situation. Specifically, during the Northern Hemisphere summer, the orientation of the East African coastline causes the southwesterly counter-trades to flow parallel to the coast (Fig. 18.19). These southwesterly winds are reinforced by the inland pull of the *Tibetan Low* to the east—the cause of onshore monsoon in South Asia. This configuration and flow draws humid air away from East Africa.

The monsoon circulation stimulates the **Somali jet** (Fig. 18.20) also known as the **Findlater jet**, an important feature of the local-scale circulation of the region.

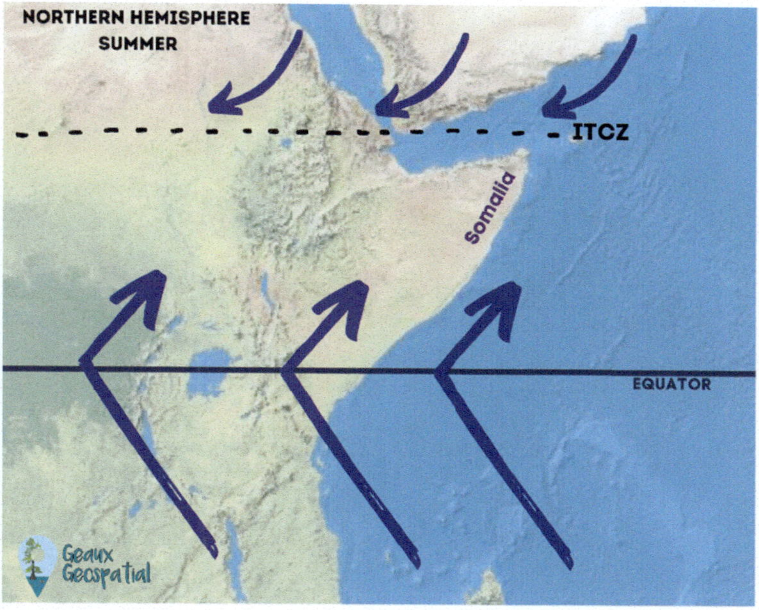

Fig. 18.19 Idealized Northern Hemisphere summer airflow over East Africa, with southeast trades south of the Equator, southwesterly counter-trades between the Equator and ITCZ, and northeast trades north of the ITCZ. *Source* Geaux Geospatial

Typically 1–1.5 km in altitude, this jet flows from the southwest, paralleling the Somali coast in association with the summer Asian monsoon system. The airflow also forces the Somali Current beneath the jet. As the Ekman spiral pushes surface water seaward toward the central ocean basin, upwelling associated with the Somali Current supports an extremely productive ecosystem.

During the Northern Hemisphere winter, the northeast trades again parallel the East African coast but in a different direction from summer. The airflow is initiated by the offshore Asian winter monsoon, with air gaining humidity over water and curving away from the coast as the northwesterly counter-trades after crossing the Equator (Fig. 18.21). At the same time, dry cT air masses from the Arabian Peninsula are advected into East Africa. This circulation results in a very dry landscape where precipitation is less than 3 cm (1 in) in both winter and summer. For example, approximately 5 cm (2 in) of precipitation falls in Berbera, Somalia, annually. Figure 18.22 shows East African locations for which climagraphs and water balance diagrams are shown.

The high temperatures, which average 35–36 °C (96–97 °F) during summer, create very high evapotranspiration rates, ensuring aridity. Farther south, Mogadishu's proximity to the cooler coastal waters keeps temperatures slightly lower, but meager rainfall totals, despite the overhead passage of the ITCZ twice per year, keep moisture stress consistent throughout the year (Fig. 18.23). Deserts extend to the Equator in East Africa as a result of the unique combination of airflow and coastal configuration.

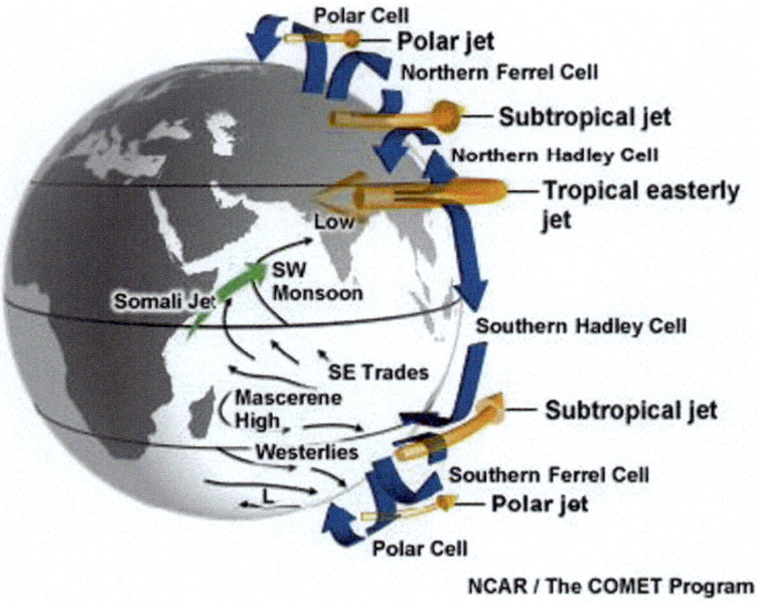

Fig. 18.20 The Somali jet (green arrow) and global scale circulation during Northern Hemisphere summer. *Source* UCAR CoMet. Permission from Tara Torres

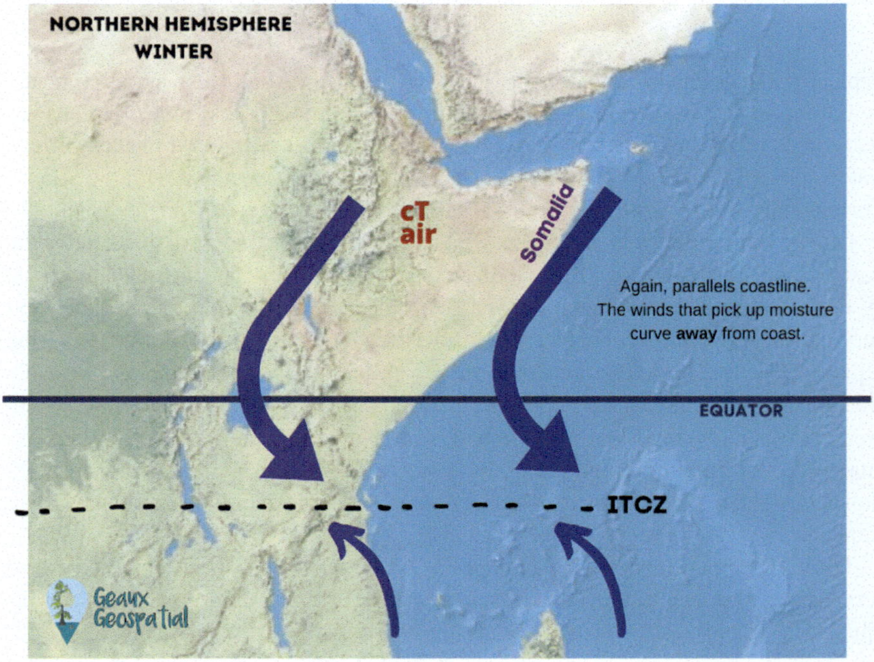

Fig. 18.21 Idealized depiction of northwesterly counter-trades and southeast trades over East Africa during winter. *Source* Geaux Geospatial

Fig. 18.22 East African cities for which climagraphs and water balance diagrams are shown. *Source* Geaux Geospatial

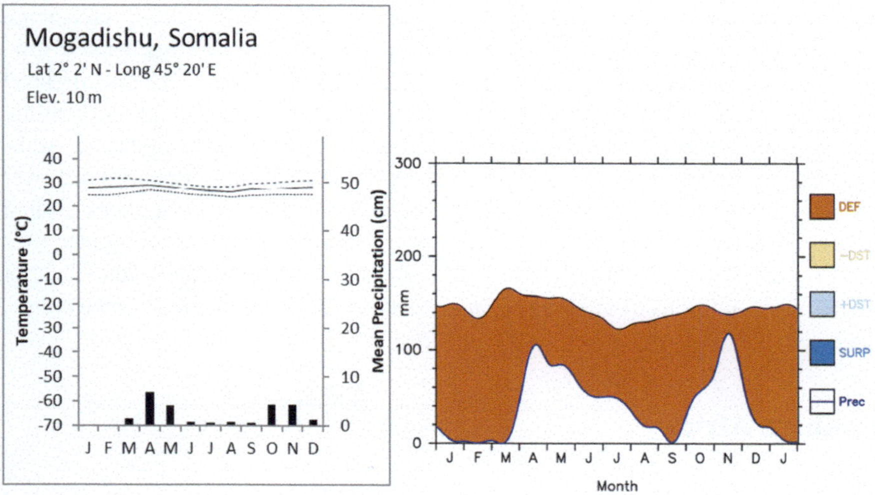

Fig. 18.23 Climagraph and water balance diagram for Mogadishu, Somalia. Data for 1981–2016 averaging period. *Source* Authors. https://weatherspark.com/, http://cyclops.deos.udel.edu/wimp/public_html/

The region is only slightly wetter near the Equator at Nairobi, Kenya (Fig. 18.24), but it becomes wetter farther inland and closer to eastern Kenya (Fig. 18.25) and especially the Ethiopian Highlands as in Addis Ababa, where high elevations reduce *potential evapotranspiration* as the climate is substantially cooler (Fig. 18.26).

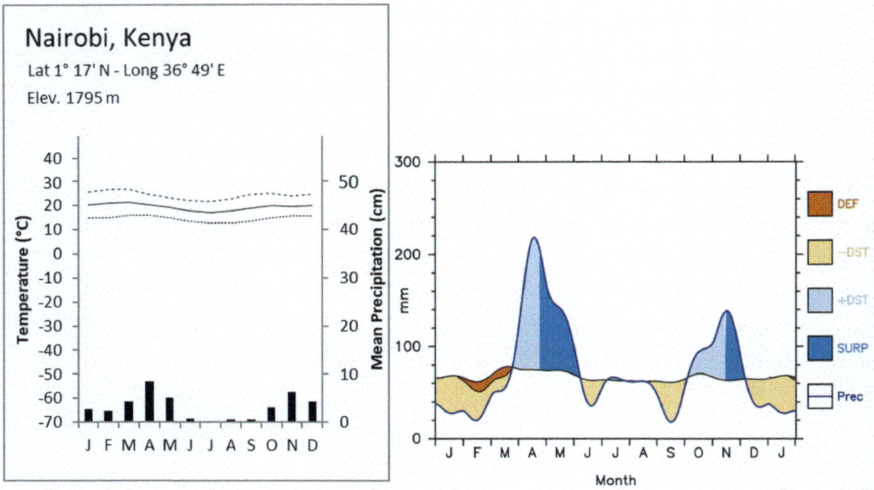

Fig. 18.24 Climagraph and water balance diagram for Nairobi, Kenya. Data for 1981–2016 averaging period. *Source* Authors. https://weatherspark.com/, http://cyclops.deos.udel.edu/wimp/public_html/

Atlantic water vapor can penetrate the region with *orographic lifting* enhancing precipitation. Wet and dry periods are also influenced by locally-generated low-level circulation induced by topography. Specifically, the **East African low-level jet** develops from an inland diurnal thermal low formed by abundant inland heating. The moisture is pushed from the coast into the interior, initiating precipitation. Additionally, the **Turkana jet** flows between the 950- and 650-mb pressure level over the Ethiopian Highlands and the Lake Turkana basin. The jet strengthens around February to March as airflow is constricted over the highlands. Channeling increases wind speeds, especially through the late night and early morning hours. This jet creates dry conditions by suppressing *mesoscale* cyclonic circulations in the area.

Southern Africa

Here the term "Southern Africa" is used to refer to the area south of the tropical climates of Africa, and it is distinguished from the Republic of South Africa, which is a nation in the extreme southern part of Africa. Southern Africa is influenced strongly by the subtropical anticyclones that are positioned in each of the ocean basins near 30° S. These pressure features are much stronger than their Northern Hemisphere counterparts. The lack of land in the hemisphere and consistency of ocean temperatures throughout the year allows uninterrupted development, strengthening, and persistence of gyres.

In extreme Southern Africa, the *mid-latitude westerlies* tend to be *zonal* throughout the Southern Hemisphere. This pattern again results from the lack of

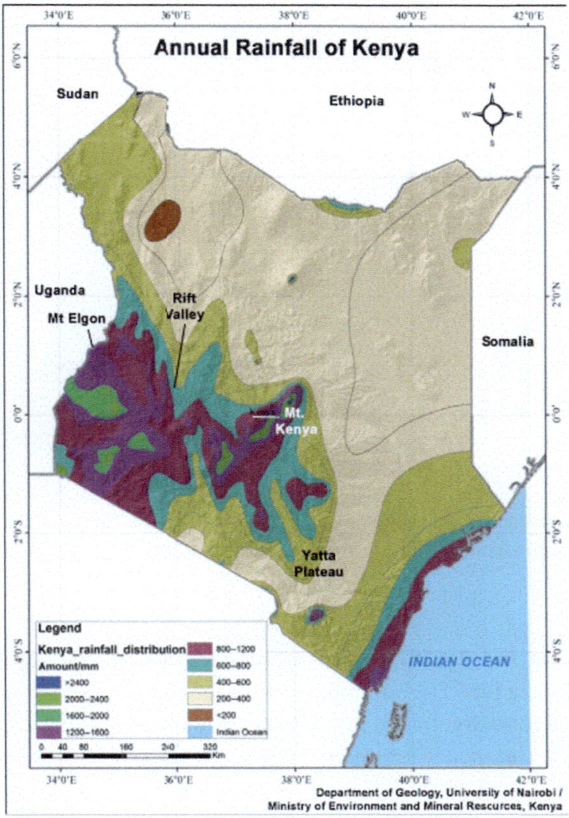

Fig. 18.25 Spatial pattern of precipitation totals in Kenya. *Source* UCAR MetEd. Permission from Tara Torres

significant land masses through the "water hemisphere." The extensive oceans create a situation of weak latitudinal thermal gradients. The smaller differences in temperatures between the high and low latitudes produce the predominantly zonal air flow.

Disturbances in Southern Africa originate from two main sources. First, the ITCZ is the site of humid air converging at lower latitudes and can initiate storms in this part of Africa during the Southern Hemisphere's summer. The ITCZ may extend as far south as Mozambique and Madagascar in summer. Second, mid-latitude wave cyclones and their associated cold fronts can force convection and precipitation at higher latitudes, primarily in winter. The *polar front* typically remains south of Africa, but trailing *cold fronts* may cross the region in the cooler months. In the Southern Hemisphere, the combined effect of the *pressure gradient force*, Coriolis effect, and *friction* causes air to circulate clockwise and inward around cyclones (Fig. 18.27). Therefore, mid-latitude wave wave cyclones appear reversed on weather maps from their Northern Hemisphere counterparts as cold air lies on the southern

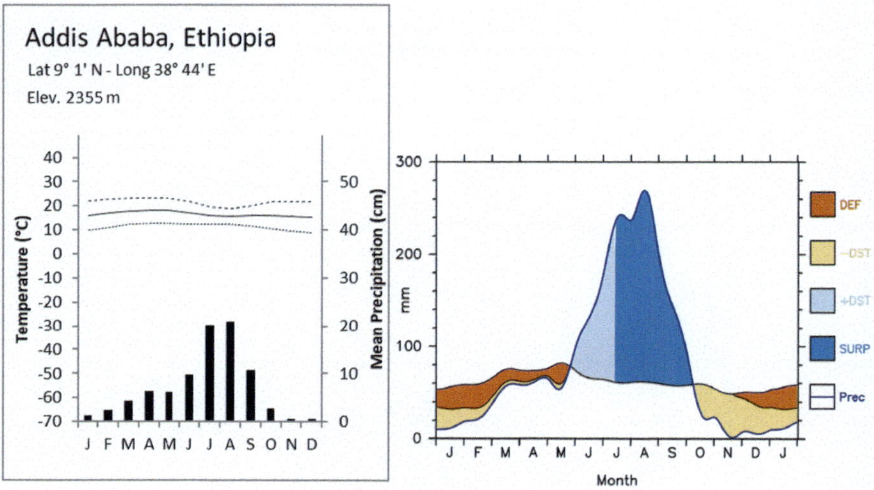

Fig. 18.26 Climagraph and water balance diagram for Addis Ababa, Ethiopia. Data for 1981–2016 averaging period. *Source* Authors. https://weatherspark.com/, http://cyclops.deos.udel.edu/wimp/public_html/

side of the systems (Fig. 18.28). The 20° S parallel of latitude roughly demarcates the boundary between the ITCZ-dominated wet summer climate to the north and the subtropical anticyclone-dominated dry summer climate to the south. Ocean currents exert significant importance in Southern African climates, both offshore and onshore (Fig. 18.29).

A north–south line bisecting Southern Africa separates the dry (on the west) and the wet (on the east) areas. The South Atlantic High produces the cold Benguela Current near the western coast, promoting low-level stability and little precipitation. BW climates extend from western parts of the Republic of South Africa westward, well out into the South Atlantic Ocean. The Namib Desert in the southwestern African nation of Namibia receives an average of less than 13 cm (5 in) of precipitation

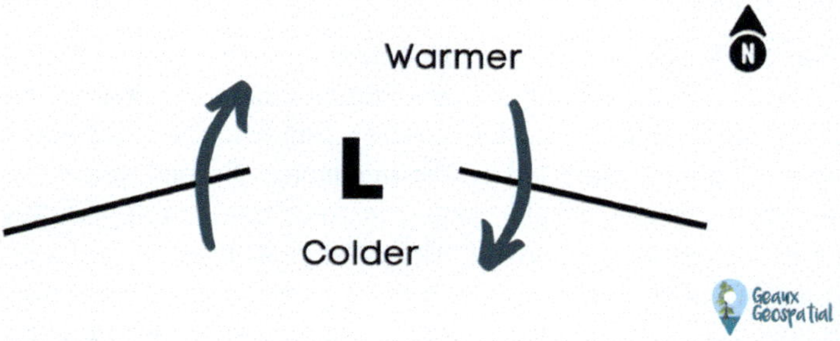

Fig. 18.27 Idealized depiction of airflow around a Southern Hemisphere cyclone. *Source* Geaux Geospatial

Fig. 18.28 Weather map depiction of an idealized Southern Hemisphere mid-latitude wave cyclone. *Source* Geaux Geospatial

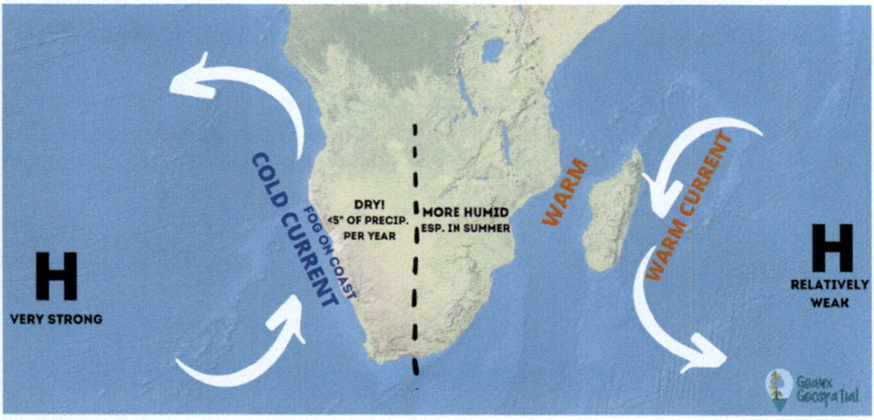

Fig. 18.29 Ocean currents off Southern Africa and associated climates. *Source* Geaux Geospatial

annually. However, significant fog occurs along the coast, allowing plants and animals to adapt by using fog droplets as their primary water source.

East of the north–south line, the South Indian Ocean High (or Mascarene High) produces the warm Agulhas Current near the eastern coast of Southern Africa. The tilt of the subtropical highs such that stronger subsidence exists to the surface on their eastern sides (i.e., western sides of continents) than on their western sides, as a result of a *subsidence inversion* (see again Fig. 15.3), means that southeastern Africa has more opportunities for precipitation. The warm current supports precipitation processes by contributing humidity and instability, leading to wetter conditions on the eastern coast than the western side of Southern Africa.

Atmospheric and hydrologic variability in Southern Africa is affected by periodic warming episodes of similar duration to ENSO in the Atlantic Ocean adjacent to southwestern Africa. These **Benguela Niño** events can cause floods in normally dry southwestern Africa, including the Namib Desert, and limit offshore nutrient availability, among other impacts. Periods of cooling waters known as **Benguela Niña** events are often associated with drought in these same areas.

Examples of the regional climate are shown for locations plotted in Fig. 18.30. The influence of the western edge of the South Indian Ocean High on nearby areas include Madagascar and eastern South Africa. Cities such as Pretoria, South Africa (Fig. 18.31), and Gaborone, Botswana (Fig. 18.32), experience a rainfall maximum in the Southern Hemisphere summer, when the South Indian Ocean High is displaced southward. Yet precipitation is insufficient to prevent deficits from occurring for much of the year, especially near Gaborone, even with some relief from reduced potential evapotranspiration in these cooler, high-elevation cities. Rainfall is especially heavy on the windward slopes of Madagascar where orographic lifting increases totals.

A small region of *Humid Subtropical* (Cfa) *climate* exists in southeastern Africa, where summer convection supplements occasional winter cold fronts. The southern tip of South Africa experiences a *Mediterranean–warm summer* (Csb) *climate*, with dry and warm summers and cooler and wetter winters. Cape Town, Republic of South Africa, has a climagraph and water balance diagram that shows these patterns (Fig. 18.33).

Africa's large latitudinal range from the deserts of the Sahara, to the areas alternately affected throughout the year by the subtropical highs and ITCZ, to the rain forests of the tropics, to the savannas of the Southern Hemisphere, to the subtropical

Fig. 18.30 Southern African cities for which climagraphs and water balance diagrams are shown. *Source* Geaux Geospatial

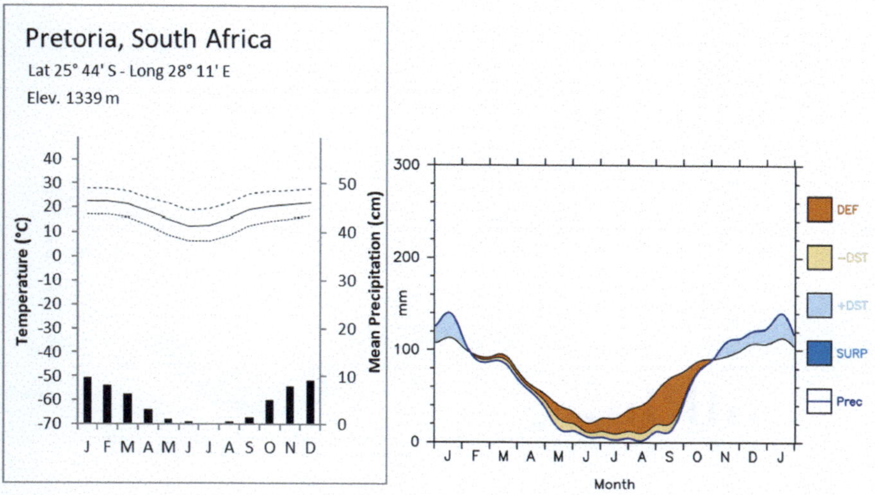

Fig. 18.31 Climagraph and water balance diagram for Pretoria, Republic of South Africa. Data for 1981–2016 averaging period. *Source* Authors. https://weatherspark.com/, http://cyclops.deos.udel.edu/wimp/public_html/

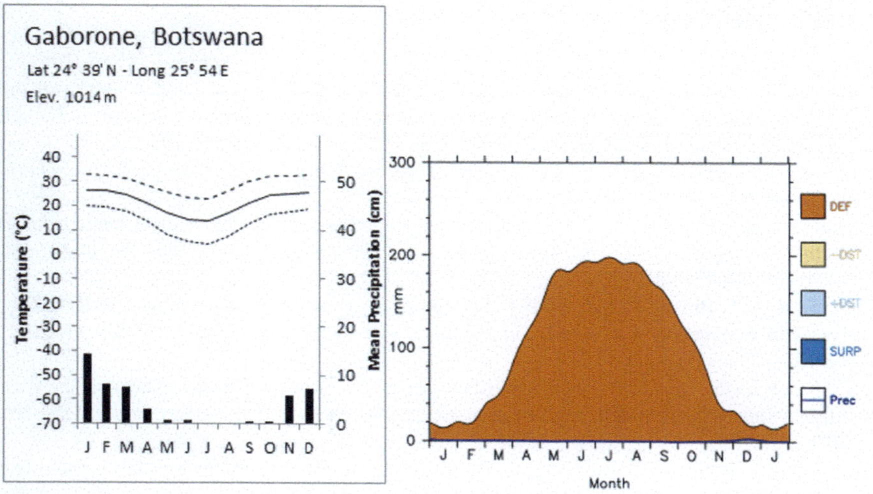

Fig. 18.32 Climagraph and water balance diagram for Gaborone, Botswana. Data for 1981–2010 averaging period. *Source* Authors. https://weatherspark.com/, http://cyclops.deos.udel.edu/wimp/public_html/

and mesothermal environments of South Africa, creates a wide variety of climate regions. The continent is influenced by both tropical and extratropical atmospheric and oceanic circulation. This diversity of influences is only matched by the world region that is the subject of the next chapter—Latin America and the Caribbean.

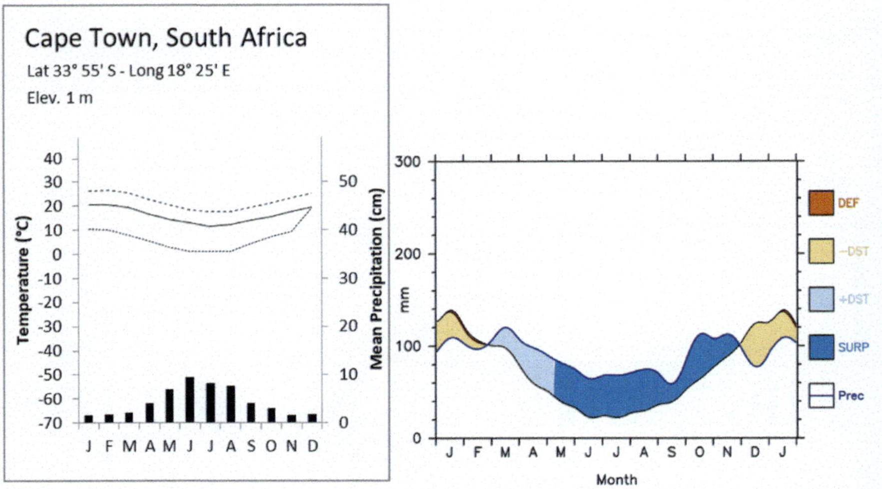

Fig. 18.33 Climagraph and water balance diagram for Cape Town, Republic of South Africa. Data for 1981–2016 averaging period. *Source* Authors. https://weatherspark.com/, http://cyclops.deos.udel.edu/wimp/public_html/

Further Reading

Camberlin, P., Janicot, S., & Poccard, I. (2001). Seasonality and atmospheric dynamics of the teleconnection between African rainfall and tropical sea-surface temperature: Atlantic vs. ENSO. *International Journal of Climatology, 21*(8), 973–1005. https://doi.org/10.1002/joc.673

Collier, A. B., & Hughes, A.R. (2011). Lightning and the African ITCZ. *Journal of Atmospheric and Solar-terrestrial Physics, 73*(16), 2392–2398.

Dunion, J. P., & Velden, C. S. (2004). The impact of the Saharan Air Layer on Atlantic tropical cyclone activity. *Bulletin of the American Meteorological Society, 85*(3), 353–366. https://doi.org/10.1175/BAMS-85-3-353

Koungue, R. A. I., Brandt, P., Lübbecke, J., Prigent, A., Martins, M. S., & Rodrigues, R. R. (2021). The 2019 Benguela Niño. *Frontiers in Marine Science, 8*. https://doi.org/10.3389/fmars.2021.800103

Lele, M. I., & Lamb, P. J. (2010). Variability of the intertropical front (ITF) and rainfall over the West African Sudan-Sahel zone. *Journal of Climate, 23*(14), 3984–4004. https://doi.org/10.1175/2010JCLI3277.1

Marchant, R., Mumbi, C., Behera, S., & Yamagata, T. (2007). The Indian Ocean Dipole—The unsung driver of climatic variability in East Africa. *African Journal of Ecology, 45*(1), 4–16. https://doi.org/10.1111/j.1365-2028.2006.00707.x

Munday, C., Washington, R., & Hart, N. (2021). African low-level jets and their importance for water vapor transport and rainfall. *Geophysical Research Letters, 48*(1). https://doi.org/10.1029/2020GL090999

Nicholson, S. E. (2018). The ITCZ and the seasonal cycle over equatorial Africa. *Bulletin of the American Meteorological Society, 99*(2), 337–348. https://doi.org/10.1175/BAMS-D-16-0287.1

Rohli, R. V., Ates, S. A., Rivera-Monroy, V. H., Polito, M. J., Midway, S. R., Castañeda-Moya, E., Gold, A. J., Uchida, E., Mangora, M. M., & Makoto, S. (2019). Inter-annual hydroclimatic variability in coastal Tanzania. *International Journal of Climatology, 39*(12), 4736–4750. https://doi.org/10.1002/joc.6103

Chapter 19
Latin America, the Caribbean, and Adjacent Oceans

Abstract The Latin America and Caribbean region spans a wide range of latitudes and climatic types. From north to south are the arid (B) climates of northern Mexico, tropical (A) climates dominant in the hurricane-prone Caribbean and much of southern Mexico and Central America and low-latitude South America, B climates in isolated areas, and mesothermal (C) climates in higher elevations and in southern South America. The Caribbean low-level jet affects weather and climate variability in the Caribbean and Central America, where it interacts with the topography to produce the veranillo. The Amazon rain forest overlaps with the inter-tropical convergence zone (ITCZ), the western side of which is held in place by the cold Peru Current. The dry season lengthens poleward, with Tropical Rain Forest (Af) climates grading into Tropical Monsoon (Am) and then Tropical Wet-dry (Aw) climates. Subsidence from the South Pacific High, coastal upwelling, and rain shadow effects of the Andes Mountains create the Atacama Desert. In mid-latitude South America, Semi-arid (BS) and mesothermal climates prevail, most commonly the Mediterranean (Cs) and Humid Subtropical (Cfa) climates. Microthermal (D) types are rare because of the lack of continentality, but South America's southern tip has a windy, Tundra (ET) climate.

Keywords Amazon rain forest · Andes Mountains · Atacama Desert · Atlantic Warm Pool · Brazil Current · Cabo (Cape) Verde tropical cyclones · Caribbean low-level jet · Easterly wave · El Niño-Southern Oscillation (ENSO) · Falkland Current · Loop Current · Peru (Humboldt) Current · Roaring Forties · Saharan air layer (SAL) · South Atlantic High · Veranillo

General

The region of Latin America and the Caribbean covers a wide range of latitudes and is physically diverse. The area includes long mountain cordilleras, lowlands, wetlands, high plateaus, peninsulas, the world's most prominent isthmus, and an array of volcanic islands. The region spans the hemispheres from the subtropics of

R. V. Rohli et al., *Atmospheric and Oceanic Circulation*,
https://doi.org/10.1007/978-3-031-66142-6_19

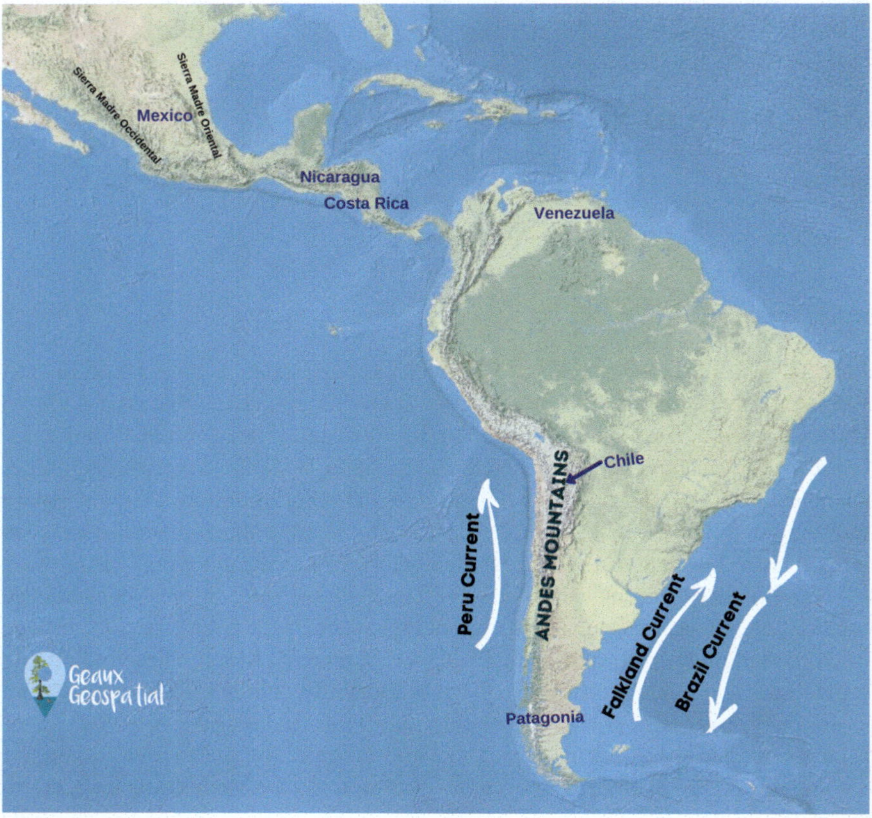

Fig. 19.1 Selected physical features and place names in Latin America, the Caribbean, and adjacent ocean region. *Source* Geaux Geospatial

the Northern Hemisphere nearly to 60 °S (Fig. 19.1). Climatologically, the region includes significant areas of aridity from the *rain shadow* of mountains, exceedingly wet areas enhanced by *orographic lifting*, influence of strong *gyres* in the adjacent oceans, and many microclimates. Because of its extensive coastlines and narrow landmasses, maritime influences and the interaction of atmospheric and oceanic circulations are especially important to the climate patterns in this region.

Topographic Influences (Site)

Tectonic activity influences the landscapes of Latin America and the Caribbean. Volcanism and uplift from converging tectonic plates have created a vast array of mountain chains, plateaus, and volcanoes throughout the mainland and volcanic islands extending out into the nearby Caribbean Sea.

The large interior Mexico plateau is bounded by two mountain ranges that align in a generally north–south direction. The Sierra Madre Occidental borders the plateau on the west and the Sierra Madre Oriental forms the eastern boundary. The Sierra Madre Occidental is an extension of the cordillera that extends from Canada, through the U.S., western Mexico, Central America, South America, and into Antarctica.

In southern Mexico and Central America, the convergence of these two mountain chains with great variety in peak elevations in the narrow land strip creates interesting gap-wind phenomena. For example, high pressure to the north of Central America can generate particularly strong easterly winds descending the slopes through the gaps that can affect the adjacent Pacific coast in the low-Sun season for up to a week. One such violent wind is the **tehuantepecer**, which occurs in southern Mexico across the Isthmus of Tehuantepec. A related wind is the **papagayo** of Central America, which generally produces fair weather on its descent. The tehuantepecers and papagayos, and the seasonal movement of the *inter-tropical convergence zone* (ITCZ) affect nearby ocean currents. Cyclonically-turning surface water between the *North Equatorial Current* and the *North Equatorial Counter-current* raises the level of the thermocline and creates the **Costa Rica Thermal Dome** offshore west of Central America. This thermal dome varies in size and shifts slightly in location, as the colder waters are drawn upward in a dome-like shape. The Costa Rica Thermal Dome, and other similar thermal domes in other ocean basins, such as the Seychelles Dome, Angola Dome, Dakar Dome, Mindanao Dome, and Sri Lanka Dome, is a nutrient-rich, biologically diverse zone supporting marine life like turtles, tuna, dolphins, and blue whales that must be protected.

The most prominent part of the cordillera of the Americas—the Andes Mountains—parallel the western coast of South America. The chain is the longest continuous continental mountain chain on Earth, exceeded only by the mid-Atlantic ridge that occupies the central basins of the North and South Atlantic Oceans. With an average height of over 4000 m (13,000 ft), the Andes Mountains affect the broad-scale atmospheric circulation and climates of South America.

East of the Andes Mountains lie the Brazilian Highlands (Fig. 19.2). The Highlands comprise approximately one-half of the country of Brazil. They were created by massive but extinct lava flows of basaltic rock. A substantial escarpment exists near the coast which consists of fluvial deposits of eroded sediments from the Highlands.

Lowland areas exist largely in narrow coastal strips throughout much of South America, similar to Africa. The abrupt elevation transition between the interior highlands and the coast creates non-navigable rivers with rapids and waterfalls, with some notable exceptions, such as the Rio de la Plata and its tributaries in Argentina, Uruguay, Brazil, Paraguay, and Bolivia. Where escarpments exist, river trade is difficult as the movement of goods into the continental interior is hindered. As in Africa, an even greater proportion of development occurred near the coasts than for most other continents. Coastal megacities contain a sizable percentage of the national populations in South America, as in Africa. Also similar to Africa, Interior regions of the continent remain less developed and sparsely populated.

The exception is the large river basin dominated by the Amazon, which extends far inland with headwaters along the eastern axis of the Andes Mountains. The basin

Fig. 19.2 Physiographic regions of Brazil. *Source* Worldmaps. https://www.freeworldmaps.net/southamerica/brazil/map.html

crosses a tropical rain forest environment, where abundant runoff from water balance *surpluses* is drained into the Amazon for nearly its entire west-to-east-length to the Atlantic Ocean.

Two other notable areas of flat grassland environments that have become important agricultural and ranching regions affect and are affected by the atmospheric circulation. The "Pampas" describes an area of coastal subtropical Argentina extending inland to cover all of Uruguay and into southernmost Brazil (Fig. 19.3a). Physically, the region is similar to the U.S. Great Plains. Like the Pampas, the "Llanos" is an area of plains in Colombia and Venezuela (Fig. 19.3b).

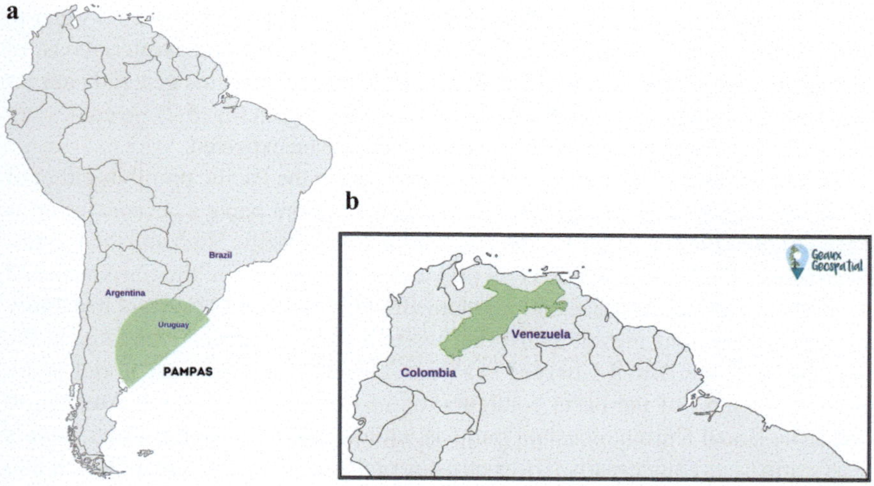

Fig. 19.3 **a** The Pampas of Argentina; and **b** the Llanos of Colombia and Venezuela. *Source* Geaux Geospatial

Circulation Influences (Situation)

As with the other tropical continents covered in the previous chapters, Latin America's atmospheric circulation is strongly influenced by the *subtropical anti-cyclones* and the ITCZ. The subtropical anticyclone known as the *South Pacific High* is extremely stable and consistent in size and strength. The cell is more permanent than the quasi-permanent high-pressure cells of other ocean basins. This consistency is caused by the uninterrupted extent of the South Pacific Ocean and the predominantly cold surface water near South America advected around the **South Pacific Gyre**. The counter-clockwise and outward flow of the South Pacific Gyre draws surface water from high latitudes near Antarctica along the western coast of South America (see again Fig. 5.2). Low water *temperatures* are reinforced by persistent cold water upwelling from the deep ocean. These factors combine to form the *Humboldt Current*—also known as the *Peru Current*—the coldest surface current on Earth. This strong current stabilizes the overlying atmosphere along nearly all of the South American western coastline. The West Coast of South America, particularly coastal Chile and Peru, is incredibly dry from the persistent high-pressure-induced *subsidence* reinforced by the cold Peru Current. The Atacama Desert, the driest desert in the world, is a direct result of these ocean–atmosphere interactions.

The other subtropical anticyclones that affect South America—the *Hawaiian-Pacific High*, the *Bermuda-Azores High*, and *South Atlantic High*—are more variable in their influence as compared to their South Pacific counterpart. These three highs act similarly to other ocean basin *anticyclones* in their summer expansion and winter contraction.

The *North Pacific Gyre* around the Hawaiian-Pacific High provides cold water to Latin America, via the remnants of the *California Current* to the western coast of Central America. However, the shape of Mexico's Pacific coastline generally curves away from the influence of the North Pacific Gyre (see again Fig. 5.2), so cold water affects Central America far less than might otherwise be expected.

The smaller area of the Atlantic Ocean compared to the Pacific prohibits either of the Atlantic subtropical anticyclones and their gyres from being as expansive or as strong as either the Hawaiian-Pacific High or the South Pacific High and their gyres. Nevertheless, the North Atlantic Gyre provides warm currents to the Caribbean and the coasts of Central America and northern Brazil. Associated low-level instability and humid conditions prevail as in other western gyre-influenced locations. In the South Atlantic, the **Brazil Current** is a warm current that affects equatorial South America. Because of the basin configuration and proximity of South America to Africa, the Brazil Current does not pull vast uninterrupted equatorial water, so it is not as warm as its counterparts. The protrusion of the easternmost point of Brazil into the Atlantic directs some tropical water into the Northern Hemisphere. Nevertheless, its *advection* of warmer waters poleward and cooler waters equatorward remains an important component of the global *thermohaline circulation.*

Like all tropical locations, the ITCZ is also an important climatic feature in Latin America. As in West Africa, the ITCZ remains in the Northern Hemisphere in Latin America and is especially fixed in place along the western coast (see again Fig. 3.12). The ITCZ is held in place because the extremely cold Humboldt Current and associated cold water upwelling stabilizes the atmosphere such that *convection* and *convergence* into the low-pressure core of the ITCZ are restricted to the air over the warmer waters in the Northern Hemisphere. The resulting northward "kink" in the ITCZ over western South America is similar in appearance to that over Africa. Thus, the Humboldt Current has a similar impact on western South America as the *Benguela Current* has on West Africa, but the northward displacement over South America is even more pronounced as the water surface remains cooler throughout the year than the Benguela Current.

Beyond the tropics, *mid-latitude westerlies* affect South America's southernmost areas, known as the Southern Cone or Patagonia. Upper-atmospheric airflow is mainly *zonal* in the Southern Hemisphere because of the lack of significant land to disrupt the westerly flow pattern, and the surface marine *isotherms* remain oriented latitudinally. However, an exception to the low-variability zonal flow regime involves a persistent *ridge* that forms over the southern Andes Mountains, with adjacent upwind and downwind *troughs*, similar to the airflow over the Rockies in North America.

In the Southern Hemisphere, ridges in the *Rossby waves* embedded in the mid-latitude westerlies resemble Northern Hemisphere troughs, while the troughs appear as Northern Hemisphere ridges, giving Northern Hemisphere dwellers the propensity to refer to Southern Hemisphere ridges and troughs as "inverted." Regardless of hemisphere, however, ridges always buckle toward higher latitudes and troughs dip toward lower latitudes. Thus, the Southern Hemisphere ridges appear as southward (toward the pole) dips in the westerly flow and Southern Hemisphere troughs appear as northward (toward the Equator) buckles in the flow.

Similar to the Northern Hemisphere situation, at a fixed pressure gradient Southern Hemisphere air flows more rapidly around ridges (*supergeostrophic* flow) and more slowly around troughs (*subgeostrophic* flow). This supergeostrophic or subgeogrophic flow occurs because air flowing around the Southern Hemisphere ridge must turn to the left to remain parallel to the isohypses, which is the direction that the *Coriolis effect* steers it, while air must turn to the right, against the Coriolis effect, as it flows around the Southern Hemisphere trough. The speed increase around the ridge is accompanied by increased Coriolis effect, as the two are directly proportional, which allows the enhanced leftward flow. Likewise, the speed decrease around the trough is associated with a weakened Coriolis effect, as the flow counteracts the Coriolis effect by turning to the right.

Thus, as in the Northern Hemisphere, the ridge-to-trough side of the Rossby wave in the Southern Hemisphere experiences air slowing and converging, which supports descending air motions and high surface air pressures. Fair weather is associated with the sinking air as in the Northern Hemisphere. Thus, the ridge-to-trough side of the Rossby wave in the Southern Hemisphere typically has fair weather beneath it.

By contrast, the trough-to-ridge side of the Rossby wave in the Southern Hemisphere has air accelerating, and therefore diverging, as in the Northern Hemisphere. This zone supports rising motions as lower atmospheric mass rushes to fill the vacuum created by this upper-level divergence. Upper-level venting supports the development of surface low-pressure systems and associated cloud and *precipitation* formation. Thus, preferred cyclogenesis areas in the Southern Hemisphere mid-latitudes are on the trough-to-ridge side of the Rossby wave, as in the Northern Hemisphere. One common example occurs in southern Patagonia, downstream from the ridge anchored in place by the Andes Mountains.

The primary air masses important to Latin America and the Caribbean include *maritime tropical* (mT) *air masses* from the Gulf of Mexico and the Southern Hemisphere sections of the Atlantic and Pacific Oceans (Fig. 19.4). In northern Mexico, *continental tropical* (cT) *air masses* exert a strong influence on associated areas, especially during summer. In equatorial areas, *maritime equatorial* (mE) *air masses* from the low-latitude oceans extend across the region. *Maritime polar* (mP) *air masses* from the Southern Ocean exert a strong influence at higher latitudes of the Southern Hemisphere. *Continental polar* (cP) *air masses* are only a minor factor in Latin America and the Caribbean because of the low latitudes in the Northern Hemisphere and the lack of a large, high-latitude land mass in the Southern Hemisphere to provide enough *continentality* to produce cold, dry air.

Mexico and Central America

Similar to several other subtropical regions, the area of Mexico near 30° N is marked by dry, subsidence-induced climates (Fig. 19.5). Northwestern Mexico and the marine environment west of Baja California are very arid as a result of extensive subsidence from the eastern side of the Hawaiian-Pacific High. The cold California

Fig. 19.4 Air masses affecting Latin America, the Caribbean and adjacent oceans. *Source* UCAR CoMet. Permission from Tara Torres

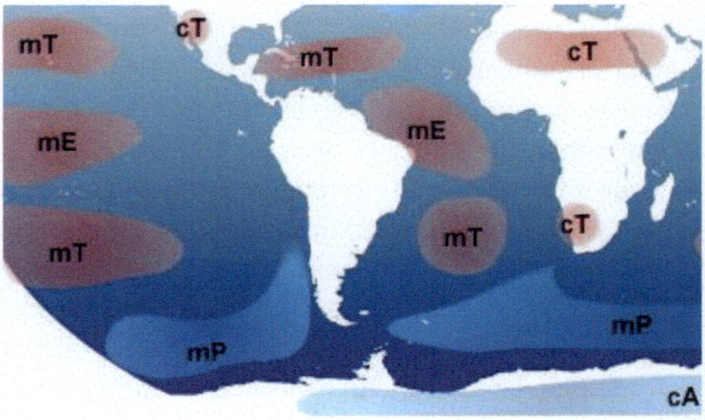

Current off the Pacific coast of Mexico reinforces low-level atmospheric stability which further restricts convective activity. The aridity of this *True Desert–hot* (BWh) *climate* area is reinforced by the *double rain shadow* effect which blocks water vapor originating from either the Gulf of Mexico to the east or the Pacific Ocean to the west from entering. The arid conditions persist westward well into the Pacific Ocean, with the areas in the path of the California Current in the *True Desert–cold* (BWk) *climate*, those farther southward in the *True Desert–hot* (BWh) *climate*, and the *Semi-arid* (BS) *climate* surrounding the desert core (see again Fig. 19.5). Northeastern Mexico is somewhat more humid as a result of direct moisture advection from the Gulf of Mexico through the lower Sierra Madre Oriental, resulting in a *Semi-arid—hot* (BSh) *climate*. Higher elevations may have the *Semi-arid—cold* (BSk) *climate*.

Figure 19.6 displays the locations in Mexico and Central America for which climagraphs and water balance diagrams are shown. The Sonoran Desert of northwestern Mexico and the southwestern U.S. is represented by the BWh climate of Mexicali (Fig. 19.7), with *deficits* for each month of the year. The somewhat less arid fringe of the Chihuahuan Desert of the southwestern U.S. and northeastern Mexico is depicted

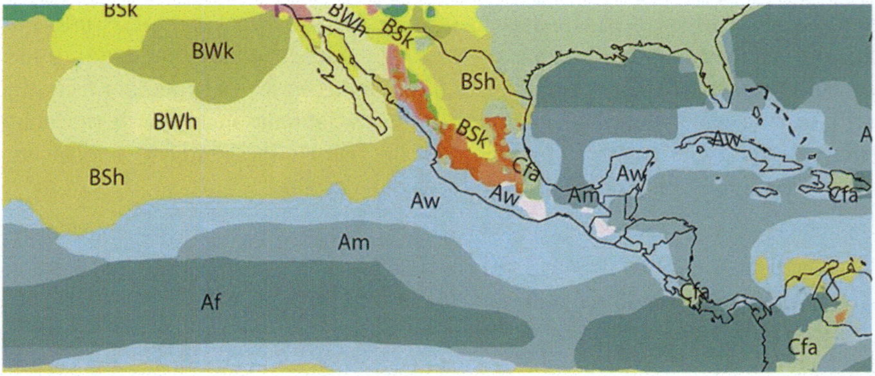

Fig. 19.5 Köppen climatic types around Central America and the Caribbean and surrounding marine environments. *Source* Modified from Reynolds, Rohli, Johnson, Waylen, and Francek, *Exploring Physical Geography*, McGraw-Hill

by Monterrey (Fig. 19.8), where precipitation from occasional *tropical cyclones* in autumn mitigates the deficits and briefly recharges the soil with water.

South of 30° N, Mexico is influenced by the ITCZ in summer. Much of southern Mexico, including the Yucatán Peninsula, is classified as a *Tropical Wet-dry* (Aw)

Fig. 19.6 Mexican and Central American cities for which climagraphs and water balance diagrams are shown. *Source* Geaux Geospatial

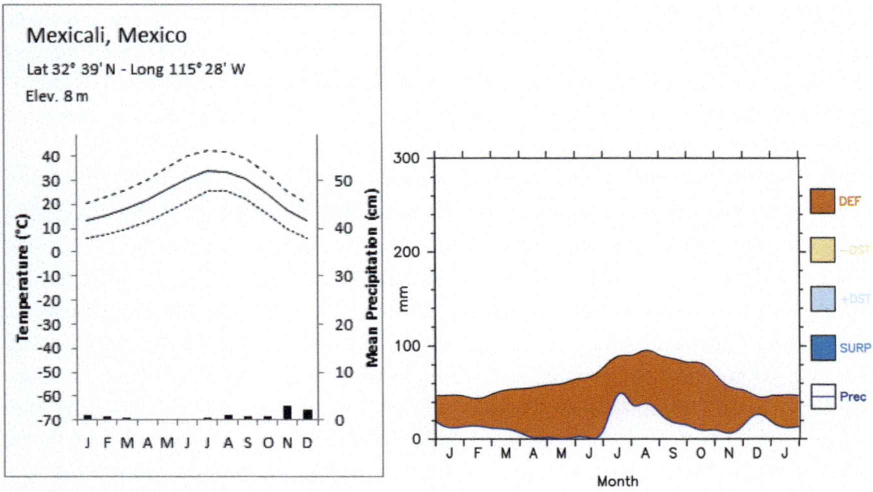

Fig. 19.7 Climagraph and water balance diagram for Mexicali, Mexico. Data for 1951–2010 averaging period. *Source* Authors. https://weatherspark.com/, http://cyclops.deos.udel.edu/wimp/public_html/

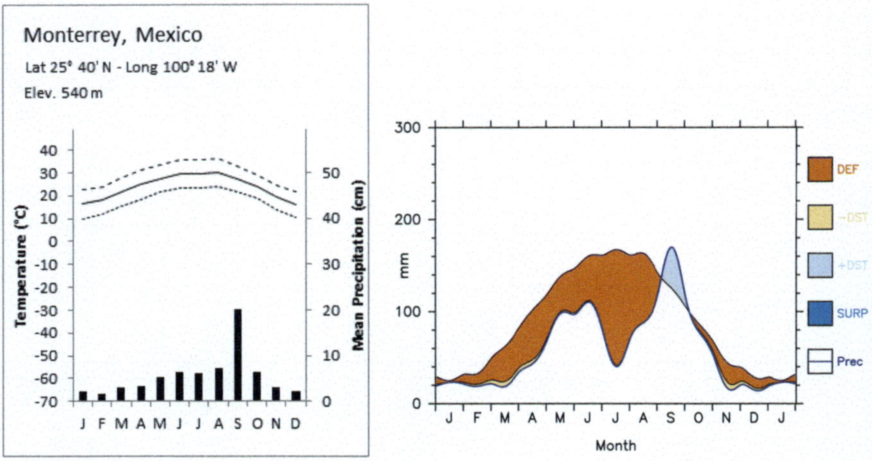

Fig. 19.8 Climagraph and water balance diagram for Monterrey, Mexico. Data for 1991–2020 averaging period. *Source* Authors. https://weatherspark.com/, http://cyclops.deos.udel.edu/wimp/public_html/

climate, also known as *Tropical Savanna climate*, with a wet, ITCZ-influenced summer and the dry winter when the ITCZ migrates southward. The higher elevations of the Sierra Madres have the same seasonal pattern of precipitation, but temperatures drop to an extent that *Mesothermal Winter Dry–hot summer* (Cwa) and *Mesothermal Winter Dry–warm summer* (Cwb) *climates* replace the Aw climates. These Cw climates are too localized to be labeled but are colored red and orange in Fig. 19.5, but the climagraph and water balance for Mexico City serve as an example (Fig. 19.9). Notice the increased summer water surplus because of the elevation and impact of the ITCZ and *northeast trades*. In some versions of the *Köppen climate types*, high-elevation locations such as Mexico City are classified into a "Highland" category to avoid the spatial complexities of locally differing climates by elevation and *slope aspect* and angle changes.

South of Mexico, the Central American nations of Belize, Guatemala, Honduras, El Salvador, Nicaragua, Costa Rica, and Panama, from north to south, have a particularly complicated topography and influence on the interplay between the air and ocean circulation. The result is an intricate collection of microclimates that is only partially apparent at the scale shown in Fig. 19.5 and that supports remarkably diverse ecosystems. A generalization is that three basic climate zones run through Central America. Along the eastern coast, the region remains wet all year, forming the *Tropical Rain Forest* (Af) *climate*. During summer, precipitation is produced by free convection and convergence associated with the ITCZ. During winter, the northeast trades drive moisture into the area from the Gulf of Mexico and the Caribbean Sea. High mountain ranges provide orographic uplift which enhances precipitation totals. Autumn precipitation originates largely from tropical disturbances over the ocean

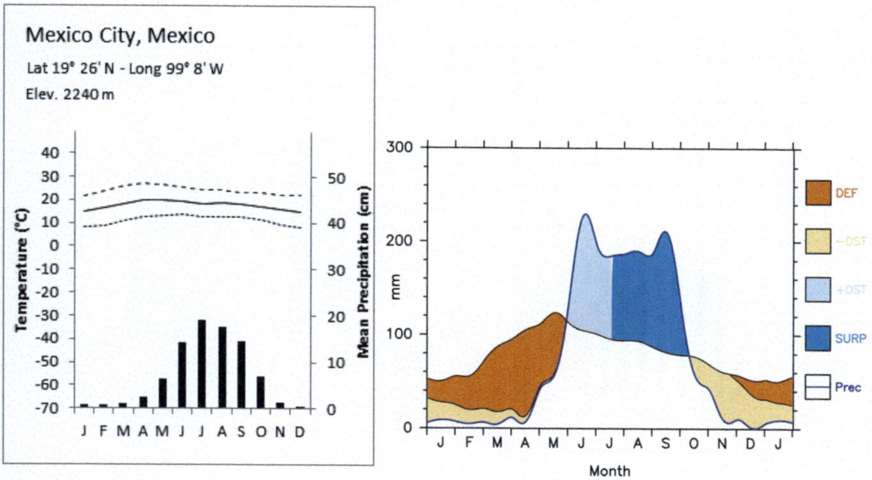

Fig. 19.9 Climagraph and water balance diagram for Mexico City, Mexico. Data for 1981–2010 averaging period. *Source* Authors. https://weatherspark.com/, http://cyclops.deos.udel.edu/wimp/public_html/

areas. These systems include intense *tropical storms* and *hurricanes*, but the majority are weaker disturbances such as *easterly waves* and *tropical depressions*.

The second zone includes the Aw climates of the Pacific coast, where wet summers and dry winters are common. During summer, afternoon precipitation is driven directly by the near-overhead ITCZ and, to some extent, by the *southwesterly counter-trades* which provide moisture advection from the Pacific Ocean. During winter, the ITCZ is replaced by the northeast trades, which advect drier air from the more arid interior. The Sierra Madre Occidental forms a rain shadow on its southwestern side, preventing humid Gulf of Mexico air from being carried to the Pacific coast via the northeast winds.

The third zone comprises highlands of the interior of this narrow land mass. This zone is much cooler than the latitude would suggest, as elevation provides relief from the high-Sun influence. Small areas of Cw and *Humid Subtropical* (Cf) *climates* are found through the central mountains. As in the two adjacent zones of Central America, precipitation peaks during summer, largely because of ITCZ proximity. A winter or spring minimum occurs as forcing mechanisms and water vapor diminish. Temperatures remain relatively consistent throughout the year but decrease at higher elevations where many of the larger cities are located. An example of a *Tropical Monsoon* (Am) *climate* in this region is San Jose, Costa Rica (1170 m (3800 ft) in elevation), which records monthly average temperatures between 22 and 24 °C (71 and 75 °F), as shown in Fig. 19.10. The small *annual temperature range* is explained by the tropical location and similar numbers of daylight hours throughout the year. Precipitation ranges from 36 cm (14 in) in September to 0.6 cm (0.2 in) in January, denoting the strong ITCZ influence and migration. In summer, surpluses are large

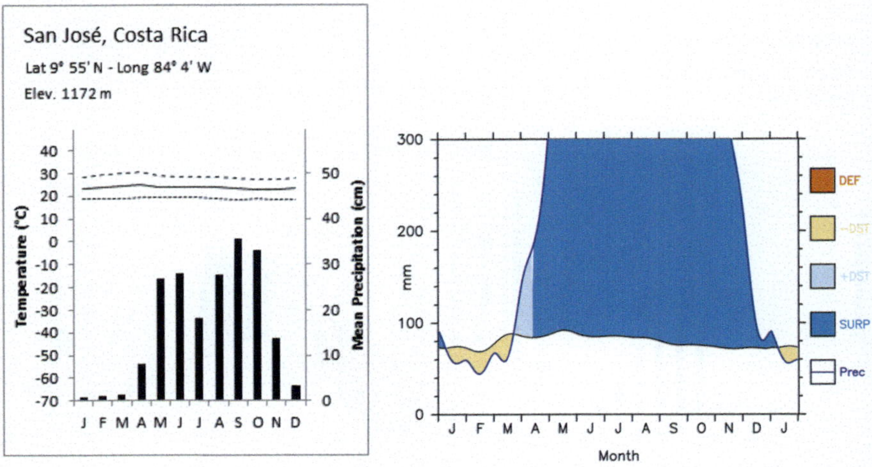

Fig. 19.10 Climagraph and water balance diagram for San Jose, Costa Rica. Data for 1961–1990 averaging period. *Source* Authors. https://weatherspark.com/, http://cyclops.deos.udel.edu/wimp/public_html/

and winter has mild decreases in water storage, but not enough to displace the tropical rain forest vegetation.

Caribbean Islands

At similar latitudes to northern Central America, precipitation in the Caribbean islands is also largely influenced by the ITCZ and the trade wind flow. Precipitation tends to peak in summer and reach a minimum in winter. Coastal and marine locations tend to have a delayed onset of the warmer and cooler, as well as the wetter and drier, seasons, as represented by the cities in the location map (Fig. 19.11) and by the climagraph for Castries, St. Lucia (Fig. 19.12).

As in Central America, local topography in the form of mountainous interiors of many of the Caribbean islands (Fig. 19.13) creates a complex array of microclimates. Depending on the location, the winter may be dry while other locations simply have a reduction in winter precipitation. The orographic effect and the northeast trades often make the northeastern sides of the islands much windier and wetter than the southwestern sides of the islands. The persistent breezes make these locations ideal for wind energy farms. Precipitation seasonality across rugged islands is much greater than across those with flatter terrain.

Water supply is a critical issue in the Caribbean. The high rates of *evapotranspiration* and the seasonality of rainfall lead to significant dry-season moisture stress. San Juan, Puerto Rico, is a representative example (Fig. 19.14). The deficits can be even larger on the flatter islands, forcing residents to turn to limited groundwater sources

Fig. 19.11 Caribbean cities for which climagraphs and water balance diagrams are shown. *Source* Geaux Geospatial

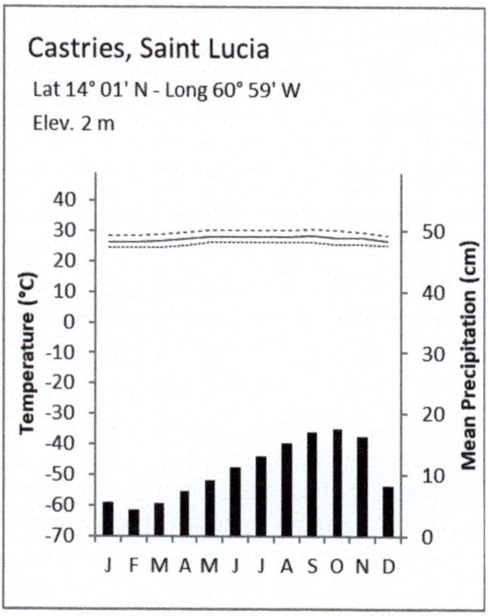

Fig. 19.12 Climagraph for Castries, St. Lucia. Data for 1981–2016 averaging period. *Source* Authors. https://weatherspark.com/

Fig. 19.13 The Pitons (Gros and Petit) and the city of Soufriere, St. Lucia. *Source* Authors

or desalination plants to meet demand. Even on the mountainous islands, steep topography increases the rate of runoff through relatively small watersheds on the windward slopes, while the southern and western sides remain in a rain shadow. The water shortages have become critical in places like St. Thomas and St. John in the Virgin Islands. During the wet season, increased runoff leads to frequent mudslides, especially in heavily deforested regions. Such a situation has been especially problematic in Haiti.

The heaviest Caribbean island rainfall and flooding is associated with tropical cyclones. On average, ten named tropical storms and six hurricanes form each year from the easterly waves that cross the low-latitude North Atlantic. The season extends from June 1 to November 30, with highest frequencies occurring in August and September. The normal parabolic track of such systems, which skirt the southern edge of the *Bermuda-Azores High* as they move westward and/or northward, often places the storms over the Caribbean Islands. Often the strongest hurricanes are *Cabo (Cape) Verde storms* that form just off the West African coast over a relatively weak *Canary Current*.

Several factors in addition to the conditions near the West African coast also affect the likelihood of tropical cyclones in the Caribbean, Mexico, Central America, and the southeastern United States. A prominent cause of Atlantic tropical cyclone variability is *El Niño-Southern Oscillation* (ENSO), as the *El Niño*, or warm, phase causes upper-air circulation shifts that tend to shear the tops off of developing tropical systems in this part of the world. However, tropical cyclones do still occur and may even be major during El Niño. Other teleconnections and regional circulation systems described later in this chapter also affect Atlantic tropical cyclone frequency and

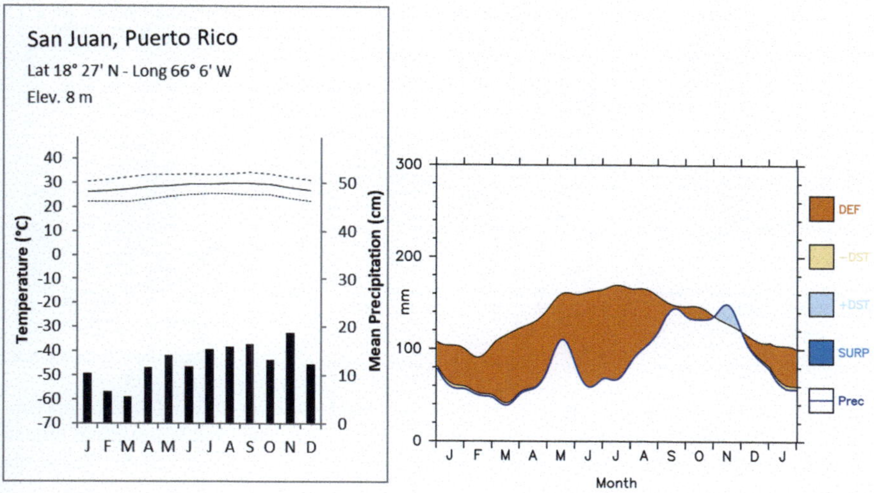

Fig. 19.14 Climagraph and water balance diagram for San Juan, Puerto Rico. Data for 1991–2020 averaging period. *Source* Authors. https://weatherspark.com/, http://cyclops.deos.udel.edu/wimp/public_html/

intensity. And in recent years, the importance of the *Saharan air layer* (SAL) as a deterrent to Atlantic tropical cyclone development has been noted.

The Caribbean Low-Level Jet

A particularly important regional circulation resulting from atmospheric and ocean circulation in the region is the **Caribbean low-level jet** (CLLJ). The CLLJ is represented as an intensification in the northeast trades in the western Caribbean basin from about 70–80° W, at about 15° N (Fig. 19.15). This intensification is associated with the clockwise, outward flow from a stronger than normal subtropical Bermuda-Azores High.

The CLLJ strength is also related to strong upper-level *pressure gradient force* over the strong sea surface temperature gradients between the eastern equatorial Pacific, where water temperatures are cool, and the tropical **Atlantic Warm Pool**. The warmer air over the Atlantic Warm pool expands upward to a much greater extent than the air over the much cooler equatorial Pacific Ocean. This "slanted" atmospheric mass, with more mass and pressure over the Atlantic than over the Pacific at a given height, causes the air to flow toward the lower *geopotential heights*. The isohypses at a given pressure level are higher over the Caribbean than over the equatorial Pacific. The relatively weak Coriolis effect in such low latitudes allows the flow to move without being deflected significantly.

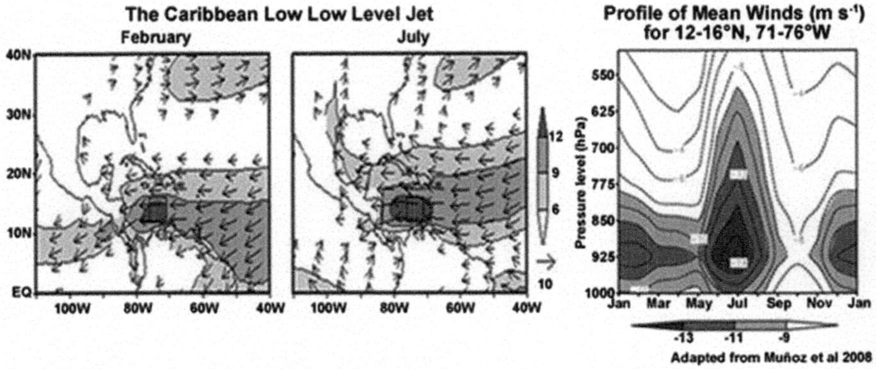

Fig. 19.15 Mean February and July position of the Caribbean low-level jet; wind vectors superimposed on shading of wind speed isolines, with the jet streak near Panama and fastest winds around 925 mb (equivalent to hPa). *Source* UCAR CoMet; adapted from Muñoz et al. (2008). We cite the Muñoz paper in this chapter reference list

The *Loop Current* can also supply energy to the CLLJ (Fig. 19.16). This warm current flows northward into the Gulf of Mexico between the Yucatán Peninsula and Cuba. It is an extension of the warm, low-latitude North Equatorial Current steered into the region by the Bermuda-Azores High. Waters warm substantially as they are funneled through the Caribbean Sea into the relatively shallow Gulf of Mexico. The waters then flow back equatorward, through the Florida Straits—between Cuba and the Florida Keys—and up the Atlantic seaboard of Florida as the *Gulf Stream*. The Loop Current provides energy and moisture to the regional atmosphere over eastern Gulf of Mexico, Cuba, and the Bahamas.

The CLLJ is usually found below the mid-*troposphere* (i.e., about 4–5 km or 2.5–3 mi) and often tracks over the areas of greatest **equivalent potential temperature**, which indexes the *enthalpy* available for weather disturbance. An elongated area of high equivalent potential temperature is known as a **theta-e ridge** (Fig. 19.17). The theta-e ridges are locations where abundant combined energy occurs from both *sensible heat* from the surface and *latent heat* released as water vapor condenses.

Maximum winds are about 16 m s^{-1} (36 mph) near the surface and occur from June to August. The CLLJ causes a precipitation maximum along coastal Central America during the June–August period, as the air over the Atlantic Warm Pool advects moisture into the zone of *unstable atmosphere*. The acceleration of the CLLJ triggers orographic precipitation on the eastern side of the Central American mountain spine. At the same time, the high-velocity winds dry out the Caribbean basin, ironically, by steering storms through the area faster than usual.

If the fast CLLJ winds ascend the eastern slopes of Central America with enough force, they can displace the southwesterly counter-trades westward while forcing surface waters away from the Pacific coast of Central America. This situation may also occur if the CLLJ squeezes through the mountain pass between Costa Rica and Nicaragua. The result is enhanced upwelling of cold, deep water along the Pacific

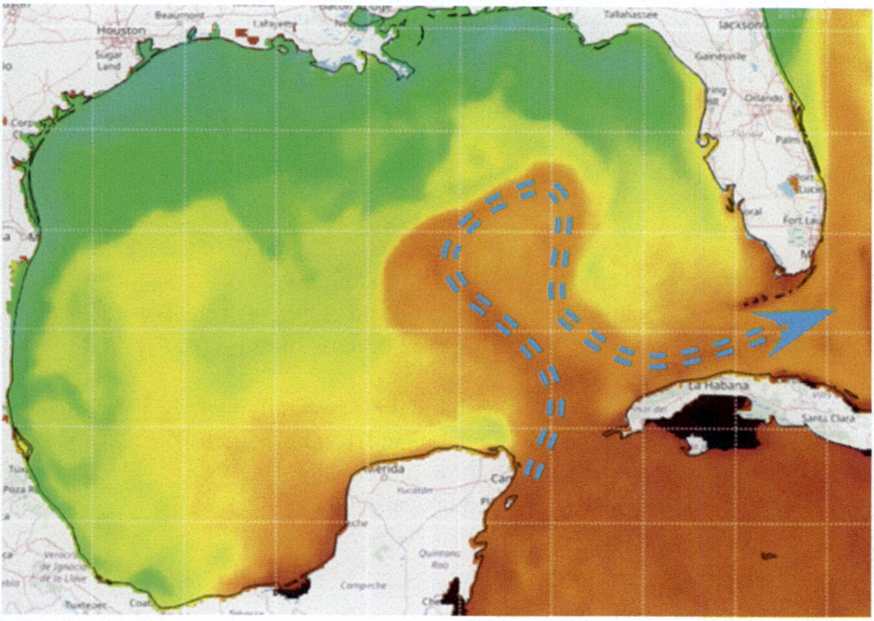

Fig. 19.16 Idealized depiction of the Gulf of Mexico Loop Current flow pattern with temperatures (hotter in orange). *Source* NOAA. www.aoml.noaa.gov/phod/gom/

coast of Central America. This pattern stabilizes the atmosphere, reducing cloud development and precipitation. The condition may even occur amid the wetter time of year. The dry spell that occurs is known as the **veranillo** ("little summer").

If the CLLJ is anomalously weak during the Atlantic hurricane season, tropical cyclone activity is supported because the lack of upper-level shear allows tropical cyclone cloudiness to remain stacked vertically. This vertical development allows the circulation inside the storm to remain organized and, if other conditions are favorable, to strengthen. The CLLJ is also believed to enhance disturbances related to interactions between the land-sea boundary and the topography in the Lake Maracaibo basin of Venezuela. These perturbations produce horizontal convergence, primarily at night when the *land breeze* is in force. One result is the **Catatumbo** lightning, perhaps making this area the most lightning-struck place in the world.

Northern and Northeastern South America

The northern part of South America has high temperatures year-round but with a slightly cooler rainy season. Temperatures are also moderated as the local summer winds spiral clockwise out of the Bermuda-Azores High and accelerate into the CLLJ. These winds tend to parallel the east–west-oriented coastline over parts of

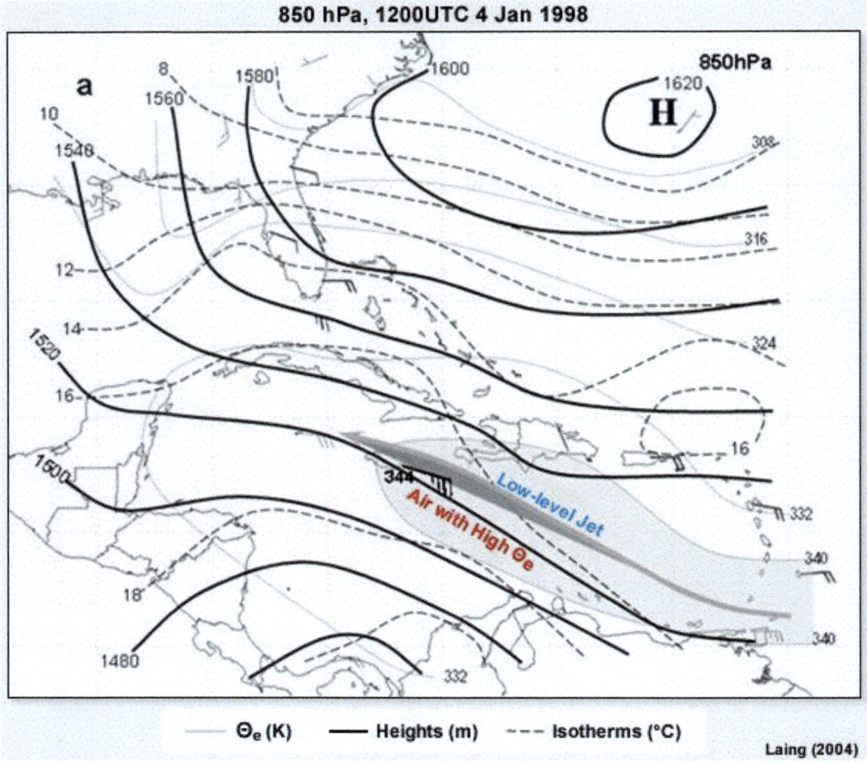

Fig. 19.17 The Caribbean low-level jet and theta-e region, with 850-mb geopotential heights. Note the ridge represented by the gray lines enclosing the shaded area. *Source* UCAR CoMet; adapted from Laing (2004). We cite the Laing paper in this chapter reference list. But it is in an American Meteorological Society journal—Journal of Hydrometeorology. Permission from Tara Torres

northern South America rather than penetrating inland, as shown in Fig. 19.18. The resulting *Ekman spiral* then pushes the near-surface water away from the coast, initiating cold water upwelling, which moderates the temperature, particularly in the Northern Hemisphere summer months. However, because the Coriolis effect is so slight in these low latitudes, the upwelling and temperature moderation are subtle.

The precipitation regime is also affected by this local circulation. The upwelling stabilizes the atmosphere, suppressing precipitation, especially along west–east-oriented local coastlines. Local areas with north–south oriented coastlines tend to be rainier. Together, the temperature and precipitation regime causes Venezuela, northern Colombia, and extensive areas of the adjacent Caribbean Sea to be dominated by an Aw climate (Fig. 19.19), as precipitation shows strong seasonality. Figure 19.20 displays the locations in northern and northeastern South America for which climagraphs and water balance diagrams are shown, including Caracas, Venezuela (Fig. 19.21), where the Aw climate is evident.

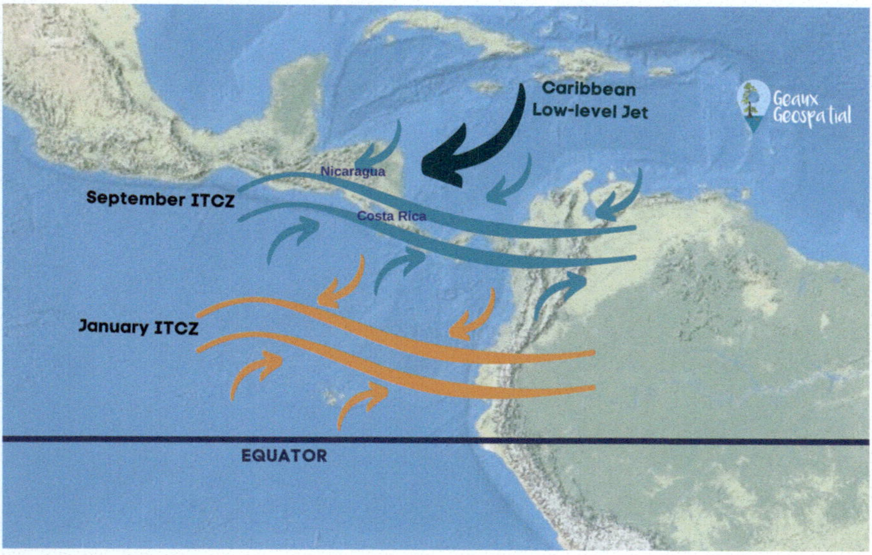

Fig. 19.18 Seasonal inter-tropical convergence zone and Caribbean low-level jet interactions in low-latitude Latin America. *Source* Geaux Geospatial

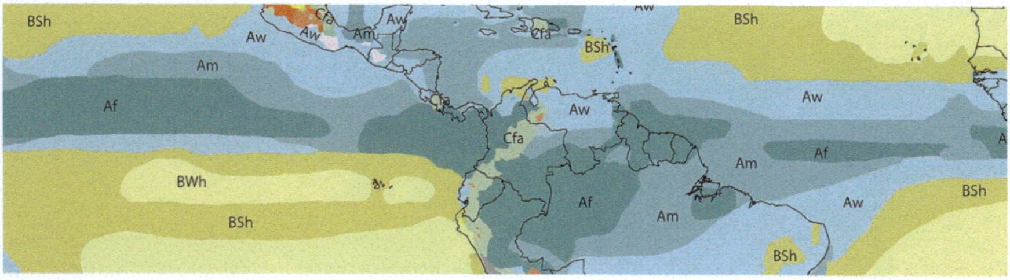

Fig. 19.19 Köppen climatic types around northern South America and surrounding marine environments. *Source* Modified from Reynolds, Rohli, Johnson, Waylen, and Francek, *Exploring Physical Geography*, McGraw-Hill

Localized parts of coastal northern Venezuela and the adjacent Caribbean are drier for a sufficiently long part of the year to be classified as an Arid (B) climate (see again Fig. 19.19). To the east of Venezuela, Af climates dominate Guyana, Suriname, the French overseas department of French Guiana, and the adjacent ocean. Only a narrow transition area of Tropical Monsoon (Am) climate separates the Aw and Af regions of northern South America, as lifting mechanisms are effective only short distances from locations where the upwelling effects suppress precipitation for a substantial part of the year. An intricate patchwork of climates too localized to appear in Fig. 19.19 occurs in rugged areas of northern South America.

East of the Andes Mountains, equatorial regions of Colombia, Ecuador, Peru, and Brazil—South America's largest country—are generally classified as Af climate

Fig. 19.20 Northern and northeastern South American cities for which climagraphs and water balance diagrams are shown. *Source* Geaux Geospatial

(see again Fig. 19.19). This largest contiguous land-covered area of Af climate on Earth receives abundant precipitation through all months even though precipitation decreases when the ITCZ moves farthest away. Rains fall overwhelmingly in the afternoon hours, on perhaps five to six days per week, in most South American Af climates. Manaus, by far the largest city in the heart of the Amazon rain forest, has temperature, precipitation, and water balance features (Fig. 19.22) typical of the Af climate.

Oceanic moisture advection causes maximum precipitation to occur along the eastern and northeastern coast. Orographic uplift along the mountain slopes and converging trade winds cause another precipitation maximum immediately to the east of the Andes. Runoff from the extremely high precipitation totals throughout the region form the river with the greatest discharge on Earth—the Amazon. Near the coast, the enhanced rainfall augments the already voluminous river flow, resulting in a discharge amount that is 12 times that of the Mississippi River in the United States.

South of the Amazon basin, the Brazilian coast juts southeastward and then southwestward, with the easternmost extent near the cities of Fortaleza, Natal, and Recife. The South Atlantic High affects these areas of eastern Brazil during the part of the

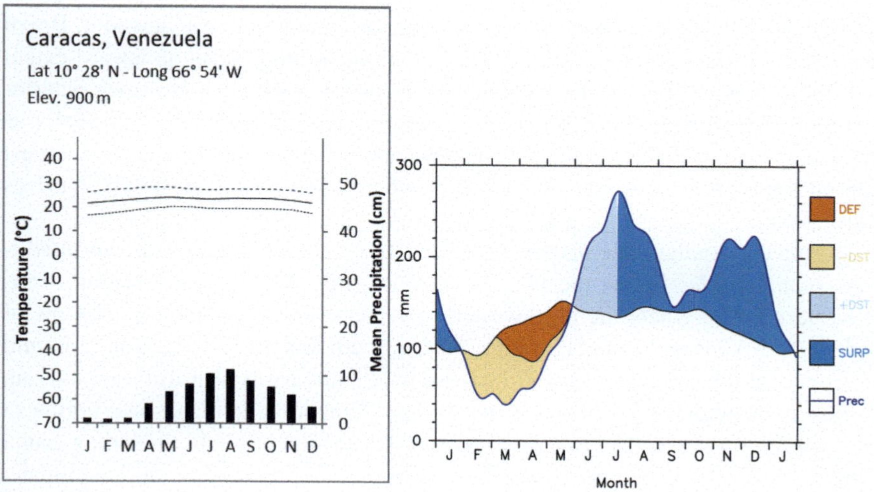

Fig. 19.21 Climagraph and water balance diagram for Caracas, Venezuela. Data for 1981–2016 averaging period. *Source* Authors. https://weatherspark.com/, http://cyclops.deos.udel.edu/wimp/public_html/

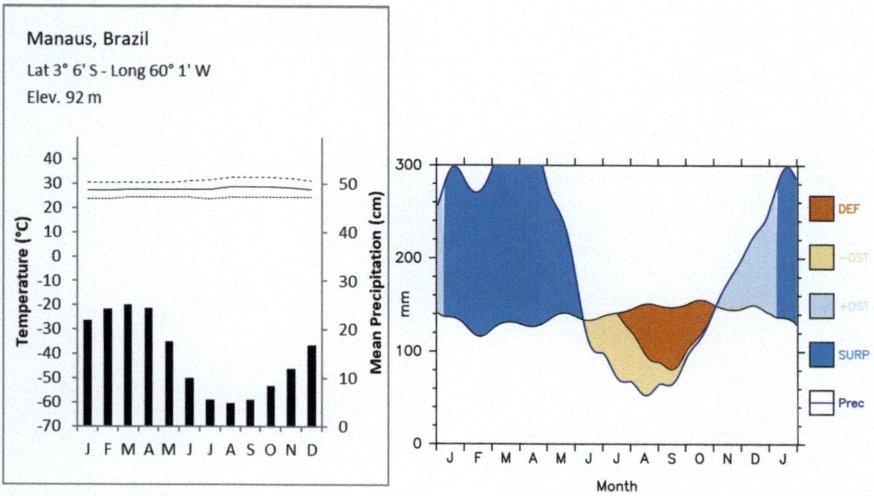

Fig. 19.22 Climagraph and water balance diagram for Manaus, Brazil. Data for 1981–2016 averaging period. *Source* Authors. https://weatherspark.com/, http://cyclops.deos.udel.edu/wimp/public_html/

year, when its counter-clockwise and outward circulation advects moisture inland. Thus, precipitation patterns across most of Brazil are seasonal, as the ITCZ and South Atlantic High shift.

In Southern Hemisphere winter (June–August), onshore winds create orographic precipitation along the immediate coast of the Brazilian Highlands (Fig. 19.23). These winter rains allow for a small strip of Af climates, including near Salvador, Brazil (Fig. 19.24), where enough rainfall also occurs in summer at the poleward extent of ITCZ influence. Yet substantial deficits still occur in summer. However, interior Southern Hemisphere Brazil is drier in Southern Hemisphere winter, especially as the distance from a moisture source increases (see again Fig. 19.23). The largest rainfall totals occur in northeastern Brazil, where convergence at the ITCZ drives convection.

During the Southern Hemisphere summer (December–February), the *northwesterly counter-trades* follow the ITCZ southward (Fig. 19.25). A dry region forms in western Brazil where the northwesterly counter-trades cross the rain shadow of highlands in Venezuela and adjacent northern South America. Advection of humid air from over the warm currents of the western Atlantic feeds plentiful seasonal rain in the eastern regions. The result is that most of Brazil, including coastal Fortaleza (Fig. 19.26) on the north coast and areas westward into Bolivia, including Santa Cruz de la Sierra, or simply Santa Cruz (Fig. 19.27), experience an Aw climate. Small areas of Cwa and Cwb climates exist in the southern Brazilian Highlands and

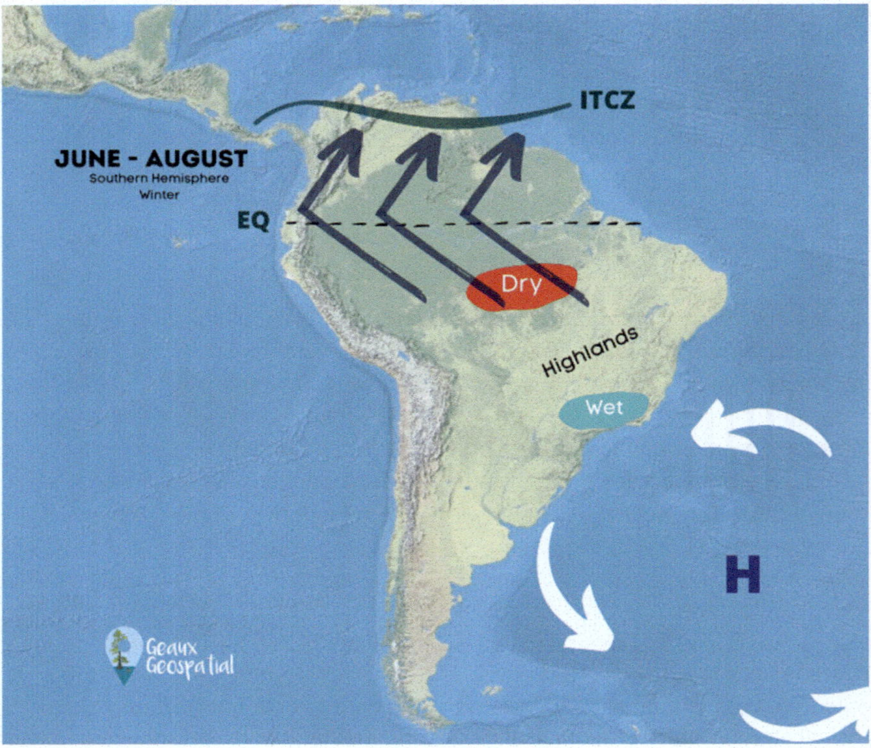

Fig. 19.23 June–August South America airflow and adjacent subtropical anticyclone. *Source* Geaux Geospatial

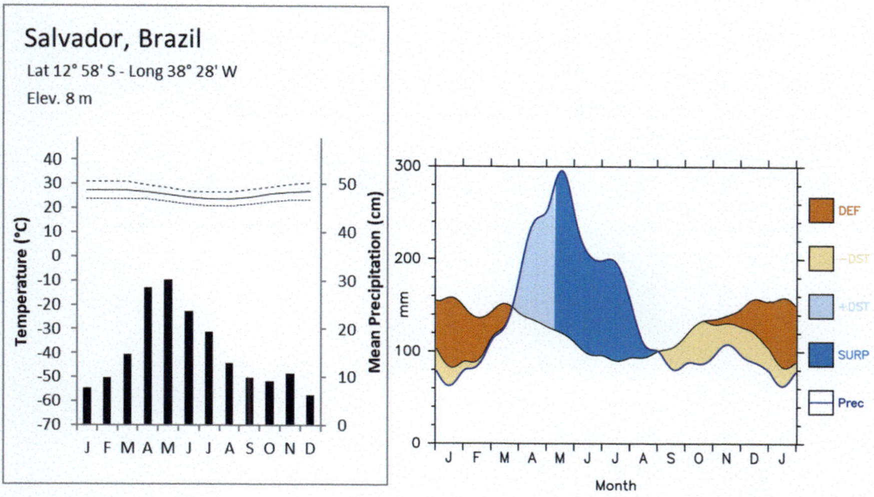

Fig. 19.24 Climagraph and water balance diagram for Salvador, Brazil. Data for 1991–2020 averaging period. *Source* Authors. https://weatherspark.com/, http://cyclops.deos.udel.edu/wimp/public_html/

on the eastern slopes of the Andes in Bolivia, Paraguay, and northern Argentina, where elevation reduces the temperature.

West of the Andes

South America west of the Andes is narrowed by the proximity of the mountains along the coastline but stretches from near the Equator to approximately 54° S. At the lowest latitudes, the ITCZ remains north of the Equator year-round over western Colombia and the adjacent ocean because of the influence of the Humboldt Current in pushing the warmest water and overlying air to the Northern Hemisphere. Therefore, western Colombia remains wet year-round. An average of approximately 500–560 cm (200–220 in) of precipitation is recorded annually, with precipitation reported on about two-thirds of all days, mostly in the late afternoon and early evening. The *southeast trades* bend back across the Equator in these areas, becoming southwesterly counter-trades north of the Equator. Orographic lifting initiates clouds and precipitation as the damp southwesterly winds encounter the western Andean slopes. Not surprisingly, this area falls into the Af climate region (see again Fig. 19.19). Poleward of the equatorial latitudes, South America west of the Andes is dominated by arid (B) and, farther south, mesothermal (C) climates (Fig. 19.28). The region of C climates is dominated by the *Mediterranean-warm summer* (Csb) *climate* and, farther poleward, the *Marine West Coast* (Cfb) *climate*. The regions are shifted somewhat north of their typical

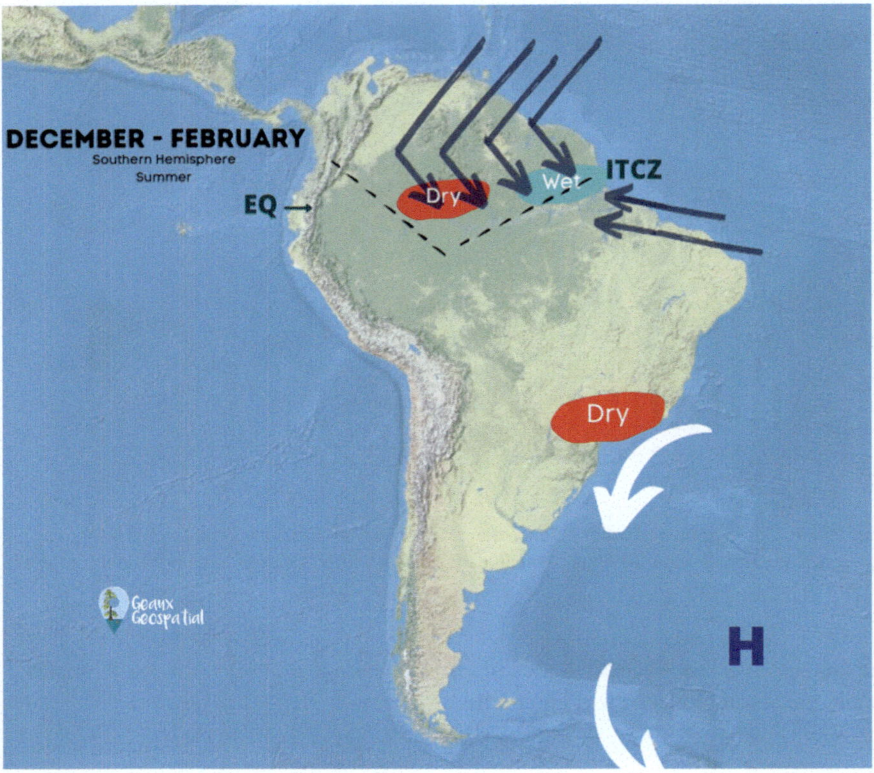

Fig. 19.25 December–February South America airflow and adjacent subtropical anticyclone. *Source* Geaux Geospatial

latitudes because of the ITCZ's position locked in place in the Northern Hemisphere over western South America.

Ecuador lies in a sharp transition zone. Despite its location on the Equator, only a small part of it is covered by the Af climate type (see again Fig. 19.19). At 5° N, precipitation is about 250 cm (100 in) per year, but at 2° S, desert conditions prevail, and diverse climate types occur in this relatively small nation. The counter-clockwise winds around the strong South Pacific High, anchored in place by the Humboldt Current that moves away from the continent just south of 5° N, promotes strong subsidence over much of southern Ecuador. Abundant *specific humidity* is usually present in this desert air, but lifting of that humid air is commonly suppressed by a *subsidence inversion*, resulting in a *stable atmosphere*. Frequent periods of *stratus* clouds and fog sustain plants and animals that can trap water droplets. The dry zone is restricted to the coast, as the Humboldt Current's influence wanes sharply with distance inland.

Peru and northern Chile experience similar conditions as southern Ecuador, but with even more robust subsidence and stability, for two reasons. First, the Humboldt Current contains even colder waters adjacent to this area than those adjacent to

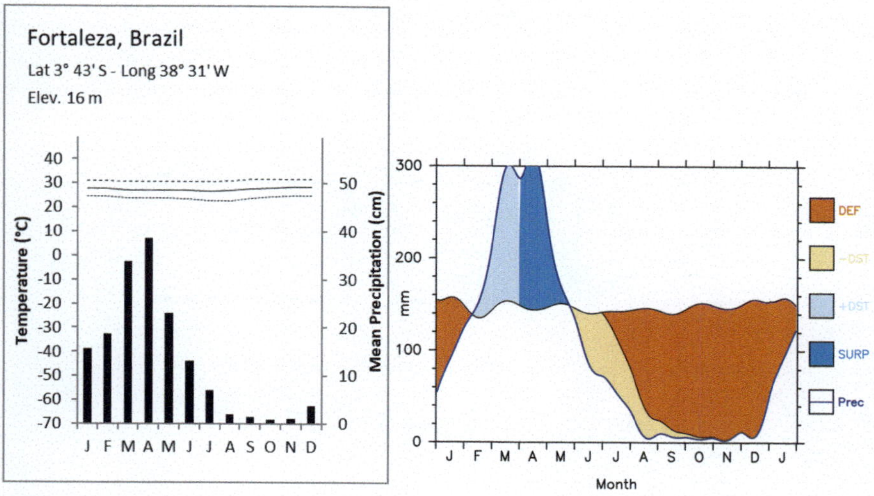

Fig. 19.26 Climagraph and water balance diagram for Fortaleza, Brazil. Data for 1991–2020 averaging period. *Source* Authors. https://weatherspark.com/, http://cyclops.deos.udel.edu/wimp/public_html/

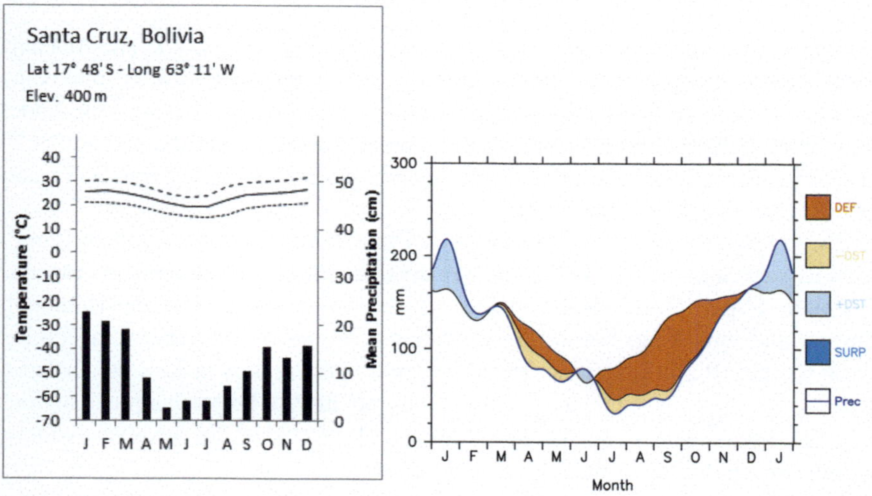

Fig. 19.27 Climagraph and water balance diagram for Santa Cruz, Bolivia. Data for 1991–2020 averaging period. *Source* Authors. https://weatherspark.com/, http://cyclops.deos.udel.edu/wimp/public_html/

Ecuador, which have had more time to warm in the lower latitudes. And second, the increased Coriolis effect at the higher latitudes of coastal Peru and northern Chile enhance the Ekman spiral, which pushes surface waters westward away from the coast more effectively, allowing more efficient upwelling as surface waters are replaced by waters from depth. These factors allow the sinking side of South Pacific High and the

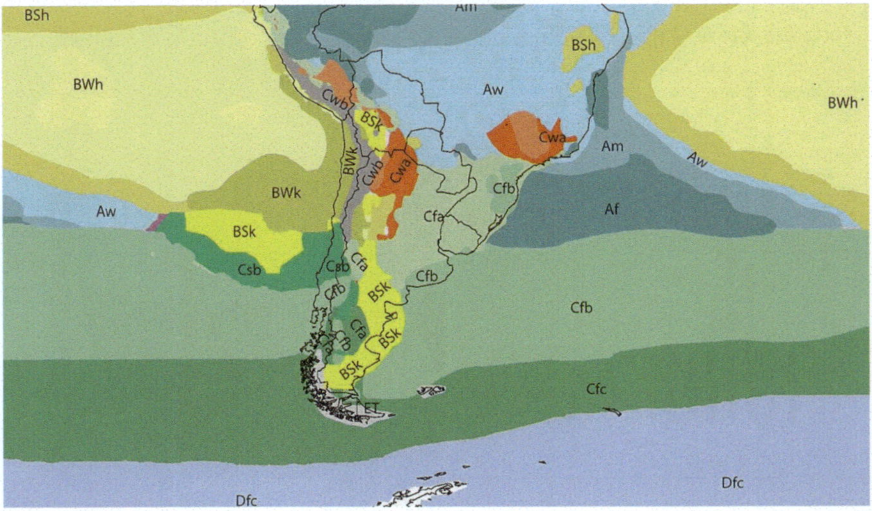

Fig. 19.28 Köppen climatic types around southern South America and surrounding marine environment. *Source* Modified from Reynolds, Rohli, Johnson, Waylen, and Francek, *Exploring Physical Geography*, McGraw-Hill

Humboldt Current to generate the conditions of the Atacama Desert. The Atacama probably gets the least precipitation on Earth, outside of Antarctica, despite being adjacent to the largest body of water on the planet. Some stations between the Andes Mountains and the coast have not recorded precipitation for decades. Figure 19.29 displays the locations in Andean South America for which climagraphs and water balance diagrams are shown.

The coastal dry region is represented by Lima, Peru. Several features of Lima's climate, depicted in Fig. 19.30, are remarkable. First, the temperature remains moderate despite the tropical latitude and clear skies, because of the Humboldt Current and upwelling. Second, precipitation is virtually non-existent, averaging about 0.6 cm (0.2 in) per year. Third, the mild temperatures make water balance deficits much less severe than might otherwise be expected in such a precipitation-deficient environment. However, the elevations of Lima and nearby areas vary widely, creating vastly different microclimates across short distances.

Farther south, precipitation is also suppressed, especially along coastal Chile. However, the high *Sun angles* provide abundant energy to evaporate the ocean water, generating a frequently-saturated environment that supports coastal fog and marine stratus clouds, even in this precipitation-starved environment, between the coastline and the low coastal mountains west of the Andes (Fig. 19.31). The combination of the mild temperatures and frequent fogs and stratus creates a desert in which the temperatures are tolerable, even toward the low latitudes. *Isotherms* parallel the coast as temperatures are influenced by proximity to the coast rather than solely by latitude. The lowest temperatures correspond to the cloudiest areas. Some "spillover" precipitation can occur at higher elevations as systems move from the east across the Andes,

Fig. 19.29 Andean South American cities for which climagraphs and water balance diagrams are shown. *Source* Geaux Geospatial

as depicted by Fig. 19.32. The South Pacific High influence wanes substantially with latitude from about 35° S southward. The Csb climate, similar to coastal California, dominates from about 35° S to 45° S. Summers (December–February) are typically dry but are even drier than normal when the South Pacific High is shifted to the south. In winter (June–August), the *polar front* is shifted to the north and *mid-latitude wave cyclones* can bring precipitation, contributing to snowpack in the highlands. Spring snowmelt from the Andes provides an important source of runoff for water supplies. The Csb climate at these latitudes is depicted by the climagraph and water balance diagram for Santiago, Chile (Fig. 19.33), where conditions border on BSk conditions. Mild temperatures, small precipitation totals (particularly in summer), and moderate to substantial deficits are all features of Santiago's climate. The rugged topography creates an intricate network of microclimates in this area that support diverse ecosystems and agriculture. From an agricultural perspective, it is advantageous to have

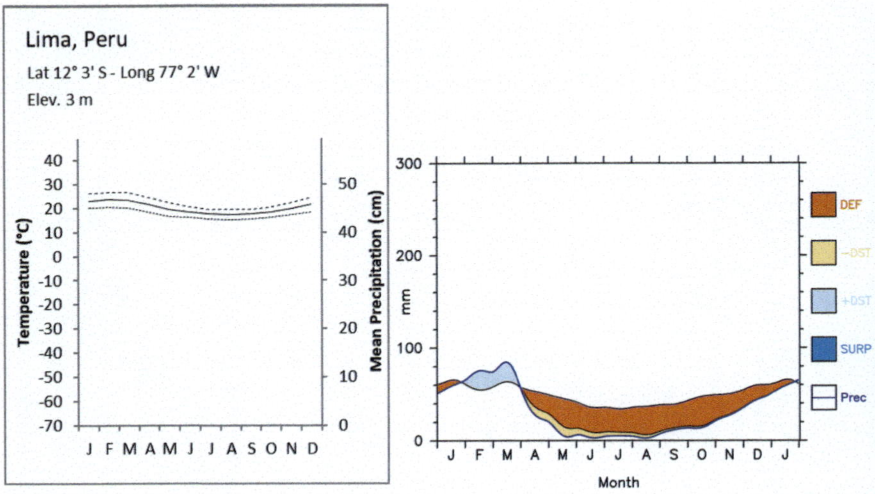

Fig. 19.30 Climagraph and water balance diagram for Lima, Peru. Data for 1981–2016 averaging period. *Source* Authors. https://weatherspark.com/, http://cyclops.deos.udel.edu/wimp/public_html/

a source of fruits and vegetables analogous to what is grown in California, such as Chilean grapes, that is available at the opposite time of year.

South of 40–45° S, the stabilizing effect of the South Pacific High decreases enough, even in summer, to allow the Cfb climate to prevail, similar to coastal Oregon, Washington, and British Columbia in North America. The mild temperatures, even at high-latitude Punta Arenas, Chile (Fig. 19.34), are caused by the mid-latitude westerlies that advect moderate *maritime polar* (mP) *air masses* from the Pacific Ocean source region.

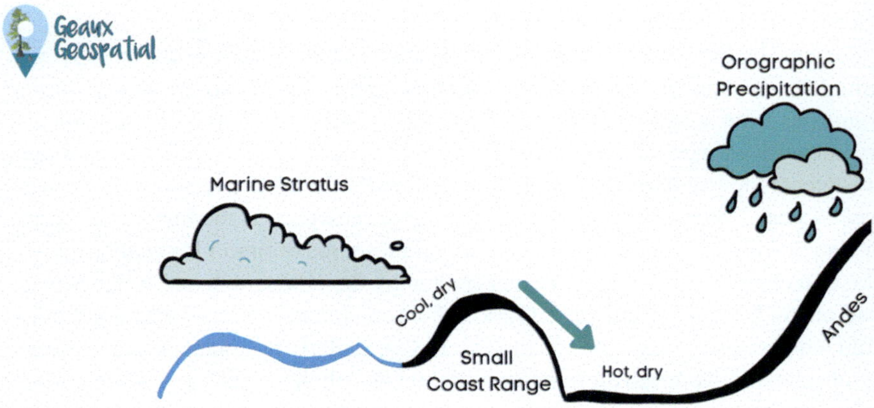

Fig. 19.31 Atacama regional fog, cloud, rain shadow, and precipitation areas. *Source* Geaux Geospatial

Fig. 19.32 Idealized depiction of Andes precipitation distribution. *Source* Geaux Geospatial

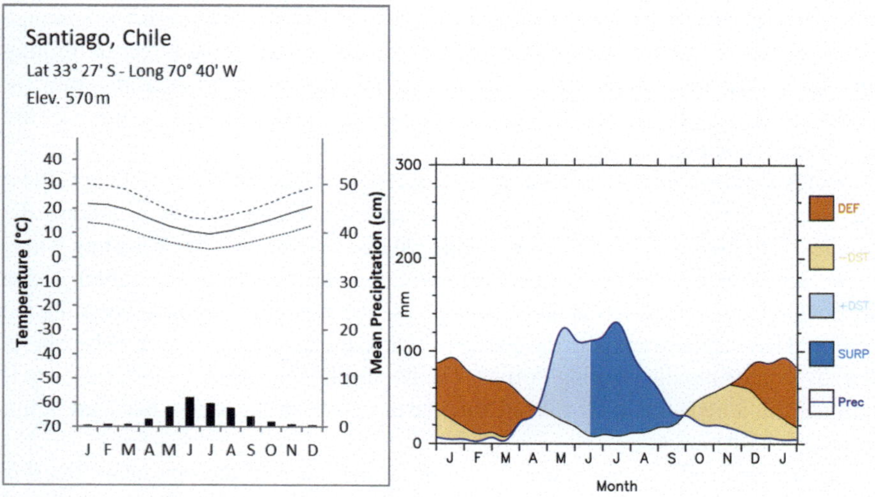

Fig. 19.33 Climagraph and water balance diagram for Santiago, Chile. Data for 1981–2016 averaging period. *Source* Authors. https://weatherspark.com/, http://cyclops.deos.udel.edu/wimp/public_html/

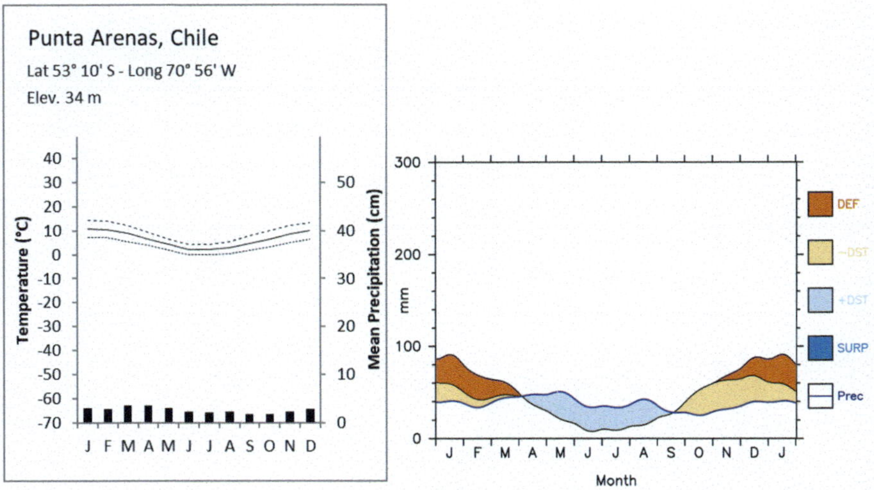

Fig. 19.34 Climagraph for Punta Arenas, Chile. Data for 1981–2016 averaging period. *Source* Authors. https://weatherspark.com/, http://cyclops.deos.udel.edu/wimp/public_html/

Precipitation occurs throughout the year from storms that are steered eastward along the polar front through the Rossby waves. In lower latitudes than Punta Arenas, precipitation usually peaks in winter, as suggested by the climagraph for Santiago (see again Fig. 19.33) as the South Pacific High has an increasing effect in limiting summer precipitation with distance equatorward. The water balance diagram for Punta Arenas suggests that both warm-season deficits and cold-season surpluses are moderate at this high latitude.

Winds in this region are known as the **Roaring Forties** (but even into the Fifties) because of their latitude, strength, and persistence. Punta Arenas is among the windiest settlements in the world. Vegetation shows evidence of the prevailing winds by the way it leans downwind, especially in the southern tip of the continent in Tierra del Fuego. Little land exists at these latitudes, so the westerlies circumnavigating the Southern Hemisphere are not slowed by *friction* to the extent that they are elsewhere, thereby supporting the Roaring Forties. The persistent and strong winds create navigational challenges for small aircraft and ocean vessels in the region between South America and Antarctica.

The *Tundra* (ET) *climate* prevails on the southern tip of the continent, but this area is not as cold as what might be expected in other ET climates. The lack of land minimizes winter *continentality* effects. Nevertheless, the area is damp, windy, and cold, with temperatures remaining fairly close to −1 to 4 °C (30°–40 °F) all year. The lack of a true summer prevents tree growth while supporting the low-to-the-ground tundra vegetation.

Southeastern South America

A narrow band of BW and BS climate zones parallels the eastern side of the Andes Mountains (see again Fig. 19.28). A rain shadow effect hinders Pacific Ocean moisture from reaching the *leeward* side of the range. In summer (December–February), the western part of northern Argentina receives some precipitation from the South Atlantic High but distance from the Atlantic Ocean limits moisture advection (Fig. 19.35). Farther to the east, the *Humid Subtropical* (Cfa) *climate* ranges across southernmost Brazil and Paraguay, northeastern Argentina, and Uruguay (see again Fig. 19.28). Figure 19.36 displays the locations in southeastern South America for which climagraphs and water balance diagrams are shown. These include São Paulo, Brazil (Fig. 19.37); Asunción, Paraguay (Fig. 19.38); and Buenos Aires, Argentina (Fig. 19.39).

Subtropical convective processes in unstable *maritime tropical* (mT) *air masses* produce summer rainfall maxima and soil moisture surpluses. Nevertheless, substantial winter deficits are evident at São Paulo (see again Fig. 19.37), which lies near

Fig. 19.35 Typical summer position of the South Atlantic High. *Source* Geaux Geospatial

Fig. 19.36 Southeastern South American cities for which climagraphs and water balance diagrams are shown. *Source* Geaux Geospatial

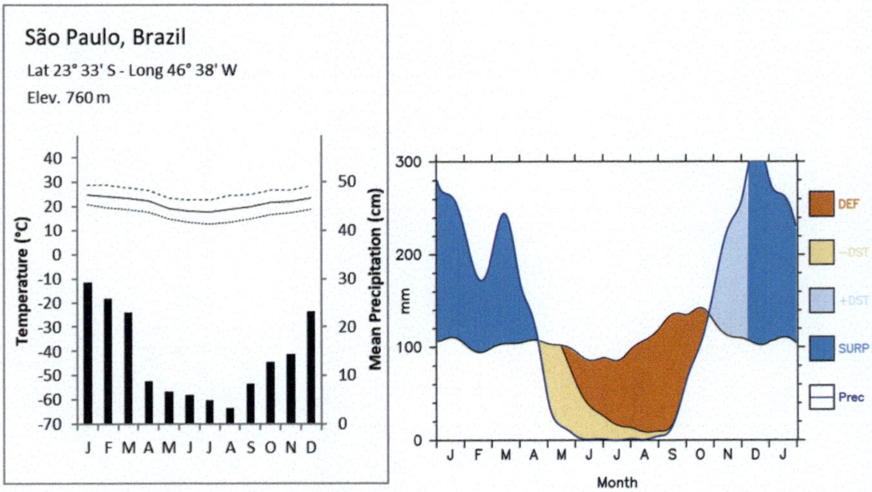

Fig. 19.37 Climagraph and water balance diagram for São Paulo, Brazil. Data for 1991–2020 averaging period. *Source* Authors. https://weatherspark.com/, http://cyclops.deos.udel.edu/wimp/public_html/

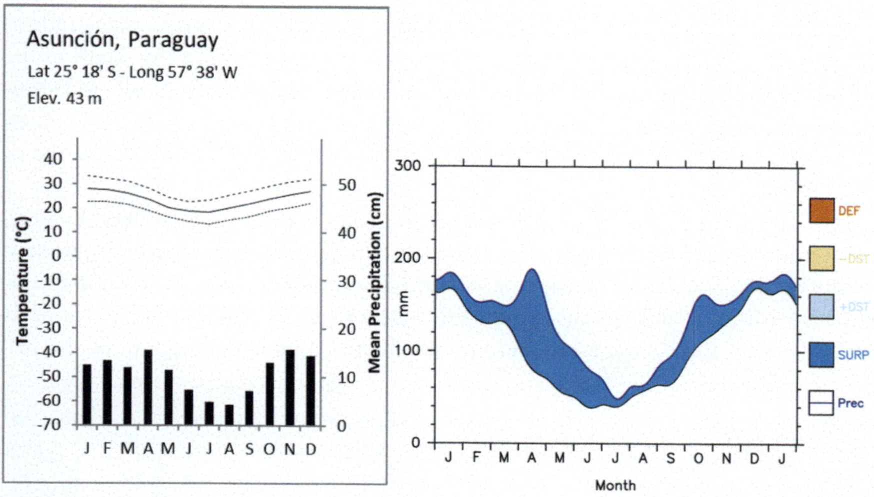

Fig. 19.38 Climagraph and water balance diagram for Asunción, Paraguay. Data for 1991–2020 averaging period. *Source* Authors. https://weatherspark.com/, http://cyclops.deos.udel.edu/wimp/public_html/

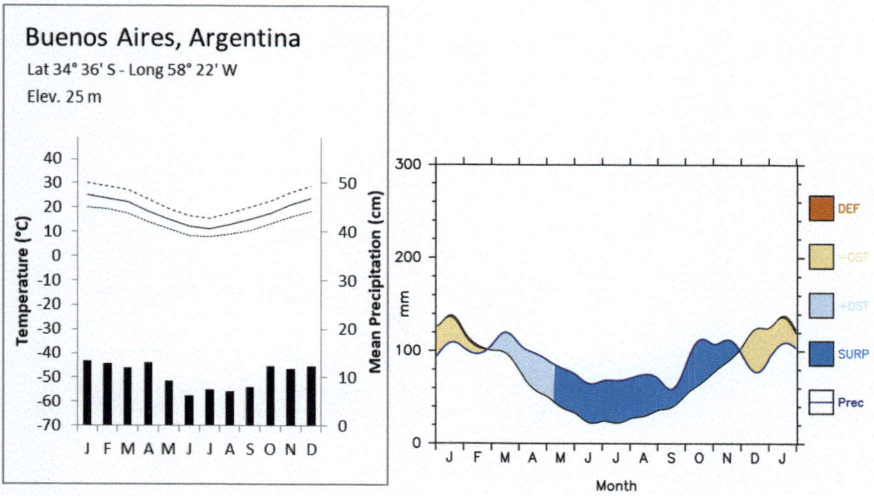

Fig. 19.39 Climagraph and water balance diagram for Buenos Aires, Argentina. Data for 1991–2020 averaging period. *Source* Authors. https://weatherspark.com/, http://cyclops.deos.udel.edu/wimp/public_html/

the Cwa/Cfa boundary. By contrast, the seasonality of the 140 cm (55 in) of precipitation per year at Asunción is so well-aligned with the annual cycle of *potential evapotranspiration* that deficits are absent in an average year (see again Fig. 19.38).

In winter, the preferred Southern Hemisphere Rossby wave positioning leads to smaller precipitation totals. The inverted ridges elongated toward the south in the Southern Hemisphere support high pressure and fair skies, as in the Northern Hemisphere. In the Southern Hemisphere, the ridge-to-trough side of the Rossby waves pushes colder mid-latitude anticyclones from the southwest toward the northeast (Fig. 19.40).

Because the Andes tend to lock a ridge in place, northern Argentina, Uruguay, and southernmost Brazil are more often under the influence of surface high-pressure systems in the Southern Hemisphere winter (June through August) when the mid-latitude westerlies and Rossby waves exert important influences. Cyclogenesis is more likely off the Pacific coast and into the South Atlantic (see again Fig. 19.40).

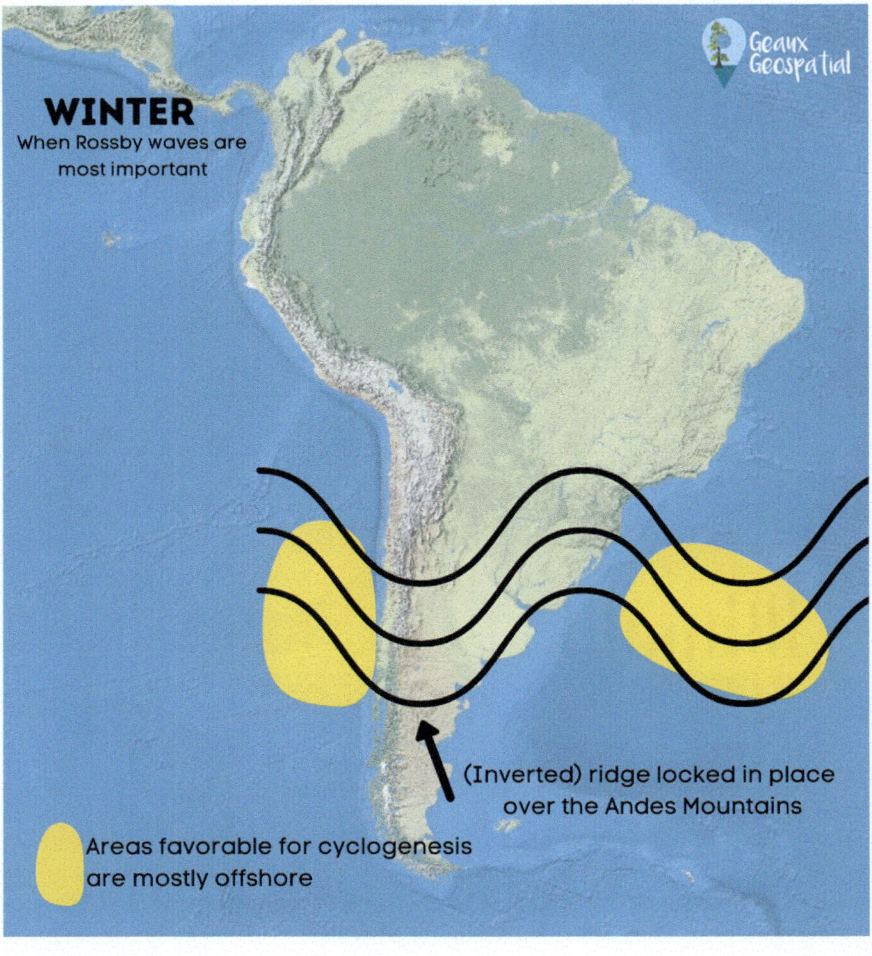

Fig. 19.40 Typical winter position of Rossby wave ridges and troughs in southern South America. *Source* Geaux Geospatial

By the time mid-latitude wave cyclones reach land, their upper-level support has typically dwindled.

In Patagonia, temperatures remain cool and subject to strong marine influences. The cold **Falkland Current** slips into the area south and east of the continent as a branch of an Antarctic ocean current, where the South Atlantic High moves away from the coast. This current has a similar role to that of the *Labrador Current* in the Northern Hemisphere, decreasing temperatures and creating frequent fogs. The narrowness of the land area in Patagonia creates little continentality.

The major Patagonian circulation features explain the aridity of the area. Patagonia is dominated by the mid-latitude westerlies year-round. As the landmass narrows, even the eastern coastal areas remain under the influence of the upper-level ridge, contributing to the aridity. The precipitation often results from fronts trailing (northward) from mid-latitude wave cyclones that push some orographic clouds over the Andes onto the eastern sides of the smaller mountains. This precipitation is also sparse because the Roaring Forties tend to steer storm systems past the area quickly. In fact, the Roaring Forties are even stronger in Patagonia than elsewhere, because in addition to the lack of land to slow them, the relatively low elevations of the Andes Mountains in that area impart less friction than higher-elevation areas. The temperature and precipitation climatology produce arid (B) climates near the Atlantic coast. Cfa climates are west of the B climates, closer to the Andes—the opposite pattern from farther north (see again Fig. 19.28).

El Niño-Southern Oscillation Effects

The proximity of the *Walker circulation* and associated ocean-air teleconnections are important for the climate of South America. The strong and persistent trade winds blowing from east to west across the equatorial Pacific associated with the Walker circulation cause sea levels to be about 15 cm (6 in) higher in the western equatorial Pacific near Australia and the *Maritime Continent* than near South America. This pattern creates the warm water mound called the *Pacific Warm Pool* in the western Pacific, while cold upwelled water occupies the eastern side near South America.

Every 3–7 years, when the Walker Circulation weakens for several months at a time, the weakened easterly trade winds allow the warm water mound in the Pacific Warm Pool to slosh back toward the east, creating an El Niño event. The water moves eastward toward South America because of gravity, at a speed of about 3 m s^{-1} (7 mph). Because the trade winds still move from east to west, albeit at a speed slower than normal, the surface waves move in opposition to the pool as it migrates eastward. This type of offsetting surface-to-depth water movement describes an oceanic (or equatorial) **Kelvin wave**.

A Kelvin wave is a large type of gravity wave—a wave in a fluid (e.g., air or ocean) that oscillates downward because of the effect of gravity, then rebounds upward as it propagates downwind or downstream, and then back downward again because of gravity, etc.—that occurs along boundaries. The flow along the boundary (either coast or Equator) has a geostrophic balance in which the pressure gradient force due to gravity and the Coriolis effect are balanced. Water molecules in the **gravity waves** move in the direction of propagation at ridges, and in the opposite direction at troughs. Therefore, when the Coriolis effect is taken into account, the amplitude of the ridges and troughs increases near the boundary. This balance becomes impossible offshore, so the waves are trapped near the boundary. Thus, in the Northern Hemisphere, the Kelvin waves propagate with the boundary on the right, and in the Southern Hemisphere they propagate with the boundary on the left.

Sea level and Kelvin waves are not the only ocean features to change with Walker circulation oscillations. The boundary between the warm surface water and the much colder water beneath it, where temperature changes very abruptly in the ocean, is known as the *thermocline* (Fig. 19.41). The thermocline is significant because the deep ocean circulation allows little water to rise up above the thermocline, since the warmer, less dense water is above colder, denser water. During normal periods, the thermocline is higher in the ocean on the eastern side of the Pacific, as coastal upwelling in the east drives the thermocline upward toward the surface. However, during the onset of an El Niño event, the warm water movement eastward across the basin deepens the thermocline in the central and then the eastern equatorial Pacific Ocean. Thus, El Niño events tend to confine the deep-ocean circulation to a deeper level, off the South American coast. This deepening of the thermocline during El Niño can have implications for global temperatures associated with the thermohaline circulation.

During an El Niño event, when the larger mass of warm water at the surface stymies cold water upwelling to the surface, nutrients settle through the ocean column as dead and decaying organisms filter downward. The lack of cold-water upwelling means that those nutrients cannot as easily remain near the surface to support an extensive oceanic biosphere. During El Niño, the process is reduced, causing ecosystem stress

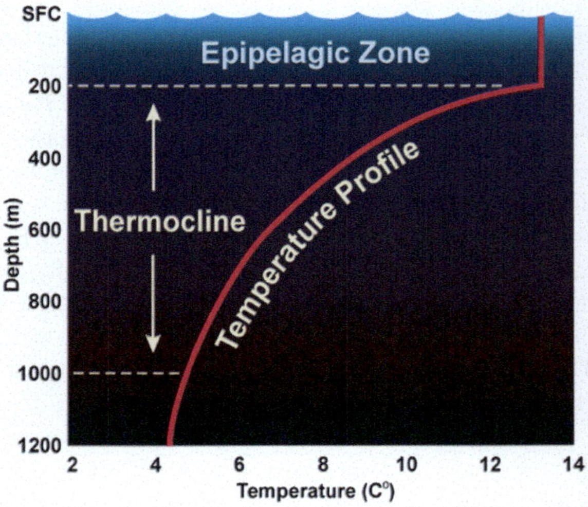

Fig. 19.41 The oceanic thermocline. *Source* NOAA. https://www.noaa.gov/jetstream/ocean/layers-of-ocean

through the top 100 m (330 ft) of the ocean. Life in this *epipelagic zone* or *photic zone*, where light initiates photosynthesis, is most affected by the lack of nutrients during El Niño. A decrease in plant life comes with a cascade of negative effects on the food chain. Prolonged El Niño events can create an environmental catastrophe and economic problems in sectors dependent on oceanic food harvests.

El Niño-associated warm anomalies in ocean temperature near South America and the cold anomaly near Asia alter the temperature and circulation patterns typical of the Walker circulation (see again Fig. 6.3). During El Niño, normally dry locations such as western South America become exceedingly wet as lower pressure occurs with the warming of the ocean surface. Along the arid western coast of South America, including along the Atacama Desert, heavy precipitation and a lack of vegetation to decrease surface runoff creates a situation conducive to mudslides and floods. East of the Andes, a northward migration and weakening of the ITCZ during El Niño often leads to drought in northeastern Brazil.

The large circulation, temperature, and precipitation departures during ENSO events cause direct impacts in South America but are also felt around the world. The important connections between atmospheric and oceanic circulation are felt in places like Peru, Chile, and Brazil. These ocean/atmosphere interactions become even more important to consider as the climate of the Earth's largest ocean basin—the Pacific—is explored in the next chapter.

Further Reading

Alvera-Azcarate, A., Barth, A., & Weisberg, R. H. (2009). The surface circulation of the Caribbean Sea and the Gulf of Mexico as inferred from satellite altimetry. *Journal of Physical Oceanography, 39*(3), 640–657. https://doi.org/10.1175/2008JPO3765.1

American Meteorological Society. (2024). Tehuantepecer. Glossary of Meteorology. Retrieved May 31, 2024. Available at: https://glossary.ametsoc.org/wiki/Tehuantepecer

Bürgesser, R. E., Nicora, M. G., & Avila, E. E. (2012). Characterization of the lightning activity of "Relámpago del Catatumbo." *Journal of Atmospheric and Solar-Terrestrial Physics, 77*, 241–247. https://doi.org/10.1016/j.jastp.2012.01.013

Fieldler, P. C. (2002). The annual cycle and biological effect of the Costa Rica dome. Deep-Sea Research Part 1: *Oceanographic Research Papers, 49*(2), 321–338. https://doi.org/10.1016/S0967-0637(01)00057-7

Laing, A. G. (2004). Cases of heavy precipitation and flash floods in the Caribbean during El Niño winters. *Journal of Hydrometeorology, 5*(4), 577–594. https://doi.org/10.1175/1525-7541(2004)005%3c0577:COHPAF%3e2.0.CO;2

Luko, C. D., da Silveira, I. C. A., Simoes-Sousa, I. T., Araujo J. M., & Tandon, A. (2021). Revisiting the Atlantic South Equatorial current. *Journal of Geophysical Research-Oceans, 126*(7). https://doi.org/10.1029/2021JC017387

Marengo, J. A., Torres, R. R., & Alves, L. M. (2017). Drought in Northeast Brazil—Past, present, and future. *Theoretical and Applied Climatology, 129*(3–4), 1189–1200. https://doi.org/10.1007/s00704-016-1840-8

Maurer, E. P., Stewart, I. T., Joseph, K., & Hidalgo, H. G. (2022). The Mesoamerican mid-summer drought: The impact of its definition on occurrences and recent changes. *Hydrology and Earth System Sciences, 26*(5), 1425–1437. https://doi.org/10.5194/hess-26-1425-2022

Montecinos, A., & Aceituno, P. (2003). Seasonality of the ENSO-related rainfall variability in central Chile and associated circulation anomalies. *Journal of Climate, 16*(2), 281–296. https://doi.org/10.1175/1520-0442(2003)016%3c0281:SOTERR%3e2.0.CO;2

Muñoz, E., Busalacchi, A. J., Nigam, S., & Ruiz-Barradas, A. (2008). Winter and summer structure of the Caribbean low-level jet. *Journal of Climate, 21*(6), 1260–1276. https://doi.org/10.1175/2007JCLI1855.1

Shi, W., & Wang, M. H. (2024). Ocean variability in the Costa Rica thermal dome region from 2012 to 2021. *Remote Sensing, 16*(8): Art. No. 1340.https://doi.org/10.3390/rs16081340

Sinclair, M. R. (1995). A climatology of cyclogenesis for the Southern Hemisphere. *Monthly Weather Review, 123*(6), 1601–1619. https://doi.org/10.1175/1520-0493(1995)123%3c1601:ACOCFT%3e2.0.CO;2

Thiel, M., Macaya, E. C., Acuña, E., Arntz, W. E., Bastias, H., Brokordt, K., Camus, P. A., Castilla, J. C., Castro, L. R., Cortés, M., Dumont, C. P., Escribano, R., Fernandez, M., Gajardo, J. A., Gaymer, C. F., Gomez, I., González, A. E., González, H. E., Haye, P. A., … Vega, J. M. A. (2007). The humboldt current system of northern and central Chile. *Oceanography and Marine Biology, 45*, 195–344.

Waylen, P., & Sadí Laporte, M. (1999). Flooding and the El Niño-Southern Oscillation phenomenon along the Pacific coast of Costa Rica. *Hydrological Processes, 13*(16), 2623–2638. https://doi.org/10.1002/(SICI)1099-1085(199911)13:16%3c2623::AID-HYP941%3e3.0.CO;2-H

Whyte, F. S., Taylor, M. A., Stephenson, T. S., & Campbell, J. D. (2008). Features of the Caribbean low level jet. *International Journal of Climatology, 28*(1), 119–128. https://doi.org/10.1002/joc.1510

Chapter 20
Australia and the Pacific

Abstract Australian and Pacific climates are the product of the moderating influence and atmospheric pressure features of the world's largest ocean. The atmospheric circulation is controlled by the Hawaiian-Pacific High, the South Indian Ocean High, and the South Pacific High. The associated tropical trade winds converge on the inter-tropical convergence zone (ITCZ), oscillating northward and southward with the seasons. Converging winds from anticyclones situated over Australia with those from the South Pacific High create the South Pacific convergence zone (SPCZ). Ocean currents, particularly the West Australian and Leeuwin Currents, bring moderating and also varying conditions to mainland Australia. Temperatures are tropical to subtropical over nearly all of the region, with excessively high temperatures in the arid Australian Outback. Coastal Australia has moderate temperatures and abundant precipitation. New Zealand has topography and a position that allows for relatively abundant precipitation and moderate temperatures. Both the ITCZ and SPCZ create precipitation for much of the Pacific islands, but variability in their position, especially for the SPCZ and associated with El Niño-Southern Oscillation (ENSO) and the Quasi-biennial Oscillation (QBO), makes precipitation inconsistent. Inhabitability of the islands is directly related to consistency of water availability.

Keywords Arid (B) climates · Atoll · Berson westerlies · Inter-tropical convergence zone (ITCZ) · Leeuwin Current · Mesothermal (C) climates · Ningaloo Niño/Niña · Quasi-biennial Oscillation (QBO) · South Equatorial Current · South Pacific convergence zone · South Pacific Gyre · South Pacific High · Tropical (A) climates · West Australian Current

General

On the western edge of the South Pacific Ocean, the atmospheric and oceanic circulation on and around the continent of Australia and the varied landscapes of New Zealand creates unique climates. The Pacific Ocean basin, which comprises one-third of the Earth's surface (Fig. 20.1), is also home to countless other islands and marine

R. V. Rohli et al., *Atmospheric and Oceanic Circulation*,
https://doi.org/10.1007/978-3-031-66142-6_20

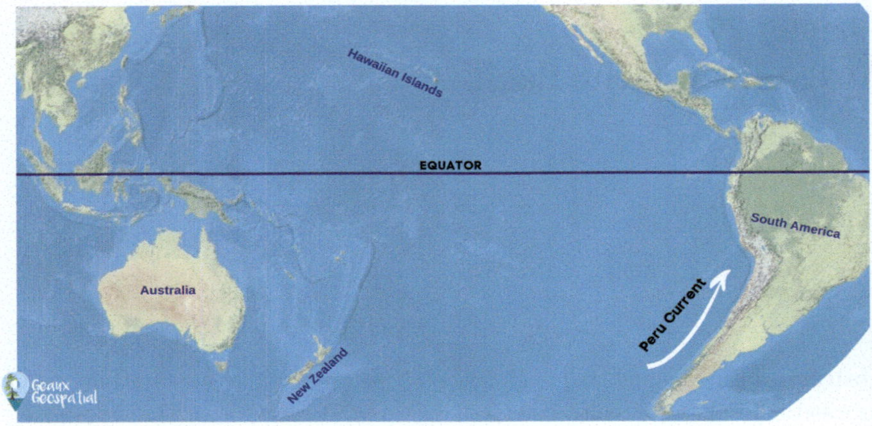

Fig. 20.1 Selected physical features and place names in Australia and the Pacific. *Source* Geaux Geospatial

ecosystems that affect and are affected by the atmospheric and oceanic circulation. The *Walker circulation* causes higher atmospheric pressure and arid conditions to persist in the eastern tropical Pacific and lower atmospheric pressure and wet conditions in the western tropical Pacific Ocean. This atmospheric and oceanic seesaw, and the transition zone between its ends, along with variability related to El Niño-Southern Oscillation (ENSO) and other modes of variability, influence the local climates of the entire area.

Topographic Influences (Site)

Australia's topography is unique compared to the other continents. It is the smallest continent, spanning latitudes 10° S to 40° S (Fig. 20.1). Australia lacks major mountain chains and high-elevation plateaus. The highest mountains are in the "Dividing Range" in eastern Australia, with the Australian Alps in the extreme southeast having the highest peaks. Elevations are approximately 300–1600 m (980–5250 ft), with the peak elevation being Mount Kosciuszko at 2228 m (7310 ft). These ranges are insufficient to be an air mass barrier or to create much *orographic lifting*. The "Outback" occupying most of non-coastal Australia consists of dry, desolate plains and plateaus. Coastal regions are much more humid and physically diverse than the interior on the flattest continent.

New Zealand consists of two major islands, simply termed the North Island and the South Island. Both are physically rugged, especially the South Island which has many peaks over 3000 m (9800 ft). The islands are composed of uplifted remnant continental crust interspersed with volcanoes and features that cooled from magma at the surface, such as lava fields and plateaus. The South Island contains the Southern

Alps, which dominates the island. Upslope *advection* of water vapor from maritime sources makes orographic *precipitation* plentiful on the South Island.

The Pacific Ocean is the largest and deepest ocean. Open water stretches across 20,000 km (12,000 mi) from the eastern coasts of Asia and Australia to the western coast of the Americas. Small islands are scattered across the Pacific, but most are in the western parts of the basin south of the Equator. The topography of these islands varies. Most formed from a variety of tectonic processes including volcanism and other forms of tectonic uplift. Consequently, many islands, including the Hawaiian chain, have high mountain peaks in the center. Other islands formed from the accumulation of coral as the mountain center eroded, forming a ring-shaped **atoll** that is usually vulnerable to sea level changes.

Circulation Influences (Situation)

The atmospheric circulation of the Pacific is controlled by three *subtropical anticyclones* and the associated tropical trade winds converging on the *inter-tropical convergence zone* (ITCZ). In the Northern Hemisphere, the winds rotating clockwise and outward from the *Hawaiian-Pacific High* direct the *northeast trades* toward the Equator. There, they meet the *southeast trade winds* which are circling counter-clockwise and outward around the *South Pacific High.* The wind and currents that rotate counter-clockwise and outward around the *South Indian Ocean High* or *Mascarene High* affect Australia, especially in that continent's winter. The winds and currents driven by the South Pacific High and *South Pacific Gyre* are important sources of variability in New Zealand and the Pacific islands.

The strength of *convection* along the ITCZ is strongly influenced by sea surface *temperatures*. The east–west oriented Walker circulation—the atmospheric response to surface water conditions—accumulates warm water and intense ITCZ activity in the western Pacific. By contrast, the cold *Humboldt Current* in the southeastern Pacific chills the overlying atmosphere in the eastern basin, supporting higher pressure, clearer skies, a much weaker ITCZ, and less precipitation. The limited water supply limits eastern island inhabitability. Across the middle of the equatorial Pacific, the trade winds normally force waters from east to west, with waters warming as they flow across the low latitudes, increasingly supporting low atmospheric pressure, abundant clouds, and precipitation characteristic of a robust ITCZ. All of the independent countries in Oceania are in the western Pacific, where an adequate freshwater supply is driven by the Walker circulation.

Rainfall from another convergence zone besides the ITCZ—the *South Pacific convergence zone* (SPCZ)—affects the western Pacific islands. The SPCZ is similar to the ITCZ in that air streams from *convergence* of airflow promote cloud formation and precipitation. Unlike the ITCZ, which appears from above as a primarily west–east line of clouds across the tropics, the SPCZ appears from above as a diagonal band of clouds protruding southeastward from the ITCZ in the western tropical Pacific near the *Maritime Continent*, as shown during a La Niña event in Fig. 20.2.

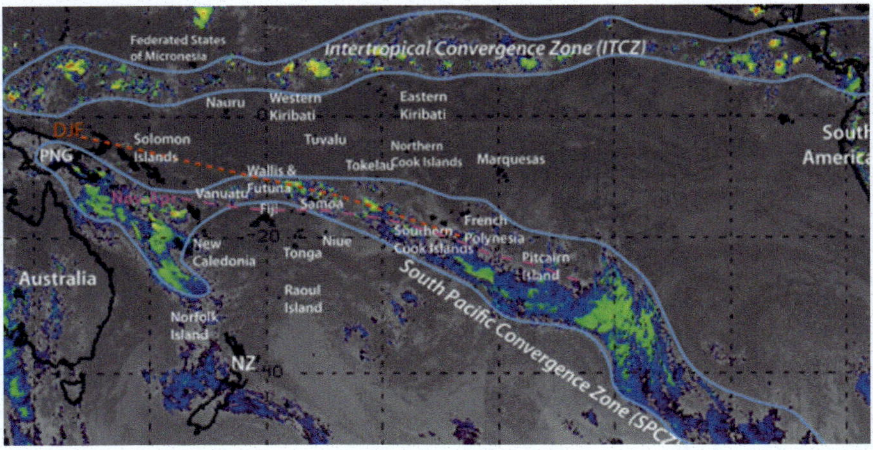

Fig. 20.2 Enhanced infrared satellite image from 28 October 2010 showing the SPCZ. *Source* NASA Global Hydrology and Climate Center; Lorrey et al. (2012)

The SPCZ forms because of the convergence of the counter-clockwise flow associated with a regional area of high pressure over Australia and the semi-permanent *South Pacific High* (Fig. 20.3). The Australian high pressure is derived from a series of regional-scale anticyclones that move generally from west to east in the *mid-latitude westerlies*, rather than from a semi-permanent subtropical anticyclone. Thus, while the northwestern side of the SPCZ remains in roughly the same position, the southeastern end fluctuates in position, like a swinging gate, depending on the position of the traveling anticyclone to its west. The orientation of the SPCZ over time is also a product of the relative intensities of high-pressure cells over Australia and the South Pacific.

Australia also has other important local marine circulation features. Warm, equatorial waters to the north and east moderate Australian temperatures. The *South Equatorial Current* flows westward near Australia's North coast, piling warm water nearby and along the Maritime Continent. Some of this warm water spills to Australia's eastern (Pacific) coast. The South Indian Ocean High forces cold waters from near Antarctica counter-clockwise toward lower latitudes, with the section of the *Indian Ocean Gyre* approaching Australia's western (Indian Ocean) coast known as the **West Australian Current**. This current continues westward, north of Australia in the Indian Ocean, as the South Equatorial Current and then feeds into the *Agulhas Current* off the southeast African coast. The weak, cold West Australian Current affects Perth and other areas on the West coast, but at times, the **Leeuwin Current** flows southward between the West Australian Current and the coast, giving western Australia a diverse set of conditions, depending on which current is dominant. The Leeuwin Current can occasionally turn eastward and provide warmth along the western half of Australia's south coast.

The broad-scale circulations affecting northern Australia most directly are the *southeast trades* and South Equatorial Current. The sinking on the eastern side

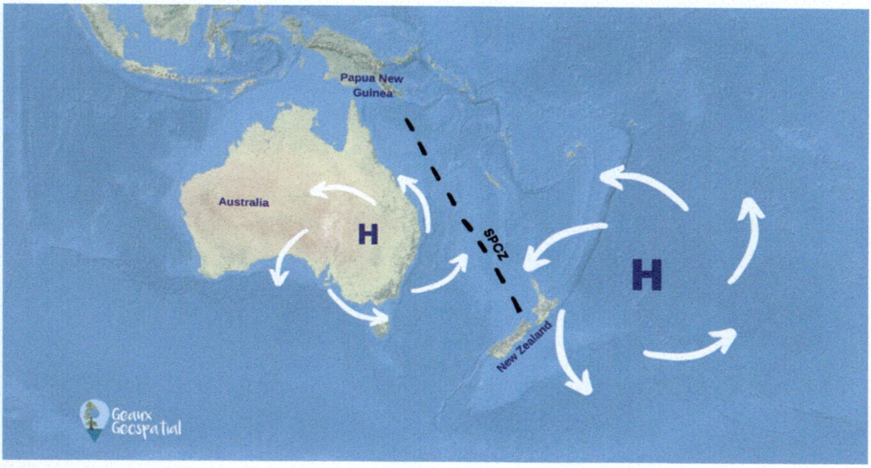

Fig. 20.3 Formation of the South Pacific convergence zone. *Source* Geaux Geospatial

of the South Indian Ocean High (i.e., Mascarene High), reinforced by the West Australian Current, dominates much of Australia. Southern Australia experiences the mid-latitude westerlies in winter.

Similar to the Northern Hemisphere mid-latitudes, migratory anticyclones and *cyclones* moving west to east, as secondary circulations embedded within the westerly wind system affect southern Australia. Because of the lack of mountain ranges which can anchor ridges and anticyclones in place, Australia experiences a "statistical high." This term describes a series of migratory high pressure cells parading from west to east in the westerlies, giving the area a propensity for high pressure but not from any single semi-permanent, quasi-stationary system. During Southern Hemisphere winter, when the *Polar Front* shifts farthest north, passing *mid-latitude wave cyclones* also bring precipitation to the South coast.

Air masses affecting the most-populated sections of the continent are primarily maritime in origin (Fig. 20.4). The northern reaches of Australia and the Maritime Continent are dominated by *maritime equatorial* (mE) *air masses*. The source regions include the Indian and western Pacific Oceans, especially during the high-Sun period of December through February. Poleward, *maritime tropical* (mT) *air masses* from the Indian and Pacific Oceans occur, especially during June through August. The southern areas of Australia are occasionally affected by *maritime polar* (mP) *air masses* originating in the Southern Ocean. Meanwhile, the vast, dry Outback produces *continental tropical* (cT) *air masses* that extend over much of the interior of the continent.

Because the de-amplified, *zonal* upper-level flow patterns provide little support for vigorous clashes of drastically different air masses, mid-latitude wave cyclones are usually weak in extratropical Australia and New Zealand. Fronts typically separate cT air from the mT or mP air rather than separating much colder, drier air from much warmer and more humid air. Thus, weak *cold fronts* trail from strong low-pressure

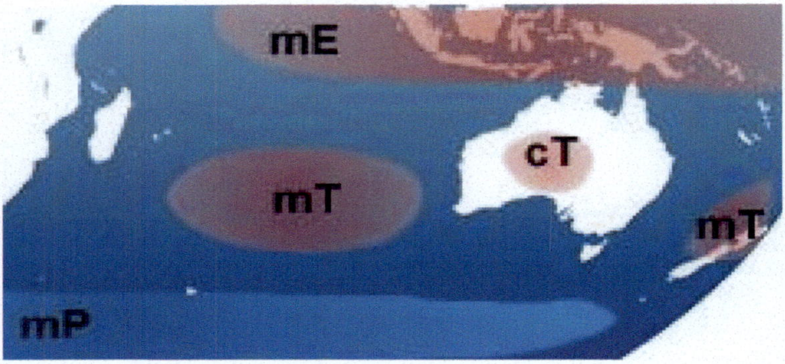

Fig. 20.4 Air masses affecting Australia and New Zealand. *Source* UCAR CoMet. Permission from Tara Torres

systems in the mid-latitudes of this region. These fronts are typically oriented in a northwest-to- southeast position (Fig. 20.5) with limited precipitation.

New Zealand spans from approximately 30°–40° S latitude, embedding it completely within the mid-latitude westerlies. Mid-latitude wave cyclones originating from the *baroclinic* zone off the eastern coast of Australia affect New Zealand throughout the year. The systems typically involve mT and mP air masses and generally produce only light precipitation, as air mass contrast is too weak to support strong fronts and vertical cloud development.

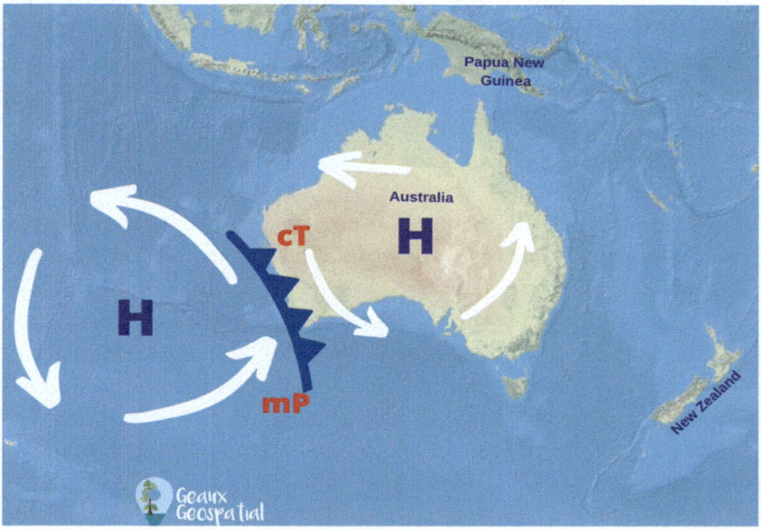

Fig. 20.5 Typical frontogenesis situation in Australia. *Source* Geaux Geospatial

Atmospheric and oceanic circulation interactions create important teleconnections, particularly ENSO, over the Pacific region and cause significant hydroclimatic variability. During a warm (El Niño) event, the Walker circulation seesaws in opposition to the normal and La Niña modes, the latter of which is an anomalously strengthened Walker circulation. On the western side of the South Pacific basin, the thermocline moves closer to the surface with the eastward migration of the warm water pool, and drought sets in over northeastern Australia. In the east Pacific, *upwelling* weakens or ceases near the coast of South America and warmer water creates increased convection over the central and eastern Pacific. The normally wet areas in the west undergo drought as increased air pressure builds with decreasing ocean temperatures. A strengthening of the high pressure over Australia swings the southern edge of the SPCZ to the northeast, further limiting rainfall over the western Pacific islands and enhancing precipitation in the central tropical Pacific.

During warm (El Niño) events, normal ecosystem processes are disrupted and stressed. The changes in ocean circulation affect fisheries, birds, coral, and marine mammals, particularly through the lack of nutrients associated with reduced upwelling in the eastern Pacific. Rain forests in northern Australia, the Maritime Continent, Southeast Asia, and western Pacific islands undergo ecosystem stress as water shortages occur. Drying vegetation increases the threat of forest fires.

Another important climate variability factor in the tropical Pacific region is the **Quasi-biennial Oscillation** (QBO). The QBO describes the alternating direction of prevailing winds in the **stratosphere**—the layer of the atmosphere directly above the weather layer (i.e., *troposphere*). Before 1908, it was assumed that upper atmospheric winds at these levels moved from east to west. This assumption was based on the observation of ash and dust moving from east to west after the massive eruption of the Krakatoa volcano in Indonesia on August 27, 1883. The winds were then named the Krakatoa easterlies. In 1908, weather balloons released over Lake Victoria in East Africa recorded westerly winds in the stratosphere, in opposition to the Krakatoa dust cloud observation. The westerly stratospheric flow became known as the **Berson westerlies**. By the 1960s, enough data became available to reveal that stratospheric winds reverse direction on an approximate 26-month cycle. Subsequent studies concluded that the winds of this QBO emanate downward from the upper regions of the stratosphere at a rate of approximately 1 km (0.6 mi) per month. Once the winds reach the *tropopause*, the winds weaken and the opposite phase occurs via the same downward propagation method. The easterly wind phase produces winds approximately twice the strength of the Berson westerlies, but the cause of this reversal—the QBO—remains uncertain.

The QBO is responsible for mixing gases in the stratosphere, which otherwise remains stagnant. The oscillation also combines with ENSO to influence zones of enhanced convection across the tropical Pacific, the strength of Asian monsoonal precipitation, and precipitation over the Sahel of Africa. *Tropical cyclone* activity in the western Pacific, western Indian, and North Atlantic oceans is strongly influenced by the QBO. Increased tropical cyclone activity in the North Atlantic Ocean occurs in association with the Berson westerlies, and the easterly QBO phase triggers notable decreases.

Australia

The temperature climatology in Australia and adjacent waters is relatively straightforward. *Isotherms* are primarily zonal across the adjacent oceans and coastal Australia, with higher temperatures occurring in lower latitudes and lower temperatures found at higher latitudes, as expected. But summer temperatures are elevated in the sunny, dry Outback. Australia generally lacks *Rossby wave* amplification that might create *meridional* flow and bring polar air to low latitudes and tropical air to higher latitudes. The low and relatively uniform topography and general lack of land surface obstructions upwind from Australia contributes to this Rossby wave pattern and temperature climatology. The maritime influence moderates temperatures from what would otherwise be expected at the latitude in question, to some extent continent-wide, but especially at the coasts. The vast majority of Australians reside near the coast where the climates are most temperate.

The marine influence on precipitation is much smaller than on temperature. Persistent clear skies from the parade of "statistical highs" make precipitation sparse beyond the coastal margins. The clear skies and high Sun angles allow the Outback to heat substantially in summer, with strong *sensible heat flux*, weak *latent heat flux*, and large *Bowen ratios* and *potential evapotranspiration*. But little precipitation falls to offset PE, leaving large water balance *deficits* away from the coast. Approximately 80% of Australia falls into the arid (B) climate category (Fig. 20.6). Representative cities in Australia and New Zealand with climagraphs and water balance diagrams are shown in Fig. 20.7.

In northern Australia, the high temperatures year-round are accompanied by strong seasonal precipitation regimes. Rainfall events occur during the high-Sun December through February period when the ITCZ extends farthest south, over northern Australia. Darwin and Cairns are representative of this coastal, *Tropical Wet-dry* (Aw) *climate*, also known as the *Tropical Savanna climate* (Figs. 20.8 and 20.9). The climate of Darwin is particularly sensitive to ENSO, as one popular ENSO index—the *Southern Oscillation index*—has traditionally used the difference between sea level pressure between Tahiti and Darwin as the indicator of the strength of El Niño and La Niña. The annual precipitation maximum is associated with a mini-monsoon effect as wind direction shifts seasonally. In addition, the *northwesterly counter-trades* result from the abrupt re-direction of the northeast trade wind flow at the Equator because of the changed direction of deflection of the *Coriolis effect* (Fig. 20.10). The northwesterly counter-trades collide with the southeast trade winds to produce convergence and copious rainfall, with annual totals near 360 cm (140 in) in the northeastern state of Queensland and the adjacent offshore Great Barrier Reef.

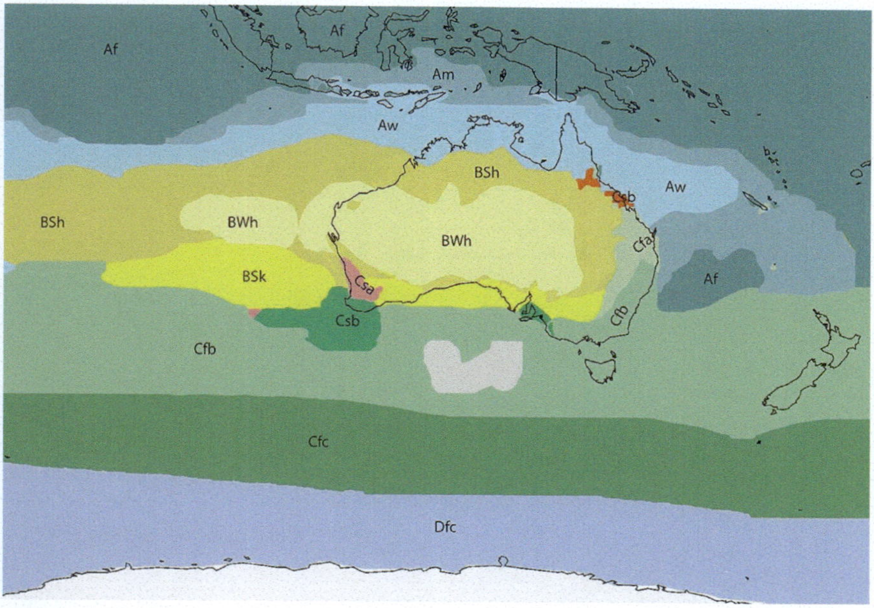

Fig. 20.6 Köppen climatic types around Australia and New Zealand. *Source* Modified from Reynolds, Rohli, Johnson, Waylen, and Francek, *Exploring Physical Geography*, McGraw-Hill

Typhoons are regular weather features of low-latitude Australia, especially along the northeastern coast, where frequencies exceed those in the North Atlantic basin. They move from east to west and are fueled by the warm western Pacific waters, with the season extending from February through May. El Niño events suppress typhoon development as the warm water pool migrates eastward. As in southern Africa, Australian atmospheric and hydrologic variability is also affected by periodic warming (**Ningaloo Niño**) and cooling (**Ningaloo Niña**) of ocean water off Australia's Indian coast.

Much of Australia's northeastern coast experiences a *Humid Subtropical* (Cfa) *climate*. Farther south, ironically, a *Marine West Coast* (Cfb) *climate* prevails despite a location on an eastern coast, as the poleward extent of the southeast trades advect relatively cool ocean water, moderating summer temperatures in this subtropical setting. As both the Cfa and Cfb indicate, precipitation is consistent throughout the year. Summers have frequent convective thunderstorms, and occasional tropical cyclones in the north. Winters along the entire eastern coast are associated with periodic mid-latitude wave cyclone impacts on both temperature and precipitation. Sydney—Australia's largest city—is representative of a Cfa climate (Fig. 20.11). Summers there are hot, winters are mild but with periodic cold spells, precipitation is distributed relatively evenly throughout the year, and water balance deficits are minimal.

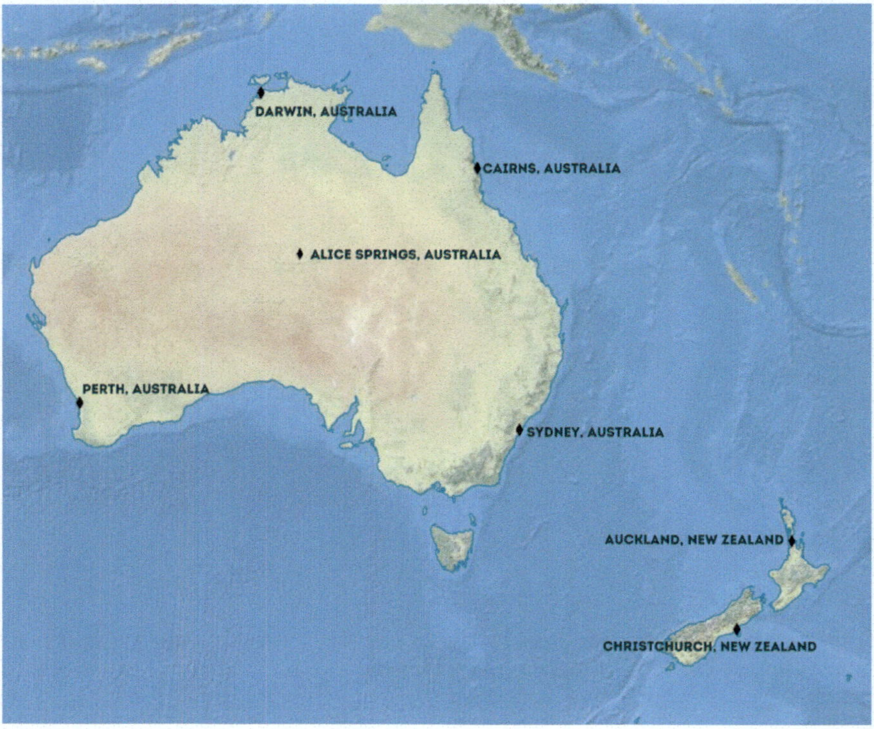

Fig. 20.7 Cities in Australia and New Zealand for which climagraphs and water balance diagrams are shown. *Source* Geaux Geospatial

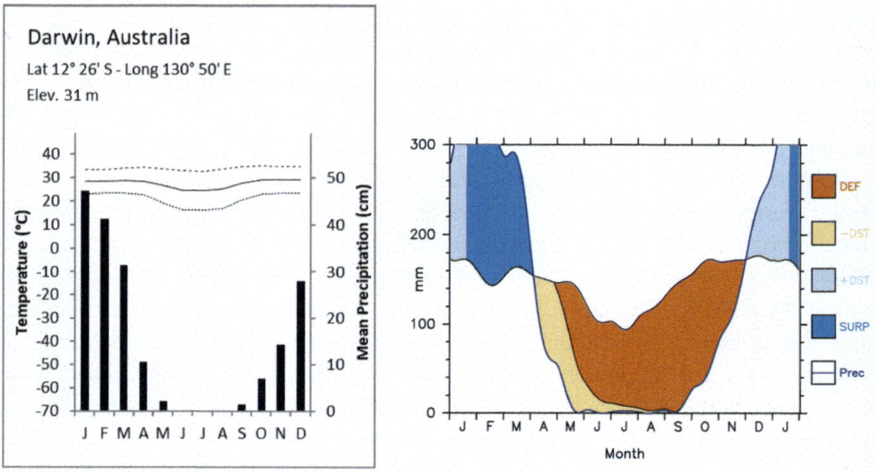

Fig. 20.8 Climagraph and water balance diagram for Darwin, Australia. Data for 1991–2020 averaging period. *Source* Authors. https://weatherspark.com/, http://cyclops.deos.udel.edu/wimp/public_html/

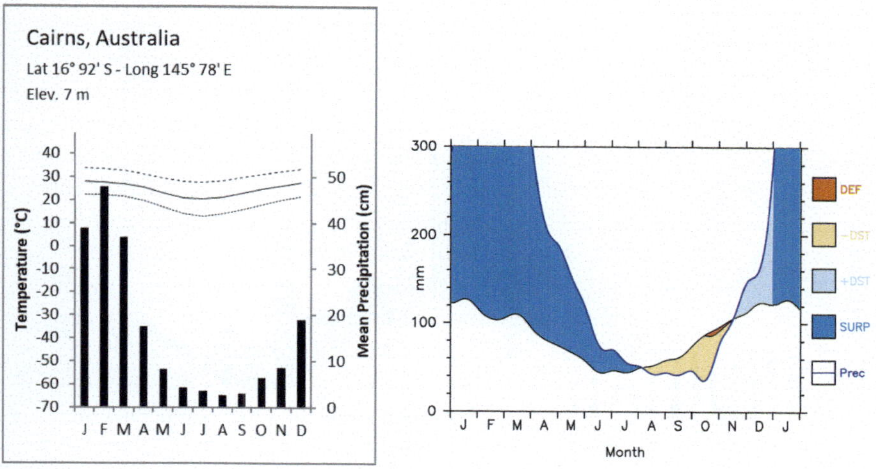

Fig. 20.9 Climagraph and water balance diagram for Cairns, Australia. Data for 1991–2020 averaging period. *Source* Authors. https://weatherspark.com/, http://cyclops.deos.udel.edu/wimp/public_html/

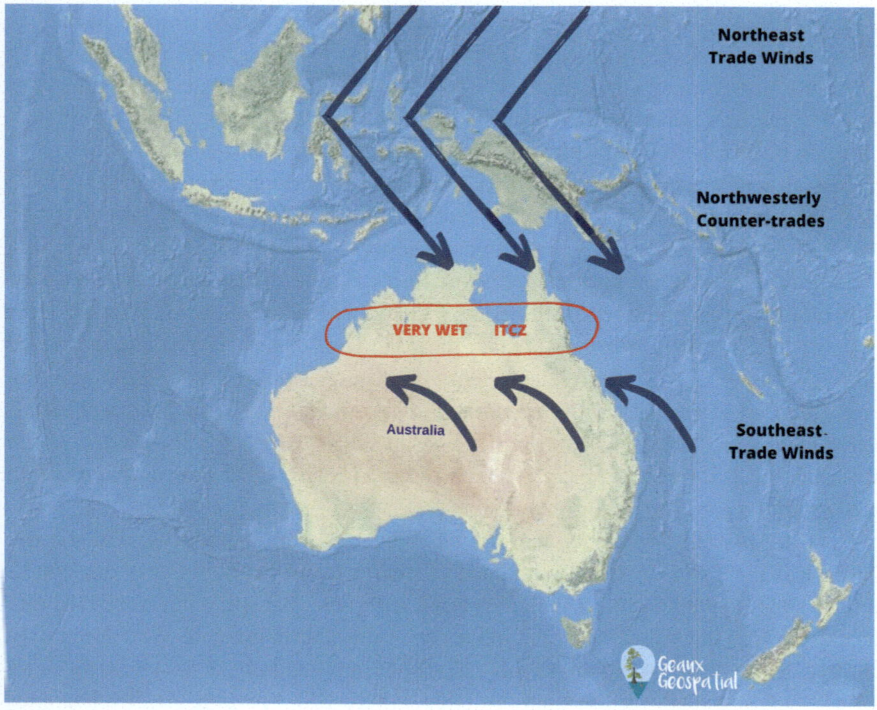

Fig. 20.10 Effect of the northwesterly counter-trades and southeast trades on northern Australia. *Source* Geaux Geospatial

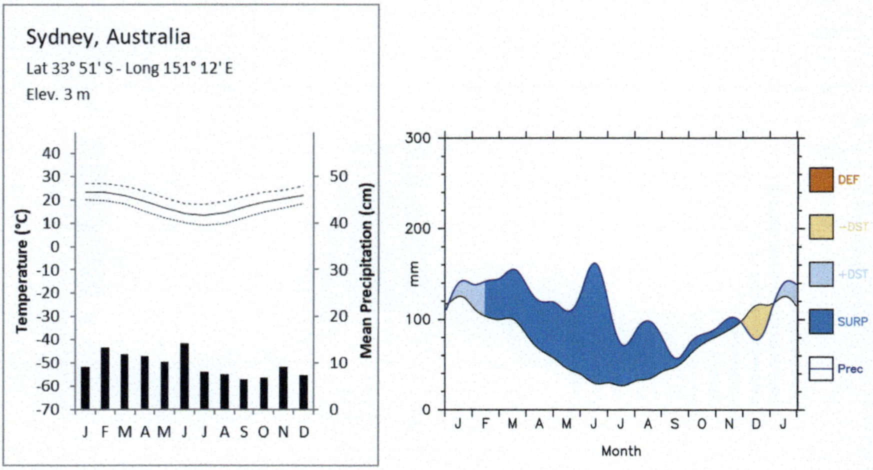

Fig. 20.11 Climagraph and water balance diagram for Sydney, Australia. Data for 1991–2020 averaging period. *Source* Authors. https://weatherspark.com/, http://cyclops.deos.udel.edu/wimp/public_html/

Off Australia's eastern coast, robust mid-latitude wave cyclones can form along a coastal *baroclinic* zone. These affect New Zealand more than Australia, with some becoming violent. In the Australian Alps in the extreme southeast, snowfall is common but variable from year-to-year.

In southern Australia, the main precipitation-generating mechanisms are the weak cold fronts associated with cold-season mid-latitude wave cyclones. The Cfb climate continues across the easternmost one-third of the South Coast, including Melbourne—Australia's largest metropolitan area—and the island Australian state of Tasmania. Most of the rest of the South Coast is a *Semi-arid—cold* (BSk) *climate*, as it sits in the transition zone between the *True Desert–hot* (BWh) *climate* of the Outback and the Csb offshore. The western edge of the South Coast, including Perth (Fig. 20.12), experiences a *Mediterranean-hot summer* (Csa) or *Mediterranean-warm summer* (Csb) *climate* in higher coastal latitudes. As in all *Mediterranean* (Cs) *climates*, summers near Perth typically have large water balance deficits, with most precipitation, soil water recharge, and *surplus* occurring in winter.

Central Australia contrasts sharply with the continental margins. The central Outback is normally under the influence of high pressure, creating very dry conditions. The Outback core is classified as a BWh climate that spills out westward into the eastern Indian Ocean, under the influence of the stabilizing cold West Australian Current, where little precipitation falls. Surrounding the Outback is *Semi-arid—hot* (BSh) *climate*, except on the southern edge, where BSk prevails. Aridity becomes more intense toward the interior, where both temperature and potential evapotranspiration increases. The overwhelming dominance of high pressure means clear skies, abundant surface *insolation* (K↓), and large Bowen ratio. The result is that daily maximum temperatures can average up to about 37 °C (98 °F) in summer. Alice

Springs experiences only about 28 cm (11 in) of precipitation per year. A *thermal low* that develops in summer causes more humid air masses to be advected in from coastal regions and produces a modicum of rainfall. Alice Springs is at the heart of the Outback; deficits prevail in every month climatologically (Fig. 20.13).

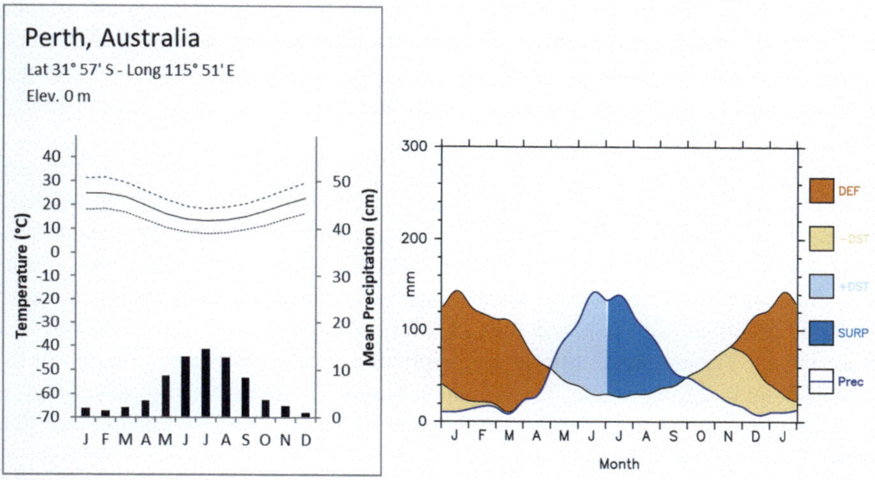

Fig. 20.12 Climagraph and water balance diagram for Perth, Australia. Data for 1991–2020 averaging period. *Source* Authors. https://weatherspark.com/, http://cyclops.deos.udel.edu/wimp/public_html/

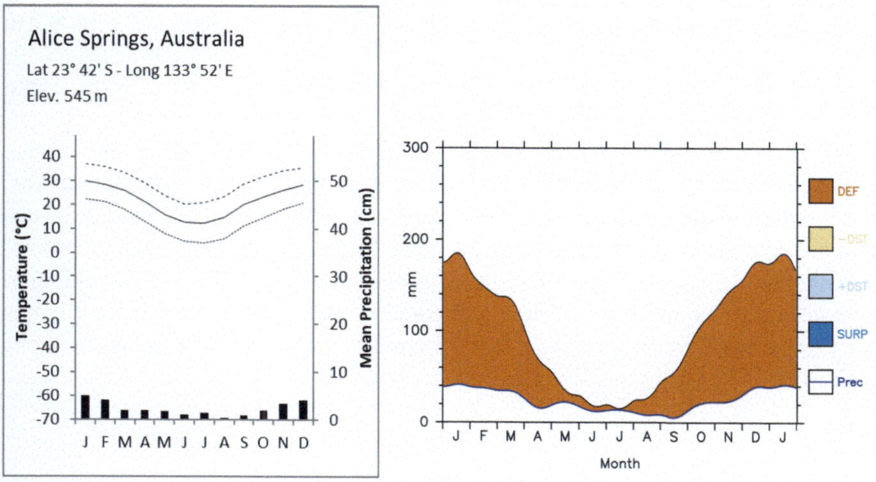

Fig. 20.13 Climagraph and water balance diagram for Alice Springs, Australia. Data for 1991–2020 averaging period. *Source* Authors. https://weatherspark.com/, http://cyclops.deos.udel.edu/wimp/public_html/

New Zealand

The extensive maritime influence moderates New Zealand's temperatures. Maximum summer temperatures range from 18 to 24 °C (64–75 °F) on the North Island and 20 to 24 °C (68–75 °F) on the South Island. Winter temperatures average approximately 9–11 °C (48–52 °F) on North Island and 4–8 °C (39–46 °F) on the South Island. Higher elevations have much lower overall temperatures, especially in the Southern Alps. Glaciers are permanent features above the snowline, which lies at an average of about 2000 m (6600 ft). Winds, including the *Roaring Forties*, are persistently strong in most of New Zealand.

The maritime influence on the mid-latitude westerlies also makes New Zealand generally wetter than any part of Australia except the northern ITCZ-affected area. Precipitation occurs throughout the year, with a winter (June to August) maximum occurring when the polar front shifts equatorward. Although many clear days occur throughout the summer, nearby warm water and abundant *humidity* ensure that moderate precipitation totals are still achieved. Violent thunderstorms are rare on either island. These temperature and precipitation conditions place virtually all of New Zealand in a Cfb climate. The climate is similar to the western coastal climates of the northwestern U.S., adjacent southwestern Canada, western Europe, and the central Pacific coast of South America.

The mountain regions of the North Island are more complex and sporadic compared to the higher, more continuous mountain chain of the South Island. This topography creates a complicated climatology, with a haphazard pattern of rainfall totals on the North Island. The western side does experience some orographic effect. However, the lack of high and continuous mountains minimizes the *rain shadow* effect. Tropical cyclones occur on the east side of the North Island during the warm season, thus masking more of the effects of rain shadows. Auckland (Fig. 20.14) represents the climate of the very heterogeneous and complicated North Island (Fig. 20.15).

Precipitation patterns are much simpler on the South Island because of the consistency of the westerly-embedded storm systems interacting with the higher mountains. The Southern Alps range parallels the coast from northeast to southwest and runs at an angle to the westerly wind flow. The resulting orographic influence on precipitation creates a very wet western side with over 500 cm (200 in) of precipitation in some locations and a much drier rain shadow side. The eastern side, in cities like Christchurch, receives only about 50 cm (20 in) per year in many locations (Fig. 20.16).

Fig. 20.14 Auckland, New Zealand, in August. *Source* Authors

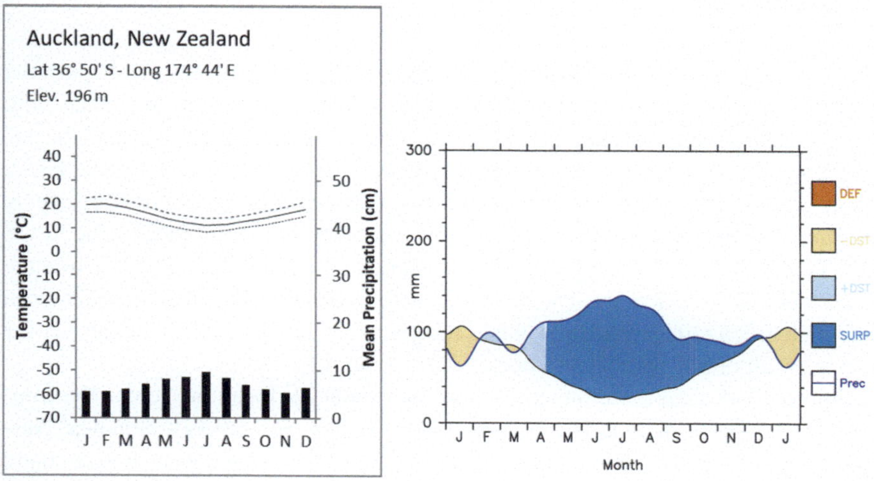

Fig. 20.15 Climagraph and water balance diagram for Auckland, New Zealand. Data for 1980–2016 averaging period. *Source* Authors. https://weatherspark.com/, http://cyclops.deos.udel.edu/wimp/public_html/

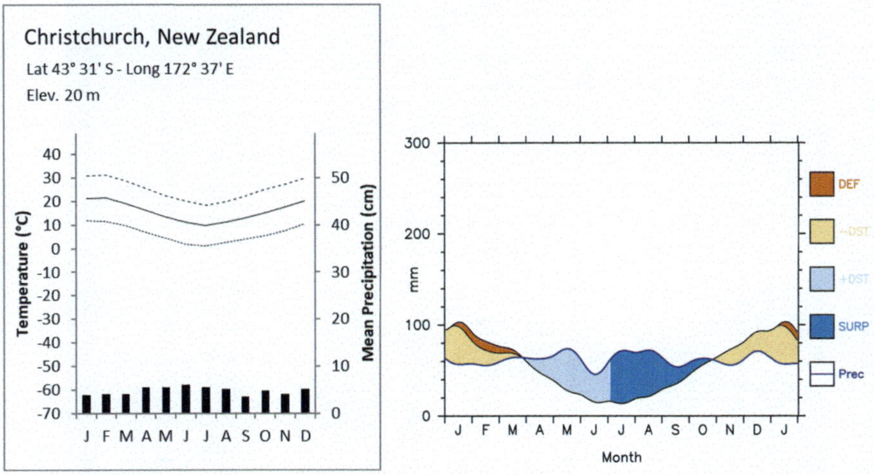

Fig. 20.16 Climagraph and water balance diagram for Christchurch, New Zealand. Data for 1991–2020 averaging period. *Source* Authors. https://weatherspark.com/, http://cyclops.deos.udel.edu/wimp/public_html/

Pacific Islands

The realm of the wet-all-year *Tropical Rain Forest* (Af) *climate* forms the shape of the letter "c" across the western tropical Pacific basin, including numerous islands, with *Tropical Monsoon* (Am) *climate* and the Aw climate progressively "inside" the "c" (see again Fig. 11.1). Precipitation amounts decrease from west to east, except during warm-event ENSO years. The westernmost islands are influenced more directly by the *Pacific Warm Pool* and the strong convection of the ITCZ and SPCZ. The SPCZ and ITCZ cause a boomerang-shaped precipitation maximum (Fig. 20.17) across the equatorial and southwestern Pacific basin, with South Pacific islands frequently in their path much rainier than they would be otherwise. BWh and BSh climates cover a wide swath of the mostly open ocean of the eastern Pacific.

Figure 20.18 displays climagraphs for representative locations across the tropical Pacific. These include Honiara, Solomon Islands (Fig. 20.19); Nouméa, New Caledonia (Fig. 20.20); Honolulu, Hawai'i (Fig. 20.21); and Mataveri, Easter Island (Fig. 20.22).

These locations demonstrate a general decrease in precipitation eastward, with consistent temperatures year-round both within and across sites. Water balance diagrams are unavailable for many of these small island locations, but water availability decreases generally from west to east. Local topography affects rainfall patterns on the islands themselves with windward slopes, usually on the east, being far

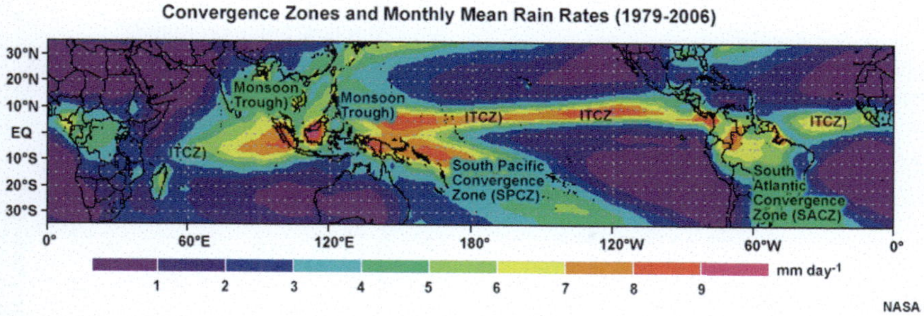

Fig. 20.17 Mean rainfall rates across the Pacific basin. *Source* UCAR CoMet. Permission from Tara Torres

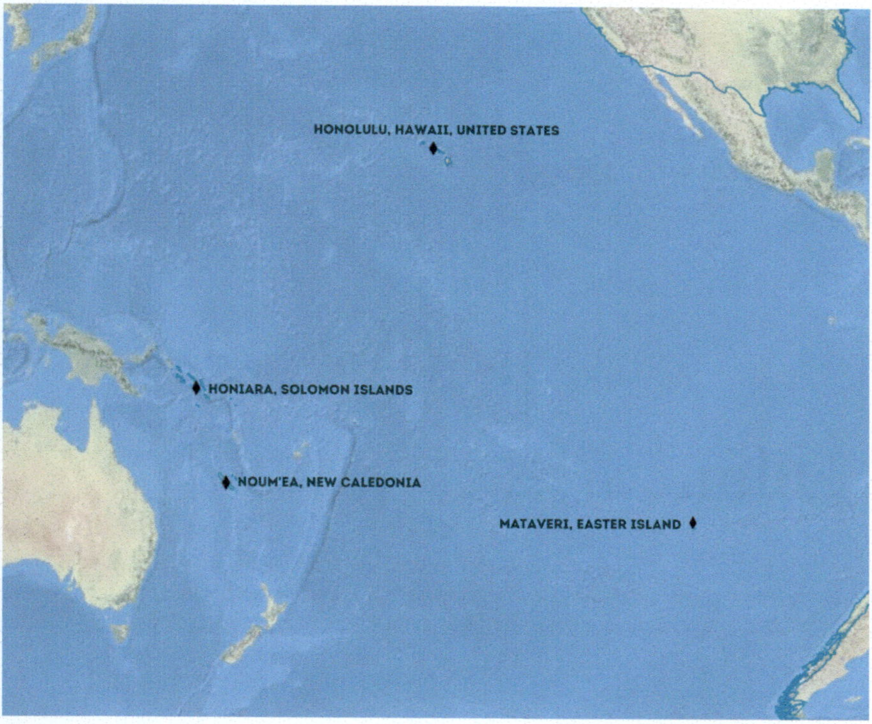

Fig. 20.18 Pacific island locations for which climagraphs and water balance diagrams are shown. *Source* Geaux Geospatial

wetter than *leeward* sides. Figure 20.23 shows average annual precipitation across the Hawaiian Islands. The windward slopes can be some of the wettest places on Earth, while the leeward sides of the islands are often B climates. Figure 20.24 depicts temperatures for Hawai‘i and suggests that the cloud cover on the windward side suppresses the daily temperature range by reflecting insolation during the hot

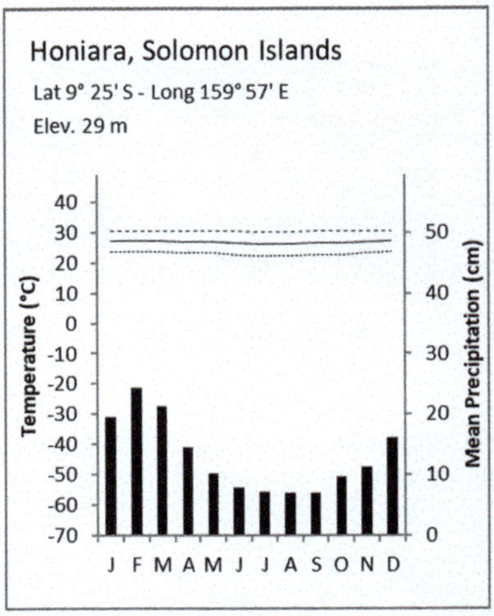

Fig. 20.19 Climagraph for Honiara, Solomon Islands. Data for 1980–2016 averaging period. *Source* Authors. https://weatherspark.com/

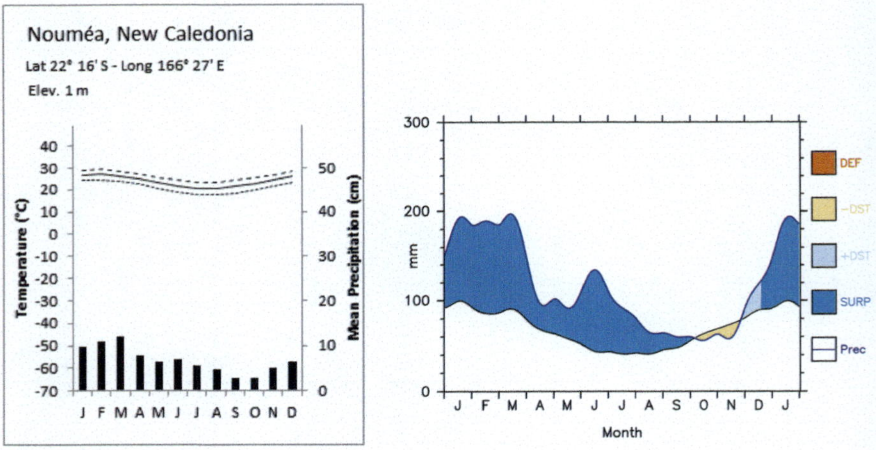

Fig. 20.20 Climagraph and water balance diagram for Nouméa, New Caledonia. Data for 1980–2016 averaging period. *Source* Authors. https://weatherspark.com/, http://cyclops.deos.udel.edu/wimp/public_html/

afternoon hours and minimizing *outgoing longwave radiation* all the time, but with its greatest effect supporting increased daily minimum temperatures during the early morning hours.

Though sparsely populated, the Pacific basin is an important component of the global climate system. Interactions between atmospheric and oceanic circulation patterns in that region not only affect Australia, New Zealand, and the Pacific islands, but have important consequences on climate variability around the world through teleconnections such as ENSO and the QBO. A discussion of the critical role of

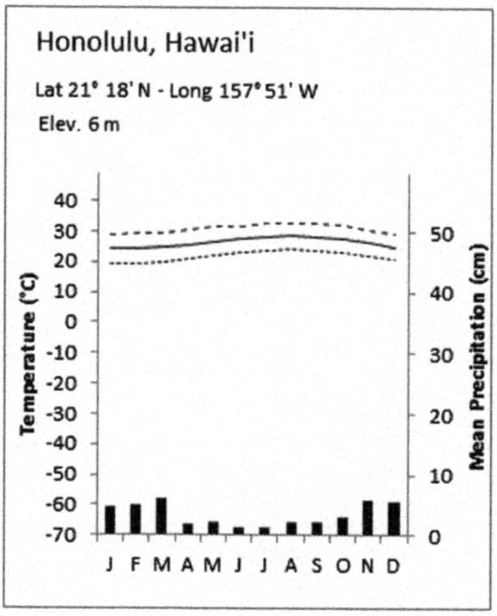

Fig. 20.21 Climagraph for Honolulu, Hawai‘i. Data for 1991–2020 averaging period. *Source* Authors. https://weatherspark.com/

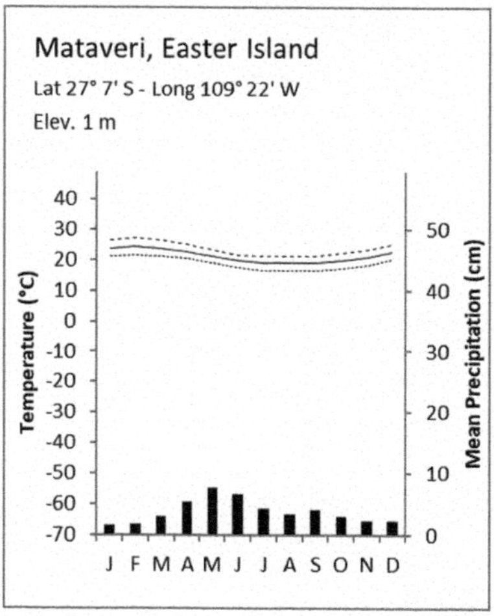

Fig. 20.22 Climagraph for Mataveri, Easter Island. Data for 1980–2016 averaging period. *Source* Authors. https://weatherspark.com/

circulation features in remote parts of the Earth continues in the next chapter, which covers the last major region—the poles.

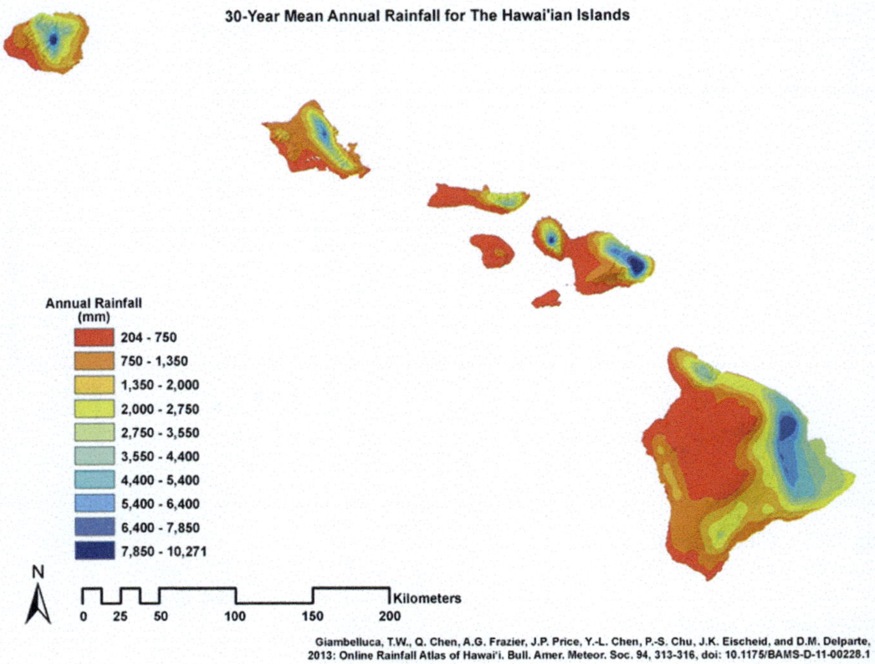

Fig. 20.23 Annual mean precipitation in the Hawaiian Islands. *Source* UCAR CoMet/NOAA

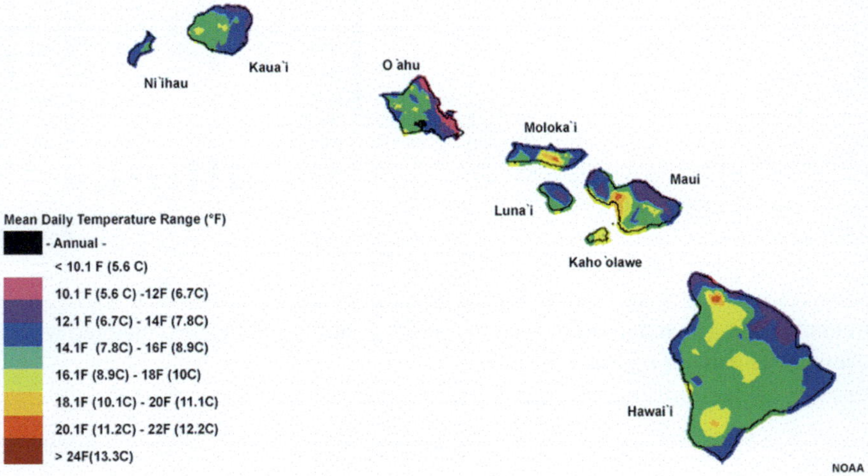

Fig. 20.24 Mean daily temperature range in the Hawaiian Islands. *Source* UCAR CoMet/NOAA

Further Reading

Baldwin, M. P., Gray, L. J., Dunkerton, T. J., Hamilton, K., Haynes, P. H., Randel, W. J., Holton, J. R., Alexander, M. J., Hirota, I., Horinouchi, T., Jones, D. B. A., Kinnersley, J. S., Marquardt, C., Soto, K., & Takahashi, M. (2001). The quasi-biennial oscillation. *Reviews of Geophysics, 39*(2), 179–229. https://doi.org/10.1029/1999RG000073

Beischer, T. A., Gregory, P., Dayal, K., Brown, J. R., Charles, A. N., Wang, W. X. D., & Brown, J. N. (2021). Scope for predicting seasonal variation of the SPCZ with ACCESS-S1. *Climate Dynamics, 56*(5–6), 1519–1540. https://doi.org/10.1007/s00382-020-05550-6

Chiew, F. H. S., Piechota, T. C., Dracup, J. A., & McMahon, T. A. (1998). El Niño Southern Oscillation and Australian rainfall, streamflow, and drought: Links and potential for forecasting. *Journal of Hydrology, 204*(1–4), 138–149. https://doi.org/10.1016/S0022-1694(97)00121-2

Hastenrath, S. (2007). Equatorial zonal circulations: Historical perspectives. *Dynamics of Atmospheres and Oceans, 43*(1–2), 16–24. https://doi.org/10.1016/j.dynatmoce.2006.04.001

Kataoka, T., Tozuka, T., Behera, S., & Yamagata, T. (2014). On the Ningaloo Niño/Niña. *Climate Dynamics, 43*, 1463–1482. https://doi.org/10.1007/s00382-013-1961-z

King, A. D., Lingaman, N. P., Alexander, L. V., Donat, M. G., Jourdain, N. C., & Maher, P. (2014). Extreme rainfall variability in Australia: Patterns, drivers, and predictability. *Journal of Climate, 27*(15), 6035–6050. https://doi.org/10.1175/JCLI-D-13-00715.1

Lorrey, A., Dalu, G., Renwick, J., Diamond, H., & Gaetani, M. (2012). Reconstructing the South Pacific convergence zone position during the Presatellite Era: A La Niña case study. *Monthly Weather Review, 140*(11), 3653–3668. https://doi.org/10.1175/MWR-D-11-00228.1

Reed, R. J., Campbell, W. J., Rasmussen, L. A., & Rogers, D. G. (1961). Evidence of a downward-propagating, annual wind reversal in the equatorial stratosphere. *Journal of Geophysical Research, 66*(3), 813–818. https://doi.org/10.1029/JZ066i003p00813

Salinger, M. J. (1980). New-Zealand climate: 1 precipitation patterns. *Monthly Weather Review, 108*(11), 1892–1904. https://doi.org/10.1175/1520-0493(1980)108%3c1892:NZCIPP%3e2.0.CO;2

Chapter 21
Polar Regions

Abstract The polar regions are important components of the atmospheric and oceanic circulation. While the Arctic region is largely capped by sea ice that is decreasing in extent, the Antarctic is covered by land and surrounded by a continuous ocean. Both hemispheres have a semi-permanent surface anticyclone near the pole, along with a circumpolar vortex aloft. The intense continentality and cryosphere gives the South Polar region special characteristics that make it an important driver of the global thermohaline circulation, particularly through subsurface ocean currents in such an oxygen-rich part of the world ocean. While Arctic temperatures are frigid and contain polar climates, temperatures are far more moderate than over the Antarctic. Likewise, precipitation is sparse in both the Arctic and Antarctic, but especially in the Antarctic, where virtually no precipitation occurs near the pole. Occasional summer mid-latitude wave cyclones affect coastal Antarctica and the Arctic. The stratospheric ozone hole has been problematic especially in the Antarctic, but it has been repairing itself in recent years due to global bans on the products that were depleting it.

Keywords Antarctic Bottom Water (AABW) · Antarctic Circumpolar Current · Antarctic Coastal Current · Antarctic Deep Water (AADW) · Antarctic Intermediate Water (AAIW) · Circumpolar vortex · Common Water · North Pacific Intermediate Water · Ozone hole · Pacific Subarctic Water (PSW) · Permafrost · Polar amplification · Polar (ET and EF) climates · Subarctic (Dfc and Dfd) climates

General

Situated at the highest latitudes, the polar regions have the harshest environments on Earth's surface. Limited daylight hours, low *Sun angles*, high surface *albedo*, and small *sensible heat fluxes* lead to extremely low winter *temperatures*. While both poles have similar radiation processes and atmospheric circulation patterns, different geographies create different levels of climate extremes in the Arctic and Antarctic.

R. V. Rohli et al., *Atmospheric and Oceanic Circulation*,
https://doi.org/10.1007/978-3-031-66142-6_21

Topographic Influences (Site)

The northern and southern polar regions have very different topographic characteristics. Approximately 75% of the Arctic is covered by ocean and only 17% of the region is land surface (Fig. 21.1). Approximately 8% of the Arctic is covered by *glaciers*, mostly in the interior of Greenland, on which a continental *ice sheet* lies.

The continental areas of the Arctic are of low elevation and largely covered by tundra vegetation. Both land and water are affected by ice in many forms which comprise the Earth's *cryosphere*. On land, this designation includes frozen soil moisture, termed **permafrost**, as deeper levels remain frozen permanently, as well as seasonal snow cover on the surface.

The Arctic Ocean is largely capped by *sea ice*, though total ice coverage is decreasing. *Seasonal sea ice* forms during winter and melts during summer. Even during winter, this ice cover is typically only 1–2 m (3–7 ft) thick. *Multi-year sea ice* lasts longer than a year, with typically double the thickness of the annual sea ice layer, being 3–4 m (7–13 ft) thick. Even in zones of ice cover, liquid water is never far from the surface. The warmer ocean adds energy into this Arctic environment

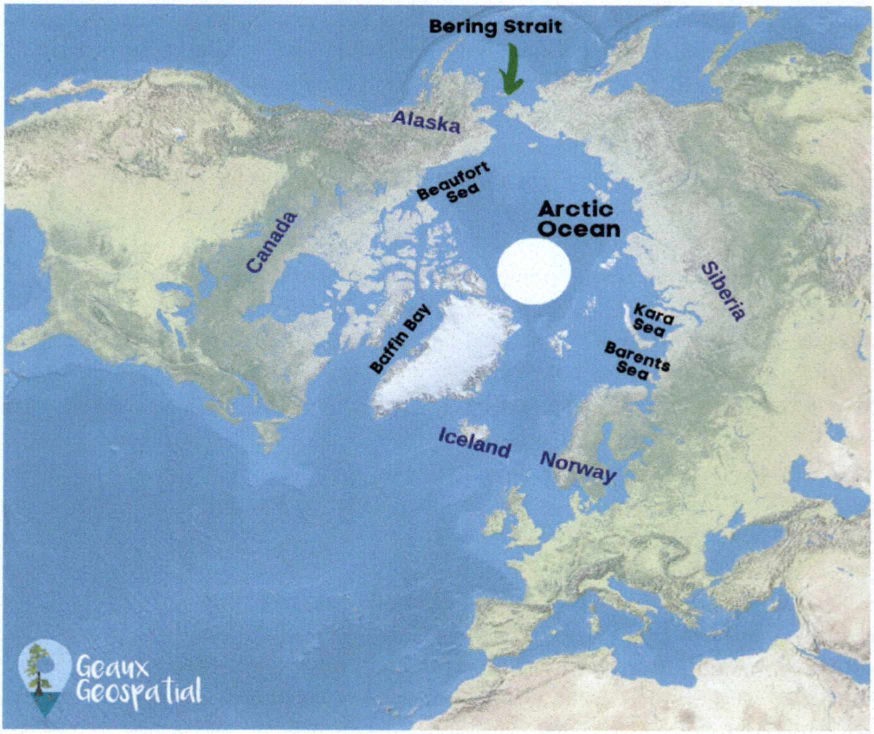

Fig. 21.1 Selected locations of the Arctic. *Source* Geaux Geospatial

and this contribution keeps temperature much higher than the land-based Antarctic region of the Southern Hemisphere, even in winter.

Most of the area south of the *Antarctic Circle* is covered by land. The geography of the land mass is skewed toward the Indian Ocean. Antarctica has the highest average elevation at approximately 2500 m (8200 ft) of all the continents. Most of the elevation consists of the glacial *ice sheet*, which averages 2160 m (7100 ft) in thickness after accumulating for the last 45 million years. Today, ice encompasses 14 million km^2 (5.4 million mi^2), an area twice the size of Australia. Ice covers approximately 95% of the continent. The exception to this ice cover is the amazing Antarctic dry valleys near McMurdo Sound. As the surrounding ocean is far warmer than the interior, elevation and ice thickness decrease abruptly near the coast.

The surface of the Antarctic *ice cap* is rough. Compression ridges form where the ice is pushed together and are a common feature of ice caps as well as sea ice. Additionally, deep cravesses form at the surface, as the speed of the moving ice exceeds that of the increased-*friction* regions that underlie the surface. As a result, the upper ice splits and fractures.

Circulation Influences (Situation)

Because of their low temperatures, the North and South Poles have "permanent" areas of high pressure at the surface, caused by the bitterly cold, dense air. In the *polar cell*, cold dense air sinks at the poles and flows equatorward toward the *polar fronts*. High pressure at the highest latitudes stymies lifting by *convection*. However, these cold-cored surface highs become lows aloft. The "extra" air that makes high pressure near the surface piles up atmospheric mass there, but since it is so cold, it compresses. Thus, with increasing height, the compression causes the *isobars* (orange lines in Fig. 21.2) to flatten and eventually become bowl-shaped (i.e., low-pressure areas) with height.

The anticyclone over the North Pole is separate from the *Siberian High* which forms over the large continental landmass of Asia during winter. The upper-air pattern above the *North Polar High* forms the core of the *circumpolar vortex*—a rotating disk of cold air that is especially strong in winter and positioned over the North Pole. The surface high and the circumpolar vortex in general are usually neither symmetrical nor centered on the pole. Instead, often the high-pressure cell stretches down over Asia and the circumpolar vortex skews toward the Pacific Ocean side of the North Pole. Energy, air masses, and transitory pressure systems are likely to move longitudinally around the "wavy" edges of the Northern Hemisphere's circumpolar vortex. Near the surface, however, the *polar easterlies* prevail. Thus, a reversal of atmospheric circulation occurs between the near-surface atmosphere and the upper-level atmosphere.

In the future, this high-amplitude pattern of the circumpolar vortex may be amplified further. Polar areas are expected to warm more than the tropics because of a positive feedback loop involving snow/ice melt, decreasing albedoes, and increasing

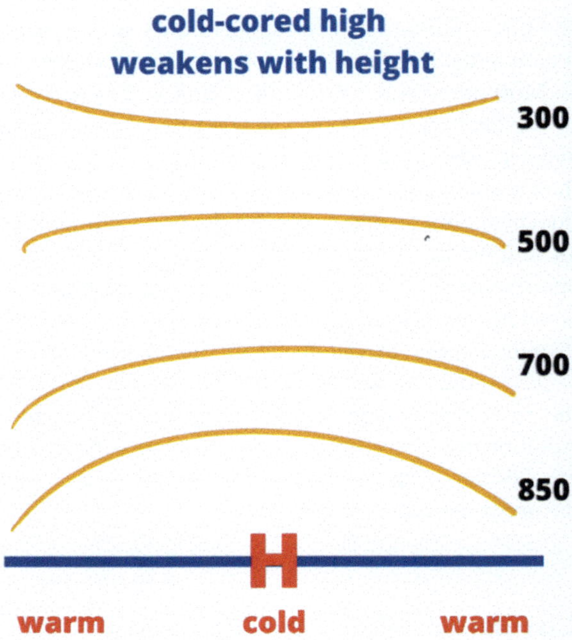

Fig. 21.2 Schematic representation of pressure pattern above a cold-cored surface anticyclone. *Source* Geaux Geospatial

radiation absorption. This *polar amplification*, sometimes known as *Arctic amplification*, would reduce the hemispheric temperature gradient. Many atmospheric scientists hypothesize that this change will enhance meridional flow at the expense of *zonal* flow across the mid-latitudes, but some disagree. Ironically, climate change could cause more frequent cold polar air outbreaks into the troughs that transport cold air equatorward.

Over Antarctica, a semi-permanent surface *anticyclone* also encircles the South Pole's *polar high*. The system is a very cold-cored high which, as with other cold-cored highs, weakens with height, eventually giving way to pressure aloft that is even lower than the areas surrounding it at the same height. The general circulation associated with the Antarctic circumpolar vortex is counter-clockwise when viewed from above, as with circulation around all high-pressure areas in the Southern Hemisphere. However, the Southern Hemisphere's circumpolar vortex is shallow because gravity confines it closely to the generally high elevation of the continent. The geography of the polar continent surrounded by the continuous Southern Ocean around the hemisphere enhances zonal flow. The counter-clockwise circulation is persistent and strong, and it hinders the exchange of energy between Antarctica and the lower latitudes.

The Siberian region in the Northern Hemisphere and Antarctica in the Southern Hemisphere are source regions for extreme cold air outbreaks brought by *continental Arctic* (cA or A) *air masses*. These exceedingly cold and dry air masses originate near the poles. In the Arctic, meridional waves of the circumpolar vortex allow cA air masses to sweep down into North America and Asia periodically. They often follow the passage of *mid-latitude wave cyclones* and can migrate as far south as the U.S. Midwest. cA air masses also influence the majority of Antarctica, where they are contained by strong zonal winds at the continental margins. cA air masses have a very small *water vapor capacity*. Such cold, dense air masses produce a *stable atmosphere* by altering the *environmental lapse rate* or even creating a *temperature inversion*, except when they overlie warmer areas like unfrozen lakes or bare soil that heat more than the surrounding areas.

Continental polar (cP) *air masses* originate in other high-latitude regions of the Northern Hemisphere, such as Canada, Alaska, and Europe. These cold but not excessively cold, dry air masses are most prominent in the polar regions during summer when temperatures increase. In the Southern Hemisphere, most polar air masses have maritime source regions. *Maritime polar* (mP) *air masses* from the Southern Ocean can interact with cA air to produce frontal *precipitation* in coastal Antarctica, especially in summer. The polar regions tend to experience moderate to light snowfall rather than heavy blizzards since the extreme cold reduces water vapor capacity which, in turn, reduces snowfall totals. Instead, the direct *deposition* and *riming* of water vapor onto the surface builds annual ice layers.

A final feature of atmospheric circulation over Antarctica is the topographically-induced local wind. Winds over the continent average 80 km hr^{-1} (50 mph), but gusts often exceed 160 km hr^{-1} (100 mph). The winds originate in the high-elevation continental interior. The extreme cold, dense air flows downhill toward the coastal areas. These dense flows are termed *katabatic winds* and are strongest in winter when the air is coldest, and near coasts, where the slopes are steepest. Extreme surface winds continually blow surface ice, creating blizzard-like blowing snow and poor visibility conditions. Winds are often strong enough to push sea ice out away from the coast, even to the extent that sea ice shelves can be detached from the continent.

Because of its distinctive temperature, salinity, and oxygen content, Antarctic ocean circulation is very important as a modulator of the global ocean *thermohaline circulation*. The **Antarctic Circumpolar Current** or **West Wind Drift** (Fig. 21.3) flows near the surface of the ocean from west to east, or clockwise as viewed from below the South Pole, around Antarctica. The strong flow isolates Antarctica by preventing the much warmer water from approaching Antarctica's coastal areas. This pattern aids in maintaining the extremely low temperatures of the continent. The Antarctic Circumpolar Current is arguably the world's most robust ocean current.

Trapped between the Antarctic Circumpolar Current and the continent is the *Antarctic Coastal Current* or *East Wind Drift*, which flows from east to west, creating a shear with the Antarctic Circumpolar Current. The Antarctic Coastal Current forms as katabatic winds from interior Antarctica race to the coast and are deflected to the left by the *Coriolis effect*.

Antarctic Bottom Water (AABW; Fig. 21.4) consists of a large mass of deep, cold water at the floor of the Southern Ocean next to Antarctica. The water forms because of extreme surface chilling, which increases water density. As dense water sinks to the ocean floor, it spreads outward through deep currents that are driven by temperature and salinity variations—a thermohaline circulation. The water has a greater oxygen content than the other deep waters of the world ocean because

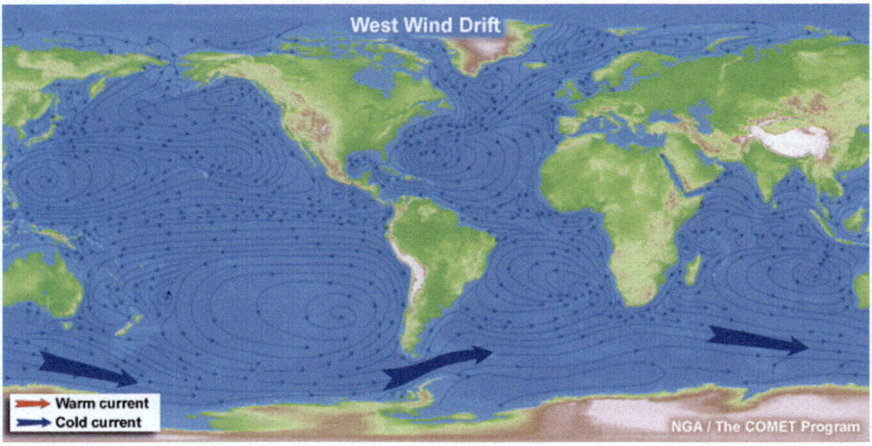

Fig. 21.3 The Antarctic Circumpolar Current or West Wind Drift. *Source* UCAR CoMet. Premission from Tara Torres

less oxygen is used in the decay of organic matter in the harsher, more lifeless AABW. Antarctic bottom water is important in the global ocean circulation because it ventilates the rest of the deep world ocean with oxygen.

Antarctic Deep Water (AADW) is another deep ocean current that flows over the AABW as it moves northward and gently upward. The flow eventually moves

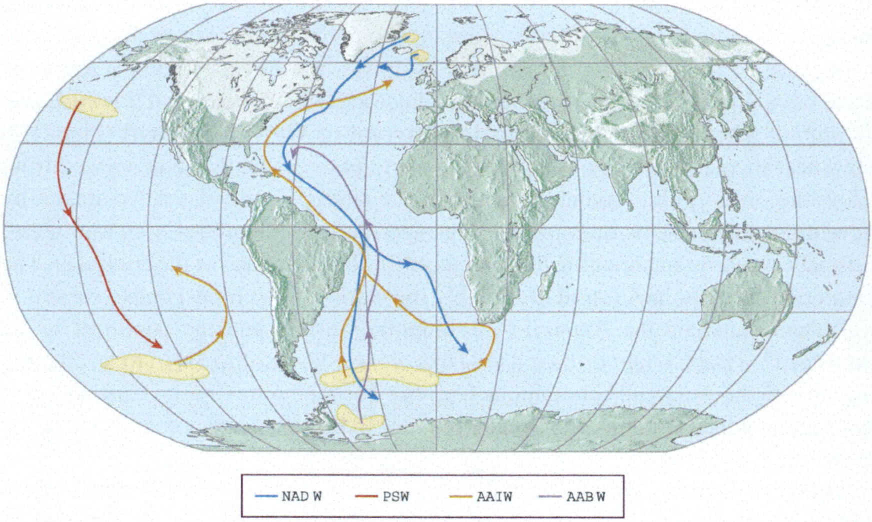

Fig. 21.4 Generalized path of the North Atlantic deep water, Pacific Subarctic Water, Antarctic Intermediate Water, and Antarctic Bottom Water. *Source* Figure 04.07 in Rohli and Vega, *Climatology* 4/e, where it was adapted from Gordon, A. L., Lamont-Doherty Geological Observatory Report 1990–91

beneath the current known as the **North Atlantic Deep Water** (NADW; see again Fig. 21.4) as part of the global ocean circulation. **Common Water** is yet another deep thermohaline current that is initiated near Antarctica, in the Pacific Ocean sector. Along with the **Pacific Subarctic Water** (PSW), Common Water occupies the lowest layers of the ocean in that area. **Antarctic Intermediate Water** (AAIW; see again Fig. 21.4) along with the **North Pacific Intermediate Water** overrides the Common Water and Pacific Subarctic Water.

Source areas for the water's properties are shown by the shaded areas of the ocean in Fig. 21.4. While seemingly confusing, all currents, much like air or other fluids, follow gradients. The gradients move substances of greater concentration to areas of lesser concentration, in fulfillment of the *Second Law of Thermodynamics*. For deep ocean waters, the gradients are in density, which are caused by differences in temperature and salt content—thermohaline properties that drive the thermohaline circulation. This flow is an important mechanism for moving energy, matter—such as oxygen, and momentum between the high and low latitudes. The polar regions are important in global circulation.

The Arctic

The lowest Arctic temperatures occur over inland Greenland and Siberia in all seasons rather than over the pole itself. *Advection* of heat from warm ocean currents is stored in the Arctic Ocean, keeping coastal temperatures somewhat moderated. Temperatures on the northern Greenland coast average about 4 °C (40 °F) in July, while inland areas are colder where ice cover is permanent. The frigid cA air masses overlying the much warmer water leave the coastal areas with a frequently *unstable atmosphere*, with rapid cooling with height and therefore a steep *environmental lapse rate*. Cloud cover is limited by the relatively small *water vapor capacity* but is often sufficient to absorb enough longwave radiation from the surface to prevent temperatures from plummeting drastically overnight.

High-latitude regions have very long daylight periods in summer but very short daylight periods in winter. Because of the extreme daylight period variation through the year, the equinox periods are close to the time of highest and lowest temperatures. Lowest temperatures typically occur in March, just before the Sun finally rises and remains above the horizon for many consecutive calendar days—up to six months at the Pole. The highest temperatures are usually in August or September, as the long continuous daylight period can keep the *net radiation* (Q*) positive until just before the daylight hours become less than the nighttime hours at the September equinox.

Even though there are long hours of daylight in summer, *shortwave radiation* (K↓) remains small to moderate because the perpetually low Sun angles cause significant *attenuation* through the long path length from the top of the atmosphere to the surface. *Albedo—reflection* of shortwave radiation, or K↑—is very high (60–80%) on the smooth, lightly-colored surfaces of ice and water throughout the region. Water has a high reflectivity when Sun angles are small.

Because the surface often remains cold, a deep temperature inversion which may extend up to 2 km (1 mi) or more is common. The persistence and thickness of the inversion increases in winter. As uplift is stymied, the only clouds that typically form are fogs and low *stratus,* which are frequent. The situation is problematic for scientists because the *condensation* can hide the surface and make fog indistinguishable from stratus cloud cover in satellite images and aerial photographs. Winds are persistently strong.

Most of the heat in the Arctic reaches the surface from below, rather than from above as in most of the rest of the Earth. The sensible heat flux is directed laterally landward from the subsurface water, before moving vertically upward, even when the ocean surface is ice-covered. Small areas of open water can be important sources of heat throughout the year.

Annual sea ice melts, with the smallest coverage of remaining ice occurring during the warmest month of August or September. Snow may linger into June and melts when the energy balance shifts quickly to sensible heating at the expense of latent heating. The snow and ice melt decreases albedo to 20–30%.

Most precipitation falls during summer as the *polar front* brings a few cyclones into the region. However, the totals are meager, with only about 25 cm (10 in) of precipitation recorded per year. Rapid weather changes occur, especially when the *Rossby wave* pattern is *meridional.* Typically, about twice as many cyclones occur during summer than in winter. However, because of the surface chilling, most cyclonic systems dissipate quickly over the region.

Climagraphs and water balance diagrams for four representative Arctic sites at locations shown in Fig. 21.5 appear in Figs. 21.6, 21.7, 21.8 and 21.9. The continental temperatures characteristic of Yellowknife, Northwest Territories, Canada (Fig. 21.6), are apparent. Likewise, the abrupt increase in surface water when the ice finally begins to melt in late spring is also evident at Yellowknife and the other three locations—Nuuk, Greenland (Fig. 21.7), Utqiagvik, Alaska (Fig. 21.8), and Nome, Alaska (Fig. 21.9). All sites except for the most extreme—Nuuk—show water *deficits* in mid- to late summer, though the actual environmental impact depends on the local status of permafrost in the soil layers. Locations in the vast *Subarctic* (Dfc, Dfd, Dwc, Dwd) *climates* containing the *taiga*—the Siberian term for *boreal forest*, across the treeline into the *Tundra* (ET) *climate* of central and eastern Eurasia, would show similar climagraphs and water balance diagrams.

The Antarctic

Because of its vast remoteness and extreme climate, scientists still know relatively little about the Antarctic. Only a few thousand research scientists and support personnel live poleward of 60° S latitude. The Amundsen expedition was the first to reach the South Pole, in 1911. With only 100 years of observation, our information about Antarctica is very recent.

Fig. 21.5 Arctic locations for which climagraphs and water balance diagrams are shown. *Source* Geaux Geospatial

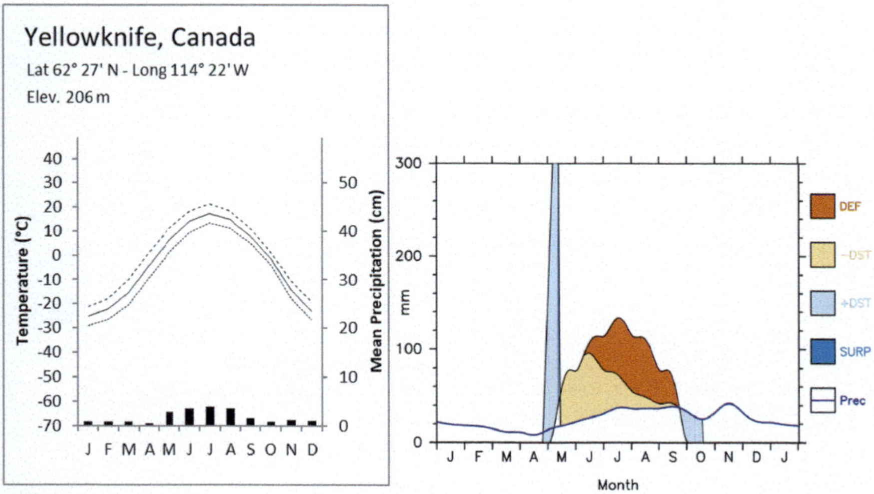

Fig. 21.6 Climagraph and water balance diagram for yellowknife, Northwest territories. Data for 1980–2016 averaging period. *Source* Authors. https://weatherspark.com/, http://cyclops.deos.udel.edu/wimp/public_html/

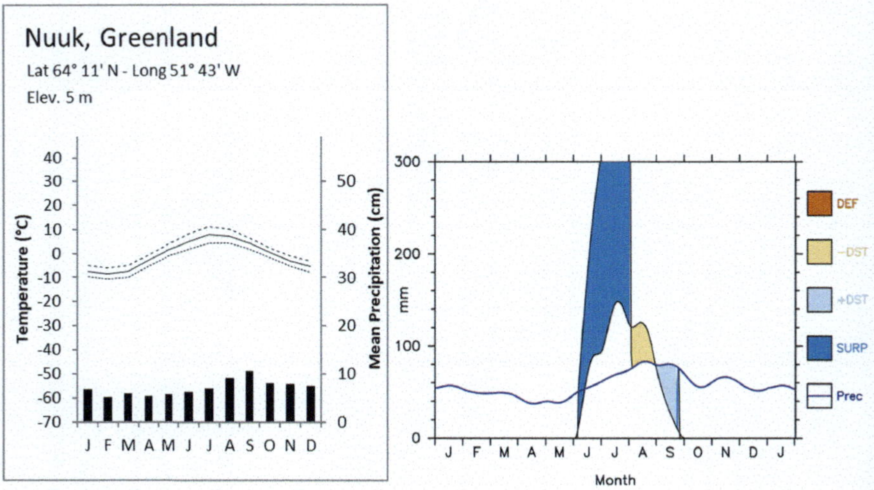

Fig. 21.7 Climagraph and water balance diagram for Nuuk, Greenland. Data for 1991–2020 averaging period. *Source* Authors. https://weatherspark.com/, http://cyclops.deos.udel.edu/wimp/public_html/

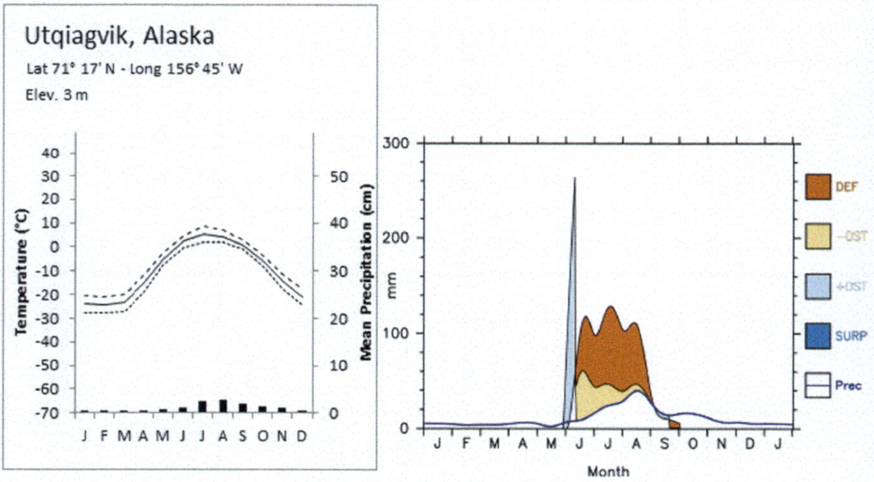

Fig. 21.8 Climagraph and water balance diagram for Utqiagvik, Alaska. Data for 1991–2020 averaging period. *Source* Authors. https://weatherspark.com/, http://cyclops.deos.udel.edu/wimp/public_html/

Temperatures over Antarctica are extremely low, especially during the winter. The lowest temperature ever recorded at the Earth's surface was −89 °C (−129 °F), at the then-Soviet station of Vostok [78° 28′ S, 106° 48′ E, elevation: 3488 m (11,444 ft)] on 26 July 1983. Temperatures are low even during summer, as high surface albedo sends a great percentage of the incoming solar radiation (85–90%) upward since

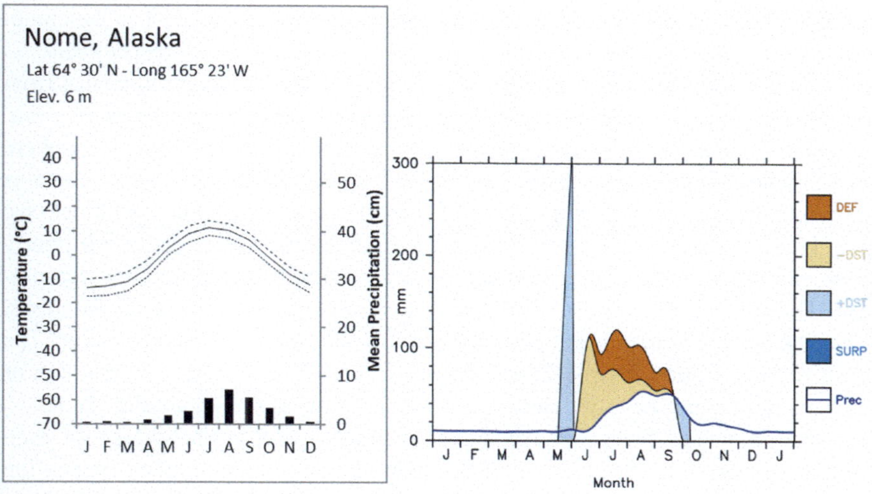

Fig. 21.9 Climagraph and water balance diagram for Nome, Alaska. Data for 1980–2016 averaging period. *Source* Authors. https://weatherspark.com/, http://cyclops.deos.udel.edu/wimp/public_html/

snow/ice is always present, except in the dry valleys. The high elevation means that *adiabatic* cooling with any rising motion already starts from a very low temperature and only gets colder as it rises. Oblique Sun angles further exacerbate the low temperatures as solar radiation has little heating power per unit area. The strongly zonal circumpolar vortex limits latitudinal heat transport. All of these factors combine with continentality to produce the coldest climate on Earth.

As expected, the lowest temperatures occur in the continental interior. Average annual temperatures in the interior are approximately −50 °C (−60 °F). Because of the strong temperature inversions over the continent, surface temperatures may be 10–18 °C (18–32 °F) colder than the overlying air. The *annual temperature range* is not as great as in Siberia, because Antarctic summers do not warm as much as Siberia warms.

The coastal regions are milder because of the energy contained in the ocean body. There, average annual temperatures are approximately −20 to −10 °C (−4 to –14 °F). The highest temperatures along the coast tend to be around −1 °C (30 °F), but a record high temperature of 21 °C (69 °F) was recorded in February 2020. Still, coastal Antarctica is typically colder than even the coldest Arctic temperatures. The summer climate experiences long daylight periods and abundant solar radiation, but the albedo is high and the Sun angles are small. As a result, there is little heating power and summers are cold. A few exceptions occur where local dark surfaces occur. Blowing snow exposing darker underlying surfaces produces a *positive feedback* loop. Exposed land warms fast, melting more snow, allowing for more area to be warmed.

In the extremely cold air, water vapor content is very low, even approaching zero during winter. The result is that Antarctica is not only the coldest but also the driest continent. The aridity is apparent for the locations shown in Fig. 21.10, in the climagraphs for coastal and inland Antarctica, in Figs. 21.11 and 21.12, respectively. As in the Arctic, the very small amount of precipitation in the Antarctic comes from mid-latitude wave cyclones associated with seasonal migration of the *polar front*. Precipitation totals are miniscule but are concentrated during Southern Hemisphere summer when the polar front retreats to higher latitudes and the warmer air contains more *humidity*, as in the Northern Hemisphere's polar sites for boreal summer (see again Figs. 21.8 and 21.9). Nevertheless, precipitation in Antarctica is still poorly understood because the high winds and extreme conditions confound its measurement. The elevated interior limits cyclone-induced precipitation to coastal locations, where annual precipitation is approximately 25–40 cm (10–15 in) per year. The higher-elevation interior typically experiences 5 cm (2 in) of total precipitation or less per year, dwindling to virtually zero at the South Pole (see again Fig. 21.12).

A final important feature of Antarctica's climate is not at the surface, or even in the *troposphere*. In 1985, a team of British researchers studying the composition of the *stratosphere* discovered the Antarctic *ozone* (O_3) hole. Later, a smaller hole was found over the Arctic. Research ultimately answered the question of why such holes form over the high latitudes. Ozone is built naturally in the atmosphere in the presence of **ultraviolet radiation** from the Sun. Diatomic oxygen (O_2) produced by plants in photosynthesis is dissociated (broken apart) in the atmosphere by sunlight. The photodissociation produces two monatomic oxygen (O) atoms. An O atom can then bond with an O_2 in the stratosphere to produce O_3. Solar radiation then acts as a catalyst to photodissociate the O_3 molecule into O_2 and O. The process continues

Fig. 21.10 Antarctic locations for which climagraphs and water balance diagrams are shown. *Source* Geaux Geospatial

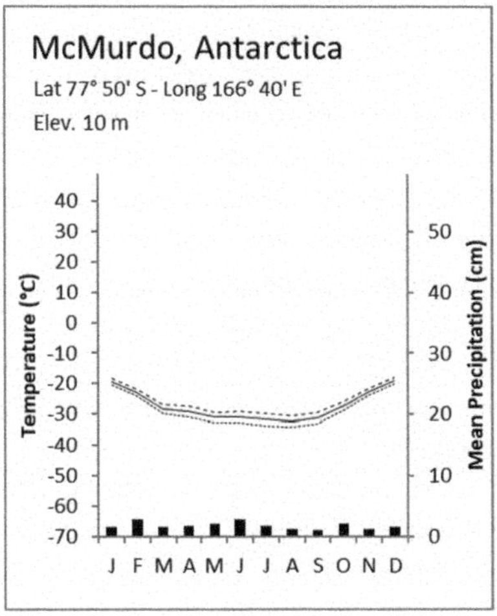

Fig. 21.11 Climagraph for McMurdo Research Station, Antarctica. Data for 1991–2020 averaging period. *Source* Authors. https://weatherspark.com/

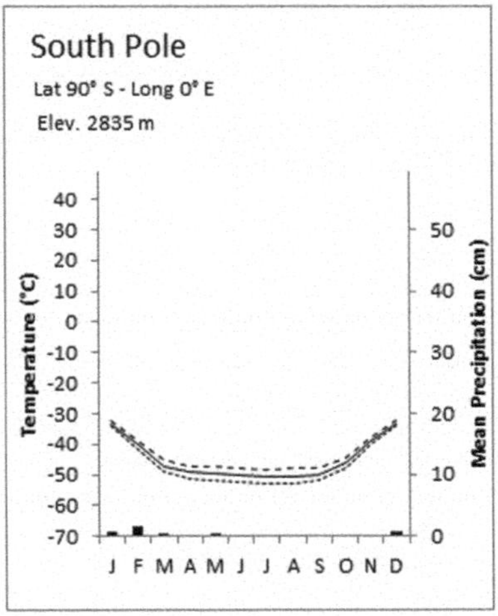

Fig. 21.12 Climagraph for the South Pole. *Source* Authors. https://weatherspark.com/

with each iteration using solar radiation in the ultraviolet radiation wavelengths of the spectrum, which are shorter, more energetic, and much more harmful to life than *visible light*. By absorbing this high-energy form of sunlight high in the stratosphere, ozone prevents dangerous levels of exposure at the surface.

Because light initiates the process, ozone is naturally depleted over the long and dark winters of the high latitudes. By the time the September equinox arrives, Antarctica will have been in darkness for many months, so no O_3 will have been generated to replace O_3 molecules that are photodissociated into O_2 molecules and O atoms. Antarctica has a larger ozone hole than the Arctic because it has a strongly zonal circumpolar vortex, compared with the more meridional flow in the North polar area. Thus, influx of ozone from lower latitudes is much less in the South polar region.

Further exacerbating the issue is that humans produced chlorofluorocarbons (CFCs) which were primarily used in the creation of aerosol sprays, refrigerants, foams, and insulation products. Scientists found that the CFCs naturally degrade and the CFC molecule dissociates. The chlorine in CFCs is particularly problematic as it drifts into the stratosphere where O can create a strong bond with it that does not photodissociate easily, rather than with O_2 to form ozone. Ozone concentrations decreased over time.

As sunlight enters the Antarctic region after its vernal equinox, the naturally formed hole would close as new ozone was produced. However, long-term data showed that the hole was growing and lasting much longer than would otherwise be expected because of human interference through CFC production. In response, the U.S. banned the manufacture of CFC products in 1996. The world followed in 2000. Nevertheless, some CFC products are still produced legally, in limited quantities, for specialized purposes. The polar ozone holes have recovered much more quickly than forecasted after the banning of CFC products. The ozone layer is still thin and holes still persist longer into the sunlight period than would occur naturally, but the rapid response offers optimism that nature repairs itself when given an opportunity.

Further Reading

Bushra, N., & Rohli, R. V. (2021). Relationship between atmospheric teleconnections and the Northern Hemisphere's circumpolar vortex. *Earth and Space Science, 8*(9): Art. No. e2021EA001802. https://doi.org/10.1029/2021EA001802

Chipperfield, M. P., Bekki, S., Dhomse, S., Harris, N. R. P., Hassler, B., Hossaini, R., Steinbrecht, W., Thieblemont, R., & Weber, M. (2017). Detecting recovery of the stratospheric ozone layer. *Nature, 549*(7671), 211–218. https://doi.org/10.1038/nature23681

Cohen, J., Screen, J. A., Furtado, J. C., Barlow, M., Whittleston, D., Coumou, D., Francis, J., Dethloff, K., Entekhabi, D., Overland, J., & Jones, J. (2014). Recent Arctic amplification and extreme mid-latitude weather. *Nature of Geoscience, 7*(9), 627–637. https://doi.org/10.1038/NGEO2234

Gordon, A. L., Lamont-Doherty Geological Observatory Report 1990–91.

Kidston, J., Scaife, A. A., Hardiman, S. C., Mitchell, D. M., Butchart, N., Baldwin, M. P., & Gray, L. J. (2015). Stratospheric influence on troposphere jet streams, storm tracks and surface weather. *Nature Geoscience, 8*(6), 433–440. https://doi.org/10.1038/NGEO2424

Orsi, A. H., Johnson, G. C., & Bullister, J. L. (1999). Circulation, mixing, and production of Antarctic bottom water. *Progress in Oceanography, 43*(1), 55–109. https://doi.org/10.1016/S0079-6611(99)00004-X

Parish, T. R., & Bromwich, D. H. (2007). Reexamination of the near-surface airflow over the Antarctic continent and implications on atmospheric circulations at high southern latitudes. *Monthly Weather Review, 135*(5), 1961–1973. https://doi.org/10.1175/MWR3374.1

Solomon, S. (1999). Stratospheric ozone depletion: A review of concepts and history. *Reviews of Geophysics, 37*(3), 275–316. https://doi.org/10.1029/1999RG900008

Part VII
Pulling It All Together

Chapter 22
Review and Applied Meteorology

Abstract The culmination of the processes that acquire and exchange energy, as described in Chap. 2, and those that move energy, matter, and momentum via the general atmospheric circulation (Chap. 3), secondary atmospheric circulations (Chap. 4), and the ocean circulation (Chap. 5), drives the climate system. Interactions between the various climate system components (Chaps. 6–9) alter the distribution of energy, matter, and/or momentum on or near the Earth's surface. These factors acting in concert make the world the way that it is. Synoptic classification methods (Chap. 10) including the Köppen climate types (Chap. 11) reveal several broad-scale patterns of climate. Those patterns described in Chaps. 12–21 are reviewed here. Various, specific, and in some cases, unique, features in world regions exchange energy, matter, and momentum in the climate system. Whether primary or secondary, their impacts are felt at different spatial scales. The example of El Niño-Southern Oscillation is used here. Uncertainty remains a major theme in both the nature and future impacts of the atmospheric and oceanic circulation variability and change. Such impacts may affect resilience and vulnerability in environmental, economic, and social systems.

Keywords Cloud seeding · Energy · Köppen climate types · Matter · Momentum · Primary impacts · Resilience · Secondary impacts · Vulnerability

Highlights of Regional Weather and Climate Patterns

The culmination of the processes that cause energy exchanges as described in Chap. 2 and involve the movement of matter and momentum via the general circulation of the atmosphere (Chap. 3), including secondary atmospheric circulations (Chap. 4) and the ocean circulation (Chap. 5) drive the climate system. Furthermore, the interactions between the various components of the climate system as described in Chaps. 6–9 alter the distribution of energy, matter, and/or momentum on or near the Earth's surface. These factors acting together make the world the way that it is.

R. V. Rohli et al., *Atmospheric and Oceanic Circulation*,
https://doi.org/10.1007/978-3-031-66142-6_22

Synoptic classification methods (Chap. 10) including the *Köppen climate types* (Chap. 11) reveal the effects of the atmospheric and oceanic circulation in the form of climate types around the world. A careful review of the extent of Köppen climate types, which correspond to mean monthly *temperature* and *precipitation* ranges of natural vegetation regions that can be extended to the marine atmosphere, reveals recurring patterns globally (Fig. 22.1, as shown previously in Fig. 11.1). This chapter summarizes the main recurring patterns. For instance, in most of the world's regions, *subtropical anticyclones* and associated *gyres* play an important role in the atmospheric and oceanic circulation and their impacts (Fig. 22.2). More comprehensive explanations for how these circulations and others occur and the climates that result are found in the chapters of Parts V and VI.

In the U.S. and Canada, several distinctive climatic realms responding to atmospheric and oceanic forcing exist. The Pacific coastal area is dry in summer, especially in the southern part. There, the *Hawaiian-Pacific High* and associated cold *California Current* frequently produce a *stable atmosphere* as cold water is overlain by much warmer air—a *temperature inversion.* The persistent high pressure suppresses lifting that might produce precipitation, both over the eastern North Pacific Ocean and the adjacent coastal land mass. However, the effect remains restricted to a narrow area inland because the coastal mountains confine the cooled near-surface air to the coastal zone. These narrow strips of land, along with vast adjacent marine areas, have a *Marine West Coast* (Cfb) *climate.* Farther south are the *Mediterranean-hot summer* (Csa) *climate* and *Mediterranean-warm summer* (Csb) *climate* of California, with their characteristic dry summers and wet winters.

A few dozens of kilometers from the Pacific coast is a linear area sandwiched between the coastal mountain ranges and much higher mountains including the Cascades in British Columbia, Washington, and Oregon, and the Sierra Nevada in California. These areas are partially blocked from precipitation-generating storms from the west by the coastal ranges and from the east by the higher mountains.

Farther east is another, larger area sandwiched between mountains—this time between these generally high mountains and the much higher Canadian and U.S. Rockies. Thus, a large area of the western U.S. and Canada are situated in this Intermontane Basin in a *double rain shadow* area, creating arid conditions and B climates. Summer temperatures in such areas are often oppressively high, especially in the lower elevations. Winter temperatures are low but often are interrupted by warmer areas as local winds can warm quickly as they descend slopes *adiabatically*. The only mechanisms for relief from the aridity are small monsoonal effects from the Gulf of California in late summer, elevation that cools enough to reduce *evapotranspiration* rates, and the occasional storm that can survive the trek over the mountains while retaining some water vapor. The northern part of this Intermontane Basin generally experiences a maximum precipitation in winter because of more frequent passages of precipitation-bearing *mid-latitude wave cyclones.* This is advantageous because winter snowpack replenishes water supplies, especially where elevations are high. In the Southwest desert area, the weak monsoon effect causes a summer precipitation maximum.

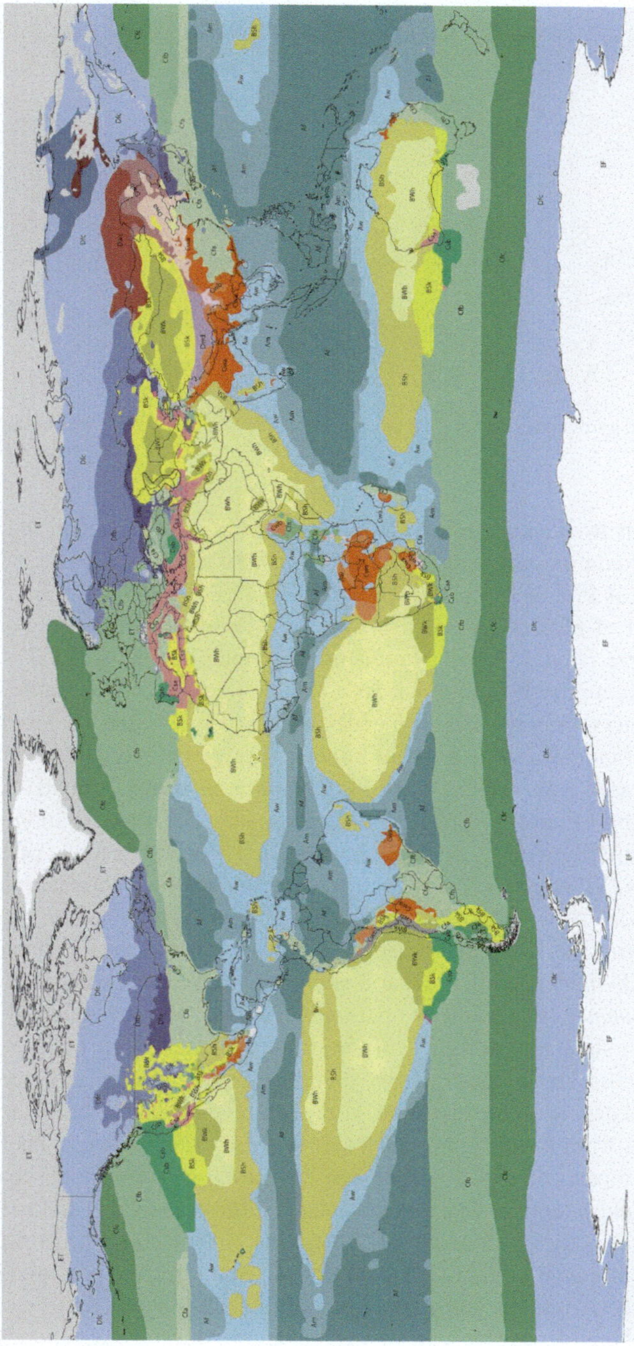

Fig. 22.1 Köppen climate types around the world. *Source* Modified from Reynolds, Rohli, Johnson, Waylen, and Francek, *Exploring Physical Geography*, McGraw-Hill

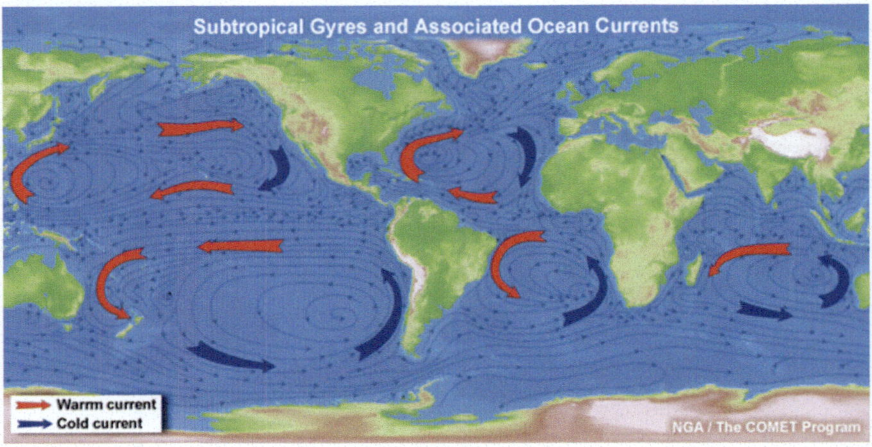

Fig. 22.2 Subtropical gyres and associated ocean currents. *Source* UCAR CoMet. Permission from Tara Torres

Just as the high Rockies form the Continental Divide for drainage, they also separate two worlds atmospherically. From the western edge of the Great Plains and Canadian Great Prairies eastward, the main source region for the humid air masses is the Gulf of Mexico. As air flows northwestward on the western side of the *Bermuda-Azores High*, it provides moisture for precipitation, especially when it makes its nearest approach in summer. As a result, almost all of the southeastern U.S. has hot summers, mild winters with occasional cold spells, and the absence of a dry season—a *Humid Subtropical* (Cfa) *climate*. The geography and orientation of the Gulf of Mexico with respect to the circulation around the Bermuda-Azores High along with the location of primary air mass contrast and the *polar front jet stream* set up a primarily west-to-east rather than south-to-north gradient of annual precipitation totals. Vigorous air mass contrast also characterizes the Heartland of the U.S. and Canada, and with it can come many forms of severe weather.

The northeastern U.S. states and adjacent Maritime Provinces of Canada have cold, uninterrupted winters and hot to warm summers, with no dry season—*Humid Continental-hot summer* (Dfa) *climate* or *Humid Continental-warm summer* (Dfb) *climate*. This area is situated near the *cyclolysis* area of preferred storm tracks regardless of the *cyclogenesis* area of the track, and therefore experiences a high percentage of cloudy days and days with precipitation. However, severe weather frequency and intensity decreases from southwest to northeast in these climate zones. The Great Lakes also create impacts on their adjacent areas by setting up late-summer to early-winter *unstable atmosphere* and late winter to early summer stability, driven by the fact that large water bodies heat and cool slower than the air above them. In most locations in the continental U.S. or Canada, however, humidity and precipitation in a given year are largely influenced by zonality/meridionality of *Rossby waves*.

Europe is heavily influenced by semi-permanent atmospheric and oceanic circulation features in the North Atlantic Ocean, particularly the *North Atlantic Drift*, which is on the eastern side of the Bermuda-Azores High—known as the Azores High in Europe—and the *Icelandic Low*. The North Atlantic Drift moderates temperatures, with winters that are very warm for the latitude, to such an extent that produces the *Humid Subtropical–mild summer* (Cfc) *climate* over much of the adjacent ocean and Iceland, though the moderating effect diminishes from west to east. Northern and western Europe experience a high percentage of days with precipitation. The Cfb climate across wide swaths of northern and western Europe and the adjacent water bodies is similar to that in northwestern U.S. and Pacific Canada, but unlike in localized areas of western North America that can experience strong *orographic effect*, no extraordinarily high precipitation totals exist in Europe. Central and eastern Europe become increasingly continental eastward in terms of temperature, with western Russia having brutally cold and long winters, in Dfa and Dfb climates. The Bermuda-Azores High keeps the summer dry south of the Alps, where Csa and Csb climates are prevalent. However, southern European winters are wet as the circulation around the *Siberian High* converges with that around the Bermuda-Azores High, leading to uplift, particularly where warm water underlies much colder air to destabilize the atmosphere. Arid climates are virtually absent in Europe.

North Africa, Southwest/Central Asia, and the adjacent waters are dominated by overwhelmingly and excessively arid climates, with a *True Desert—hot* (BWh) *climate* in most of the region and other B climates in most of the remainder. The *subsidence* side of the Bermuda-Azores High, reinforced by the temperature inversion from the cold *Canary Current* underlying warmer, subtropical air, suppresses precipitation across Saharan Africa and the eastern tropical North Atlantic Ocean. The *Walker circulation* supports even stronger sinking motion across the subtropical latitudes of Africa and especially over the Arabian Peninsula of Southwest Asia than in most other places on Earth. Central Asia suffers from distance to humid source regions, as the moisture in mid-latitude wave cyclones is largely wrung out before finally reaching the region. *Rain shadow* effects of several mountain ranges exacerbate the aridity. Across the entire swath from Morocco to Kyrgyzstan, the lack of cloud cover makes summer temperatures excessively high, while winter temperatures range from occasionally cold in coastal north Africa to characteristically frigid in central Asia.

By far the largest continent in terms of both land area and population, Asia contains the greatest diversity of responses to the atmospheric and oceanic circulation, with all of the major climate and air mass types. The other defining environmental characteristic of East and South Asia is the strong and distinctive monsoon circulation. In East Asia, winter is dry because of the sinking motion and clockwise circulation from inland, caused by the enormous, strong, and persistent Siberian High. The *polar front jet stream* bifurcates around the world's highest mountains—the Himalaya, steering the mid-latitude wave cyclones with it. In summer, rain in East Asia, particularly China, is brought by the opposite phase of the monsoon, with winds from the southwest and, later in the season, the southeast. Typhoons are an interesting juxtaposition between major environmental hazards in coastal areas and important precipitation-generating mechanisms inland. Coastal areas, including all of Japan, humid climates, while in areas farther north and inland, extratropical winter-dry Köppen types (Cw or Dw) or arid (B) climates prevail. The Gobi Desert of northern China and adjacent southern Mongolia is the driest area of the continent.

South and Southeast Asia is also influenced heavily by a monsoon circulation. Despite being shielded from the dryness that accompanies the coldest air masses by the massive Himalaya Mountains, the winter monsoon season nevertheless brings substantially dry air. By late spring or early summer, the monsoon reverses direction. Abundant moisture *advection*, mostly in the form of the southwest monsoon, occurs in most locations on the Indian subcontinent and Southeast Asia. Large seasonal differences in precipitation and surface water balances characterize the region. The result is generally a *Tropical Wet-dry* (Aw) *climate*, also known as *Tropical Savanna climate*, and *Mesothermal Winter Dry–hot summer* (Cwa) *climate* in most of India, with Southeast Asia characterized by all three tropical (A) climate types (*Tropical Rain Forest* (Af) *climate*, *Tropical Monsoon* (Am) *climate*, and Aw climate), and several *mesothermal* (C) types.

The hottest time of year in much of India and adjacent areas tends to be in March through May, before the wet monsoon is established. The only major places in the region that are wet in winter are on the northeast sides of the islands and in extreme Southeast Asia, which is too far north for the *inter-tropical convergence zone* (ITCZ) to provide precipitation at any time of year and where the *northeast trades* are most effective in winter months. However, interesting patterns of spatial and temporal variability exist because of complex and unique topography. The wettest regions, such as northeastern India and the western sides of southern India and southeastern Asia, experience tremendous rainfall totals during the summer phase of the monsoon. While the winter monsoon brings drier air to most of South Asia, a few places in the region may receive their only substantial precipitation of the year in winter. For instance, Pakistan and northwestern India experience winter frontal precipitation, the south-central Indian Peninsula has additional precipitation as the double rain shadow effect that hinders onshore moisture during the summer monsoon is disrupted by winter fronts, and the eastern coast of southern India and Southeast Asia are affected by humid northeast winds over water. The strength of the monsoon also varies over interannual time scales, particularly because of the mode of the *Indian Ocean Dipole* (IOD), and at intra-annual time scales, such as from *Madden–Julian Oscillation* (MJO) variability.

Sub-Saharan Africa has the full range of tropical (A), arid (B), and mesothermal (C) climate types. Thus, temperatures are high year-round, with some nocturnal and winter cooling in the arid subtropical margins. The sinking air from the Bermuda-Azores High not only causes the Sahara to exist but also influences northern sub-Saharan Africa. The Aw climate remains nearer to the Equator than elsewhere in the Northern Hemisphere, because the ITCZ is tilted vertically, reducing the *convergence* that produces the vertical movement allowing for cloud growth. The result is that the ITCZ produces a quasi-front with warmer air being pushed diagonally upward, from north to south. Precipitation in the Af climate of Africa is therefore substantially less than in the same climate type elsewhere in the world. A small monsoon effect in coastal West Africa enhances this precipitation. East Africa lacks a similar wet area because the winds parallel the coast, leading to coastal *upwelling* that produces a temperature inversion, thereby suppressing upward motion, cloud cover, and precipitation. The Aw climate south of the Equator gives way to Cwa or *Mesothermal Winter Dry–warm summer* (Cwb) *climate* in higher elevations. The southernmost section of Africa, and vast sections of the Southern Ocean south of Africa around the hemisphere, have Cfa, Cfb, Csa, and Csb climates, which allow for more winter cooling and influence of mid-latitude wave cyclones.

The relatively small section of the African continent that is south of the Equator experiences the Southern Hemisphere circulation features caused by a leftward deflection of the *Coriolis effect*: clockwise and inward rotation around cyclones, and counter-clockwise and outward circulation around anticyclones. This effect generates the *southeast trades* in Southern Hemisphere Africa, the *southwesterly counter-trades* in tropical Northern Hemisphere Africa in the Northern Hemisphere summer, and the *northwesterly counter-trades* in tropical Southern Hemisphere Africa in the Southern Hemisphere summer.

The region of Latin America, the Caribbean basin, and other adjacent water bodies has a wide array of conditions. Northern Mexico and adjacent oceans are largely arid as a continuation of the U.S. Desert Southwest. Much of the Caribbean and western coast of Central America has an Aw climate because the ITCZ provides rainfall in summer but the region lies in a rain shadow from the humid northeast trades and/or is under the effect of subsidence from the subtropical anticyclone in winter. The Af and Am climates are more common, at least at the local scale, along the eastern coast of Central America. The *Caribbean low-level jet* is an important factor in the climate variability of this area. Northern coastal South America has some desiccating effects from coastal upwelling, but farther south is the rainiest area of the region—the Amazon basin, where precipitation exceeds evapotranspiration to such an extent that runoff gives the Amazon River by far the most discharge of any major world river. West of the Andes Mountains, South America has a mirror image of the array of narrow climatic zones as in North America, with the brutally dry Atacama Desert equatorward of the Csa and Csb climates. Eastern South America has a subtropical climate that is dry in Southern Hemisphere winter as the ITCZ moves to the Northern Hemisphere and subtropical anticyclones have a desiccating effect overhead. The Southern Cone of South America, known as Patagonia, is dry from the Andes to the east coast because the area is in the Andean rain shadow and is locked under a ridge in the *Rossby wave*, which is inverted in the Southern Hemisphere compared to the way ridges appear in the Northern Hemisphere. Nevertheless, storms do form offshore of the extreme southern section of the continent, around the Falkland Islands, with trailing cold fronts bringing some precipitation.

Australia and the rest of Oceania is mostly a product of the Pacific Ocean. Australia is dominated by arid climates, but unlike other places with arid climates, the dryness is produced by a series of moving highs to a stronger extent than by the effect of a persistent, semi-permanent, quasi-stationary subtropical anticyclone. Northern Australia is rainy because of the ITCZ, with some monsoon and *tropical cyclone* impacts. Cyclogenesis off Australia's southeast coast gives New Zealand rain as the *mid-latitude westerlies* push these storms eastward. The inhabitability of tropical and subtropical Pacific islands is determined mainly by water availability which results from orographic effects of the trade winds and counter-trades, and precipitation overhead of the ITCZ and *South Pacific convergence zone* (SPCZ). *El Niño-Southern Oscillation* (ENSO) causes more disruption to the "normal" conditions in this part of the world and South America than elsewhere, but the impacts of ENSO are felt globally.

The polar regions have atmospheric circulation that keeps them largely confined within the influence of the *circumpolar vortex* through summer and winter. Even though the number of daylight hours varies excessively through the course of the year, the environment remains cold year-round. Radiant energy amounts are generally too low to evaporate enough surface water to generate high precipitation totals. Nevertheless, because the environmental demand for water is also low, these landscapes take on the appearance of relatively humid environments, at least where green plants can have enough warmth to persist. The North and South polar regions are very different in that the former is covered by water, which means that *sea ice* dominates, while the latter is covered by land, which means that land ice in the form of ice sheets covers the vast majority of Antarctica. The South polar region is much colder, and virtually no precipitation occurs at the South Pole.

The spatial differences in climate are ironic in a sense, because they repeat around the world yet places nearby may have more differences than places half a world apart. Several pairs of cities represented by climagraphs in Chaps. 12–21 are near each other. Consider the similarities and differences in climate, and the causes of each, between, for example, Washington, District of Columbia and Philadelphia, Pennsylvania (see again Figs. 13.8 and 13.9); Hangzhou and Shanghai, China (see again Figs. 16.27 and 16.28); and Mumbai and Ahmedabad, India (see again Figs. 17.17 and 17.19). To what extent and in which ways do the factors described in Chapter 8 come into play in causing differences in the climates among the pairs of cities?

It is important to remember that all of the forcing mechanisms in the atmospheric and oceanic circulation that give rise to characteristic climates around the world are changing through time, rapidly in some cases. Furthermore, these changes are manifested at different spatial scales, with different time scales of responses based on scale. And in turn, those climatic changes at various spatial and temporal scales also feed back on the forcing mechanisms and land covers by impacting all aspects of atmospheric and oceanic circulation, including all components of the *climate system*. How will these phenomena change in the future? Volumes have been written on the topic, yet there is still so much more to be learned. The climate system is complicated.

Coming Full Circle: Re-evaluating the Climate System and Weather and Climate Impacts on Society

It should now be much clearer than it was at the end of Chap. 1 that atmospheric and oceanic circulation, and the climates and environments that result from them, involve more than just the *atmosphere*. Spend some time studying the items in Fig. 22.3 and think even more deeply than you did at the end of Chap. 1 about how the atmosphere interacts with other "spheres" in the climate system. It is now apparent that the Earth-ocean–atmosphere feedback system is complicated, so this time list as many concrete examples from around the world as you can.

Also in Chap. 1, some of the major *primary impacts* and *secondary impacts* of climate were described. Now that the processes that affect the atmospheric and oceanic circulation were covered, along with the expression of those processes in the form of weather and climatic landscapes and seascapes around the world, it may be useful to revisit those impacts through the lens of what has been covered in previous chapters. All the while, consider not just that the circulation impacts the climate, but how it does so, why it is important, and how and why it might be changing over time. Likewise, remember the relevance of the atmospheric and oceanic circulation processes to your field of specialization. Try to think of your own hypotheses of possible climate impacts in your field, considered within a climate change framework. Be creative.

Primary effects of the atmospheric circulation on climate are sectors of the economy that are affected most directly by climate. They include agriculture, fisheries, water resources, energy, transportation, and construction. Secondary effects are only indirectly related to climate. They include the economy itself, political upheaval, tourism, and migrations. What other areas can you add to the list?

Regardless of the particular sector or impact, the weather and climate elements under consideration might include temperature and temperature variability, precipitation and its variability, moisture index, solar radiation, wind patterns, growing season and variability of it, and risk of damaging storms or other weather/climate

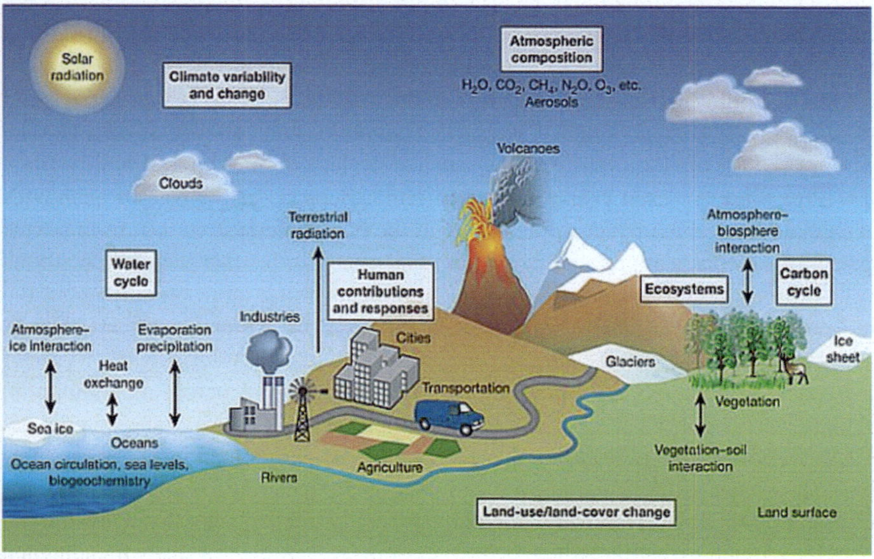

Fig. 22.3 Schematic of processes occurring in the climate system. *Source* NOAA

extremes. Each sector is more sensitive to the effects of certain atmospheric or environmental variables and changes in those variables than they are to others. And each sector is unique in its sensitivity to each weather and climate element. Even within a sector, each actor within the sector has different sensitivity to different variables. For example, some types of agriculture are more sensitive to an unseasonable drought than others.

Whether primary or secondary, these and other impacts are felt at different spatial scales. Using agriculture as the example again, a local-scale freeze might devastate citrus in a county of northern Florida, while at the regional-scale, a peach crop could be destroyed in the southeastern United States. At the national-to-international scale, weather and climate could cause a coffee crop failure in a major producing country. Agricultural implications of changing weather and climate could then ripple into various other areas. For example, at the business level, food supply would be affected. How much of a subsidy would one country need to give another, and what implications would such decisions have for food security? Notice how these questions necessarily involve interplay between climate effects and other effects. Large nations like the U.S. may be somewhat insulated from these kinds of problems because of its size and variety of climates but they are still vulnerable.

Uncertainty remains a major theme in both the nature of the changes to the atmospheric and oceanic circulation and in the nature of the future impacts. Will our **vulnerability**—the impacts or consequences that we suffer as a proportion of the magnitude of the event—to weather and climate extremes increase or decrease in the future? Will technology rescue us or trap us? Vulnerability may also depend on the scale of the problem, with more *resilience* in the future to minor hurdles but more vulnerability to major hurdles. What can be done at the individual, community, state, national, and international levels to build resilience to climate extremes? What should be done? What argument would an opponent of your ideas make to refute them? How would you refute those arguments against your idea?

One example that has recurred throughout earlier chapters is the relationship between ENSO extremes and ecological, economic, and/or societal effects. The anchovy industry in Peru is affected because of the lack of nutrient availability when upwelling slows or even ceases during El Niño. Because of the role of anchovies as chicken feed, this impact ripples through the economic system too. In this way, agricultural impacts are related to those to fisheries and, for that matter, forestry as well. But beyond these kinds of specialized impacts, think more broadly about the impacts of El Niño and *La Niña*, after studying Fig. 22.4, which is similar to Fig. 6.7, and Figs. 22.5, 22.6 and 22.7.

Because of the importance of water, its limited supply, and increasing population, water resources may be another of the most important impacts in a changing climate. Weather/climate affects water supply (precipitation) and demand (evapotranspiration). How large should a dam, levee, drainage system, etc. be? These questions are often answered via hydrodynamic modeling (actually tracing water through aquifers, etc.). How much irrigation will be necessary or even possible? Places with adequate summer rain but too little winter precipitation (second letter of "w" in the Köppen system) may be impacted minimally in the "average year" because rain falls to the

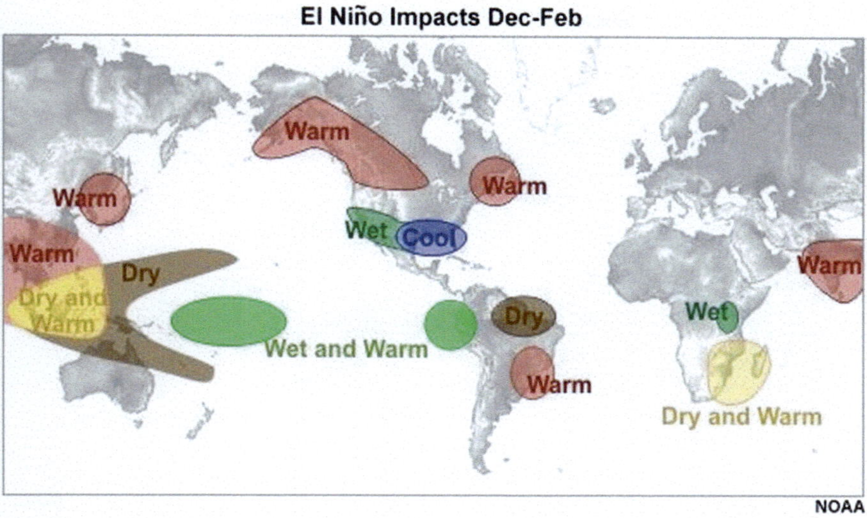

Fig. 22.4 Major El Niño impacts from December through February. *Source* NOAA/UCAR CoMet. Permission from Tara Torres

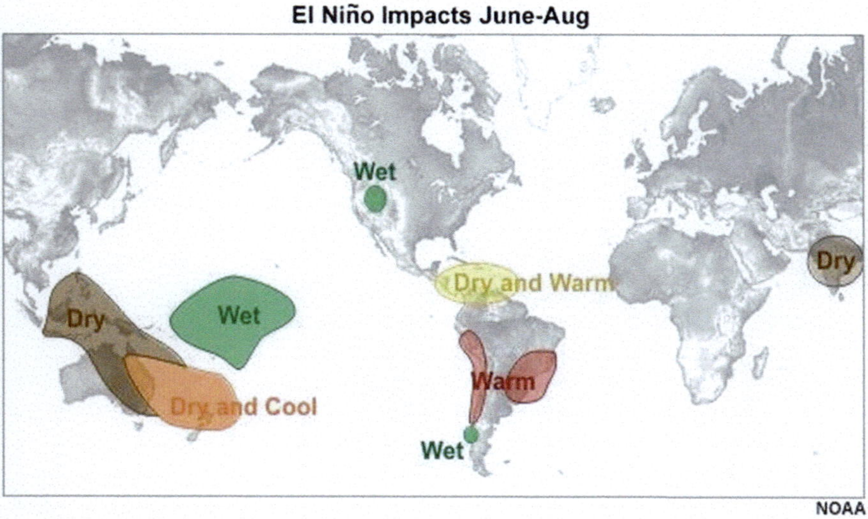

Fig. 22.5 Major El Niño impacts from June through August. *Source* NOAA/UCAR CoMet. Permission from Tara Torres

ground at the time when the *latent heat flux* is moving water from the soil to the atmosphere. However, when the rains do not arrive on time, a problem exists. All too often, such problems happen after people have been lured into depending on the water supply for their livelihoods during the "average" year. In *Mediterranean* (Csa or Csb) *climates*, rains do not come at the time of year when they are needed, so the summer

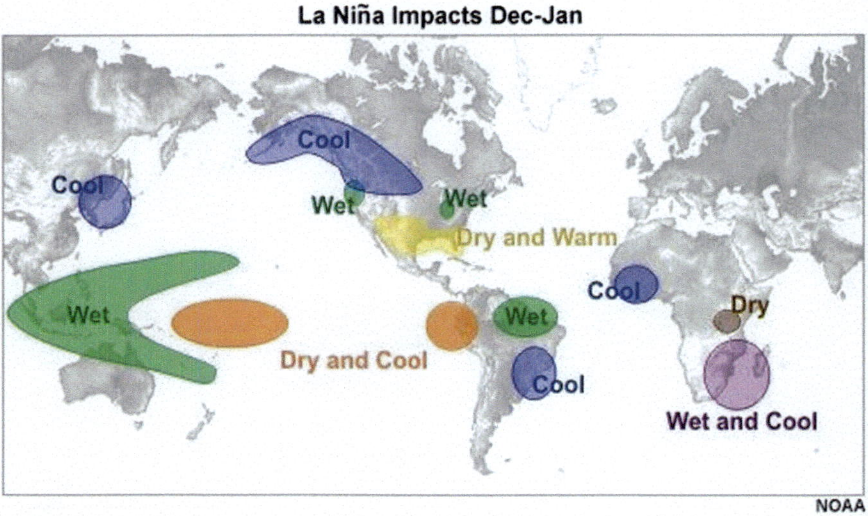

Fig. 22.6 Major La Niña impacts from December through January. *Source* NOAA/UCAR CoMet. Permission from Tara Torres

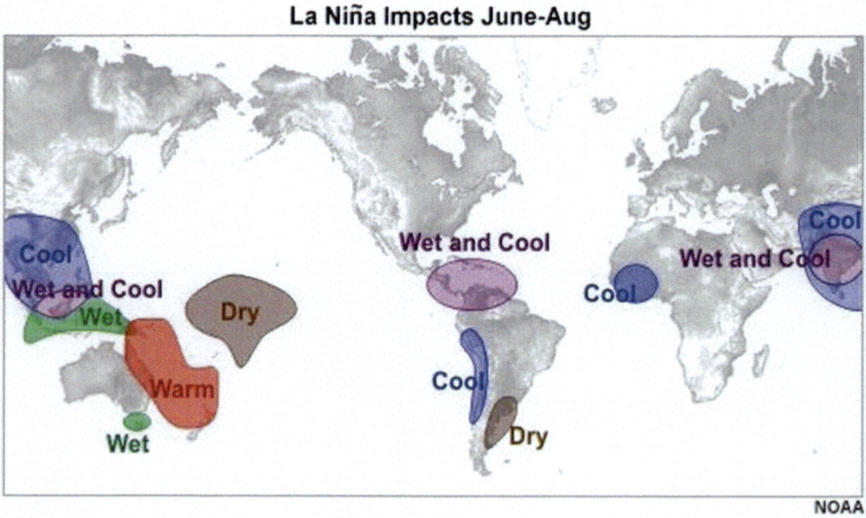

Fig. 22.7 Major La Niña impacts from June through August. *Source* NOAA/UCAR CoMet. Permission from Tara Torres

water supply relies on groundwater and snowpack. The California water crisis will only worsen over time as more people are competing for the same resources, but with intermittent periods of wetter conditions that lure people into activities requiring more intensive use of water. On the other hand, flooding is also a major concern, especially in climate types with the second letter of "f" (wet year-round).

Based on the importance of energy in our daily lives, and in our planning for the future, it is not surprising that discussion of energy impacts in the context of weather/climate and changes therein has become an increasingly important topic. Climate affects energy supply (solar, wind, and hydropower, primarily) and demand (heating and cooling of buildings, including homes, primarily). Again, the concept of variability is vital. The limitation of the given power source is determined by the lowest available amount of solar, wind, or stream discharge, not by the mean or the maximum. Deserts are often the best candidates to make use of solar power, especially for densely populated areas nearby. Mountains, coasts, and places influenced by the trade wind systems and near the *polar front* are the most suitable places for wind power. Places with abundant precipitation and hilly terrain are best for hydropower.

We have experienced relatively high variability in temperatures in the last half century. Winters in the late 1970s were full of cold extremes that affected energy supply and also transportation (shipping, airlines, trucking, etc.), while winters in recent years have been dominated by warm extremes. Which temperature extremes have the more disadvantageous impacts? Why? What kinds of temperature extremes have the most deleterious impacts in summer? What does the consensus of experts say about the future of temperature extremes, and how might these relate to energy supply and availability, and spinoff impacts from it?

Construction is also an industry that is impacted intricately by atmospheric and oceanic circulation, weather and climate resulting from it, and changing weather/climate patterns. Precipitation and construction are enemies. Cement cannot harden correctly below a certain temperature threshold. When our current infrastructure was built, people failed to consider the fact that climates change, and now the changing conditions have caused our infrastructure to be more vulnerable than they should be. An example is the failure to construct homes and other buildings high enough to mitigate the flood risk.

Political impacts of the atmospheric and oceanic circulation and the climate resulting from it are also all around us. For example, injecting particles in the atmosphere in the hope of causing water vapor to condense around them and precipitate to produce rain—**cloud seeding**—can cause political problems. Starting in the 1960s, the Cuban government has claimed that the U.S. was conducting seeding experiments to try to steer a hurricane away from Florida and toward Cuba. But even domestic political problems can come from weather and climate changes. The U.S. states of Idaho and Montana have had disputes during times of drought. In a famous court case, Montana claimed that Idaho seeded clouds that would have brought precipitation to Montana. Whose rainwater was it when a drought occurred? The states of Georgia and Tennessee also have frequent disputes regarding water rights in the Chattahoochee River. These kinds of political problems inevitably turn into economic and social justice issues. With growing populations situated in environmentally vulnerable areas amid a changing climate, such problems are likely to increase substantially in the future. How will we manage such issues, legally, ecologically, logistically, and ethically? How do we set priorities in so many cases in which a price tag cannot be put on the various implications? Do we always select the option that helps the greatest number of people and harms the fewest? Is such a strategy always ethical? Who determines what is fair? Is it fair that this entity or these entities determine fairness?

Of course, health impacts also need to be managed, and many of these same questions apply to health amid a changing climate. In fact, with an aging population, such as exists in most of the more developed world, such questions become even more complicated. Despite popular misconceptions, it is unclear whether more people die from exposure to cold weather or from exposure to hot weather. The true answer likely depends on the location and what "counts" as a "heat" or "cold" death, and whether such death rate calculations adjust for the fact that more people are out and about, and therefore exposed to weather hazards, in extreme heat than in extreme cold. Nevertheless, heat stress, especially in *urban heat islands*, has become an increasing concern. More people are killed in heat waves in poleward areas of the mid-latitudes than in subtropical areas, as they are less likely to have the means of dealing with the extreme heat. This is particularly true in mid-latitude climates for the disadvantaged, who are more likely to live in homes with poor ventilation designed to keep heat inside. Other health impacts of weather and climate involve the relationship between sunshine and cancer, *ozone* depletion and cancer, and warm, wet weather that allows insect-borne disease vectors and pests to proliferate faster.

None of the questions raised in this chapter are easy to answer. Earth and environmental scientists are the most qualified people to begin to confront these types of challenges. An improved awareness, understanding, and healthy respect for the amazing atmospheric and oceanic circulation that sets the foundation for the cascading environmental, economic, and social landscapes that make our world the way that it is, the way that it was, and the way that it will be, is a great start.

Further Reading

Basu, R. (2009). High ambient temperature and mortality: A review of epidemiological studies from 2001 to 2008. *Environmental Health, 8*(40). https://doi.org/10.1186/1476-069X-8-4

Eisner, S., Florke, M., Chamorro, A., Daggupati, P., Donnelly, C., Huang, J., Hundecha, Y., Koch, H., Kalugin, A., Krylenko, I., Mishra, V., Piniewski, M., Samaniego, L., Seidou, O., Wallner, M., & Krysanova, V. (2017). An ensemble analysis of climate change impacts on streamflow seasonality across 11 large river basins. *Climatic Change, 141*(3), 401–417. https://doi.org/10.1007/s10584-016-1844-5

Jones, L., & Tanner, T. (2017). 'Subjective resilience': Using perceptions to quantify household resilience to climate extremes and disasters. *Regional Environmental Change, 17*(1), 229–243. https://doi.org/10.1007/s10113-016-0995-2

Malhi, G. S., Kaur, M., & Kaushik, P. (2021). Impact of climate change on agriculture and its mitigation strategies: A review. *Sustainability, 13*(3). https://doi.org/10.3390/su13031318

Perry, A., & Thompson, R. D. (Ed.). (1997). *Applied climatology: Principles and practice.* Routledge, 384p., 978-0415141000.

Schaeffer, R., Szklo, A. S., de Lucena, A. F. P., Borba, B. S. M. C., Nogueira, L. P. P., Fleming, F. P., Troccoli, A., Harrison, M., & Boulahya, M. S. (2012). Energy sector vulnerability to climate change: A review. *Energy, 38*(1), 1–12. https://doi.org/10.1016/j.energy.2011.11.056

Glossary

30–60 Day Oscillation See Madden–Julian Oscillation
40–50 Day Oscillation See Madden–Julian Oscillation
A climate The generic term for tropical climate types (11)
Ablation Melting or sublimation (9)
Absorption Energy acquired by a substance (2)
Accumulation Buildup of snow in the formation of glaciers (9)
Adiabatically A change in temperature that occurs solely by thermodynamic effects, without an energy exchange (2)
Advection fog A cloud at the surface that occurs when warm, moist air passes over a cold water body and chills to the dew point temperature (5)
Advection The lateral transport of energy and matter (2)
Aerosol A solid or liquid particle suspended in the air; also known as a particulate (1)
African easterly jet A low tropospheric jet stream over West Africa (18)
Agulhas Current Warm ocean current in the western Indian Ocean basin (18)
Air mass thunderstorm Single-celled convection induced thunderstorms (12)
Air mass Air with uniform weather characteristics over large areas (temperature, stability, humidity, etc.) across a wide region (4)
Alaska Current The warm ocean current of the eastern North Pacific Ocean (12)
Albedo The percentage of incident shortwave radiation that is reflected off of an object, which is color- and roughness-dependent (2)
Alberta Clipper A weak mid-latitude wave cyclone originating in lee of the Canadian Rocky Mountains (12)
Aleutian Low The semi-permanent subpolar low over the North Pacific Ocean (3)
Alpine glacier See mountain glacier
Amplitude The degree of amplification of a wave, measured from the depth or most equatorward extent of a trough to the crest or most poleward extent of a ridge (3)
Anabatic wind A valley breeze (8)
Anemometer Device that measures wind speed and direction (1)

R. V. Rohli et al., *Atmospheric and Oceanic Circulation*,
https://doi.org/10.1007/978-3-031-66142-6

Annual temperature range The difference between the temperature in the warmest month and that of the coldest month for the same place (2)

Antarctic amplification See polar amplification

Antarctic Bottom Water (AABW) A large mass of deep, cold water at the bottom of the Southern Ocean next to Antarctica (21)

Antarctic Circle The 66.5°S parallel of latitude and the latitude at which 24 consecutive daylight hours occurs on the December solstice (2)

Antarctic Circumpolar Current The west-to-east cold saline ocean circulation around Antarctica but equatorward of the Antarctic Coastal Current; also known as West Wind Drift (21)

Antarctic Coastal Current The east-to-west cold, saline ocean circulation flowing along the coast of Antarctica between the Antarctic Circumpolar Current and the coast; also known as East Wind Drift (5)

Antarctic Deep Water (AADW) The deep ocean current that flows over the AABW as it moves northward (21)

Antarctic Intermediate Water (AAIW) The deepwater thermohaline current that overrides the Common Water (21)

Antarctic Oscillation (AAO) A seesaw between the strength of the South Polar High and mid-level geopotential heights in the Southern Hemisphere mid-latitudes; also known as Southern Annular Mode (6)

Anticyclone A spinning enclosed area of surface high pressure (3)

Anticyclonic shear The leeward side of an upper-level ridge initiating anticyclonic spin (7)

Arctic (A or cA) air mass A large area of extremely cold and dry air from a high-latitude continental source region (4)

Arctic amplification See polar amplification

Arctic Circle The 66.5°N parallel of latitude and the latitude at which 24 consecutive daylight hours occurs on the June solstice (2)

Arctic Oscillation (AO) A seesaw between the North Polar High and mid-level geopotential heights in the Northern Hemisphere mid-latitudes; also known as Northern Annular Mode (6)

Arctic sea smoke See steam fog

Arid Dry, as defined by prominent water balance deficits throughout the year (11)

Artificial intelligence (AI) The use of computers to solve problems by building models that classify or predict future information (10)

Artificial neural network (ANN) A type of machine learning model that attempts to mimic the human brain by relaying on/off signals throughout a collection of interconnected elements working together (10)

Atlantic meridional overturning circulation (AMOC) A major component of the global thermohaline circulation in the Atlantic Ocean that has been observed to be weakening in recent years (5)

Atlantic Multidecadal Oscillation (AMO) A water temperature oscillation in the North Atlantic Ocean with a period of about 70 years (6)

Atlantic Multidecadal Variability (AMV) See Atlantic Multidecadal Oscillation

Atlantic Warm Pool Warm water region of the tropical Atlantic Ocean (19)

Atmosphere The "system component" that consists of Earth's environment above the surface, comprised of gases, and solid and liquid particles (1)

Atmospheric river A narrow, persistent "conveyor belt" of abundant, persistent moisture in air (6)

Atmospheric window The range of wavelengths of radiant energy from Earth that can escape to space at a given place or time (2)

Atoll A ring-shaped island composed of coral, usually formed from eroded mountain slopes (20)

Attenuation A reduction of the radiation incident on a surface because of absorption, scattering, and reflection of radiation before it reaches the surface (2)

Automated (objective) classification system Flow pattern analysis and classification using empirical computing systems (10)

Autumnal equinox One of the two days each year when every place on Earth receives a 12 hours of daylight and 12 hours of darkness; corresponds to the September equinox in the Northern Hemisphere and to the March equinox in the Southern Hemisphere (2)

B climates The generic term for arid climate types (11)

Backdoor front A cold frontal passage which moves from the east or northeast to the west or southwest, against the normal west-to-east steering flow (4)

Baiu phenomenon See Mei-yu phenomenon (16)

Baroclinic zone A location of strong thermal gradients (9)

Benguela Current The cold ocean current in the eastern South Atlantic basin (18)

Benguela Niña A periodic warming of the southwestern tropical surface Atlantic Ocean that causes atmospheric variability in Southern Africa (18)

Benguela Niño A periodic cooling of the southwestern tropical surface Atlantic Ocean that causes atmospheric variability in Southern Africa (18)

Bergen School Norwegian school of meteorology reaching prominence in the early 20th century (1)

Bermuda-Azores High Quasi-stationary subtropical high pressure is over the North Atlantic Ocean (3)

Berson westerlies Westerly stratospheric flow circumnavigating the global tropics (20)

Biome An area classified by the species in that area (11)

Biosphere The layer of life in the Earth-atmosphere system (1)

Blackbody Hypothetical object that absorbs and emits radiation uniformly and with 100% efficiency (2)

Blocking pattern An upper-level longwave configuration that prohibits the movement of migratory pressure cells (4)

Boreal forest Evergreen high-latitude forest (11)

Bowen ratio The ratio of sensible heating to latent heating (2)

Brazil Current A warm ocean current on the western side of the South Atlantic Ocean (19)

C climates (mesothermal) The generic term for subtropical, mid-latitude, mild climate types (11)

Cabo (Cape) Verde storm A tropical cyclone that forms from clusters of thunderstorms over West Africa and strengthens as it tracks across the eastern north Atlantic basin (18)

California Current The major cold ocean current in the eastern North Pacific Ocean (5)

Calving The break-off of ice from a glacier (9)

Canary Current The major cold ocean current in the eastern North Atlantic basin (18)

Canopy The upper vegetation layer of a forest (8)

Cape Verde storm See Cabo Verde storm

Caribbean low-level jet The region of intensified low-level tropical easterly winds carrying moisture from the Caribbean Sea into eastern Central America (19)

Catatumbo Lightning of Venezuela in the Lake Maracaibo basin (19)

Centers of action High-variability regions of atmospheric teleconnections (12)

Centrifugal force A force initiated toward the outside of a moving object's curved trajectory and is proportional to the speed and radius of curvature of the object (3)

Chaparral A type of short, woody, hardy shrubs specially adapted to dry soils and high temperatures (13)

Chinook winds Warm and dry air flow initiated by adiabatic heating of air descending mountains in the western United States (8)

Circumpolar vortex Geostrophic winds that circumnavigate the poles at upper levels (3)

Climagraph A line and bar chart that depicts average temperature and precipitation, respectively, for a given place; also known as a climograph (13)

Climate system The Earth-ocean–atmosphere system composed of components and processes/functions that can be measured (1)

Climatography Average and extreme weather data, and very simple statistics on the data, devoid of explanations about "why" or "how" the data occurred (1)

Cloud seeding Human-induced precipitation enhancement (22)

Cloud street An atmospheric equivalent to the Langmuir circulation in the ocean (9)

Cluster analysis An automated classification system using correlations to key days (10)

Cold front The boundary created by colder air displacing warmer air (4)

Cold-air funnel A phenomenon that occurs over large lakes whereby funnels form beneath showers or weak thunderstorms when the air aloft is especially cold (7)

Colorado Wave Cyclone A traveling mid-latitude low-pressure center originating on the leeward side of the Rocky Mountains over or near Colorado (12)

Common Water The deep thermohaline current initiated near Antarctica in the Pacific Ocean (21)

Condensation A phase change from vapor (or gaseous) to a liquid state (1)

Condensation nuclei Microscopic solid particles that encourage the formation of a liquid water droplet by providing a tiny surface around which water vapor can undergo condensation or deposition (8)

Conditional instability A stability situation in which an unstable situation is reached if a parcel is forced aloft to a certain height (2)

Conduction The transfer of energy in solids (1)

Conductive heat flux Flow of energy through solids (1)

Confluence Lateral motion from two sources that converges into one flow into a narrower area, resulting in a speeding up of the flow (3)

Conservation of potential vorticity The constant nature of the trajectory of a fluid parcel that remains unchanged in a vortex (4)

Continental polar (cP) air mass A large area of cold/cool, dry air from a high-latitude continental source region (4)

Continental tropical (cT) air mass A large area of hot/warm and dry air from a low-latitude, dry continental source region (4)

Continentality The concept of seasonal heating and cooling of continental interiors (2)

Convection The transfer of energy in fluids (1)

Coriolis effect The deflection of moving objects in the atmosphere-ocean resulting from Earth rotation (3)

Coriolis parameter The product of twice the Earth's angular velocity and the sine of the latitude, which, when multiplied by wind speed, describes the magnitude of the Coriolis effect (3)

Costa Rica thermal dome An offshore zone west of Central America that varies in size and shifts slightly in location, where colder waters are drawn upward in a dome-like shape (19)

Counter-radiation (L↓) Longwave radiation absorbed and re-emitted back downward to Earth's surface by the atmosphere (2)

Counter-trades See northwesterly counter-trades, or southwesterly counter-trades

Cryosphere Ice on or near Earth's surface, including ice caps, ice sheets, floating **sea ice**, frozen inland water bodies, ice in soils, and glaciers

Cumulonimbus A towering cumulus cloud that precipitates (2)

Current meter Device that measures water motion (1)

Cutoff low Spinning eddy that is cut off from the main meandering airflow (4)

Cyclogenesis The initial formation of a cyclone (4)

Cyclolysis The dissipation of a cyclone (4)

Cyclone A spinning enclosed area of surface low pressure (3)

Cyclonic shear The leeward side of an upper-level trough that initiates cyclonic spin (7)

D climates (microthermal) The generic term for cold, harsh, mid-latitude climate types (11)

December solstice The day with the shortest daylight period in the Northern Hemisphere and longest daylight period in the Southern Hemisphere, corresponding to winter solstice in the Northern Hemisphere and summer solstice in the Southern Hemisphere (2)

Decision tree (DT) "Bin" phenomenon characterized by various types of data (10)

Deep learning (DL) A subclass of various ML models in which the models are constructed over numerous layers (10)

Deficit A water shortage at a location (11)
Deposition A phase change from a vapor to a solid state (1)
Depth of frictional influence The ocean depth at which a vector first points 180° from the overlying wind vector; also known as depth of frictional resistance (5)
Depth of frictional resistance See depth of frictional influence
Depth of the oceanic mixed layer See layer of frictional influence
Desert pavement Hard, rocky surface of deserts (15)
Desertification Expansion of deserts (18)
Dew point depression The difference between the dew point and the surface air temperature (2)
Dew point temperature The temperature at which saturation occurs (2)
Diffluence Lateral motion diverges into two separate flow into a wider area, resulting in a slowing of the flow (3)
Dishpan experiments Experimental studies of fluid motion in the 1930s (3)
Dixie Alley The high-fatality tornado region of the Gulf South of the United States (12)
Doldrums An alternate name for the intertropical convergence zone (ITCZ) (3)
Doppler radar The radar system capable of determining direction and speed of airflow (12)
Double lake breeze Lake breeze from two directions over land sandwiched between two lakes (7)
Double rain shadow An Intermontane Basin desert (12)
Dry line A moisture density boundary that forms a quasi-frontal boundary, regardless of whether a temperature gradient exists (12)
Dynamics Movement and/or change, especially across space (2)
E **(Arctic/Antarctic) climates** The generic term for bitter cold, high-latitude climate types (11)
East African low-level jet A low-level moisture jet extending from the coast into the interior of Africa (18)
East Wind Drift See Antarctic Coastal Current
Easterly wave A curved airflow from east to west embedded in the low-latitude trade wind flow and capable of developing into tropical cyclones (4)
Eastern-boundary Current A wide, shallow, slower-moving current in the subtropical to middle latitudes of the eastern sides of ocean basins (5)
Eddy A local-to-mesoscale whirling, spin-off circulation (3)
Eigenvector analysis An objective automated classification technique (10)
Ekman spiral A turning of ocean water with depth due to the Coriolis effect (1)
El Niño A warming of eastern equatorial Pacific Ocean temperatures with a corresponding cooling of western equatorial Pacific Ocean temperatures (1)
Electromagnetic radiation Energy in wave form that contains both electrical and magnetic properties (2)
Electromagnetic spectrum The full assemblage of all possible electromagnetic waves by wavelength (2)
Electromagnetic wave A traveling oscillation of a combined electric and magnetic field (2)

Empirical classification system A climate classification based on similar values of weather elements (11)
Empirical orthogonal function A type of eigenvector analysis (10)
Energy The capacity or power to do work (1)
Energy balance An accounting of all incoming versus outgoing energy in a system (2)
Enhanced Fujita (EF) scale An intensity rating system for tornadoes based on fastest wind gust, ranging from EF0 to the theoretical EF5
ENSO The combined effect of El Niño/La Niña in the ocean and the Southern Oscillation in the atmosphere (6)
Enthalpy The total energy content of the atmosphere (14)
Environmental determinism The notion that weather, climate, and other aspects of the environment determine society (1)
Environmental lapse rate (γ) Rate of temperature change with height in a static atmosphere (2)
Environmental possibilism The notion that the environment sets broad limits but does not determine human achievement (1)
Epipelagic zone The highest layer of the ocean, defined by the distance that light can penetrate; also known as the photic zone (5)
Equator 0° latitude, dividing Earth into the Northern and Southern Hemispheres (2)
Equatorial (E or mE) air mass Hot and humid air from a tropical source region (4)
Equatorial trough See intertropical convergence zone (3)
Equilibrium line The boundary between the zone of accumulation and the zone of ablation on a glacier; also known as firn line (9)
Equivalent potential temperature An index representing the combined sensible and latent energy (i.e., enthalpy) (19)
Evaporation A phase change of a substance from a liquid to a gaseous state (1)
Evapotranspiration Total surface water evaporated and transpired (2)
Extraterrestrial shortwave radiation (K_{EX}) Shortwave solar radiation that reaches the "top" of the atmosphere (2)
Eye Central area of a well-developed tropical cyclone (12)
Falkland Current The cold ocean current south and east of South America (19)
Ferrel cell Theoretical cells in each hemisphere between the Hadley and polar cells (3)
Fetch The distance over which air flows over a uniform surface (typically water) (7)
Findlater jet See Somali jet (18)
Firn line See equilibrium line
Fluid A state where particles easily move and change position without separation of the mass and easily yields to pressure (2)
Foehn winds Warm and dry air flow initiated by adiabatic heating of air descending mountains of the European Alps (8)
Fog hollow A local landscape depression that accumulates frost or fog because of cold, dense air drainage at night; also known as a frost hollow (8)

Friction A force of opposition (3)

Frost hollow See fog hollow

Frost point temperature A below-zero dew point temperature; air temperature at which freezing and deposition, rather than condensation and sublimation, begins (2)

General circulation See primary circulation (3)

Genetic classification system Climate classification system in which types are determined by a major climate forcing feature (11)

Geopotential height The height above the surface of a particular atmospheric pressure (3)

Geostrophic wind Airflow that results from a balance between the pressure gradient force and the Coriolis effect (3)

Gradient wind A wind in a curved flow situation resulting from a balance between the pressure gradient force, Coriolis effect, and centrifugal force (3)

Gravity wave A wave in a fluid (e.g., air or ocean) that oscillates downward because of the effect of gravity, then rebounds upward as it propagates downwind or downstream, and then back downward again because of gravity, etc. (19)

Greenhouse effect Warming of the air associated with the absorption of longwave radiation from the surface and re-emission of longwave radiation back downward to the surface by certain gases (2)

Greenhouse gas Water vapor, carbon dioxide, methane, or other gases that decrease the amount of longwave radiation escaping the Earth-system (2)

Ground heat flux Substrate heating into a land surface (2)

Gulf Low A mid-latitude wave cyclone originating in the Gulf of Mexico (12)

Gulf of Alaska Low A mid-latitude wave cyclone originating in the Gulf of Alaska (12)

Gulf of Mexico low-level jet (LLJ) A stream of fast-moving air from the Gulf of Mexico into the southern Great Plains (12)

Gulf Stream The major warm surface ocean current along the Atlantic coast of North America (2)

Gust front Temperature density boundary and leading edge of a thunderstorm downdraft (12)

Gyre Any large, permanent, ocean basin-wide circular anticyclonic circulation in the mid-latitudes of each ocean basin (5)

Hadley cell A hemispheric convection cell located approximately between 0° and 30°N and between 0° and 30°S latitude, but oscillating in position with the seasons (3)

Harmattan wind Hot and dry northeasterly trade wind over North Africa (18)

Hatteras Low A mid-latitude wave cyclone originating off the east coast of the United States (12)

Hawaiian High See Hawaiian-Pacific High (3)

Hawaiian-Pacific High The major, quasi-stationary subtropical high-pressure system over the North Pacific Ocean; also known as the Hawaiian High or the Pacific High (3)

Heat burst A mesoscale feature that involves sinking air without the intense pressure gradient and strong winds, sometimes accompanied by swift increase in the dew point temperature (4)
Heat dome Excessively hot conditions caused by subsiding air in a stationary high pressure region, often beneath a ridge (4)
Horse latitudes Alternate name for the subtropical high-pressure belt (3)
Humboldt Current The major cold current in the eastern South Pacific Ocean; also known as Peru Current (1)
Humid Continental—cool summer (Dfd) climate A type of microthermal climate with a cool summer and no dry season; along with the Dfc climate type, is known as Subarctic climate (11)
Humid Continental—hot summer (Dfa) climate A type of microthermal climate with a cool summer and no dry season (11)
Humid Continental—mild summer (Dfc) climate A type of microthermal climate with a mild summer and no dry season; along with the Dfd climate type, is known as Subarctic climate (11)
Humid Continental—warm summer (Dfb) climate A type of microthermal climate with a cool summer and no dry season (11)
Humid Subtropical (Cfa) climate A type of C climate with a relatively consistent and abundant annual precipitation and hot summers (11)
Humidity Atmospheric moisture in the gaseous state, which has absorbed latent energy when it evaporated (1)
Hurricane A strong tropical storm occurring over the North Atlantic Ocean (4)
Hydrosphere The climate system component comprised of liquid water (1)
Hypothesis A testable statement (1)
Ice cap A large, continuous ice sheet at the poles (9)
Ice Cap (EF) climate A type of E climate in which conditions are frozen year-round (11)
Ice sheet A mass of ice greater than 50,000 km^2 (30,000 mi^2) in area (9)
Iceberg A floating broken chunk of glacial ice (9)
Icelandic Low The major, semi-permanent subpolar low over the North Atlantic Ocean (3)
Ideal Gas Law An expression that pressure in a gas is equal to the product of the density and Kelvin temperature of the gas, and the dry gas constant (3)
Indian Ocean Dipole (IOD) The coupled anomalies of sea surface temperature between the western and eastern Indian Oceans (6)
Indian Ocean Gyre The counter-clockwise surface ocean circulation in the South Indian Ocean (18)
Indonesian Throughflow The migration of near-surface water from the Maritime Continent into the Indian Ocean (6)
Inertial flow The balance of forces between the Coriolis effect manifested as the Ekman spiral and gravity (5)
Insolation (K↓) Solar radiation received at the Earth's surface (2)
Instability See unstable atmosphere (2)

Inter-tropical convergence zone (ITCZ) The low-pressure belt along the equatorial edge of each Hadley cell; also known as the equatorial trough (3)

Isobar A lines of constant pressure (3)

Isobaric surface A constant-pressure level in the air (3)

Isohypse A line on a map touching all places that have the same geopotential height (3)

Isotherm A line on a map touching all places that have a given temperature (11)

Jet streak The fastest part of a jet stream (3)

Joule The metric unit of work or energy or heat (2)

June solstice The day with the longest daylight period in the Northern Hemisphere and the shortest daylight period in the Southern Hemisphere, corresponding to summer solstice in the Northern Hemisphere and winter solstice in the Southern Hemisphere (2)

Katabatic wind A mountain breeze; a drainage of cold, dense air from high elevations (8)

Kelvin temperature An absolute temperature scale with zero degrees at absolute zero, representing a theoretical absence of kinetic energy (2)

Kelvin wave A flow of surface waves propagating in opposition to a migrating pool of water (19)

Kinetic energy Energy of motion at the molecular level (2)

Kirchofer technique A correlation based classification technique (10)

Köppen climate types A type of classification of daily weather using maps (10)

Kuroshio Current The major warm current of the western North Pacific Ocean (5)

La Niña A strengthening of "normal" water temperatures across the equatorial Pacific Ocean (1)

Labrador Current The cold current that flows equatorward to fill the location between the departing North Atlantic Drift and the coastline of eastern North America (6)

Lake breeze Airflow directed from a lake to land (7)

Lake effect snow Leeward snow formed as cold air is advected over water bodies (7)

Land breeze Airflow directed from land to a position over an ocean or large lake (7)

Land breeze front The leading edge of a land breeze over water, and/or the convergence of multiple land breezes from different directions (8)

Langmuir circulation A corkscrew-shaped circulation that moves upward and downward in the mixed layer of the oceans, parallel to and pushed along by a steady, prevailing wind (9)

Latent heat flux (Q_E) The transport of latent energy contained within its elements water molecules in solid, liquid, and gaseous states, both vertically and laterally (1)

Latent heat Work or energy contained within molecules or released from molecules in the phase changes of water (1)

Latent heating Energy or work done in the phase change of water from a gas to a liquid (2)

Latitude The north/south location on Earth, measured in angular degrees of the angle formed by the location, the center of the Earth, and the Equator (1)

Law An absolute and predictable truth with no known variations (1)

Layer of frictional influence The layer from the surface to the depth of frictional influence; also known as depth of the oceanic mixed layer (5)

Leeuwin Current A warm ocean current that flows southward intermittently along the coast of western Australia (20)

Leeward The downwind side (4)

Limnology The scientific study of lakes (7)

Lithosphere The solid Earth comprised of crust and the upper, rigid mantle (1)

Local scale A small, limited spatial scale (3)

Long wave See Rossby wave (3)

Longwave radiation Electromagnetic radiation having wavelength exceeding 4 μm (2)

Longitude The east/west location on Earth relative to Greenwich, England (1)

Loop Current A Caribbean Sea and Gulf of Mexico extension of the Gulf Stream (12)

Machine learning (ML) A subclass of artificial intelligence in which the models "learn" characteristics from an input set of previously observed data (10)

Madden–Julian Oscillation (MJO) Cloud bands that cross the tropical Pacific Ocean from west to east; also known as the 30–60 day oscillation or 40–50 day oscillation (17)

Manual classification Analysis and categorization of weather maps by hand (10)

March equinox The date in March when every line of latitude receives a 12-h day and night (2)

Marine West Coast (Cfb) climate A mesothermal climate, moderated by its location downwind of maritime air masses, with warm but not hot summer and no dry season (11)

Maritime Continent The many islands of and near Indonesia (6)

Maritime polar (mP) air mass A large area of cold/cool and humid air from a high-latitude oceanic source region (4)

Maritime tropical (mT) air mass A large area of warm/hot and humid air from a low-latitude oceanic source region (4)

Mascarene High See South Indian Ocean High

Mass balance (glacier) The balance of ice mass between the upper elevation zone accumulation the lower zone of ablation (melting) below with the firn (snow) line denoting the mass balance point

Matter A substance made up of various particles that occupies physical space (1)

Mediterranean—hot summer (Csa) climate A type of mesothermal climate with a hot, dry summer (11)

Mediterranean—warm summer (Csb) climate A type of mesothermal climate with warm or hot, dry summers (11)

Mediterranean (Cs) climate A type of mesothermal climate with warm or hot, dry summers (11)

Mei-yu A quasi-persistent, nearly stationary, east–west-oriented weak baroclinic zone in the lower troposphere from the coast of China eastward into the Pacific, south of Japan, most common in spring; also known as Baiu (16)

Melting A phase change from a solid to a liquid state (9)

Meridional Amplified Rossby wave flow, with ridges and troughs occupying a wide range of latitudes (3)

Mesocyclone A low-pressure, rotating cell and updraft region of a severe thunderstorm (12)

Mesoscale Phenomena occurring at a medium spatial scale (1); also see regional scale

Mesoscale convective complex (MCC) A line of thunderstorms in advance of a cold front; also known as a squall line (12)

Mesoscale convective system (MCS) A globular cluster of interacting severe thunderstorms (12)

Mesoscale ocean eddy A secondary, rotating circulation in the ocean with a diameter of a few kilometers up to hundreds of kilometers (5)

Mesothermal Winter Dry (Cw) climate A type of mesothermal (C) climate with a winter dry period, often because of a monsoon effect; can either have hot summer (Cwa) or warm summer (Cwb) (11)

Mesothermal A subtropical mid-latitude temperate (C) climate (11)

Meteorologist A scientist who studies the atmosphere (1)

Micrometer One-millionth of a meter (2)

Microscale Phenomena, including in the atmosphere and ocean, that occur on small spatial and temporal dimensions (1)

Microthermal A harsh mid-latitude or Subarctic (D) climate (11)

Microthermal Winter Dry (Dw) climate A type of D climate with a winter dry period, often because of a monsoon effect; can either have hot summer (Dwa), warm summer (Dwb), mild summer (Dwc), or cool summer (Dwd), the latter two of which are types of Subarctic climates (11)

Mid-Atlantic Bight The gradual bend of the shoreline that protects the open bay north of Cape Hatteras, North Carolina (12)

Mid-latitude wave cyclone A frontal storm system that includes a low-pressure center, and trailing cold, warm, stationary, and/or occluded fronts that is steered by the mid-latitude westerlies (1)

Mid-latitude westerlies The primary circulation in the mid-latitudes, consisting of air moving generally from west to east but with ridges and troughs embedded; includes the Rossby waves and polar front jet stream (3)

Millibar (mb) One one-thousandth of a bar of pressure, equal to 100 pascals (3)

Mindanao Current A major tropical ocean current near the Philippines (16)

Modoki A type of El Niño event in which anomalously high sea surface temperatures extend only part of the distance to the eastern side of the Pacific basin adjacent to the Americas (6)

Momentum The quantity of motion of a moving body measured as a product of its mass and velocity (1)

Momentum flux The transfer of the combined effect of mass and velocity in the air and ocean by processes involving wind and ocean currents (1)

Monochromatic emittance Amount of radiant energy emitted at a given wavelength (2)

Mountain (alpine) glacier The smallest glaciers, formed in high-elevation areas (9)

Mountain breeze A katabatic wind initiated during nighttime hours (8)

Multicell thunderstorm A severe thunderstorm containing multiple up- and downdrafts (12)

Multi-year sea ice Ice floating in the ocean that continues building thicker, year after year, but still not more than 3–4 m (10–13 ft) thick (9)

Negative feedback A process that decreases an initial condition (9)

Net longwave radiation (L^*) The difference between longwave radiation reaching the surface ($L\downarrow$) and longwave radiation emitted from the surface ($L\uparrow$) (2)

Net radiation (Q^*) The actual amount of radiation a place on Earth's surface receives at a given instant (2)

Net shortwave radiation (K^*) The difference in quantity of shortwave radiation that is absorbed vs. reflected (i.e., $K\downarrow - K\uparrow$) (2)

Nimbostratus Precipitating stratus clouds (7)

Ningaloo Niña A periodic cooling in the surface ocean water west of Australia that contributes to Australian atmospheric and hydrologic variability (20)

Ningaloo Niño A periodic warming in the surface ocean water west of Australia that contributes to Australian atmospheric and hydrologic variability (20)

Niño 3.4 index A statistically-standardized representation of the sea surface temperature in the so-called Regions 3 and 4 of the central equatorial Pacific Ocean (6)

Nor'easter A type of intense winter storm of the U.S. Atlantic coast and East Asian Pacific coast (9)

North American monsoon The summertime thermal low development and associated thunderstorm period in the southwestern United States (13)

North Atlantic Drift A northeastward extension of the warm Gulf stream current in the northern Atlantic Ocean (5)

North Atlantic Gyre The clockwise surface ocean circulation in the North Atlantic Ocean (6)

North Atlantic Oscillation (NAO) A atmosphere-focused teleconnection manifested as a seesaw in sea-level pressure between the Bermuda-Azores High and Icelandic Low (6)

North Equatorial Current A low-latitude ocean current driven by the Northern Hemisphere trade winds (5)

North Equatorial Counter-current The major eastward-flowing ocean current slightly north of the Equator (5)

North Pacific Current The northern part of the North Pacific Gyre (12)

North Pacific Gyre The clockwise-rotating ocean circulation in the North Pacific Ocean (5)

North Pacific Intermediate Water The deep thermohaline current that overrides Pacific Subarctic Water (21)

Northeast trade winds See northeast trades

Northeast trades The consistent belt of winds that blow from northeast to southwest, generally between 0 and 30°N globally, but oscillating latitudinally with the direct rays of the Sun; also known as the northeast trade winds (3)

Northern Annular Mode See Arctic oscillation

Northwesterly counter-trades The northwesterly airflow in the low-latitude Southern Hemisphere, resulting from a bending of the northeast trades in the Northern Hemisphere being deflected to the left after crossing the Equator (18)

Occlusion The elevation of warmer air when a cold front "catches up with" the warm front entirely around a low-pressure core (4)

Oceanographer A scientist who study the oceans (1)

Omega block An upper-level atmospheric blocking pattern that resembles the Greek capital letter "omega" (4)

Orographic lifting Air ascending a mountain slope or other high topographic barrier (8)

Oyashio Current The major cold ocean current off the east coast of Asia (16)

Ozone A triatomic oxygen molecule that absorbs ultraviolet radiation effectively (1)

Pacific-North American (PNA) pattern An air-ocean teleconnection that characterizes the long-wave flow over North America in the middle (500-mb) atmosphere (6)

Pacific Decadal Oscillation (PDO) An important Pacific Ocean teleconnection similar to ENSO but on multidecadal time scales, of 20-to-30 years; also known as Pacific Multidecadal Variability (PDV) (6)

Pacific Decadal Variability (PDV) See Pacific Decadal Oscillation

Pacific High See Hawaiian-Pacific High (3)

Pacific Subarctic Water (PSW) The deep thermohaline current in the lowest layers of the Pacific Ocean (21)

Pacific Warm Pool The area of hot water that is pushed westward across the equatorial Pacific Ocean by the trade winds (5)

Papagayo A fair-weather, easterly wind of the low-Sun season that descends the Pacific slopes of Central America (19)

Particulate See aerosol (1)

Permafrost Permanently frozen soil and subsoil (21)

Peru Current See Humboldt Current

Photic zone See epipelagic zone

Pineapple Express Storm systems that migrate across the North Pacific from Hawai'i to California, especially during an El Niño event (6)

Planck's Law Relationship between the amount of blackbody emittance of radiant energy and temperature at any wavelength (2)

Planetary boundary layer The layer of air immediately adjacent the surface, where which friction is important (5)

Polar amplification Increased high-latitude temperatures due to circulation changes; also known as Arctic amplification or Antarctic amplification (9)

Polar cell High-pressure area near the poles, with associated low-pressure area near 60° of latitude in the same hemisphere, polar easterlies at the surface, and upper-level westerly circulation (3)

Polar easterlies Surface air in the polar cells that flow from east to west (3)

Polar front The boundary between the cold polar cell and the warmer mid-latitudes (3)

Polar front jet stream A fast core of west-to-east-moving upper air stream that separates warm and cold air (1)

Polar front low See subpolar low

Polar High The semi-permanent near-surface high-pressure system situated over each pole (3)

Polar westerlies Airflow aloft from west to east over the polar cells (3)

Pollution front The diurnal reversal of wind direction on slopes which brings air pollution upslope by day and downslope by night (8)

Positive feedback A process that continues and exacerbates an initial condition (9)

Positive vorticity advection Cyclonic vorticity region generally corresponding to a trough (3)

Potential evapotranspiration (PE) The amount of evaporation that could occur if water were freely available (11)

Precipitation Falling matter, usually liquid or frozen water (1)

Pressure gradient force The force that determines initial wind speed and direction resulting from the distribution of high and low pressure centers (3)

Primary circulation The prevailing, global-scale atmospheric and oceanic flow patterns (3)

Primary impacts Components of the natural or human systems that variability or changes to the atmospheric and oceanic circulation, and the resulting climate, impact directly (1)

Principal components analysis (PCA) A type of eigenvector analysis (10)

Quasi-biennial Oscillation (QBO) A seesaw of wind direction between west-to-east and east-to-west in the atmospheric layer directly above the troposphere (20)

Radiant energy Solar radiation (1)

Radiation fog A cloud near the ground that occurs as longwave radiation escapes to space in clear conditions and allows near surface air to chill to the dew point (7)

Radiometer A device that measures radiant energy (1)

Rain shadow A dry region on the leeward side of mountains or high physiographic barriers formed as ascent causes precipitation on the opposite, windward side (8)

Random forest (RF) A type of DT algorithm (10)

Reflection Energy transmitted back into a single direction by a substance (2)

Regional scale Also known as mesoscale, a spatial scale smaller than synoptic (3)

Relative humidity The ratio of the specific humidity to the water vapor capacity at that temperature, normally expressed as a percentage (2)

Resilience The ability to adapt (1)
Revolution Revolving, moving in a Circle or near-circle about a central axis (2)
Rex block An upper-level blocking pattern in the atmosphere that features adjacent cutoff high- and low-pressure centers (4)
Ridge An elongated zone of high pressure (3)
Riming Deposition of water vapor directly onto a solid surface, as ice in the growth of a hailstone or glacier, skipping the liquid phase (9)
Roaring Forties Strong winds of the Southern Hemisphere around 40°S latitude (19)
Rossby wave A large-scale, upper-level atmospheric circulation of the mid-latitudes, consisting of a bend poleward (ridge) and equatorward (trough); also known as long wave (3)
Saharan air layer (SAL) A flow of dry air from northwestern Africa to the Caribbean Sea and northern South America that has numerous impacts, including cooling the Atlantic Ocean and Caribbean Sea sufficiently to prevent or weaken Atlantic tropical cyclones (18)
Santa Ana winds The warm and dry airflow initiated by adiabatic warming of air descending mountains in southern California (8)
Saturated (wet/humid) adiabatic lapse rate (Γ_s or SALR) Temperature change on average of 4 C° km^{-1} in rising saturated air (2)
Scattering Deflection of radiant energy in all directions upon interaction with a substance (2)
Scientific Method A series of steps including empirical observations and hypothesis testing in the process of discovery (1)
Sea breeze Airflow from a water body to adjacent land areas (8)
Sea breeze front The leading edge of a sea breeze over land (8)
Sea ice Floating, frozen ocean water (1)
Seasonal sea ice Ice floating in the ocean that melts in summer and is approximately 0.3–1.2 m (1–4 ft) deep (9)
Second Law of Thermodynamics The principle that pervades all of natural science that energy flows from areas of higher concentration to areas of lesser concentration (2)
Secondary circulation Any short-lived and local-to-regional-scale flow pattern (3)
Seiche A standing wave of oscillating water in an enclosed area (7)
Self-organizing map (SOM) A type of competitive-learning-based ANN technique that is most useful in synoptic typing (10)
Semi-arid—cold (BSk) climate A type of low-temperature, high-latitude B climate that supports grasses (11)
Semi-arid—hot (BSh) climate A type of high-temperature, low-latitude B climate that supports grasses (11)
Semi-arid (BS) climate A type of B climate with adequate precipitation to support grasses (11)
Sensible heat flux (Q_H) The flow of measurable heat (1)
Sensible heat Energy detectable as "hot" as measured by temperature (1)

September equinox The date when every line of latitude receives a 12-h day/night (2)
Short wave Any small upper-level ridge and trough configuration embedded within the longwave ridge or trough (12)
Shortwave radiation Wavelengths of electromagnetic radiation shorter than 4 μm (2)
Siberian High The large, low-Sun-season, high pressure cell over Asia (3)
Site The local atmospheric and oceanic conditions at a place (1)
Situation Broader-scale factors acting upon, and exerting implications on, the atmospheric or oceanic circulation at a place (1)
Slope aspect The compass point direction that a slope faces (8)
Snow band Linearly-oriented storm producing lake effect snow formed when the winds are aligned parallel to the major axis of a lake (7)
Somali Current The major warm current in the western Indian Ocean (18)
Somali jet Southwesterly flow paralleling the Somali coast in association with the summer Asian monsoon system; also known as the Findlater jet (18)
Source region The place where an air mass forms and obtains its characteristics (4)
South Atlantic Bight The gradual bend of the shoreline that protects the open bay south of Cape Hatteras North Carolina (12)
South Atlantic Gyre The counter-clockwise surface ocean circulation in the South Atlantic Ocean (18)
South Atlantic High The semi-permanent Southern Hemisphere subtropical anticyclone in the Atlantic Ocean (3)
South Equatorial Counter-current An eastward-flowing ocean current slightly south of the Equator (5)
South Equatorial Current A low-latitude ocean current driven by the Southern Hemisphere trade winds (5)
South Indian Ocean High The semi-permanent anticyclone of the Indian Ocean basin; also known as the Mascarene High (18)
South Pacific convergence zone (SPCZ) A convergence zone extending as an arm protruding diagonally from northwest to southeast across the tropical and subtropical South Pacific Ocean (5)
South Pacific Gyre The counter-clockwise-rotating ocean current in the South Pacific Ocean (19)
South Pacific High The Southern Hemisphere subtropical anticyclone in the Pacific Ocean (3)
Southeast trade winds See southeast trades
Southeast trades The consistent belt of winds that blow from southeast to northwest, generally between (0° and 30°S latitude globally, but oscillating latitudinally with the direct rays of the Sun, also known as the southeast trade winds (3)
Southern Annular Mode See Antarctic Oscillation
Southern Oscillation A seesaw of atmospheric pressures across the equatorial Pacific Ocean (1)
Southern Oscillation index An index formulated by subtracting the sea level pressure at Darwin, Australia, from that at Tahiti (20)

Southwesterly counter-trades Southwesterly airflow in the low-latitude Northern Hemisphere, resulting from a bending of the southeast trades in the Southern Hemisphere being deflected to the right after crossing the Equator (17)

Spatial variability, extremes, and change Change over space (1)

Specific heat The amount of heat (or energy) needed to produce a temperature change of 1 Celsius degree (1.8 F°) (2)

Specific humidity Atmospheric moisture content (i.e., atmospheric humidity) measured in grams of water vapor per kilogram of air, or g kg^{-1} (2)

Squall line See mesoscale convective systems (MCS) (12)

Stability A measure of the tendency for a local parcel of air to rise (2)

Stable atmosphere A stability situation in which air locally does not have a tendency to rise (2)

Stable season A period when a lake is cooler than the adjacent land (7)

Stable stratification A fluid state whereby a less dense fluid resides above a more dense fluid (7)

Stationary front A front trailing from a mid-latitude wave cyclone that has stalled temporarily (4)

Steam fog A type of fog that forms when cold air overlies much warmer water; also known as Arctic sea smoke (7)

Stefan-Boltzmann's Law A mathematical relationship describing the total amount of radiation at all wavelengths emitted by a blackbody, as a function of temperature (2)

Stomate A plant leaf aperture through which evapotranspiration occurs (8)

Storm surge A tropical-cyclone-induced wave of water accumulating at a coast (12)

Stratosphere The second layer of the atmosphere from the surface, characterized by a temperature inversion and the ozone layer (20)

Stratus Low, thick, wide, blanket-like clouds (3)

Streamline An indication of airflow, typically used in low latitudes instead of isohypses (4)

Structuration theory The idea that there are overlying structures that enable us as a society to accomplish certain things, but that these structures also constrain us (1)

Subarctic (Df) climate See Humid Continental—mild summer (Dfc) Humid Continental—cool summer (Dfd) climate, or Microthermal Winter Dry (Dwc, Dwd) climates (11)

Subgeostrophic Air moving at slower speeds than geostrophic winds (3)

Sublimation A phase change from solid to vapor, skipping the liquid phase (2)

Subpolar low A semi-permanent, enclosed area of particularly low air pressure located near 60°N and 60°S, but oscillating in strength and position with the seasons; also known as polar front low (3)

Substrate heat flux/heating (Q_S) Transfer of energy in a liquid or solid (1)

Subtropical high-pressure belt The semi-permanent chain of enclosed high pressure and sinking air near 30°N and 30°S, but oscillating in position with the seasons (3)

Subtropical jet stream The fast-moving generally west-to-east airflow at about 30°N and 30°S (3)
Summer solstice See June solstice or December solstice
Supercell A strong, single-celled severe thunderstorm (12)
Supergeostrophic Air moving at faster speeds than the speed of geostrophic flow (3)
Surplus Excess water in the soil at a location (11)
Sverdrup transport The enhanced Ekman transport on the eastern side of a subtropical gyre (5)
Synoptic climatology The study of the relationship between atmospheric circulation patterns and the climate at the surface (10)
Synoptic scale A spatial scale on the order of 1,000 km (600 mi) (1)
Taiga Evergreen boreal forest of Siberia (11)
Techno-fix era The phase of the history of science dominated by the notion that people can and will conquer any difficulties that climate or other constraints place by using science and technology (1)
Tehuantepecer A violent, easterly wind of the low-Sun season that descends the Pacific slopes of Central America (19)
Teleconnection A correlation between energy, matter, or momentum across a large part of the Earth's surface (6)
Temperate Any subtropical, mild, mid-latitude climate (11)
Temperature Index of the presence and distribution of sensible heat, measured by various types of thermometers (1)
Temporal variability, extremes, and change Change over time (1)
Theory A consensus explanation of a physical process verified through hypotheses testing (1)
Thermal bar An area of sharp density differences and intense mixing in a fluid (7)
Thermal conductivity The rate of penetration of heat from the surface into a substance needed to change the temperature one meter away by one Kelvin (7)
Thermal low Any heat-induced low pressure system (12)
Thermally direct Thermodynamic (convection) circulation (3)
Thermally indirect A dynamically-driven circulation (3)
Thermodynamic Movement by heat (2)
Thermohaline circulation The globally contiguous, three-dimensional ocean current driven by temperature and/or salinity differences (5)
Thermometer A device that measures temperature (1)
Theta-e ridge An elongated area of high equivalent potential temperature (19)
Tibetan Low A thermal low that develops over Central Asia during summer (16)
Tornado Alley High tornado-frequency region of the U.S. extending from central Texas to the Dakotas (12)
Trade wind inversion Temperatures increasing with increasing height in this trade wind zone (3)
Transpiration Water evaporated through plants (2)
Treeline A temperature limit to tree growth (11)

Tropic of Cancer The 23.5°N parallel of latitude and the farthest poleward extent of the direct ray of sunshine, which occurs at solar noon on or near June 21 (2)

Tropic of Capricorn The 23.5°S parallel of latitude and the farthest poleward extent of the direct ray of sunshine, which occurs at solar noon on or near December 21 (2)

Tropical cyclone A strong storm of tropical origin, also called hurricanes and typhoons (6)

Tropical depression A weak storm system that develops from an easterly wave (4)

Tropical Monsoon (Am) climate A climate that is rainy for most of the year and dry for a few months, but with water balance deficits that are sufficiently small that tropical rain forest vegetation survives the dry season (11)

Tropical Rain Forest (Af) climate The wettest A climate (11)

Tropical Savanna (Aw) climate See tropical wet-dry climate (11)

Tropical storm A storm system that is stronger than a tropical depression but weaker than a hurricane or typhoon, which develops from an easterly wave (4)

Tropical Wet-dry (Aw) climate A climate in which approximately half of the year, corresponding to the low-Sun season, is either dry or wet (11)

Tropopause The top of the troposphere (3)

Troposphere The lowest major layer of the atmosphere, where nearly all weather and climate occurs (3)

Trough An elongated zone of low pressure (3)

True Desert—cold (BWk) climate A type of cool to cold, very arid B climate (11)

True Desert—hot (BWh) climate A type of warm, somewhat arid B climate (11)

True Desert (BW) climate A family of B climates with persistent and intense aridity year-round (11)

Tundra (ET) climate A type of polar climate marked by stunted vegetation (11)

Turbulence Airflow characterized by sharp, abrupt fluctuations in air speed and direction (1)

Turkana jet A jet between the 950- and 650-mb pressure levels over Ethiopia, the East African highlands, and the Lake Turkana basin (18)

Typhoon A strong tropical storm occurring over the Pacific Ocean (6)

Ultraviolet radiation Radiant energy in a band of shorter wavelengths than visible light and which is more energetic than visible light and much more harmful to life (21)

Understory The lower vegetation or floor of a forest (8)

Unsaturated adiabatic lapse rate (Γ_u), or UALR A rate of temperature change of $\pm$ 1 C° per 100 m in rising or sinking unsaturated air (2)

Unstable atmosphere A stability situation in which air locally has a tendency to rise spontaneously (2)

Unstable season A period when a lake is warmer than the adjacent land (7)

Updraft Vigorous upward vertical air motion (2)

Upwelling Upward motion in water (1)

Urban canyon A linear set of vertical walls formed by the adjacent buildings on each side of city streets (8)

Urban heat island A local area of higher temperatures centered on cities compared with their surrounding rural hinterlands (8)

Urban precipitation island A local region of precipitation maximum on the leeward side of a city caused by extra condensation nuclei from urban pollutants (8)

Valley breeze An anabatic wind; upslope winds that originate in valleys during daylight hours (8)

Valley inversion A temperature inversion that traps relatively cold air in valleys by katabatic winds from two or more surrounding slopes, common at night (8)

Veranillo A dry period occurring on the western coast of Central America as the Caribbean low-level jet descends the western sides of mountain slopes, displaying surface ocean water and causing upwelling and stable atmospheric conditions; literally, "little summer" (19)

Vernal equinox One of the two days each year when every place on Earth receives a hours of daylight and 12 hours of darkness; corresponds to the March equinox in the Northern Hemisphere and to the September equinox in the Southern Hemisphere (2)

Vertical wind shear A sharp increase in wind speed with increasing height (12)

Visible light Electromagnetic radiation having a wavelength between 0.04 and 0.07 μm and apparent to the human eye as the colors of the rainbow (2)

Vulnerability Impacts or consequences from a disaster, as a proportion of the magnitude of the event (22)

Wake low A mesoscale low-pressure system on the "back" side of a mid-latitude wave cyclone's precipitation shield, where air sinks because of its increased density as it is cooled by precipitation, warms adiabatically on its decent, and then rises nearby (4)

Walker circulation The broad-scale atmospheric circulation between the Maritime Continent and equatorial South America (6)

Warm front The boundary created by warmer air displacing colder air (4)

Water balance A system that accounts for all water input and output for a location (11)

Water vapor capacity The maximum amount of humidity that can occur before the air becomes saturated (2)

Watt The metric unit of power (2)

Wavelength The distance between an adjacent crest and trough in an oscillation (2)

West Australian Current A cold current on the eastern side of the Indian Ocean Gyre (20)

West Wind Drift See Antarctic Circumpolar Current

Western-boundary intensification A poleward-moving western ocean basin current that accelerates along with the poleward increase in Coriolis effect (5)

Wien's Law An equation that determines the electromagnetic wavelength of peak emittance of a blackbody as a function of temperature (2)

Winter solstice See December solstice or June solstice

Younger Dryas A period of abrupt global or nearly-global cooling, from about 13,000 years ago to about 11,000 years ago (5)

Zonal A de-amplified Rossby wave flow (3)

Index

R. V. Rohli et al., *Atmospheric and Oceanic Circulation*,
https://doi.org/10.1007/978-3-031-66142-6

B

G

M

T

GPSR Compliance

The European Union's (EU) General Product Safety Regulation (GPSR) is a set of rules that requires consumer products to be safe and our obligations to ensure this.

If you have any concerns about our products, you can contact us on ProductSafety@springernature.com

In case Publisher is established outside the EU, the EU authorized representative is:

Springer Nature Customer Service Center GmbH
Europaplatz 3
69115 Heidelberg, Germany

The manufacturer's authorised representative in the EU is Springer Nature Customer Service Centre GmbH, Europaplatz 3, 69115 Heidelberg, Germany. If you have any concerns regarding our products, please contact ProductSafety@springernature.com

Printed and bound by CPI Group (UK) Ltd, Croydon, CR0 4YY
30/09/2026
02225233-0015